CNS Neurotransmitters and Neuromodulators

Acetylcholine

Edited by

Trevor W. Stone, Ph.D., D.Sc.
Professor of Pharmacology
Department of Pharmacology
University of Glasgow
Glasgow, Scotland

CRC Press
Boca Raton Ann Arbor London Tokyo

Library of Congress Cataloging-in-Publication Data

CNS neurotransmitters and neuromodulators : acetylcholine / edited by
 Trevor W. Stone.
     p.   cm.
   Includes bibliographical references and index.
   ISBN 0-8493-7630-0
   1. Acetylcholine--Physiological effect.   2. Cholinergic
 mechanisms.   I. Stone, T. W.
 QP364.7.C58   1994
 612.8′042--dc20   94-26310
                                                                                                         CIP

PREFACE

The rate at which neuroscience research is growing makes it increasingly difficult for active scientists to keep abreast of topics not in their immediate sphere of interest. Despite this, there is undoubtedly much to learn from workers in closely related fields. This group of review volumes, *CNS Neurotransmitters and Neuromodulators,* is intended to provide an overview in certain areas of neuroscience, covering aspects from molecular to behavioral. These books should be valuable reference and background material for anyone working in the field of neuroscience, and will hopefully encourage discussion between groups of various disciplines, which will make possible major advances in knowledge.

Trevor W. Stone, Ph.D., D.Sc.
Glasgow, Scotland

THE EDITOR

Trevor W. Stone, Ph.D, D.Sc., is a Professor of Pharmacology at the University of Glasgow in Scotland. Professor Stone graduated in 1969 from the School of Pharmacy at London University and proceeded to the University at Aberdeen in Scotland to take a Ph.D. degree under the supervision of Professor J. Laurence Malcolm. Professor Stone was appointed to a Lectureship in Physiology at Aberdeen, where he remained until 1977 when he was appointed to a senior Lectureship and subsequent Professorship in Neurosciences at St. George's Medical School in London. He is a member of the British Physiological and Pharmacological Societies, the European and International Neuroscience Research Societies, The American Society for Neuroscience, The Royal Society of Medicine in London, and the New York Academy of Sciences.

Professor Stone has held research appointments at the National Institute of Mental Health in Washington, D.C. and has been Visiting Professor of Pharmacology at the University of Auckland, New Zealand and at the Gulbenkian Institute of Science, Portugal.

Professor Stone has presented invited lectures at international meetings and has published more than 400 research papers and communications. In 1983 Professor Stone was awarded the degree of Doctor of Science by the University of London for his work on the physiology and pharmacology of the nervous system. His current research interests include the pharmacology of synaptic transmission in the nervous system, particularly with respect to amino acids and purines, the interactions between synaptic transmitters, and the role of amino acids in neurological disorders.

CONTRIBUTORS

Vera Adam-Vizi, M.D., Ph.D., D.Sc.
Department of Biochemistry II
Semmelweis University of Medicine
Budapest, Hungary

Jesse Baumgold, Ph.D.
Departments of Radiology and Pharmacology
The George Washington University Medical
    Center
Washington, D.C.

Hendrik W. G. M. Boddeke, Ph.D.
Preclinical Research
Sandoz Pharma Ltd.
Basel, Switzerland

Peter H. Boeijinga, Ph.D.
Preclinical Research
Sandoz Pharma Ltd.
Basel, Switzerland

Jennifer A. Court, Ph.D.
MRC Neurochemical Pathology Unit
Newcastle General Hospital
Newcastle-upon-Tyne, England

Esam E. El-Fakahany, Ph.D.
Division of Neuroscience Research in
    Psychiatry
University of Minnesota Medical School
Minneapolis, Minnesota

Carola Eva, Ph.D.
Istituto di Farmacologia e Terapia
    Sperimentale
Facoltà di Medicina
Università di Torino
Torino, Italy

Rosalee Grette Lydon
Department of Psychology
Dalhousie University
Halifax, Nova Scotia, Canada

Ricardo Martínez-Murillo, Ph.D.
Department of Chemical Neuroanatomy
Cajal Institute
C.S.I.C.
Madrid, Spain

Hylan C. Moises, Ph.D.
Department of Physiology
University of Michigan Medical School
Ann Arbor, Michigan

Eugenia Monferini, D.Chem.
Department of Biochemistry and Molecular
    Pharmacolgy
Boehringer Ingelheim Italia
Milano, Italy

Elaine K. Perry, D.Sc.
MRC Neurochemical Pathology Unit
Newcastle General Hospital
Newcastle-upon-Tyne, England

James H. Pirch, Ph.D.
Department of Pharmacology
Texas Tech University
    Health Sciences Center
Lubbock, Texas

José Rodrigo, Ph.D.
Department of Chemical Neuroanatomy
Cajal Institute
C.S.I.C.
Madrid, Spain

Steffen Roßner
Department of Neurochemistry
Paul Flechsig Institute for Brain Research
University of Leipzig
Leipzig, Germany

Jennifer M. Rusted, Ph.D.
Laboratory of Experimental Psychology
University of Sussex
Brighton, Sussex, England

Reinhard Schliebs, Ph.D., D.Sc.
Department of Neurochemistry
Paul Flechsig Institute for Brain Research
University of Leipzig
Leipzig, Germany

David M. Warburton, Ph.D.
Department of Psychology
University of Reading
Reading, Birkshire, England

Lynn Wecker, Ph.D.
Department of Pharmacology and Therapeutics
University of South Florida College of
   Medicine
Tampa, Florida

Mark D. Womble, Ph.D.
Department of Physiology
University of Michigan Medical
   School
Ann Arbor, Michigan

Z. Jian Yu, Ph.D.
Department of Pharmacology and Therapeutics
University of South Florida College of
   Medicine
Tampa, Florida

TABLE OF CONTENTS

Chapter 1

The Localization of Cholinergic Neurons and Markers in the CNS

Ricardo Martínez-Murillo and José Rodrigo

CONTENTS

Summary: Convincing evidence that acetylcholine (ACh) is a neurotransmitter contained in a variety of central nervous system pathways has been available for some years (see Kasa[1] and Wainer et al.,[2] for a broad overview), ACh occupying a historically senior position among neurotransmitter candidates. ACh was first described by Loewi (1921) as "Vagustoff" in the frog heart. However, only in the last few years has the field of neuroanatomy allowed, after a growth in research technology, the unequivocal visualization of cholinergic structures. Convincing conclusions on the cholinergic neuroanatomy of the central nervous system were gained by using histochemical tools, including AChE pharmacohistochemistry, immunocytochemistry, and *in situ* hybridization procedures. In addition, the anatomic technology has benefited from an interdisciplinary strategy, and it is now possible to combine behavioral or physiological experiments with tracing and/or lesioning of neuronal pathways and with histochemical characterization of the pathways involved. These methodologies will be discussed briefly below.

I. INTRODUCTION

The cholinergic system in the central nervous system (CNS) consists of several neuronal components which are localized throughout the rostrocaudal extent of the brain and spinal cord. Since there was not any available histochemical procedure for direct visualization of ACh, the demonstration of the ACh inactivating enzyme — acetyl cholinesterase (AChE) — was first employed, in combination with lesion-induced procedures, for an earliest anatomic clarification of the CNS cholinergic system in a classical series of experiments.[3-5] These studies are of exceptional relevance in the history of chemical neuroanatomy in that they were among the first to be developed for the general purpose of identifying chemically distinctive classes of cells. These experiments[3-5] were based on the hypothesis that cholinergic neurons contain detectable levels of AChE in the cell body and processes. Shute and Lewis's consequential conclusion was the identification of two major cholinergic afferents to the forebrain, termed dorsal and ventral tegmental pathways, arising in the pontomesencephalic region. The ventral tegmental AChE pathway described by Shute and Lewis represents the dopamine-containing projection system that originates in the substantia nigra and ventral tegmental area project to the striatum, nucleus accumbens, olfactory tubercle, and frontal cortex.[6-8] It is well known that the AChE ventral tegmental pathway consistently lacks the ACh synthesizing enzyme, choline acetyltransferase (ChAT), considered as an unquestionable marker for cholinergic structures.[9] Therefore, although the histochemical description concerning the ventral tegmental AChE-containing pathway to the forebrain is precise, the conclusion concerning the cholinergic nature of this projection was erroneous. Nevertheless, the cholinergic nature of some other pathways described as containing heavy AChE staining by Shute and Lewis[4] and Lewis and Shute,[5] has been subsequently confirmed, for example, the septo-hippocampal pathway[10] and the dorsal tegmental projection to the forebrain.[11]

In general, the use of AChE histochemistry for examining cholinergic neurons has resulted in conflicting results. In the cerebral cortex, for example, some investigators have not found stained cell bodies following AChE histochemistry, concluding that the cortical mantle does not contain intrinsic cholinergic perikarya.[12-15] Others, however, have reported strong staining in a population of AChE-positive cortical cells.[16,17] However, these cells were found to be non-cholinergic.[18]

Extensive basic and clinical literature has linked ACh with memory function. These studies give information on how the central cholinergic system is involved in functions representing the highest levels of integration, including cortical arousal, sleep-wakefulness cycles, selective attention, learning, memory, and discrimination processes (see Arendt et al.,[19] Dekker et al.,[20] and Jones,[21] for a broad overview), as well as in cortical sensory processing.[22] These facts inspired an effort to better define the role that the central cholinergic system plays in human cognitive processes. To date, however, the precise neuronal circuits involved and the physiological basis of the control of memory are still relatively poorly understood. Impairment of central cholinergic function has been associated with normal aging processes and in a number of neurological diseases associated with amnesia or dementia including Alzheimer's, Parkinson's and Korsakoff's diseases; dementia of the Alzheimer's type; postalcoholic dementia; and others.[19,23-25] Cholinergic neurons have also been implicated in delirium in demented patients[26] and schizophrenia.[27] A decay in neurochemical markers linked to cholinergic function (ChAT, AChE, and muscarinic binding sites) in the cerebral cortex was found to be correlated with the grade of impairment of cognitive function and with the extent of damage to the cholinergic cortical projection system in the basal forebrain.[25,28-30] It is, however, unclear as to whether these cholinergic changes are the cause or result of the disease.[31-33] Because of central cholinergic vulnerability in neurological disorders underlying cognitive deterioration, major research to establish the role of cholinergic neurons in these processes was attempted with the aim of providing potential therapeutic trials in patients exhibiting cognitive deficits.

Approximately 70% of the cholinergic innervation of the neocortex, hippocampus, amygdala, and olfactory bulb arises from neurons in the basal forebrain,[34] through which cholinergic transmission may modulate the activity of these telencephalic areas. It has been reported that 30% of neocortical neurons is cholinoceptive, and for most of these application of ACh increases the rate of firing.[35] In addition, it seems that ACh released from cholinergic fibers of the basalocortical pathway influences the morphological development of the cerebral cortex[36] and exerts a facilitatory, permissive role, in processes of synaptic plasticity during a critical postnatal period[37,38] and during adulthood.[39] While there is ample evidence for a role of ACh in the control of cerebral blood flow, it appears that the cortical cholinergic afferents originating in the basal forebrain are not involved in such an event.[40,41] It is presumed that ACh released from intrinsic cortical cholinergic cells may mediate the cerebrovascular action.

The proven implication of central cholinergic transmission in higher brain processes, as well as in neocortical development and plasticity, has made understanding of the morphology and distribution of cholinergic neurons the subject of considerable attention and continued reinvestigation and reappraisal. To elucidate the structural organization, cytochemistry, and functional connections of the central cholinergic system in different species will provide a new viewpoint relevant to its functional significance in normal and diseased brains, particularly the pathophysiological implications of its degeneration in Alzheimer's disease and the dementia accompanying Parkinson's disease.

In this chapter we are concerned with the current status of the most conspicuous features of the neuroanatomy of the cholinergic neurons in the CNS of vertebrates, assessed from light and ultrastructural histochemical studies.

II. CONSIDERATIONS ON THE METHODOLOGIES FOR VISUALIZING CHOLINERGIC STRUCTURES

The different indicators of cholinergic activity such as concentrations of ACh,[42,43] AChE, and ChAT activities;[16,44–47] high-affinity choline uptake;[42] and muscarinic receptor sites[48–51] can be used for the localization of cholinergic structures, but these do not provide morphological peculiarities of cholinergic neurons.

A. AChE HISTOCHEMISTRY

While not all AChE-positive cells are cholinergic, most cholinergic cells are AChE positive.[12] Following this criterion, AChE histochemical staining was employed in studies of the organization of the basal forebrain cholinergic system in normal subjects and after diverse experimental conditions.[16,44,52–54] In the cerebral cortex and hippocampus, most histochemically demonstrable AChE depends on the integrity of the nucleus basalis magnocellularis and medial septum-diagonal band complex, respectively. These structures are the main source of cholinergic input to these regions. Comparative studies have concluded that the pattern of staining of nerve fibers identified after ChAT immunocytochemistry and AChE histochemistry in the cerebral cortex and hippocampus are analogous.[10,15,55] However, it is noteworthy that ChAT-immunoreactive neurons intrinsic to the rodent cortex consistently lack AChE.[18]

B. AChE PHARMACOHISTOCHEMISTRY

For reasons outlined above, the use of AChE histochemistry alone does not guarantee the cholinergic nature of stained elements. Instead, the use of a pharmacohistochemical procedure to reveal AChE-synthesizing neurons[56,57] was considered as an alternative marker for mapping domains of the mammalian cholinergic system in several regions of the forebrain.[12,14,15,34,57–60] AChE pharmacohistochemistry incorporates a modification of the histochemical approach for demonstrating AChE by pretreating experimental animals with an irreversible AChE inhibitor, diisopropyl phosphorofluoridate (DFP) (introduced by Fukuda and Koelle[61]) before processing the brain for AChE histochemistry.[56,57,62] Biochemical and histochemical studies[63,64] have suggested that rapid regeneration of AChE in the CNS following DFP intoxication occurs predominantly in the cell soma of cholinergic neurons. Following the pharmacohistochemical procedure, compared to most other neurons cholinergic neurons consistently show "intense staining" reflecting newly synthesized AChE, 4 to 8 h after DFP treatment. However, in some circumstances, it is hard to decide whether particular neurons meet the criterion of "intense staining". It is also remarkable that noncholinergic cells, such as those in the locus ceruleus and thalamus, also express intense staining after AChE pharmacohistochemistry.[65–67] However, the pharmacohistochemical procedure offers several advantages. Because of its relative simplicity, AChE pharmacohistochemistry can be used to obtain morphological data about particular cholinergic cell populations known to exhibit intense AChE staining following this procedure. In addition, AChE pharmacohistochemistry can be applied in combination with immunocytochemistry (Figure 1-1) or tract-tracing procedures to simultaneously examine the distribution of AChE perikarya and that of particular antigens or terminal fields, in the same sections, using both light and electron microscopes.[60,68]

In summary, AChE histochemistry can be considered a rather limited procedure for labeling CNS neurons that use ACh as a neurotransmitter at their synapses. Therefore, alternative methods have been sought.

With the discovery of new methodologies to specifically stain cholinergic structures, the anatomy of the central cholinergic pathways has been properly delineated. These include (1) immunocytochemical detection of ACh; (2) immunocytochemical detection of the rate-limiting enzyme required for the synthesis of the neurotransmitter ACh, ChAT; (3) in situ hybridization procedures; and (4) cytochemical markers of cholinergic neurons other than ChAT. In addition, several other studies have combined ChAT immunocytochemistry or AChE pharmacohistochemistry procedures with chemical and mechanical lesion-induced techniques, appropriate track tracing techniques, and Golgi impregnation.[4,12,33,66,69–78]

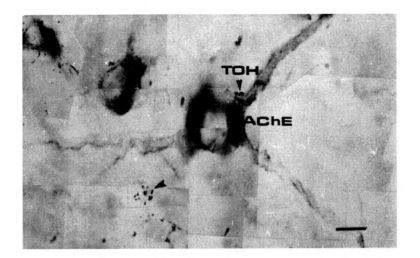

Figure 1-1   Oil immersion photomicrograph simultaneously showing in the same section details of the distribution of tyrosine hydroxylase (TOH) immunoreactive fibers (arrowheads) and intensely stained AChE-positive neurons in the region of the nucleus basalis of Meynert in the rat. (Bar = 10 μm.)

Finally, functional studies have been conducted by using intracellular horseradish peroxidase techniques in combination with antidromic identification which allows correlation of physiological properties of neurons with their morphology.[79]

C. IMMUNOCYTOCHEMISTRY OF ACh

Following a two-step immunogen synthesis procedure, Geffard et al.[80] reported the synthesis of an antibody to ACh and its use as a specific marker for cholinergic neurons in the CNS. This procedure allows a direct visualization of ACh. These authors reported that the detection of endogenous ACh requires particular conditions of the fixation protocol which can be also used for the fixation of norepinephrine, dopamine, and serotonin in the rat CNS. However, since then, ACh antibodies have to our knowledge never been used by other research groups.

D. IMMUNOCYTOCHEMISTRY OF ChAT

The immunocytochemical localization of ChAT constitutes a major improvement in the search for specific tools for a reliable identification of cholinergic neurons. Most of the difficulties involving the identification of cholinergic cells were overcome through immunohistochemical studies using those antibodies. Since ChAT is regarded as being restricted to cholinergic neurons, the immunohistochemical identification of neurons containing ChAT is considered to be a reliable means for recognizing the cholinergic nature of stained structures and providing conclusive information on the anatomy of cholinergic systems. Early studies on this line[81–83] gave rise to controversy because the incomplete purity of the enzyme preparations used to generate the immunological probe questioned the specificity of the staining subsequently obtained. These difficulties have been solved by the use of ChAT purified to near homogeneity, allowing production of monospecific antisera and/or monoclonal antibodies to the enzyme.[84–86] By using these antibodies, considerable progress has been made in defining the morphology and distribution of central cholinergic structures.

In immunocytochemical studies, it should be taken into account that the primary antibody may cross-react with unrelated proteins sharing similar epitopes. This is probable in earlier immunocytochemical studies considering the fact that ChAT is very difficult to purify and polyclonal antibodies raised against impure ChAT preparations in earlier studies possibly cross-reacted with antigens other than ChAT. The application of hybridoma technology yields highly specific antibodies which can be selectively screened by appropriate assay techniques. Therefore, monoclonal antibodies are a sensitive tool which specifically recognizes cholinergic nerve cells (Figure 1-2A; and later, Figures 1-7D–F, 1-8, 1-9, and 1-11–13).

E. *IN SITU* HYBRIDIZATION OF ChAT mRNA

The isolation of ChAT cDNA clones has made it possible to study more precisely the distribution of central cholinergic neurons by the detection of endogenous ChAT mRNA based on *in situ* hybridization

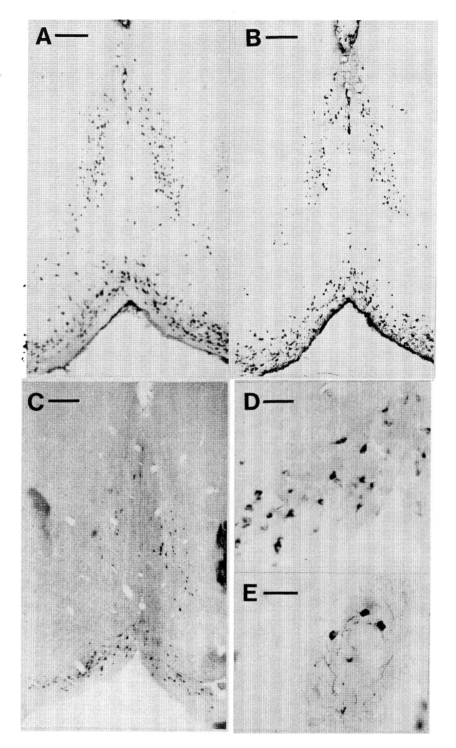

Figure 1-2 Cholinergic neurons visualized by (A) immunoperoxidase staining of ChAT; (B) immunoperoxidase staining of LNGFR; and (C) NADPH-diaphorase histochemistry, in the medial septum complex-diagonal band nuclei in a rostrocaudal level of the rat forebrain. Notice comparing (A) with (B) and (C) the same distribution of stained structures. Also notice comparing (A) or (B) with (C) that the number of stained structures following the NADPH-diaphorase histochemical procedure is remarkably smaller. (D) Higher power magnification of NADPH-diaphorase-positive neurons in the diagonal band of Broca. (E) Illustrates the aspect of ChAT-positive neurons in the isla magna of the islands of Calleja complex. (Bars = 200 μm [A]–[C], 50 μm [D], [E].)

procedure.[87–92] This approach provides unequivocal data on the location of neurons that synthesize ChAT. Because ChAT mRNA constitutes a conclusive marker for cholinergic neurons, hybridocytochemistry can be accepted as a decisive approach for revealing cholinergic somata. The specificity of this procedure is based on the high degree of specificity afforded by DNA and RNA probes. Although ChAT immunocytochemistry provided details of distal dendrites and axons, hybridocytochemistry does not. The use of nonradiolabeled probes allows, however, a good definition of neuronal perikarya similar to that provided under the light microscope by ChAT immunocytochemistry.

The use of hybridocytochemistry confirmed previous results using ChAT immunocytochemistry, except that intrinsic neurons synthesizing ChAT were not detected in particular regions of the forebrain, including the cerebral cortex and hippocampus. These results questioned the specificity of ChAT immunocytochemistry.[87,88,92] However, a more recent study by Lauterborn et al.,[90] described the presence of lightly stained ChAT synthesizing neurons in the cerebral cortex. A variety of explanations may account for the topographic differences seen through the application of ChAT-immunocytochemistry and ChAT mRNA hybridocytochemistry. It it very likely that cholinergic somata in the cerebral cortex and hippocampus express very low levels of ChAT mRNA. In this regard, it should be noted that in the rat ChAT-positive cells in the cerebral cortex and hippocampus appear to stain less intensely than other types of ChAT-positive neurons following ChAT immunocytochemistry. It is also likely that some variant of ChAT bearing the same epitope as that recognized by the monoclonal antibody may be present in these polemic neurons.

F. CYTOCHEMICAL MARKERS OF CHOLINERGIC NEURONS OTHER THAN ChAT

The various groups of cholinergic neurons have been shown to display individualized cytochemical markers, some of which are species specific.

Current information indicates that the nerve growth factor (NGF) is produced in the brain and appears to play an important regulatory role in the activity of cholinergic neurons (see Kiss et al.,[93] for a broad overview). The effect of NGF is thought to be mediated by specific receptors, which can be present in a high- (HNGFR) as well as a low-affinity (LNGFR) form. In the basal forebrain of both human and rat, the topography of LNGFR-positive neurons is strikingly similar to that of ChAT-containing neurons,[93–97] while cholinergic neurons of the upper brainstem show no correlation, indicating that LNGFRs, are restricted to cholinergic cell populations in the basal forebrain. Taken together, these results suggest immunocytochemistry of LNGFR as a choice procedure for visualizing cholinergic neurons in the basal forebrain (Figure 1-2B). By comparison, it should be noted that most cholinergic neurons in the striatum lack LNGFR immunoreactivity.

In the monkey, many neurons in the septum-basal forebrain cholinergic complex are galanin positive. In the human, however, almost none of the magnocellular cholinergic basal forebrain neurons have been shown to contain galanin.[98–101] In the rat, many cholinergic neurons in the medial septum and diagonal band complex, but not those in the substantia innominata and nucleus basalis of Meynert, contain galanin.[100,102–105]

Besides ChAT, LNGFR, and galanin, cholinergic neurons contain other cytochemical markers such as reduced nicotinamide adenine dinucleotide phosphate (NADPH)-diaphorase[104,106–108] and the vitamin D-dependent calcium-binding protein calbindin-D28k.[109–111] However, considerable regional and species-specific variations have been reported.

Cholinergic neurons in the midbrain and pons are characterized by their ability to utilize the reduced cofactor NADPH. In the rat, monkey, baboon, and man, the use of NADPH-diaphorase histochemical approach provides a simple, reliable method to selectively stain virtually all cholinergic neurons in the pontine reticular formation.[108,112–114] Characteristically, between 20 and 30% of basal forebrain cholinergic neurons in the rat contain NADPH-diaphorase activity (Figures 1-2C, D). In striking contrast, NADPH-diaphorase never occurs in basal forebrain cholinergic neurons in the monkey, baboon, and human.[112] Finally, none of the cholinergic neurons in the striatum of these species exhibit NADPH-diaphorase activity.[107,112]

In the rat, calbindin-positive neuronal population in the basal forebrain characteristically lacks cholinergic markers. In the monkey, baboon, and human, most cholinergic neurons in the basal forebrain contain calbindin.[112] In contrast, in the upper brain stem of the rat, monkey, baboon, and human, calbindin- and ChAT-positive neuronal populations are completely separated.[112] Characteristically none of the cholinergic neurons in the striatum display calbindin immunoreactivity.

Table 1-1 **Central cholinergic neuronal somata**

Local Circuit Cholinergic Neurons

Cholinergic neurons in the caudate-putamen, nucleus accumbens, main and accessory olfactory bulbs, anterior olfactory nucleus, olfactory tubercle, hippocampus, cerebral cortex, basolateral hypothalamus, and spinal cord

Cholinergic Projection Neurons

Magnocellular basal nucleus (MBN) (Ch1-Ch4): cholinergic neurons in the medial septal nucleus, nucleus of the diagonal band of Broca, substantia innominata, and nucleus basalis magnocellularis

Pontine cholinergic system (Ch5-Ch6): cholinergic neurons in the pedunculopontine tegmental and laterodorsal tegmental nuclei

Cholinergic motor neurons in the spinal cord, cholinergic neurons of cranial nerves 3–7, 9–12

Cholinergic neurons of the sympathetic nervous system

Cholinergic neurons of the parasympathetic nervous system in the spinal cord

---

In the cerebral cortex, cholinergic neurons exhibit vasoactive intestinal polypeptide (VIP) immunoreactivity.[115]

In the rodent upper brain stem about half of the cholinergic neurons contain substance P (SP).[114,116] In addition, a very small number of cholinergic neurons in the basal forebrain exhibit γ-aminobutyric acid (GABA) uptake or glutamic acid decarboxylase immunoreactivity.[117–119] In striking contrast, the basal forebrain cholinergic neurons were found to be distinctly different from GABA-synthesizing neurons in a recent study.[120]

III. ORGANIZATION AND MAJOR PROJECTIONS OF CHOLINERGIC NEURONS OF THE BRAIN

Several groups of investigators have provided data on the morphology and distribution of cholinergic perikarya in the CNS, following different approaches. The use of ChAT immunocytochemistry[57,58,69,71–74,121,122] and AChE pharmacohistochemistry[12,66,70,123,124] allowed the conclusion that central cholinergic perikarya in the mammalian CNS form five major groups in: (1) the cerebral cortex and hippocampus, (2) the striatum, (3) the magnocellular basal nucleus, (4) the pontomesencephalic tegmentum, and (5) the cranial nerve motor nuclei and motor neurons of the spinal cord. Although many much of these results have been obtained in experiments on the rat brain, there is evidence that these cell groupings are remarkably stable across many vertebrate species, including the primate, providing a new structural viewpoint relevant to the mammalian CNS in general. While the cholinergic system displays a complex anatomic expansion and shows extensive chemical differentiation as the phylogenetic scale is ascended, the organization and presumably the function of cholinergic systems appear to be conserved.[125]

Central cholinergic somata occur in many areas of the CNS. These cells are intermingled among numerous noncholinergic cells and frequently distributed in close proximity to several ascending and descending, more or less well-defined fiber systems, many of which make direct synaptic contacts with cholinergic and/or noncholinergic structures in the region. Cholinergic neurons do not form a single neuronal system or a continuous neuronal network. Instead, certain distinct groups have been mapped out. Following the initial description by Cozzari and Hartman,[81] several other reports appeared describing partial or complete maps illustrating the morphology and distribution of ChAT-containing structures in the rat,[57,69,71,85,121,122] cat,[126–128] monkey,[72,129,130] baboon,[123] mouse,[131] and human.[132,133]

Central cholinergic neurons can be subdivided into two groups, projection neurons and interneurons (see Table 1). In addition, the cholinergic projection cells in the forebrain and upper brain stem can be divided up in different groups (Ch1-Ch6 of Mesulam et al.[72,73]) according to where they project. The topography of ChAT-containing central neuronal somata in the rat brain is represented schematically in Figures 1-3 and 1-4.

A. CHOLINERGIC NEURONS IN THE TELENCEPHALON
1. Projection Neurons

Loosely organized groups of multipolar neurons are located in the basal forebrain complex giving rise to long direct projections to a wide range of cortical and subcortical structures.[12] These projection neurons

8

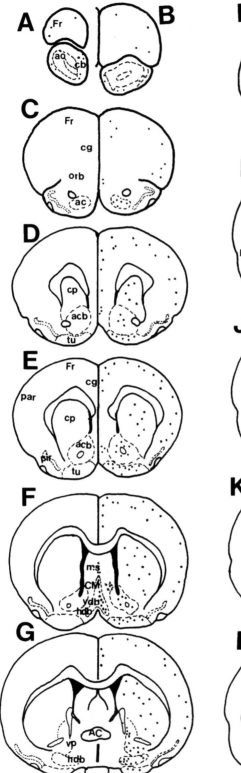
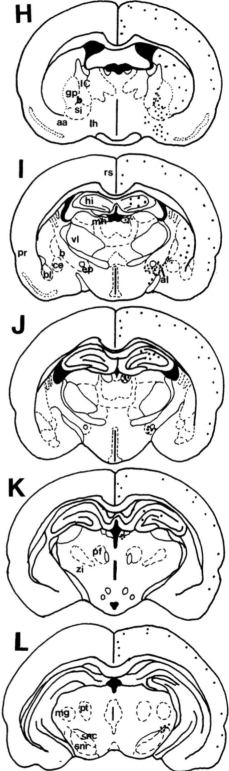

Figure 1-3

Figures 1-3 and 1-4 Schematic drawing of coronal sections illustrating the distribution of cholinergic neurons (dots) revealed by immunoperoxidase staining of ChAT at different rostrocaudal levels of the rat central nervous system. Diagrams (A)-(T) illustrate coronal sections at progressively more caudal levels of the brain, according to the stereotaxic atlas of Paxinos and Watson.[248] Cervical (U) and thoracic (V) spinal cord levels are also shown. Abbreviations: aa, anterior amygdalar area; AC, anterior commissurae; acb, nucleus accumbens; al, ansa lenticularis; ao, anterior olfactory nucleus; amb, nucleus ambiguus; b, nucleus basalis; bl, basolateral amygdala; cb, olfactory bulb; ce, central amygdaloid nucleus-lateral; cc, central canal; cen, cochlear efferent nucleus; cg, cingulate cortex; CM, isla magna of the islands of Calleja complex; cp, caudate putamen; dr, dorsal raphe nucleus; dh, dorsal horn; hi, hippocampus; im, intermediolateral horn of the spinal cord; ep, entopeduncular nucleus; Fr, frontal cortex; gp, globus pallidus; hdb, horizontal limb nucleus of the diagonal band of Broca; IC, internal capsule; ic, inferior colliculus; ip, interpeduncular nucleus; lh, lateral hypothalamus; ldt, laterodorsal tegmental nucleus; mg, medial geniculate; mh, medial habenular nucleus; mr, median raphe nucleus; ms, medial septum; orb, orbitofrontal cortex; ppt, pedunculopontine tegmental nucleus; par, parietal cortex; pb, parabigeminal nucleus; pf, parafascicular nucleus; ph, prepositus hypoglossal nucleus; pir, piriform cortex; pn, pons; pr, perirhinal cortex; pt, pretectum; rs, retrosplenial cortex; sal, salivatory nucleus; si, substantia innominata; snc, substantia nigra pars compacta; snr, substantia nigra pars reticulata; tu, olfactory tubercle; vdb, vertical limb nucleus of the diagonal band of Broca; ven, vestibular efferent nucleus; vh, ventral horn; vl, ventrolateral thalamus; vp, ventral pallidus; zi, zona incerta; 3,4,5,6,7,10,12, cranial nerve nuclei; and 7a, accessory nuclei of cranial nerve 7.

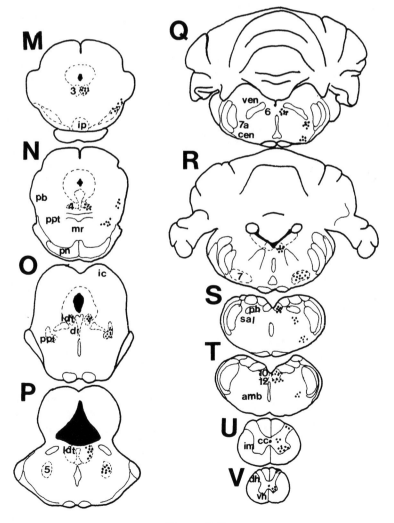

Figure 1-4

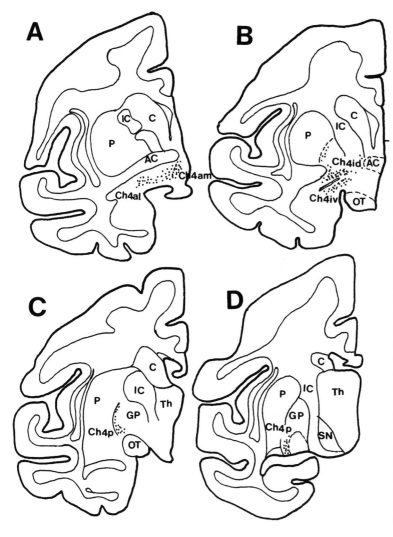

Figure 1-5 Schematic drawing showing the location of Ch4 sectors in the rhesus monkey, according to a description by Mesulam and Mufson,[249] Drawings (A)–(D) illustrate coronal sections from rostral to caudal levels of the brain. Abbreviations: AC, anterior commissure; C, caudate; Ch4al, anterolateral sector of the Ch4 group; Ch4am, anteromedial sector of the Ch4 group; Ch4id, Ch4iv, intermediate sector dorsal (d) and ventral (v) of the Ch4 group; Ch4p, posterior sector of the Ch4 group; GP, globus pallidus; IC, internal capsule; OT, optic tract; P, putamen; SN, substantia nigra; and Th, thalamus.

constitute a continuous band of large isodendritic cells; rather than being arranged as a series of discrete nuclear groups,[57,134] they start rostrally at the anterior end of the olfactory tubercle and extend caudally as far posteriorly as the cerebral peduncle, a neuroanatomic contiguum referred to collectively as the magnocellular basal nucleus (MBN)[54,135] or cholinergic basal nuclear complex.[136] Cholinergic corticopetal neurons in the basal forebrain are situated medially and basally to the basal ganglia complex forming the major cholinergic cell agglomeration of the forebrain. This sheath leads to further cholinergic neurons in the entopeduncular nucleus[137] and occasionally in the amygdala. In the rat beginning rostrally, the basal forebrain cholinergic complex occupies the medial septum; the vertical and horizontal limbs of the diagonal band of Broca including the nucleus preopticus magnocellularis, the substantia innominata, the basolateral hypothalamus, and the ventromedial region of the globus pallidus; and adjacent portions of the internal capsule and the so-called nucleus of the ansa lenticularis (Figure 1-3F–J). The basolateral hypothalamic cholinergic cellular group and the nucleus of the ansa lenticularis are considered cholinergic diencephalic structures (see Section III.B). In the primate, the picture appears to differ somewhat in that at least the cholinergic cells traversing the substantia innominata appear to be divided into medial and lateral columns rather than a single stream of cells coursing caudally[138,139] (Figure 1-5A).

Cholinergic cells in the basal forebrain provide the major cholinergic afferent fibers to the entire neocortex, the hippocampus, the amygdala, and the olfactory bulb.[4,12,15,16,34,70,72,105,140–142] One problem in the study of basal forebrain cholinergic neurons, in general, has been the lack of consensus concerning the nomenclature for anatomic regions that contain the cholinergic basal nuclear complex. Although there is not a sharp border between cholinergic projection neuronal groups in the forebrain, an attempt at classification was made based on their connectivity pattern. On these grounds, cholinergic projection neurons in the forebrain were subdivided into four major groups or sectors termed Ch1-Ch4 in the rat, monkey, and human.[72,73,143] A schematic representation of the major cholinergic pathways is represented later in Figure 1-10. It should be noted, however, that anatomic, neurochemical, and pathological evidence has emphasized the similarities between these sectors, leading to the suggestion that the distinction is topographical rather than functional. The Ch1 sector consists of ChAT-positive neurons contained within the medial septal nucleus. These vertically oriented neurons, which are the smallest in the Ch1-Ch6 cell groups, are concentrated along the midline of the septum. A few cholinergic cells are also located laterally along the outer edge of the nucleus. These neurons provide substantial cholinergic projections to the hippocampus. The Ch2 sector consists of ChAT-positive neurons within the vertical limb nucleus of the diagonal band of Broca. This cell group is continuous with the one in the medial septal nucleus and provides cholinergic projection to the hippocampus. Taken together, the Ch1-Ch2 sectors provide the major component of cholinergic projection to the hippocampal formation. The Ch3 cholinergic cell group occupies the horizontal limb nucleus of the diagonal band of Broca. The cells in the Ch3 sector, principally in the lateral part of the nucleus, project to the olfactory bulb. A portion of cholinergic cells in the medial part of the horizontal limb of the diagonal band of Broca is in continuity with the Ch2 sector anteriorly and with the Ch4 dorsally. These neurons project to piriform, cingulate, and entorhinal cortices. Therefore, a proportion of these cells may be members of Ch4 sector. The Ch4 cell group is defined on the basis of the neocortical projections of its cellular constituents. Cholinergic neurons of the Ch4 sector are distributed in the substantia innominata and the magnocellular preoptic nucleus. Moreover, a collection of Ch4 cholinergic neurons can be found in the innermost portions of the globus pallidus and in adjacent portions of the internal capsule.[144–147] The latter collection of cholinergic cells constitutes the main body of the termed nucleus basalis magnocellularis of Meynert which constitutes the main source of cholinergic input to neocortex and basolateral amygdala. In the rat, additional cholinergic corticopetal neurons have been reported in the anterior amygdala.[147] However, this is not the case in primates. A proportion of the cholinergic input to the thalamus was reported from the basal forebrain in both rodents and humans,[148–151] particularly from the nucleus basalis magnocellularis.[152]

The Ch4 group is much more extensive in primates than in other species and has been subdivided into four distinct components (Figure 1-5), each with a preferential set of cortical connections:[70,72,129,136,139,143,153] (1) the anteromedial sector (Ch4am) that provides innervation to medial cortical areas; (2) the anterolateral compartment (Ch4al) which innervates ventral orbital, frontal, parietal opercular regions, and the amygdala; (3) the intermediate sector (Ch4i-d, v) to lateral frontal, parietal, peristriate, and temporal regions including the insula; and (4) the posterior sector (Ch4p) to superior temporal and temporopolar areas.

By comparison, cholinergic neurons of the MBN in man constituted an irregularly shaped cell group. These cells constitute two main clusters: (1) the medial septal/diagonal band nuclei and (2) the nucleus basalis (Figure 1-6). The former cell group lies along the ventromedial surface of the rostral basal forebrain, and contains the typical magnocellular AChE-rich neurons of the MBN intermixed with many small neurons, some of which also exhibit AChE. In striking contrast, the nucleus basalis is located more dorsally and caudally and contains primarily magnocellular neurons. First descriptions on the MBN were made by Kolliker (see translation in Gorry[154]). More recently, Mesulam et al.,[72] provided data comparing the distribution of neurons in the MBN of the human with their histochemical and cytoarchitectonic parcellation of these neurons in the monkey. These authors delineated the cholinergic sectors Ch1-Ch4 in the human basal forebrain, which are equivalent to those previously identified in monkey. Although it appears that the cholinergic neurons in the human anterior portion of the Ch4 sector cluster into medial and lateral portions, it is not, however, possible to distinguish anterolateral and anteromedial cell groups[54] described in the monkey.[72] More recently distribution of the MBN in man has been described,[54,138] suggesting that the topographic organization of the magnocellular basal projection to the cerebral cortex in other species probably exists in man as well.

The nucleus basalis in the rat is much less densely populated by cholinergic neurons than the equivalent region in man. As described above, the rat nucleus basalis consists of scattered clusters of cholinergic neurons in the substantia innominata, and along the ventral and medial borders of the globus pallidus and adjacent portions of the internal capsule (Figures 1-3H; 1-7A–D). In humans, by comparison, there are many neurons of the nucleus basalis along the borders of the globus pallidus, but only a small number of cholinergic neurons are located within it (Figure 1-6B–E).

The projections originating in the basal forebrain toward the cereberal cortex and hippocampus are topographically organized.[147,155,156] Each cortical area is innervated by a characteristic subset of MBN neurons, always located in close association with descending cortical fibers. In the rat, three different

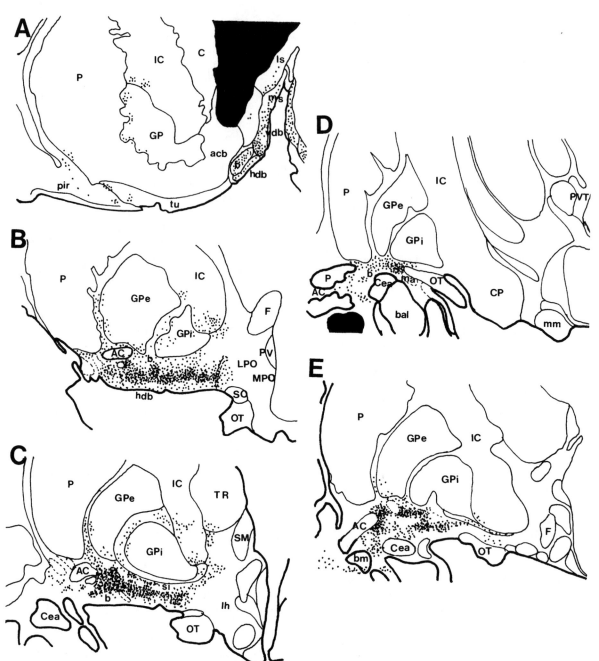

Figure 1-6 Schematic drawing showing the distribution of MBN neurons in man, according to a description by Saper and Chelimsky.[54] Panels (A)–(E) illustrate sections from rostral to caudal levels of the brain. Abbreviations: acb, nucleus accumbens; b, nucleus basalis; bal, basal accessory nucleus of the amygdala lateral pars; bm, basal nucleus of the amygdala; C, caudate; Cea, central nucleus of the amygdala; Cp, cerebral peduncle; F, column of the fornix; GP, globus pallidus; GPe,i, globus pallidus external and internal segments; hdb, horizontal limb nucleus of the diagonal band of Broca; IC, internal capsule; lh, lateral hypothalamus; LPO, lateral preoptic area; ls, lateral septum; ma, medial amygdaloid nucleus; mm, medial mammilary nucleus; MPO, medial preoptic area; ms, medial septum; OT, optic tract; P, putamen; pir, piriform cortex; PV, periventricular hypothalamic nucleus; PVT, periventricular nucleus of the thalamus; SM, stria medularis; SO, supraoptic nucleus; TR, thalamic reticular nucleus; tu, olfactory tubercle; and vdb, vertical limb nucleus of the diagonal band of Broca.

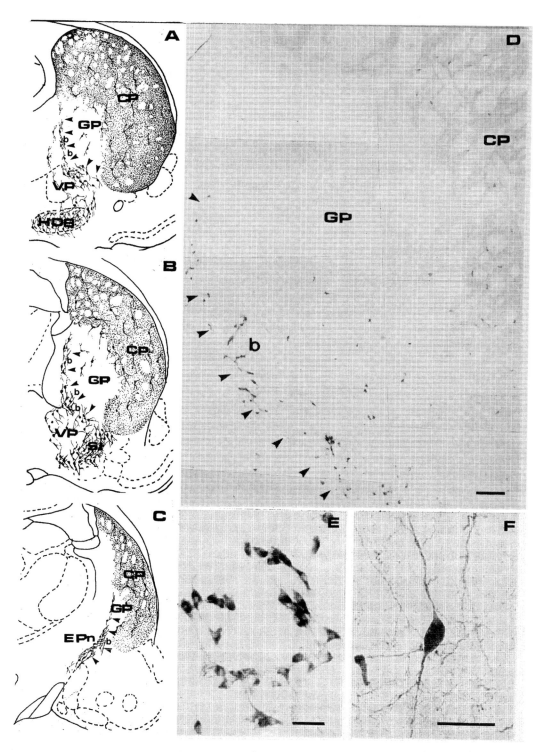

Figure 1-7 (A)–(C) Schematic drawings and (D) light micrograph representing the distribution of ChAT-positive neurons in several rostrocaudal levels of the rat forebrain. Notice the specific distribution of cholinergic neurons in the ventromedial division of the globus pallidus (GP). (E) Also note the characteristic clusters formed by cholinergic neurons in the nucleus basalis (b). (F) represents a cholinergic neuron in the nucleus basalis. Abbreviations: CP, caudate-putamen; EPn, entopeduncular nucleus; SI, substantia innominata; VP, ventral pallidum; and HDB, horizontal limb nucleus of the diagonal band of Broca. (Bars = 100 μm [D], 50 μm [E],[F].)

subsystems were identified by Eckenstein et al.,[157] originating from particular neuronal groups in the basal forebrain and innervating different cortical areas. The first arises from neurons in the septal area and diagonal band of Broca and innervates the cingulate and retrosplenial cortex. The second originates in the lateral portions of the horizontal limb of the diagonal band of Broca and innervates the entorhinal and olfactory cortex. The third starts from neurons in the globus pallidus, fans out laterally through the neostriatum toward the external capsule, and innervates all other cortical areas.

Although the cholinergic projection from the basal forebrain as a whole is widespread, the projection of individual cells is limited to a very restricted area within the cortex, not more that 1 to 1.5 mm in diameter.[147,158] However, there is evidence suggesting that cortical projection neurons with highly specific field projection may have an overlapping distribution and/or may give off collateralized axons that innervate divergent terminal fields,[34] being capable of simultaneously influencing widespread neocortical zones. Cholinergic neurons in the MBN probably receive inputs from the same cortical areas which they innervate.[147] However, this latter finding has not been further supported by more recent studies.[159] In the hippocampus direct feedback projections of Ammon's horn to the medial septum-diagonal band complex show a topographic organization.[160]

Despite the current emphasis on magnocellular cholinergic projection neurons, it is well known that a large proportion of large corticopetal neurons found in the basal forebrain in rodents is noncholinergic.[102,118,120,161–163] In the globus pallidus of the rat, a substantial population of small cells was reported to have axonal projections to the cerebral cortex and occasionally these cells also gave off axonal collaterals to the paraventricular thalamic nucleus.[164] These noncholinergic cells are intermingled in various degrees with cholinergic neurons. In striking contrast, it has been reported that in primates about 90% of the neurons projecting to the neocortex are cholinergic.[72] In recent years, GABAergic neurons have been identified in the basal forebrain and were found to be of importance in mechanisms of cortical activation. Because of the fact that in the basal forebrain the distribution of GABAergic neurons resembles that of cholinergic projection neurons, it has been suggested that these GABAergic neurons be considered as constituents of the MBN.[120]

Neurons in the MBN receive inputs from a variety of sources throughout the neuroaxis. The wide range of structures that project to the MBN can be divided in two groups which are linked by many ascending or descending connections: (1) those from limbic forebrain structures and (2) those from the hypothalamus and the brain stem. Following tracer injections in the MBN, labeling was seen in a variety of brain stem, thalamic, and hypothalamic nuclei as well as in the basal nuclei, the nucleus accumbens, the amygdaloid complex, the dorsolateral region of the frontal cortex, and the hippocampus (see Carnes et al.,[159] Cullinan and Zaborszky,[165] Haring and Wang,[166] and Záborszky and Cullinan,[167] for a broad overview). Afferents from the contralateral MBN have also been identified.[159,168,169]

Finally, in the rat frontal cortex (Krieg's area 10), cholinergic pyramidal neurons also have been described using monoclonal antibody to ChAT.[170] These authors suggested that cholinergic pyramidal cells in the frontal cortex may project to the contralateral hemisphere, having an important role for the function of memory and learning.

2. Local Circuit Neurons

These are widely distributed through telencephalic and diencephalic structures (Figure 1-3A–K). In the forebrain these are localized principally in dopamine-rich regions including the caudate-putamen and all compartments of the ventral striatum: the nucleus accumbens, the olfactory tubercle as well as island of Calleja complex (Figure 1-2E). In the cat, large striatal, apparently cholinergic, cortical projection neurons have been retrogradely labeled from the cortex, although they appear to be very occasional.[171–173] A few intrinsic cholinergic neurons are also located in the entopeduncular nucleus, the anterior amygdalar area, the main and accessory olfactory bulbs, and the anterior olfactory nucleus (Figure 1-8I).

The cholinergic extrinsic and intrinsic neuronal components of the cerebral cortex have been reported.[18,71,157] While the extrinsic component in both rodents and humans is thought to be axons derived mainly from large multipolar cells in the MBN, the intrinsic components are thought to be local bipolar-shaped circuit neurons, presumed as being inhibitory[174] (Figure 1-8G). In striking contrast, cortical cholinergic neurons in the adult primate brain have not been reported.[129,143] Although much of cholinergic innervation of the hippocampus is derived from MBN, an intrinsic source from local circuit cholinergic neurons in the rat has also been described.[2,10,18,175–179] ChAT-positive cells were found in the caudal and ventral hippocampus (Figure 1-8H).

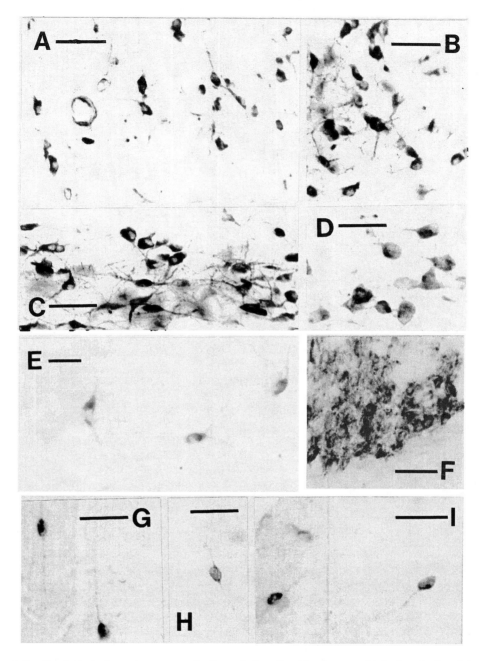

Figure 1-8 Central cholinergic neurons in the rat visualized by immunohistochemical staining for ChAT in the (A) medial septum, (B) vertical and (C) horizontal limbs of the diagonal band of Broca, (D) substantia innominata, (E) neostriatum, (F) medial habenular nucleus, (G) cerebral cortex, (H) hippocampus, and (I) anterior olfactory nucleus. (Bars = 50 μm [A]–[D], [F], 25 μm [E], [G]–[I].)

It is noteworthy that the detection of cholinergic neurons in the neocortex and hippocampus by using ChAT-immunocytochemistry has been controversial, with different groups reporting positive[74,178,180–182] or negative[69,71,85,121,183] results. It has been hypothesized that the differences in these results may be a function of the sensitivity of antibodies to ChAT and that the use of antibodies with better sensitivity may show that controversial cholinergic neurons can be detected readily.[74] In this regard, it should be noted that in the rat ChAT-positive cells in the cerebral cortex appear to stain less intensely than other types of ChAT-positive neurons.

B. CHOLINERGIC NEURONS IN THE DIENCEPHALON

The thalamus proper, the metathalamus, and the thalamic reticular nucleus do not bear cholinergic somata. Cholinergic cell bodies are located in the rostral basolateral hypothalamic area adjacent to the substantia innominata[69,184,185] (Figure 1-3H). These cells often appear as a loosely clustered group dorsolateral to the supraoptic nucleus and appear to project locally to the magnocellular neuroendocrine cells of the supraoptic nucleus. Most cells within this group are multipolar and range from 20 to 32 μm in diameter.[48] Another cholinergic group is located in the so-called nucleus of the ansa lenticularis (Figure 1-3I). Finally, small intrinsic cholinergic somata are located in the epithalamic medial habenular nucleus (Figures 1-3I–K and 1-8F).

C. CHOLINERGIC NEURONS IN THE MESENCEPHALON, PONS, AND MEDULLA

1. Midbrain and Pons

Following the nomenclature used by Mesulam et al.,[73] cholinergic projection neurons in the rostral brain stem form two groups of medium-sized neurons, Ch5 and Ch6. These cells form a continuous column arising rostrally from caudal aspects of the substantia nigra rostrally to rostral domains of the locus ceruleus caudally.[73,129,186–190] These cells appear to be mainly involved in wakefulness and sleep (see Jones,[21] for a broad overview). The neurons of the Ch5 cell group are within the limits of the pedunculopontine nucleus (Figure 1-4N, O). A few are also located in the cuneiform and parabrachial nuclei. The most rostral elements of the Ch5 cell group appear in the reticular part of the substantia nigra[113,191,192] (Figure 1-4M). Cells of the Ch5 cell group are intermingled with many noncholinergic neurons in the cuneiform and pedunculopontine nuclei of the pontomesencephalic reticular formation. The Ch6 group remains within the periventricular gray, most of these cells within the boundaries of the laterodorsal tegmental nucleus (Figure 1-4O, P). Cholinergic neurons in the pedunculopontine and laterodorsal tegmental nuclei are located in close association with several fiber tracts, including the superior cerebellar peduncle, lateral lemniscus, dorsotegmental tract, and medial longitudinal fasciculus.[77]

The Ch5-Ch6 cholinergic groups provide direct ascending projections that terminate in a number of target structures in the midbrain, diencephalon and telencephalon, including the superior colliculus, anterior pretectal area, interstitial magnocellular nucleus of the posterior commissure, lateral habenular nucleus, thalamus, magnocellular preoptic nucleus, lateral mammillary nucleus, basal forebrain, olfactory bulb, and medial prefrontal cortex[73,75,150,193–195] as well as to the pontine nuclei[196] (see Figure 1-10). As many as 80% of projection neurons in the pedunculopontine tegmental nucleus are cholinergic.[75] The widespread cholinergic pathway from the Ch5-Ch6 sector to the thalamus constitutes an important component of the ascending reticular activating system. On this basis, the basal forebrain and mesopontine tegmental cholinergic systems may influence cortical activity through both direct and indirect pathways that could be activated simultaneously. Projections toward the pontine nuclei may serve to modulate transmission of cerebellar afferent information in accordance with the behavioral state of the animal. It has been reported that single cholinergic mesopontine tegmental neurons project to both the pontine reticular formation and the thalamus in the rat.[197] The latter suggests that mesopontine cholinergic neurons simultaneously modulating neuronal activity in the pontine reticular formation and the thalamus, may regulate different aspects of rapid eye movement sleep. Cholinergic neurons in the substantia nigra of the rat may be projection neurons of the pontomesencephalotegmental cholinergic complex (Ch5-Ch6) ectopically located, which may play an important role in the regulation of the forebrain activation and locomotion.[113] Descending cholinergic projections from the Ch5-Ch6 groups are widely distributed. The target areas include the motor nuclei of cranial nerves 5, 7, and 12; the vestibular nuclei; the spinal nucleus of the fifth cranial nerve; the deep cerebellar nuclei; the pontine nuclei; the locus ceruleus; the raphe magnus; the median raphe nucleus; the medullary reticular nucleus; the oral and caudal pontine reticular nuclei; and the inferior olive. It is also of interest that these cholinergic cells send concurrent projections to more than one hindbrain site, and in some cases exhibit ascending projections as well.[77] Characteristically, none of the pedunculopontine or laterodorsal tegmental cholinergic cells project to the cerebellar cortex or spinal cord.[77,198] While the function of descending cholinergic projections may be related to integration of visceral information, descending noncholinergic projections arising in the pedunculopontine to innervate the spinal cord and cerebellum may have both motor and limbic functions.

The pontomesencephalotegmental cholinergic complex receives inputs from various sources (see Semba and Fibiger,[199] for a broad overview). The Ch6 group receives afferents from the prefrontal cortex and lateral habenula, whereas in the bed nucleus of stria terminalis and the central nucleus of the amygdala more cells are labeled from the Ch5 group. The basal forebrain also sends projections toward

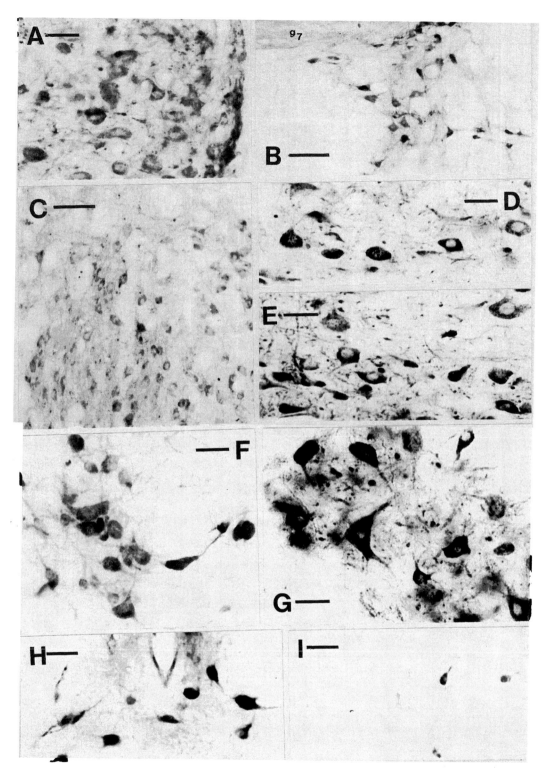

Figure 1-9 Illustration of cholinergic neurons in cranial nerve (A) 3, (B) 6, (C) 7, (D) 10, (E) 12; (F) in the nucleus ambiguous; (G) in motor neurons of the anterior horn; (H) around the central canal; and (I) in the dorsal horn of the spinal cord. Abbreviation: g_7, genu facial nerve. (Bars = 30 μm [A], [D]–[G], 100 μm [B], [C], 25 μm [H], [I]).

18

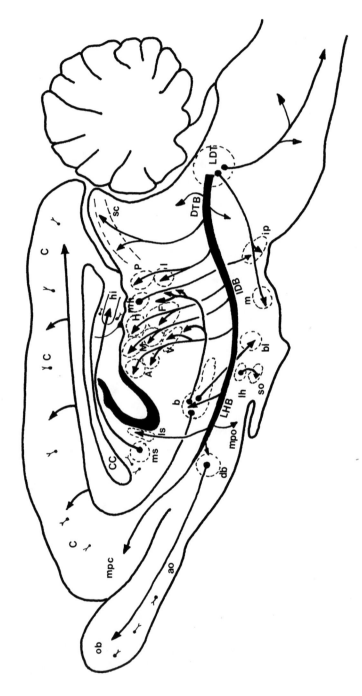

Figure 1-10 A schematic representation of principal cholinergic cell groups and projections in a sagittal view of the rat CNS. Abbreviations: A, anterior thalamic nucleus; ao, anterior olfactory nucleus; b, nucleus basalis; bl, basolateral amygdala; C, cerebral cortex; CC, corpus callosum; db, diagonal band; DTB, dorsal tegmental bundle; F, parafascicular thalamic nucleus; H, lateral habenular nucleus; hi, hippocampus; l, interstitial magnocellular nucleus of the posterior commissure; IDB, intermediate diencephalic nuclei; ip, interpeduncular nucleus; LDT, laterodorsal tegmental nucleus; lh, lateral hypothalamus; LHB, lateral hypothalamic bundle; ls, lateral septum; m, lateral mammillary nucleus; M, medial thalamic nucleus; mpc, medial prefrontal cortex; mpo, magnocellular preoptic nucleus; ms, medial septum; ob, olfactory bulb; P, anterior pretectal area; sc, superior colliculus; so, supraoptic nucleus; and V, ventromedial thalamic nucleus.

the Ch5-Ch6 group, with many projection neurons located in the lateral hypothalamus. Projections from the zona incerta, the midbrain central gray, and the accessory motor nuclei in the midbrain as well as from eye movement-related structures in the lower brain stem toward the Ch5-Ch6 group have also been reported. Finally, scarce input to the Ch5-Ch6 group was identified from neurons in somatosensory and other sensory relay nuclei in the brain stem and spinal cord.

Cholinergic neurons constitute the main body of motor nuclei associated with cranial nerves 3 to 7 and 9 to 12 (Figures 1-4M, N, P–T; and 1-9). In addition, parasympathetic nuclei related with these nuclei contained cholinergic somata. Cochlear and vestibular nuclei as well as the orofacial nuclei of cranial nerves 5, 7, and 12; the accessory nucleus of 7; the nucleus ambiguus; the nucleus solitarius; and the prepositus hypoglossal and the superior and inferior salivatory nuclei characteristically contain cholinergic cell bodies.

2. Spinal Cord

Neurons containing ChAT are localized in the ventral horns (Figures 1-4U, V; 1-9G) at all levels occupying the medial, central, and lateral motor columns and in the intermediolateral cell column at thoracic and lumbar levels. In addition, cholinergic cells are located in the central gray matter around the central canal (Figure 1-9H). The intermediate gray matter contains medium to large ChAT-positive neurons. At autonomic spinal levels, cholinergic neurons are intermingled with other reactive cells identified as preganglionic sympathetic or parasympathetic neurons which are localized in the intermediolateral nucleus in the lateral horn, dorsal to the central canal, in the intercalated nucleus, and in the funicular intermediolateral neurons of the white matter lateral to the principal intermediolateral nucleus.[200] In the dorsal horn, ChAT-positive somata are also located,[200,201] principally in laminae III-V (Figure 1-9I).

D. CHOLINERGIC NEURONS IN THE CEREBELLUM

Very recently, a number of studies have reported cerebellar neurons immunoreactive to ChAT.[126,202,203] Neurons containing ChAT were reported in each cerebellar nucleus. In addition, Golgi cells of the granular layer exhibiting ChAT-immunostaining were also described. However, disagreement with the latter has been reported in recent studies,[92,204] stating that intrinsic cerebellar neurons are devoid of ChAT immunostaining or ChAT mRNA hybridization signal.

IV. CYTOLOGICAL CHARACTERISTICS OF CENTRAL CHOLINERGIC NEURONS

Several reports have described with detail the light and electron microscopic appearance of central cholinergic cells.

A. LIGHT MICROSCOPY

While cholinergic cells in the same region usually exhibit a similar appearance, they vary commonly in appearance from region to region, these differences being occasionally quite prominent. Based on differences and similarities among central cholinergic neurons in different regions of the rat brain, Sofroniew et al.[74] have summarized central cholinergic neurons into four major groups according to their morphologies: (1) very large motor neurons (25 to 45 μm) as found in the spinal cord ventral horn and cranial nerve nuclei; (2) large forebrain neurons (18 to 25 μm) as in the striatum and MBN, including the nucleus of the diagonal band (horizontal and vertical limbs), the magnocellular preoptic nucleus, the substantia innominata, and the globus pallidus; (3) medium neurons (14 to 20 μm) as in certain forebrain and brainstem regions; and (4) small neurons (8 to 16 μm) as in the cerebral cortex, hippocampus, olfactory bulb, anterior olfactory nucleus, and dorsal horn. The distribution of these neurons is presented graphically in Figures 1-3 and 1-4, and pictorially in Figures 1-2, 1-8, 1-9, 1-12A–C). Light microscopic features of some of the most prominent cholinergic neurons in the groups mentioned above are briefly described below.

MBN neurons exhibit different morphologies. In the nucleus basalis, ChAT-positive cells display, in general, a fusiform or roughly pear-shaped perikaryon generally wider than the one of striatal cholinergic neurons, with a larger round or oval nucleus (Figure 1-7E, F). Typically, two to four dendrites arise from the perikaryon. These processes that are generally very long, varicose, spine-poor, and sparsely branched most often appear to overlap with one another and take the direction of nearby fiber bundles. The initial segment of the axon emerged from the perikaryon or from the proximal part of a large caliber dendritic trunk. Frequently, one to four thin axon collaterals arose. Characteristically, small clusters of short collaterals emerge from the main axon near the perikaryon.[205,206] In the horizontal limb of the diagonal band of Broca (Figure 1-8C), cholinergic neurons have been categorized in three groups depending on the shape of the cell body:[207] (1) the first group includes cells with fusiform or spindle-shaped soma, (2) the second group comprises large multipolar neurons with triangular or polygonal perikarya, and (3) the third group consists of medium-sized cells with round or oval somata (12 to 18 μm in diameter) and a small number (one or two) of short and sparsely branching dendrites. In the medial septum, ChAT-positive neurons exhibit oval- or bipolar-shaped perikaryon (Figure 1-8A). In the lateral septum, a small cluster of ChAT-positive cells recently have been described.[208] These cells are small and multipolar and exhibited less staining than ChAT-positive neurons in the medial septum.

The morphology of ChAT-positive neurons in the striatum has been extensively studied.[209–211] These cells are rather evenly dispersed through the anterior-posterior extent of the striatum and frequently exhibit an elongated or fusiform perikaryon (Figure 1-8E). The soma display a variety of shapes; most commonly they are oval, but elongated or multipolar somata can be also detected. Usually, two to three dendritic processes arise from each pole of the soma with no preferential orientation. Primary dendrites have smooth contours, radiating from cell bodies in all directions and usually branched several times. In general, these cells average 27 × 13 μm in their major and minor somal diameters. Cholinergic neurons in the striatum account for a small proportion (1.7%) of the total number of neurons in this nucleus.

Cholinergic cells in the cerebral cortex labeled after ChAT immunocytochemistry show a widespread distribution.[18,157,181] Cingulate, insular, retrosplenial, striate, and motor cortices have more cholinergic neurons than the somatosensory, entorhinal, or pyriform. Most of these cells occupy layers II and III. They are also present, but in smaller numbers in layers III to VI. Most of these cells are small (10 μm in maximum diameter), are vertically oriented with typical bipolar dendritic patterns, and appear to belong to the spine-free bipolar class (Figure 1-8G). These cells have a single apical dendrite directed toward the cortical surface. A basal process, which occasionally gives off some collaterals, can be followed into deep cortical layers. Another type of ChAT-positive cell is also small, but is more round and multipolar, and its processes do not follow any particular direction.

In the hippocampus, small (8.9 ± 0.25 μm in diameter), round, or ovoid and multipolar ChAT-positive neurons are localized in the caudal and ventral hippocampus. The majority of these cells is located within the subiculum. Within the CA fields of Ammon's horn and dentate gyrus, the cells are located most often in the stratum lacunosum moleculare and at the border of the stratum lacunosum moleculare and the stratum radiatum. Numerous ChAT-labeled cells are also located in the molecular layer of the dentate gyrus. In general, these cells resemble intrinsic cholinergic neocortical neurons (Figure 1-8H).

In the main olfactory bulb, small neurons (13 to 9 μm in major and minor diameters) labeled with antibodies to ChAT are sparse and particularly localized in the external and internal plexiform and internal granular layers exhibiting the aspect of tufted and deep short axon cells. Small cholinergic neurons also occur in the accessory olfactory bulb and in the anterior olfactory nucleus (Figures 1-3A, B; 1-8I). These cells are commonly bipolar in form with thin smooth dendrites, characteristics which are associated to intrinsic, local circuit neurons.[212]

The distribution and soma shapes of cholinergic neurons in the nucleus accumbens and olfactory tubercle are similar to those in the striatum. Cholinergic neurons in the island of Calleja complex are generally small sized (Figures 1-2E; 1-3C-G).

Cholinergic neurons in the pontomesencephalotegmental cholinergic complex (Ch5-Ch6), including those in the substantia nigra,[113] are large (20 to 30 μm in the longest cross-sectional diameter) and multipolar. The cell somata are of different shapes including round, spindle, or oval. One to three primary dendrites often emerge from the poles of the perikarya. The primary dendrites radiate from the somata in many directions and frequently exhibit several branches.

Motor neurons are large and multipolar and exhibit the features so very well described in many classical histological communications (Figures 1-4M–V; 1-9A–E, F).

B. ELECTRON MICROSCOPY

Besides light microscopic information, several studies have given a detailed description of the ultrastructural morphology of the central cholinergic neurons.[113,135,178,192,195,201,207,209,213–219] Cholinergic projection neurons exhibit, in general, very similar ultrastructural characteristics despite relative differences in size and shape. After ChAT immunocytochemistry, cholinergic structures are identified by the presence of an

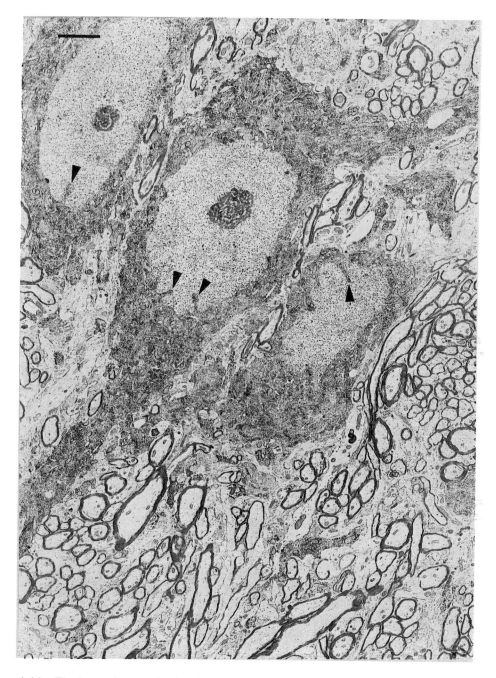

Figure 1-11 Electron micrograph of a cluster of three adjacent ChAT-positive neurons detected in the nucleus basalis of the rat. The neurons have abundant immunoreactive cytoplasm rich in organelles. The nuclei exhibit occasional nuclear infoldings (arrowheads) and prominent nucleolus. The synaptic input to the cell soma and proximal dendrites of these cells was studied in serial sections and is illustrated in Figure 1-13C, D. (Bar = 5 μm.)

electron-dense peroxidase reaction product in different cellular structures including cell bodies, dendrites, and terminal boutons (Figures 1-11, 1-12D, 1-13).

Following ChAT immunocytochemistry, the reaction product in the cell body is generally detected attached to the outer surface of mitochondria, the rough endoplasmic reticulum, and in the internal face of the plasma membrane. The cell nucleus shows one or more indentations and generally occupies a

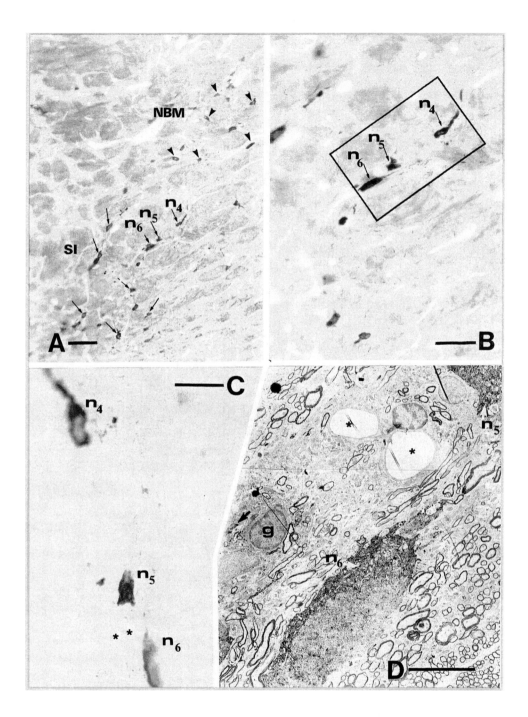

Figure 1-12 Correlated light (A), (C) and electron microscopy (D) of the same ChAT-positive neuron (n_6) identified in the substantia innominata. (A) Light micrograph taken from a flat-embedded ChAT-immunostained section through the basal forebrain that was postfixed in 1% osmium tetroxide for 1 h and subsequently processed for electron microscopy. (B) A higher power magnification than what is observed in (A). (C) Micrograph of a 2-μm thick section obtained through the boxed area shown in (B). (D) Electron micrograph showing aspects of the cholinergic n_6 neuron. Notice in (D) the presence of a ChAT-positive terminal bouton (arrow) in the proximity of a glial cell (g). (Bars = 100 μm [A], 40 μm [B], 25 μm [C], and 10 μm [D].)

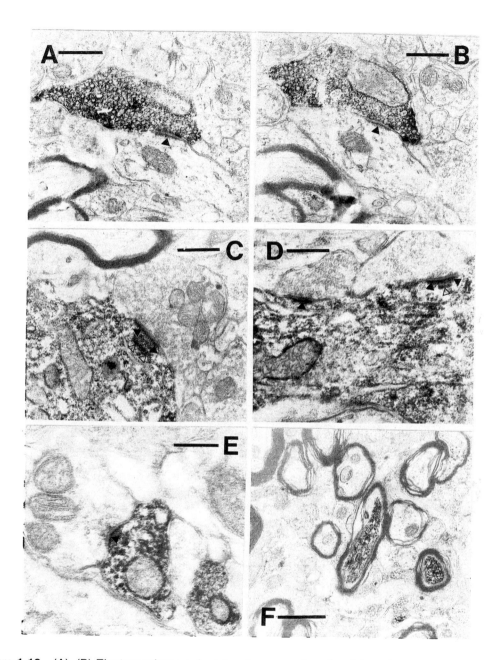

Figure 1-13 (A), (B) Electron micrographs taken from a series of serial sections through the same terminal bouton shown in Figure 1-11D. The ChAT-positive terminal forms a symmetrical synaptic contact (arrowhead) with an unlabeled dendritic profile. (C)–(E) Electron micrographs of axosomatic (C) and axodendritic (D), (E) synaptic contacts (arrowheads) established between unlabeled axon terminals and ChAT-positive postsynaptic structures in the region of the nucleus basalis. Notice in (C) and (D) that the asymmetry of synaptic contacts is well defined by the presence of subjunctional dense bodies (open arrows). Also note in (E) that the presence of an immunoperoxidase reaction product associated with the postsynaptic membrane makes difficult a definition of certain types of synaptic contact, either symmetrical or asymmetrical. (F) Illustrates ChAT-positive myelinic fibers in the region of the nucleus basalis. (Bars = 0.5 μm [A]–[D], 0.25 μm [E], and 1 μm [F].)

central position being surrounded by abundant cytoplasm rich in organelles including lysosomes, mitochondria, rough endoplasmic reticulum, and Golgi apparatus. Large lipofuscin granules can also be detected. The proximal dendrites are generally thick and contain the same organelles (Figure 1-11). The outer membrane of the cell soma and proximal dendrites are largely covered by glia. Myelinated and small unmyelinated nonterminal axons often contact with the plasmalemma, but without specialized junctions. Axosomatic synaptic contacts on cholinergic projection neurons, particularly to those in the basal forebrain and upper brainstem, are notably sparse, which is distinctly different from those on noncholinergic neighboring neurons that show abundant synaptic contacts on their cell bodies. The synaptic contact onto cholinergic structures is sometimes difficult to categorize as symmetrical or asymmetrical because of the presence of an immunoperoxidase reaction product associated with the postsynaptic membrane (Figure 1-13E). In more lightly stained material, most of axosomatic contacts are of the asymmetrical type. The presence of subjunctional dense bodies associated with the postsynaptic membrane in synaptic contacts onto the cell soma and proximal dendrites is often detected, allowing the delineation of asymmetrical types (Figure 1-13C, D). In general, the unlabeled terminals associated with junctions displaying these asymmetrical subsynaptic specializations contain mitochondria and numerous clear, mostly round, synaptic vesicles varying in size. Dense-cored vesicles are sometimes detected presynaptically in axosomatic or axodendritic synapses. The most commonly detected synaptic contacts are found on various diameters of ChAT-positive distal dendrites scattered throughout the neuropil immediately surrounding the area containing ChAT-positive neuronal somata.[60,207,214,220] Almost 80% of the postsynaptic cholinergic dendrites are of medium to small size with a diameter measured perpendicular to the synaptic complex of 0.2 to 1 µm. The immunoreactive distal dendrites contain numerous mitochondria and microtubules. Both symmetrical and asymmetrical synaptic specializations from unlabeled terminals on these labeled dendrites can be observed (Figure 1-13D, E). The immunonegative terminals generally contain mitochondria and either numerous round or pleomorphic clear vesicles. One of the shapes of synaptic vesicles mentioned above predominates depending on the type of synaptic contact. Asymmetrical synaptic contacts on distal dendrites are often defined by the presence of subjunctional dense bodies associated with the postsynaptic membrane and of clear, mostly round, synaptic vesicles within the presynaptic structures. On occasion, dense-cored vesicles can be associated with presynaptic structures in asymmetrical types. Symmetrical synaptic types display pleomorphic, mostly flattened, clear synaptic vesicles presynaptically. Characteristically, most labeled distal dendrites are postsynaptic to one or two unlabeled terminals seen in a single plane of a section (Figure 1-13E). The apposed neuronal processes are isolated from the surrounding neuropil by glial structures, which further support functional relationships.

Striatal cholinergic interneurons exhibit similar fine structural characteristics.[209,211] Cholinergic neurons in the neocortex and hippocampus are considerably smaller with a thin rim of perikaryal cytoplasm and exhibit frequent axosomatic synaptic contacts.[181]

C. CHOLINERGIC TERMINALS

The existence of cholinergic terminals was assessed under the light and electron microscopes in different brain regions including the cerebral cortex, hippocampus, amygdala, interpeduncular nucleus, basal forebrain, striatum, thalamus, hypothalamus, substantia nigra, and spinal cord,[48,113,181,192,200,211,212,220–228] (among others). The similarity between cholinergic synaptic boutons in different areas is remarkable. They are usually small, contained, round, clear vesicles and form predominantly symmetrical synaptic specializations although asymmetrical contacts established by cholinergic terminals were also reported, i.e., those in the cerebral cortex, basal forebrain (Figure 1-13A, B), and spinal cord.[71,181,220] Cholinergic terminals make frequent synaptic contact with noncholinergic structures. Terminals containing ChAT were also detected in asymmetrical synaptic contact with cholinergic structures in the basal forebrain,[213,219] although such an input was not found significant. Axo-axonic synaptic contacts between cholinergic terminals and nonimmunoreactive terminals in the dorsal horn have been described.[201]

V. TRANSMITTER-SPECIFIC AFFERENTS TO CENTRAL CHOLINERGIC NEURONS

Cholinergic neurons receive inputs from a variety of sources throughout the neuroaxis. Many studies have analyzed the morphological aspects, the chemical characteristics, and the origin of some of the various inputs that central cholinergic neurons receive. On the basis of these investigations, clear improvements have been made in the knowledge of mechanisms of modulation of central cholinergic activity. Light microscopic neurochemical information concerning neurotransmitters present in the synaptic boutons contacting the cholinergic neurons remains relatively limited. The reason for the latter is the anatomic complexity of the central cholinergic systems which have cellular elements intermingled among numerous noncholinergic cells.[120,229,230] The use of double-staining procedures at the ultrastructural level, which allows simultaneous determination of the cholinergic nature of the postsynaptic target and the chemical nature of the afferent system or its origin,[231,232] has provided substantial morphocytochemical data for the understanding of the functional modulation of the central cholinergic activity. These studies have identified various transmitter-specific afferent systems to central cholinergic neurons which include GABA, ACh, glutamate/aspartate, substance P, enkephalin, galanin, somatostatin, Neuropeptide Y, and dopamine;[60,68,112,113,159,192,214,216,217,220,228,233–243] however, further morphocytochemical and pharmacological studies are required to elucidate the relationships between cholinergic and noncholinergic neighboring cells so as to ascertain specific circuits in which cholinergic neurons are involved. Neighboring neurons to cholinergic projection neurons may be either projection neurons affecting the cholinergic innervation of particular targets by sending efferents that parallel those of cholinergic neurons, or local circuit cells forming interneuronal associative populations that may in turn innervate cholinergic projection neurons.

It is suggested that small GABAergic neurons may represent interneurons in the basal forebrain and upper brain stem that may exert an inhibitory influence on adjacent projection cholinergic neurons.[113,120,220,228,237] In addition, medium to large GABAergic neurons may also exert an inhibitory influence outside the MBN as projection neurons, potentially in parallel with cholinergic neurons, to certain regions of the neocortex. The latter is supported by the fact that cholinergic-GABAergic interactions have been suggested in the neocortex.[181,211]

In the MBN, it is interesting to consider that ChAT-positive terminals were seen establishing symmetrical synaptic contacts with nonimmunoreactive distal dendrites and, to a lesser extent, asymmetrical synaptic contacts both onto nonimmunoreactive distal dendritic spines and shafts and onto immunoreactive dendrites.[220] According to the correlation of morphological and physiological properties of synapses,[244,245] it can be speculated that symmetrical (Gray's type II) contacts could indicate an inhibitory action of ACh on target neuronal structures, while asymmetrical (Gray's type I) contacts could be excitatory. The fact that asymmetrical synaptic contacts are established between immunoreactive boutons and cholinergic dendrites suggests a monosynaptic excitatory cholinergic interaction. On the other hand, symmetrical contacts established between cholinergic terminals and immunonegative dendrites could lead to disinhibitory effects of cholinergic neurons through a polysynaptic link via inhibition of GABAergic interneurons. This assumption is consistent with the reported[79,246] excitatory and inhibitory action of ACh and GABA, respectively, on cortically projecting cholinergic cells in the MBN. However, direct evidence of a cholinergic input to the GABAergic neurons in the MBN remains to be elucidated.

The existence of local synaptic interactions between cholinergic basalocortical neurons has been suggested by certain authors.[205,206] Therefore, the cholinergic terminals detected in the MBN may be originated from basalocortical cholinergic neurons, although additional possible sources from the midbrain cannot be ruled out.[247] The latter would suggest that brain stem cholinergic neurons may regulate the activity of the neocortex through a polisynaptic pathway via modulation of the thalamocortical projection, discussed above, and through a possible cholinocholinergic relay in the basal forebrain.[220]

ACKNOWLEDGMENTS

The studies on which parts of this review are based were made possible by grants from FIS (92/0269) and CAM (C179/91), Spain.

REFERENCES

1. Kasa P., The cholinergic systems in brain and spinal cord, *Progress in Neurobiology,* 26, 211, 1986.
2. Wainer, B. H., Levey, A. I., Mufson, E. J., and Mesulam M.- M., Cholinergic systems in mammalian brain identified using antibodies against choline acetyltransferase, *Neurochemistry International,* 6, 163, 1984.
3. Shute, C. C. D. and Lewis, P. R., Cholinesterase-containing pathways of the hindbrain: afferent cerebellar and centrifugal cochlear fibers, *Nature (London),* 205, 242, 1965.
4. Shute, C. C. D. and Lewis, P. R., The ascending cholinergic reticular system, neocortical, olfactory and subcortical projections, *Brain,* 90, 497, 1967.

5. Lewis, P. R. and Shute, C. C. D., The cholinergic limbic system: projections to hippocampal formation, medial cortex, nuclei of the ascending cholinergic reticular system, and the subfornical organ and supraoptic crest, *Brain*, 90, 521, 1967.

6. Lehman, J. and Fibiger, H. C., Acetylcholinesterase in the substantia nigra and caudate-putamen of the rat: properties and localization in dopaminergic neurons, *Journal of Neurochemistry*, 30, 615, 1978.

7. Moore, R. Y. and Bloom, F. E., Central catecholamine neurons: anatomy and physiology of the dopamine systems, *Annals Review of Neuroscience*, 1, 129, 1978.

8. Ungerstedt, U., Stereotaxic mapping of the monoamine pathways in the rat brain, *Acta Physiologica Scandinavica*, 95, 1, 1971.

9. Fonnum, F., Review of the recent progress in synthesis, storage, and release of acetylcholine, in *Cholinergic Mechanisms*, Waser, P.G., Ed., Raven Press, New York, 1975, 145.

10. Matthews, D. A., Salvaterra, P. M., Crawford, G. D., Houser, C. R., and Vaughn, J.E., An immuno-cytochemical study of choline acetyltransferase containing neurons and axon terminals in normal and partially deafferented hippocampal formation, *Brain Research*, 402, 30, 1987.

11. Hoover, D. B. and Jacobowitz, D. M., Neurochemical and histochemical studies of the effect of a lesion of the nucleus cuneiformis on the cholinergic innervation of discrete areas of the rat brain, *Brain Research*, 170, 113, 1979.

12. Fibiger, H. C., The organization and some projections of cholinergic neurons of the mammalian forebrain, *Brain Research Reviews*, 4, 327, 1982.

13. Koelle, G. B., The histochemical localization of cholinesterase in the central nervous system of the rat, *Journal of Comparative Neurology*, 100, 211, 1954.

14. Lehmann, J. and Fibiger, H. C., Acetylcholinesterase and the cholinergic neuron, *Life Science*, 25, 1939, 1979.

15. Lehmann, J., Nagy, J. I., Atmadja, S., and Fibiger, H. C., The nucleus basalis magnocellularis: the origin of a cholinergic projection to the neocortex of the rat, *Neuroscience*, 5, 1161, 1980.

16. Johnston, M. V., McKinney, M., and Coyle, J. T., Neocortical cholinergic innervation: a description of extrinsic and intrinsic components in the rat, *Experimental Brain Research*, 43, 159, 1981.

17. Krnjevic, K. and Silver, A., An histochemical study of cholinergic fibers in the cerebral cortex, *Journal of Anatomy*, 99, 711, 1965.

18. Levey, A. I., Wainer, B., Rye, D. B., Mufson, E. J., and Mesulam M.-M., Choline acetyltransferase-immunoreactive neurons intrinsic to rodent cortex and distinction from acetylcholinesterase-positive neurons, *Neuroscience*, 13, 341, 1984.

19. Arendt, T., Allen, Y., Marchbanks, R. M., Schugens, M. M., Sinden, J., Lantos, P. L., and Grey, J. A., Cholinergic system and memory in the rat: effects of chronic ethanol, embryonic basal forebrain transplants and excitotoxic lesions of cholinergic basal forebrain projection system, *Neuroscience*, 33, 433, 1989.

20. Dekker, J. A. M., Connor, D. J., and Thal, L. J., The role of cholinergic projections from the nucleus basalis in memory, *Neuroscience and Behavioral Reviews*, 15, 299, 1991.

21. Jones, B. E., The role of noradrenergic locus ceruleus neurons and neighbouring cholinergic neurons of the pontomesencephalic tegmentum in sleep-wake states, *Progress in Brain Research*, 88, 533, 1991.

22. Ma, C. F., Höfmann, F., Coyle, J. T., and Juliano S. L., Lesions of the basal forebrain alter stimulus-evoked metabolic activity in mouse somatosensory cortex, *Journal of Comparative Neurology*, 288, 414, 1989.

23. Bartus, R. T., Dean, R. L., Beer, B., and Lippa, A. S., The cholinergic hypothesis of geriatric memory dysfunction, a critical review, *Science*, 217, 407, 1982.

24. Bigl, V., Arendt, T., and Biesold, D., The nucleus basalis of Meynert during aging and in dementia disorders, in *Brain Cholinergic Systems*, Steriade, M. and Biesold, D., Eds., Oxford University Press, Oxford, 1990, 364.

25. Coyle, J. T., Price, D. L., and De Long, M. R., Alzheimer's disease, a disorder of cortical cholinergic innervation, *Science*, 219, 1184, 1983.

26. Ross, C. A., CNS arousal systems: possible role in delirium, *International Psychogeriatry*, 3, 353, 1991.

27. Karson, C. N., Casanova, M. F., Kleinman, J. E., and Griffin, W. S., Choline acetyltransferase in schizophrenia, *American Journal of Psychiatry*, 150, 454, 1993.

28. Candy, J. M., Perry, R. H., Perry, E. K., Irving, D., Blessed, G., Fairbairn, A. F., and Tomlinson, R. L., Pathological changes in the nucleus of Meynert in Alzheimer's and Parkinson's diseases, *Journal of Neurological Science*, 54, 277, 1983.

29. Leherici, S., Hirsch, E. C., Cervera-Pierot, C., Hersh, L. B., Bakchine, S., Piette, F., Duickaerts, C., Hauw, J. J., Javoy-Agid, F., and Agid, Y., Heterogeneity and selectivity of the degeneration of cholinergic neurons in the basal forebrain of patients with Alzheimer's disease, *Journal of Comparative Neurology,* 330, 15, 1993.

30. Whitehouse, P. J., Price, D. L., Struble, R. G., Clark, A. W., Coyle, J. T., and DeLong, M. R., Alzheimer's disease and senile dementia, loss of neurons in the basal forebrain, *Science,* 215, 1237, 1982.

31. Pearson, R. C. A., Sofroniew, M. V., Cuello, A. C., Powell, T. P. S., Eckenstein, F., Esiri, M. M., and Wilcock, G. K., Persistence of cholinergic neurones in the basal nucleus in a brain with senile dementia of the Alzheimer's type demonstrated by immunohistochemical staining of choline acetyltransferase, *Brain Research,* 289, 375, 1983.

32. Perry E. K. and Perry, R. H., The cholinergic system in Alzheimer disease, *Trends in Neuroscience,* 5, 261, 1982.

33. Sofroniew, M. V., Pearson, R. C. A., Eckenstein, F., Cuello, A. C., and Powell T. P. S., Retrograde changes in cholinergic neurons in the basal forebrain of the rat following cortical damage, *Brain Research,* 289, 370, 1983.

34. McKinney, M., Coyle, J. T., and Hedreen, J. C., Topographic analysis of the innervation of the rat neocortex and hippocampus by the basal forebrain cholinergic system, *Journal of Comparative Neurology,* 217, 103, 1983.

35. Lamour, Y., Dutar, P., and Jobert, A., Excitatory effect of acetylcholine on different types of neurons in the first somatosensory neocortex of the rat, laminar distribution and pharmacological characteristics, *Neuroscience,* 7, 1483, 1982.

36. Hohmann, C. F., Brooks, A. R., and Coyle, J. T., Neonatal lesion of the basal forebrain cholinergic neurons result in abnormal cortical development, *Developmental Brain Research,* 42, 253, 1988.

37. Bear, M. F. and Singer, W., Modulation of visual cortical plasticity by acetylcholine and noradrenaline, *Nature (London),* 320, 172, 1986.

38. Henderson, Z., Early development of the nucleus basalis-cortical projection but late expression of its cholinergic function, *Neuroscience,* 44, 311, 1991.

39. Webster, H. H., Rasmusson, D. D., Dikes, R. W., Schliebes, R., Schober, W., Bruckner, G., and Biesold, D., Long-term enhancement of evoked potentials in raccoon somatosensory cortex following co-activation of the nucleus basalis of Meynert complex and cutaneous receptors, *Brain Research,* 545, 292, 1991.

40. Namba, H., Irie, T., Fukusi K., Yamasaki, T. and Hasegawa, S., Lesion of the nucleus basalis does not affect cerebral cortical blood flow in rats, *Neuroscience Research,* 12, 463, 1991.

41. Scremin, O. U., Torres, C., Scremin, A. M. E., O'Neal, M., Heuser, D., and Blisard, K.S., Role of nucleus basalis in cholinergic control of cortical blood flow, *Journal of Neuroscience Research,* 28, 382, 1991.

42. Kuhar, M. J., Dehaven, R. N., Yamamura, H. I., Rommelspacher, H., and Simon, J. R., Further evidence for cholinergic habenulo-interpeduncular neurons: pharmacologic and functional evidences, *Brain Research,* 97, 265, 1975.

43. Sethy, V. H., Roth, R. H., Kuhar, M. J., and Van Woert, M. H., Choline and acetylcholine: regional distribution and effect of degeneration of cholinergic nerve terminals in the rat hippocampus, *Neuropharmacology,* 12, 819, 1973.

44. Dunnet, S. B., Whishaw, I. Q., Jones, G. H., and Bunch, S. T., Behavioral, biochemical and histochemical effects of different neurotoxic amino acids injected into nucleus basalis magnocellularis of the rat, *Neuroscience,* 20, 653, 1987.

45. Everitt, B. J., Robbins, T. W., Evenden, J. L., Marston, H. M., Jones, G. H., and Sirkia T., The effect ox exitotoxic lesions of the substantia innominata, ventral and dorsal globus pallidus on the acquisition and retention of a conditional discrimination: implication for cholinergic hypothesis of learning and memory, *Neuroscience,* 22, 441, 1987.

46. Nieoullon, A., Scarfone, E., Kerkerian, L., Errami, M., and Dusticier, N., Changes in choline acetyltransferase, glutamic acid decarboxylase, high-affinity glutamate uptake and dopaminergic activity induced by kainic acid lesion of the thalamostriatal neurons, *Neuroscience Letters,* 58, 299, 1985.

47. Santos-Benito, F. F., González, J. L., and de la Torre, F., Choline acetyltransferase activity in the rat brain cortex homogenate, synaptosomes, and capillaries after lesioning the nucleus basalis magnocellularis, *Journal of Neurochemistry,* 50, 395, 1988.

48. Meeker, R. B., Swanson, D. J., and Hayward, J. N., Local synaptic organization of cholinergic neurons in the basolateral hypothalamus, *Journal of Comparative Neurology,* 276, 157, 1988.

49. Rotter, A., Birdsall, N. J. M., Burgen, A. S. V., Field, P. M., Hulme, E.C., and Raisman, G., Muscarinic receptors in the central nervous system of the rat. I. Technique for autoradiographic localization of binding of [3H]propylbenzilycholine mustard and its distribution in the forebrain, *Brain Research Reviews,* 1, 141, 1979.

50. Rotter, A., Birdsall, N. J. M., Field, P. M., and Raisman, G., Muscarinic receptors in the central nervous system of the rat. II. Distribution of binding of [3H]propylbenzilylcholine mustard in the midbrain and hindbrain, *Brain Research Reviews,* 1, 167, 1979.

51. Yamamura, H. I. and Snyder, S. H., Muscarinic cholinergic binding in rat brain, *Proceedings of the National Academy of Sciences, U.S.A.,* 71, 1725, 1974.

52. Bear, M. F., Carnes, K. M., and Ebner, F. F., An investigation of cholinergic circuitry in cat striate cortex using acetylcholinesterase histochemistry, *Journal of Comparative Neurology,* 234, 411, 1985.

53. Robbins, T. W., Everitt, B. J., Ryan, C. N., Marston, H. M., Jones, G. H., and Page, K. J., Comparative effects of quisqualic and ibotenic-induced lesions of the substantia innominata and globus pallidus on the acquisition of a conditional visual discrimination: differential effects on cholinergic mechanisms, *Neuroscience,* 28, 337, 1989.

54. Saper, C. B. and Chelimsky, T. C., A cytoarchitectonic and histochemical study of nucleus basalis and associated cell groups in the normal human brain, *Neuroscience,* 13, 1023, 1984.

55. Gasser, U. E., Van Deusen, E. B., and Dravid, A. R ., Homologous cholinergic afferents spared by partial fimbrial lesions contribute to the recovery of hippocampal cholinergic enzymes in adult rats, *Brain Research,* 367, 368, 1986.

56. Butcher, L. L., Marchand, R., Parent, A., and Poirier, L. J., Morphological characteristics of acetylcholinesterase-containing neurons in the CNS of DFP-treated monkeys. III. Brain stem and spinal cord, *Journal of Neurological Science,* 32, 169, 1977.

57. Satoh, K., Armstrong, D. M., and Fibiger, H. C., A comparison of the distribution of central cholinergic neurons as demonstrated by acetylcholinesterase pharmacohistochemistry and choline acetyltransferase immunohistochemistry, *Brain Research Bulletin,* 11, 693, 1983.

58. Eckenstein, F. and Sofroniew, M. W., Identification of central cholinergic neurons containing both choline acetyltransferase and acetylcholinesterase and of central neurons containing only acetylcholinesterase, *Journal of Neuroscience,* 3, 2286, 1983.

59. Levey, A. I., Wainer, B. H., Mufson, E. J., and Mesulam, M.-M., Colocalization of acetylcholinesterase and choline acetyltransferase in the rat cerebrum, *Neuroscience,* 9, 9, 1983b.

60. Martínez-Murillo, R., Blasco, I., Alvarez, F. J., Villalba, R., Solano, M. L., Montero-Caballero, M. I., and Rodrigo, J., Distribution of enkephalin-immunoreactive nerve fibers and terminals in the region of the nucleus basalis magnocellularis of the rat, a light and electron microscopic study, *Journal of Neurocytology,* 17, 361, 1988.

61. Fukuda, T. and Koelle, G. B., The cytological localization of intracellular acetylcholinesterase, *Journal of Biology and Biochemical Cytology,* 5, 433, 1959.

62. Lehmann, J., Fibiger, H. C., and Butcher, L.L., The localization of acetylcholinesterase in the corpus striatum and substantia nigra of the rat following kainic acid lesions of the corpus striatum. A biochemical and histochemical study, *Neuroscience,* 4, 217, 1979.

63. Austin, L. and James, K. A. C., Rates of regeneration of acetylcholinesterase in rat brain subcellular fractions following DFP inhibition, *Journal of Neurochemistry,* 17, 705, 1970.

64. Michalek, H., Bisso, G. M., and Manegus, A., Comparative studies on rat brain soluble acetylcholinesterase and its molecular forms during intoxication by DFP and paraoxon, *Cholinergic Mechanisms,* Pepeu, G. and Ladinsky, H., Eds., Plenum Press, New York, 1981, pp. 847–852.

65. Albanese, A. and Butcher, L. L., Acetylcholinesterase and catecholamine distribution in the locus ceruleus of the rat, *Brain Research Bulletin,* 5, 127, 1980.

66. Butcher, L. L. and Woolf, N. J., Histochemical distribution of acetylcholinesterase in the central nervous system, clues to the localization of cholinergic neurons, in *Handbook of Chemical Neuroanatomy, Vol.3, Classical Transmitters and Transmitter Receptors in the CNS,* Björklund, A., Hökfelt, T. and Kuhar, M.J., Eds., Elsevier Science Publishers, Amsterdam, 1984, 1.

67. Lewis, P. R. and Schon, F. E. G., The localization of acetylcholinesterase in the locus ceruleus of the normal rat and after 6-hydroxydopamine treatment, *Journal of Anatomy,* 120, 373, 1975.

68. Grove, E. A., Domesick, V. B., and Nauta, W. J. H., Light microscope evidence of striatal input to intrapallidal neurons of cholinergic cell group Ch4 in the rat: a study employing the anterograde tracer *Phaseolus vulgaris* leucoagglutinin (PHA-L), *Brain Research,* 367, 379, 1986.

69. Armstrong, D. M., Saper, C. B., Levey, A. I., Wainer, B. H., and Terry, R. D., Distribution of cholinergic neurons in rat brain, demonstrated by the immunohistochemical localization of choline acetyltransferase, *Journal of Comparative Neurology,* 216, 53, 1983.

70. Bigl, V., Woolf, N. J., and Butcher, L. L., Cholinergic projections from the basal forebrain to frontal, parietal, temporal, occipital and cingulate cortices, a combined fluorescent tracer and acetylcholinesterase analysis, *Brain Research Bulletin,* 8, 727, 1982.

71. Houser, C. R., Crawford, G. D., Barber, R. P., Salvaterra, P. M., and Vaughn, J. E., Organization and morphological characteristics of cholinergic neurons: an immunocytochemical study with a monoclonal antibody to choline acetyltransferase, *Brain Research,* 266, 97, 1983.

72. Mesulam, M.-M., Mufson, E. J., Levey, A. I., and Wainer, B. H., Cholinergic innervation of cortex by basal forebrain, cytochemistry and cortical connections of the septal area, diagonal band nuclei, nucleus basalis, substantia innominata, and hypothalamus in the rhesus monkey, *Journal of Comparative Neurology,* 214, 170, 1983.

73. Mesulam, M.-M., Mufson, E. J., Wainer, B. H., and Levey, A. I., Central cholinergic pathways in the rat, an overview based on an alternative nomenclature, Ch1-Ch6, *Neuroscience,* 10,1185, 1983.

74. Sofroniew, M. V., Campbell, P. E., Cuello, A. C., and Eckenstein, F., Central cholinergic neurons visualized by immunohistochemical detection of choline acetyltransferase, in *The Rat Nervous System, Vol. 1, Forebrain and Midbrain*, Paxinos, G., Ed., Academic Press, Sydney, 1985, 471.

75. Sofroniew, M. V., Priestley, J. V., Consolazione, A., Eckenstein, F., and Cuello, A. C., Cholinergic projections from midbrain and pons to the thalamus in the rat, identified by combined retrograde tracing and choline acetyltransferase immunohistochemistry, *Brain Research,* 329, 213, 1985.

76. Woolf, N. J. and Butcher, L. L., Cholinergic systems in the rat brain. III. Projections from the pontomesencephalic tegmentum to the thalamus, tectum, basal ganglia, and basal forebrain, *Brain Research Bulletin,* 16, 603, 1986.

77. Woolf, N. J. and Butcher, L. L., Cholinergic systems in the rat brain. IV. Descending projections of the pontomesencephalic tegmentum. *Brain Research Bulletin* 23, 519, 1989.

78. Woolf, N. J., Eckenstein, F., and Butcher, L. L., Cholinergic systems in the rat brain. I. Projections to the limbic telencephalon, *Brain Research Bulletin,* 13, 751, 1984.

79. Lamour, Y., Dutar, P., and Jobert, A., Basal forebrain neurons projecting to the rat frontoparietal cortex: electrophysiological and pharmacological properties, *Brain Research,* 362, 122, 1986.

80. Geffard, M., McRae-Degueurce, A., and Souan, M. L., Immunocytochemical detection of acetylcholine in rat central nervous system, *Science,* 229, 77, 1985.

81. Cozzari, C. and Hartman, B. K., Preparation of antibodies specific to choline acetyltransferase from bovine caudate nucleus and immunohistochemical localization of the enzyme, *Proceedings of the National Academy of Science U.S.A.*, 77, 7453, 1980.

82. Eng, L. F., Uyeda, L. P., Chao, L. P., and Wolfgram, F., Antibody to bovine choline acetyltransferase and immunofluorescent localisation of the enzyme in neurones, *Nature (London),* 250, 243, 1974.

83. McGeer, P. L., McGeer, E. G., Singh, V. H., and Chase, W. H., Choline acetyltransferase localization in the central nervous system by immunohistochemistry, *Brain Research,* 81, 373, 1974.

84. Eckenstein, F. and Thoenen, H., Production of specific antisera and monoclonal antibodies to choline acetyltransferase, characterization and use for identification of cholinergic neurons, *EMBO Journal,* 1, 363, 1982.

85. Kimura, H., McGeer, P. L., and Peng, J. H., Choline acetyltransferase-containing neurons in the rat brain, in *Handbook of Chemical Neuroanatomy, Vol.3, Classical Transmitters and Transmitter receptors in the CNS, Part II*, Björklund A., Hökfelt T., and Kuhar, M.J., Eds., Elsevier Science Publishers, Amsterdam, 1984, 51.

86. Levey, A. I., Armstrong, D. M., Atweh, S. F., Terry, R. D., and Wainer, B. H., Monoclonal antibodies to choline acetyltransferase: specificity and immunohistochemistry, *Journal of Neuroscience,* 3, 1, 1983.

87. Butcher, L. L., Oh, J. D., Woolf, N. J., Edwards, R. H., and Roghani, A., Organization of central cholinergic neurons revealed by combined *in situ* hybridization histochemistry and choline-*o*-acetyltransferase immunocytochemistry, *Neurochemistry International,* 21, 429, 1992.

88. Ibáñez, C. F., Ernfors, P., and Persson, H., Developmental and regional expression of choline acetyltransferase mRNA in the rat central nervous system, *Journal of Neuroscience Research,* 29, 163, 1991.

89. Ishii, K., Oda, Y., Ichikawa, T., and Deguchi, T., Complementary DNAs for choline acetyltransferase from spinal cords of rats and mouse, nucleotide sequences, expression in mammalian cells, and *in situ* hybridization, *Molecular Brain Research,* 7, 151, 1990.

90. Lauterborn, J. C., Isacson, P. J., Montalvo, R., and Gall, C. M., *In situ* hybridization localization of choline acetyltransferase mRNA in adult rat brain and spinal cord, *Brain Res.,* 17, 59, 1993.

91. Mori, N., Tajima, Y., Sakaguchi, H., Vanderberg, D. J., Nawa, H., and Salvaterra, P. M., Partial cloning of the rat choline acetyltransferase gene and *in situ* localization of its transcripts in the cell body of cholinergic neurons in the brainstem and spinal cord, *Brain Res.,* 17, 101, 1993.

92. Oh, J. D., Woolf, N. J., Roghani, A., Edwards, R. H., and Butcher, L. L., Cholinergic neurons in the rat central nervous system demonstrated by *in situ* hybridization of choline acetyltransferase mRNA, *Neuroscience,* 47, 807, 1992.

93. Kiss, J., McGovern, J., and Patel, A. J., Immunohistochemical localization of cells containing nerve growth factor receptors in the different regions of the adult rat forebrain, *Neuroscience,* 27, 731, 1988.

94. Hefti, F., Hartikka, J., Salvatierra, A., Weiner, W. J., and Mash, D. C., Localization of nerve growth factor receptors in cholinergic neurons of the human basal forebrain, *Neuroscience Letters,* 69, 37, 1986.

95. Martínez-Murillo, R., Fernández, T., Alguacil, M. M., Aguado, F., Achaval, M., Bovolenta, P., Rodrigo, J., and Nieto-Sampedro, M., Subcellular localization of nerve growth factor receptors in identified cells of the rat nucleus basalis magnocellularis: an immunocytochemical study, *Neuroscience,* 42, 463, 1991.

96. Mufson, E. J., Presley, L. N., and Kordower, J. H., Nerve growth factor receptor immunoreactivity within the nucleus basalis (Ch4) in Parkinson's disease: reduced cell numbers and colocalization with cholinergic neurons, *Brain Research,* 539, 19, 1991.

97. Springer, J. E., Koh, S., Tayrien, M. W., and Loy, R., Basal forebrain magnocellular neurons stain for nerve growth factor receptor: correlation with cholinergic cell bodies and effects of axotomy, *Journal of Neuroscience Research,* 17, 11, 1987.

98. Chan-Palay, V., Neurons with galanin innervate cholinergic cells in the human basal forebrain and galanin and acetylcholine coexist, *Brain Research Bulletin,* 21, 465, 1988.

99. Kordower, J. H. and Mufson, E., Galanin-like immunoreactivity within the primate basal forebrain: differential staining patterns between human and monkeys, *Journal of Comparative Neurology,* 294, 281, 1990.

100. Melander, T. and Staines, W. A., Galanin-like peptide coexists in putative cholinergic somata of the septum-basal forebrain complex and in acetylcholinesterase containing fibers and varicosities within the hippocampus in the owl monkey (*Aotus trivirgatus*), *Neuroscience Letters,* 68, 17, 1986.

101. Walker, L. C., Koliatsos, V. E., Kitt, C. A., Richardson, R. T., Rokaeus, A., and Price, D. L., Peptidergic neurons in the basal forebrain magnocellular complex of the rhesus monkey, *Journal of Comparative Neurology,* 280, 272, 1989.

102. Melander, T., Staines, W. A., Hökfelt, T., Rokaeus, A., Salvaterra, P. M., and Wainer, B. H., Galanin-like immunoreactivity in cholinergic neurons of the septum basal forebrain complex projecting to the hippocampus of the rat, *Brain Research,* 360, 130, 1985.

103. Melander, T., Hökfelt, T., and Rokaeus, A., Distribution of galanin-like immunoreactivity in the rat central nervous system, *Journal of Comparative Neurology,* 248, 475, 1986.

104. Pasqualotto, B. A. and Vincent, S. R., Galanin and NADPH- diaphorase coexistence in cholinergic neurons of the rat basal forebrain, *Brain Research,* 551, 78, 1991.

105. Senut, M. C., Menetrey, D., and Lamour, Y., Cholinergic and peptidergic projections from the medial septum and the nucleus of the diagonal band of Broca to dorsal hippocampus, cingulate cortex and olfactory bulb, a combined wheat germ agglutinin-apohorseradish peroxidase-gold immunohistochemical study, *Neuroscience,* 30, 385, 1989.

106. Knowall, N. W., Ferrante, R. J., Beal, M. F., Richardson, E. P., Jr., Sofroniew, M. V., Cuello, A. C., and Martin, J. B., Neuropeptide T, somatostatin and reduced nicotinamide adenine dinucleotide phosphate diaphorase in the human striatum: a combined immunocytochemical and enzyme histochemical study, *Neuroscience,* 20, 817, 1987.

107. Vincent, S. R., Johansson, O., Hökfelt, T., Skirboll, L., Elde, R. P., Terenius, L., Kimmel, J., and Goldstein, M., NADPH-diaphorase: a selective histochemical marker for striatal neurons containing both somatostatin- and avian pancreatic polypeptide (APP)-like immunoreactivity, *Journal of Comparative Neurology,* 217, 252, 1983.

108. Vincent, S. R., Satoh, K., Armstrong, D. M., and Fibiger, H. C., NADPH-diaphorase: a selective histochemical marker for cholinergic neurons of the pontine reticular formation, *Neuroscience Letters,* 43, 31, 1983.

109. Celio, M. R. and Norman, A. W., Nucleus basalis Meynert neurons contain the vitamin D-induced calcium binding protein (Calbindin-D 28K), *Anatomy and Embryology (Berlin),* 173, 143, 1985.

110. Chang, H. T. and Kuo, H., Relationship of Calbindin D-28K and cholinergic neurons in the nucleus basalis of Meynert of the monkey and the rat, *Brain Research,* 549, 141, 1991.

111. Ichimiya, Y., Emson, P. C., Mountjoy, C. Q., Lawson, D. E. M., and Iizuka, R., Calbindin-immunoreactive cholinergic neurons in the nucleus basalis of Meynert in Alzheimer's type dementia, *Brain Research,* 499, 402, 1989.

112. Geula, C., Schatz, C. R., and Mesulam, M.-M., Differential localization of NADPH-diaphorase and calbindin-D$_{28k}$ within the cholinergic neurons of the basal forebrain, striatum and brainstem in the rat, monkey, baboon and human, *Neuroscience,* 54, 461, 1993.

113. Martínez-Murillo, R., Villalba, R., Montero-Caballero, M. I., and Rodrigo, J., Cholinergic somata and terminals in the rat substantia nigra, an immunocytochemical study with optical and electron microscopic techniques, *Journal of Comparative Neurology,* 281, 397, 1989.

114. Vincent, S. R., Satoh, K., Armstrong, D. M., Panula, P., Vale, W., and Fibiger, H. C., Neuropeptides and NADPH-diaphorase activity in the ascending cholinergic reticular system of the rat, *Neuroscience,* 17, 167, 1986.

115. Cuello, A. C. and Sofroniew, M. V., The anatomy of the CNS cholinergic neurons, *Trends in Neuroscience,* March, 74, 1984

116. Standaert, D. G., Saper, C. B., Rye, D. B., and Wainer, B. H., Colocalization of atriopeptin-like immunoreactivity with choline acetyltransferase- and substance P-like immunoreactivity in the pedunculopontine and laterodorsal tegmental nuclei in the rat, *Brain Research,* 382, 163, 1986.

117. Arimatsu, Y. and Yamamoto, M., Colocalization of NGF receptor immunoreactivity and [³H]GABA uptake activity in developing rat septum/diagonal band neurons *in vitro, Neuroscience Letters,* 99, 39, 1989.

118. Brashear, H. R., Zaborsky, L., and Heimer, L., Distribution of GABAergic and cholinergic neurons in the rat diagonal band, *Neuroscience,* 17, 439, 1986.

119. Fisher, R. S. and Levine, M. S., Transmitter cosynthesis by corticopetal basal forebrain neurons, *Brain Research,* 491, 163, 1989.

120. Gritti, I., Mainville, L., and Jones, B. E., Codistribution of GABA- with acetylcholine-synthesizing neurons in the basal forebrain of the rat, *Journal of Comparative Neurology,* 329, 438, 1993.

121. Sofroniew, M. V., Eckenstein, F., Thoenen, H., and Cuello, A. C., Topography of choline acetyltransferase-containing neurons in the forebrain of the rat, *Neuroscience Letters,* 33, 7, 1982.

122. Tago, H., McGeer, P. L., McGeer, E. G., Akiyama, H., and Hersh, L. B., Distribution of choline acetyltransferase immunopositive structures in the rat brainstem, *Brain Research,* 495, 271, 1989.

123. Satoh, K. and Fibiger, H. C., Distribution of central cholinergic neurons in the baboon (*Papio papio*). I. General morphology, *Journal of Comparative Neurology,* 236, 197, 1985.

124. Woolf, N. J., Cholinergic systems in mammalian brain and spinal cord, *Progress in Neurobiology,* 37, 475, 1991.

125. Powers, A. S. and Reiner, A., The distribution of cholinergic neurons in the central nervous system of turtles, *Brain and Behavioral Evolution,* 41, 326, 1993.

126. Ikeda, M., Houtani, T., Ueyama, T., and Sugimoto, T., Choline acetyltransferase immunoreactivity in the cat cerebellum, *Neuroscience,* 45, 671, 1991.

127. Jones, B. E. and Beaudet, A., Distribution of acetylcholine and catecholamine neurons in the cat brainstem, a choline acetyltransferase and tyrosine hydroxylase immunohistochemical study, *Journal of Comparative Neuroanatomy,* 261, 15, 1987.

128. Vincent, S. R. and Reiner, P. B., The immunohistochemical localization of choline acetyltransferase in the cat brain, *Brain Research Bulletin,* 18, 371, 1987.

129. Mesulam, M.-M., Mufson, E. J., Levey, A. I., and Wainer, B. H., Atlas of cholinergic neurons in the forebrain and upper brainstem of the macaque based on monoclonal choline acetyltransferase immunohistochemistry and acetylcholinesterase histochemistry, *Neuroscience,* 12, 669, 1984.

130. Walker, L. C., Tigges, M., and Tigges, J., Ultrastructure of neurons in the nucleus basalis of Meynert in squirrel monkey, *Journal of Comparative Neurology,* 217, 158, 1983.

131. Mufson, E. J. and Cunningham, M. G., Observations on choline acetyltransferase containing structures in the CD-1 mouse brain, *Neuroscience Letters,* 84, 7, 1988.

132. Mizukawa, K., McGeer, P. L., Tago, H., Peng, J. H., McGeer, E. G., and Kimura, H., The cholinergic system of the human hindbrain studied by choline acetyltransferase immunohistochemistry and acetylcholinesterase histochemistry, *Brain Research,* 379, 39, 1986.

133. Nagai, T., Pearson, T., Peng, J. H., McGeer, E. G., and McGeer, P. L., Immunohistochemical staining of the human forebrain with monoclonal antibody to human choline acetyltransferase, *Brain Research,* 265, 300, 1983.

134. Schwaber, J. S., Rogers, W. T., Satoh, K., and Fibiger, H. C., Distribution and organization of cholinergic neurons in the rat forebrain demonstrated by computer-aided data acquisition and three-dimensional reconstruction, *Journal of Comparative Neurology,* 263, 309, 1987.

135. Armstrong, D. M., Ultrastructural characterization of choline acetyltranferase-containing neurons in the basal forebrain of rat: evidence for a cholinergic innervation of intracerebral blood vessels, *Journal of Comparative Neurology,* 264, 421, 1986.

136. Butcher, L. L. and Semba, K., Reassessing the cholinergic basal forebrain, nomenclature, schemata, and concepts, *Trends in Neuroscience,* 12, 483, 1989.

137. Moriizumi, T. and Hattori, T., Choline acetyltransferase-immunoreactive neurons in the rat entopeduncular nucleus, *Neuroscience,* 46, 721, 1992.

138. Hedreen, J. C., Struble, R. G., Whitehouse, P. J., and Price, D. L., Topography of the magnocellular basal forebrain system in human brain, *Journal of Neuropathology and Experimental Neurology,* 43, 1, 1984.

139. Mesulam, M.-M., Mufson, E. J., and Wainer, B. H., Three-dimensional representation and cortical projection topography of the nucleus basalis (Ch4) in the macaque. Concurrent demonstration of choline acetyltransferase and retrograde transport with a stabilized tetramethylbenzidine method for horseradish peroxidase, *Brain Research,* 367, 301, 1986.

140. Divac, I., Cortical projections of the magnocellular nuclei of the basal forebrain. A reinvestigation, *Neuroscience,* 6, 983, 1981.

141. Macrides, F., Davis, B. J., Youngs, W. M., Nadi, S. N., and Margolis, F. L., Cholinergic and catecholaminergic afferents to the olfactory bulb in the hamster, a neuroanatomical, biochemical and histochemical investigation, *Journal of Comparative Neurology,* 203, 495, 1981.

142. Struble, R. G., Lehmann, J., Mitchell, S. J., McKinney, M., Price, D. L., Coyle, J. T., and DeLong, M. R., Basal forebrain neurons provide major cholinergic innervation of primate neocortex, *Neuroscience Letters,* 66, 215, 1986.

143. Mesulam, M.-M. and Geula, G., Nucleus basalis (Ch4) and cortical cholinergic innervation in the human brain, observations based on the distribution of acetylcholinesterase and choline acetyltransferase, *Journal of Comparative Neurology,* 275, 216, 1988.

144. Divac, I., Magnocellular nuclei of the basal forebrain project to neocortex, brainstem and olfactory bulb. Review of some functional correlates, *Brain Research,* 93, 385, 1975.

145. Mesulam, M.-M. and van Hoesen, G. W., Acetylcholinesterase-rich projections from the basal forebrain of the rhesus monkey to neocortex, *Brain Research,* 109, 152, 1976.

146. Reinoso-Suarez, F., Llamas, A., and Avendaño, C., Pallido- cortical projections in the cat studies by means of the horseradish peroxidase retrograde transport technique, *Neuroscience Letters,* 29, 225, 1982.

147. Saper, C. B., Organization of cerebral cortical afferent system in the rat. II. Magnocellular basal nucleus, *Journal of Comparative Neurology,* 222, 313, 1984.

148. Chen, S. and Bentivoglio, M., Nerve growth factor receptor-containing cholinergic neurons of the basal forebrain project to the thalamic reticular nucleus in the rat, *Brain Research,* 26, 207, 1993.

149. Heckers, S., Geula, C. and Mesulam, M.-M., Cholinergic innervation of the human thalamus: dual origin and differential nuclear distribution, *Journal of Comparative Neurology,* 325, 68, 1992.

150. Jourdain, A., Semba, K., and Fibiger, H. C., Basal forebrain and mesopontine tegmental projections to the reticular thalamic nucleus, an axonal collateralization and immunohistochemical study in the rat, *Brain Research,* 505, 55, 1989.

151. Metherate, R. and Ashe, J. H., Nucleus basalis stimulation facilitates thalamocortical synaptic transmission in the rat auditory cortex, *Synapse,* 14, 132, 1993.

152. Levey, A. I., Hallanger, A. E., and Wainer, B. H., Cholinergic nucleus basalis neurons may influence the cortex via the thalamus, *Neuroscience Letters,* 74, 7, 1987.

153. Carlsen, J. L., Záborszky, L., and Heimer, L., Cholinergic projections from the basal forebrain to the basolateral amygdaloid complex: a combined retrograde fluorescent and immunohistochemical study, *Journal of Comparative Neurology,* 234, 155, 1985.

154. Gorry, J. D., Studies on the comparative anatomy of the ganglion basale of Meynert, *Acta Anatomica,* 55, 51, 1963.

155. Lamour, Y., Dutar, P., and Jobert, A., Topographic organization of basal forebrain neurons projecting to the rat cerebral cortex, *Neuroscience Letters,* 34, 117, 1982.

156. McKinney, M., Coyle J. T., and Hedreen, J. C., Topographic analysis of the innervation of the rat neocortex and hippocampus by the basal forebrain cholinergic system, *Journal of Comparative Neurology,* 217, 103, 1983.

157. Eckenstein, F., Baughman, R. W., and Quin, J., An anatomical study of cholinergic innervation in the rat cerebral cortex, *Neuroscience,* 25, 457, 1988.

158. Price, J. L. and Stern, R., Individual cells in the nucleus basalis-diagonal band complex have restricted projections to the cerebral cortex in the rat, *Brain Research,* 269, 352, 1983.

159. Carnes, K. M., Fuller, T. A., and Price, J. L., Sources of presumptive glutamatergic/aspartatergic afferents to the magnocellular basal forebrain in the rat, *Journal of Comparative Neurology,* 302, 824, 1990.

160. Gaykema, R. P., van del Kuil, J., Hersh, C. B., and Luiten, P. G., Pattern of direct projections from the hippocampus to the medial septum-diagonal band complex: anterograde tracing with *Phaseolus vulgaris* leucoagglutinin combined with immunocytochemistry of choline acetyltransferase, *Neuroscience,* 43, 349, 1991.

161. Bennett-Clarke, C. A., Cortical projections from somatostatin neurons of the basal forebrain in the rat, *Society of Neuroscience Abstract,* 81, 8, 1986.

162. Köller, C., Chan-Palay, V., and Wu, J. Y., Septal neurons containing glutamic acid decarboxylase immunoreactivity project to the hippocampal region in the rat brain, *Anatomy and Embryology (Berlin),* 169, 41, 1984.

163. Rye, D. B., Wainer, B. H., Mesulam, M.-M., Mufson, E. J., and Saper C. B., Cortical projections arising from the basal forebrain: a study of cholinergic and non-cholinergic components employing combined retrograde tracing and immunohistochemical localization of choline acetyltransferase, *Neuroscience,* 13, 627, 1984.

164. Van der Kooy, D. and Kolb, B., Non-cholinergic globus pallidus cells that project to the cortex but not to the subthalamic nucleus in rat, *Neuroscience Letters,* 57, 113, 1985.

165. Cullinan, W. E. and Zaborszky, L., Organization of ascending hypothalamic projections to the rostral forebrain with special reference to the innervation of cholinergic projection neurons, *Journal Comparative Neurology,* 306, 631, 1991.

166. Haring, J. H. and Wang, R. Y., The identification of some sources of afferent input to the rat nucleus basalis magnocellularis by retrograde transport of horseradish peroxidase, *Brain Research,* 366, 152, 1986.

167. Záborszky, L. and Cullinan, W. E., Projections from the nucleus accumbens to cholinergic neurons of the ventral pallidus: a correlated light and electron microscopic double-immunolabeling study in the rat, *Brain Research,* 570, 92, 1992.

168. Gaykema, R. P., van Weeghel R., Hersh, L. B., and Luiten, G. M., Prefrontal cortical projections to the cholinergic neurons in the basal forebrain, *Journal of Comparative Neurology,* 303, 563, 1991.

169. Semba, K., Reiner, P. B., McGeer, E. G., and Fibiger, H. C., Non-cholinergic basal forebrain neurons project to the contralateral basal forebrain in the rat, *Neuroscience Letters,* 84, 23, 1988.

170. Nishimura, Y., Natory, M., and Mato, M., Choline acetyltransferase immunopositive pyramidal neurons in the rat frontal cortex, *Brain Research,* 440, 144, 1988.

171. Jayaraman, A., Anatomical evidence for cortical projections from the striatum in the cat, *Brain Research,* 195, 29, 1984.

172. Parent, A., Boucher, R., and O'Reilly-Fromentin, J., Acetylcholinesterase-containing neurons in cat pallidal complex: morphological characteristics and projection towards the neocortex, *Brain Research,* 230, 356, 1981.

173. Royce, C. J. and Laine, E., Efferent connections of the caudate nucleus including cortical projections of the striatum and other basal ganglia: an autoradiographic and horseradish peroxidase study in the cat, *Journal of Comparative Neurology,* 226, 28, 1984.

174. Sillito, A. M. and Kemp, J. A., Cholinergic modulation of the functional organization of the cat visual cortex, *Brain Research,* 289, 143, 1983.

175. Blaker, S. N., Armstrong, D. M., and Gage, F. H., Cholinergic neurons within the rat hippocampus: response to fimbria-fornix transection, *Journal of Comparative Neurology*, 272, 127, 1988.

176. Clarke, D. J., Cholinergic innervation of the rat dentate gyrus: an immunocytochemical and electron microscopical study, *Brain Research*, 360, 349, 1985.

177. Clarke, D. J. and Dunnett, S. B., Ultrastructural organization of choline acetyltransferase immunoreactive fibers innervating the neocortex from embryonic ventral forebrain graft, *Journal of Comparative Neurology*, 252, 483, 1986.

178. Frotscher, M. and Léránth, C., Cholinergic innervation of the rat hippocampus as revealed by choline acetyltransferase immunocytochemistry, a combined light and electron microscopic study, *Journal of Comparative Neurology*, 239, 237, 1985.

179. Frotscher, M., Schlander, M., and Léránth, C., Cholinergic neurons in the hippocampus: a combined light- and electron-microscopic immunocytochemistry study in the rat, *Cell and Tissue Research*, 246, 293, 1986.

180. Brady, D. R. and Vaughn, J. E., A comparison of the localization of choline acetyltransferase and glutamate decarboxylase immunoreactivity in rat cerebral cortex, *Neuroscience*, 24, 1009, 1988.

181. Houser, C. R., Crawford, G. D., Salvaterra, P. M., and Vaughn, J. E., Immunocytochemical localization of choline acetyltransferase in rat cerebral cortex, a study of cholinergic neurons and synapses, *Journal of Comparative Neurology*, 234, 17, 1985.

182. Ichikawa, T. and Hirata, Y., Organization of choline acetyltransferase-containing structures in the forebrain of the rat, *Journal of Neuroscience*, 6, 281, 1986.

183. Eckenstein, F. and Thoenen, H., Cholinergic neurons in the rat cerebral cortex demonstrated by immunohistochemical localization of choline acetyltransferase, *Neuroscience Letters*, 36, 211, 1983.

184. Hatton, G. I., Ho, Y. W., and Mason, W. T., Synaptic activation of phasic bursting in rat supraoptic nucleus neurones recorded in hypothalamic slides, *Journal of Physiology (London)*, 345, 297, 1983.

185. Mason, W. T., Ho, Y. W., Eckenstein, F., and Hatton, G. I., Mapping of cholinergic neurons associated with rat supraoptic nucleus: combined immunocytochemical and histochemical identification, *Brain Research Bulletin*, 11, 617, 1983.

186. Hallanger, A. E., Levey, A. Y., Lee, H. J., Rye, D. B., and Wainer, B. H., The origin of cholinergic and other subcortical afferents to the thalamus in the rat, *Journal of Comparative Neurology*, 262, 105, 1987.

187. Hallanger, A. E. and Wainer, B. H., Ascending projections from the pedunculopontine tegmental nucleus and adjacent mesopontine tegmentum in the rat, *Journal of Comparative Neurology*, 274, 483, 1988.

188. Sugimoto, T., Mizukawa, T., Hattori, T., Konishi, T., Kaneco, J., and Mizuno, N., Cholinergic neurons in the nucleus tegmenti pedunculopontinus pars compacta and the caudoputamen in the rat: a light and electron microscope immunohistochemical study using a monoclonal antibody to choline acetyltransferase, *Neuroscience Letters*, 51, 113, 1984.

189. Lee, H. J., Rye, D. B., Hallanger, A. E., Levey, A. I., and Wainer, B. H., Cholinergic vs. non-cholinergic efferents from the mesopontine tegmentum to the extrapyramidal motor system nuclei, *Journal of Comparative Neurology*, 275, 469, 1988.

190. Mesulam, M.-M., Geula, C., Bothwell, M. A., and Hersh, L. B., Human reticular formation: cholinergic neurons of the pedunculopontine and laterodorsal tegmental nuclei and some cytochemical comparisons to forebrain cholinergic neurons, *Journal of Comparative Neurology*, 281, 611, 1989.

191. Gould, E. and Butcher, L. L., Cholinergic neurons in the rat substantia nigra, *Neuroscience Letters*, 63, 315, 1986.

192. Martínez-Murillo, R., Villalba, R. M., and Rodrigo, J., Electron microscopic localization of cholinergic terminals in the rat substantia nigra, an immunocytochemical study, *Neuroscience Letters*, 96, 121, 1989.

193. Satoh, K. and Fibiger, H. C., Cholinergic neurons of the laterodorsal tegmental nucleus: efferent and afferent connections, *Journal of Comparative Neurology*, 253, 277, 1986.

194. Semba, K., Reiner, P. B., McGeer, E. G., and Fibiger, H. C., Brainstem afferents to the magnocellular basal forebrain studied by axonal transport, immunohistochemistry, and electrophysiology in the rat, *Journal of Comparative Neurology*, 267, 433, 1988.

195. Sugimoto, T. and Hattori, T., Organization and afferent projections of nucleus tegmenti pedunculopontinus pars compacta with special reference to its cholinergic aspects, *Neuroscience*, 11, 931, 1984.

196. Aas, J.-E., Brodal, P., Baughman, R. W., and Storm- Mathisen J., Projections to the pontine nuclei from choline acetyltransferase-like immunoreactive neurons in the brainstem of the cat, *Journal of Comparative Neurology,* 300, 183, 1990.

197. Semba, K., Reinier, P. B., and Fibiger, H. C., Single mesopontine tegmental neurons project to both the pontine reticular formation and the thalamus in the rat, *Neuroscience,* 38, 643, 1990.

198. Goldsmith, M. and van der Kooy, D., Separate non- cholinergic descending projections and cholinergic ascending projections from the nucleus tegmenti pedunculopontinus, *Brain Research,* 445, 386, 1988.

199. Semba, K. and Fibiger, H. C., Afferent connections of the laterodorsal and the pedunculopontine tegmental nuclei in the rat: a retro- and antero-grade transport and immunohistochemical study, *Journal of Comparative Neurology,* 323, 847, 1992.

200. Barber, R. P., Phelps, P. E., Houser, C. R., Crawford, G. D., Salvaterra, P.M., and Vaughn, J.E., The morphology and distribution of neurons containing choline acetyltransferase in the adult rat spinal cord, an immunocytochemical study, *Journal of Comparative Neuroanatomy,* 229, 329, 1984.

201. Ribeiro-da-Silva, A. and Cuello, A. C., Choline acetyltransferase-immunoreactive profiles are presynaptic to primary sensory fibers in the rat superficial dorsal horn, *Journal of Comparative Neurology,* 295, 370, 1990.

202. Illing, R.-B., A subtype of cerebellar Golgi cells may be cholinergic, *Brain Research,* 522, 267, 1990.

203. Ojima, H., Kawajiri, S.-I., and Yamasaki, T., Cholinergic innervation of the rat cerebellum, qualitative and quantitative analysis of elements immunoreactive to a monoclonal antibody against choline acetyltransferase, *Journal of Comparative Neuroanatomy,* 290, 41, 1989.

204. Barmack, N. H., Baughman, R. W., and Eckenstein, F., Cholinergic innervation of the cerebellum of the rat, rabbit, cat, and monkey as revealed by choline acetyltransferase activity and immunocytochemistry, *Journal of Comparative Neurology,* 317, 233, 1992.

205. Kristt, D. A., Mc Gowan, K. A., Jr., McKinnon, N. M., and Solomon, J., Basal forebrain innervation of rodent neocortex, studies using acetylcholinesterase histochemistry, Golgi and lesion strategies, *Brain Research,* 337, 19, 1985.

206. Semba, K., Reiner, P. B., McGeer, E. G., and Fibiger, H. C., Morphology of cortically projecting basal forebrain neurons in the rat as revealed by intracellular iontophoresis of horseradish peroxidase, *Neuroscience,* 20, 637, 1987.

207. Dinopoulos, A., Parnavelas, J. G., and Eckenstein, F., Morphological characterization of cholinergic neurons in the horizontal limb of the diagonal band of Broca in the basal forebrain of the rat, *Journal of Neurocytology,* 15, 619, 1986.

208. Kimura, H., Tago, H., Akiyama, H., Hersh, L. B., Tooyama, I., and McGeer, P. L., Choline acetyltransferase immunopositive neurons in the lateral septum, *Brain Research,* 533, 165, 1990.

209. Bolam, J. P., Wainer, B. H., and Smith, A. D., Characterization of cholinergic neurons in the rat neostriatum. A combination of choline acetyltransferase immunocytochemistry, Golgi-impregnation and electron microscopy, *Neuroscience,* 12, 711, 1984.

210. Geula, C., Tokuno, H., Hersh, L., and Mesulam, M.-M., Human striatal cholinergic neurons in development, aging and Alzheimer's disease, *Brain Research,* 508, 310, 1990.

211. Phelps, P. E., Houser, C. R., and Vaughn, J. E., Immunocytochemical localization of choline acetyltransferase within the rat neostriatum: a correlated light and electron microscopic study of cholinergic neurons and synapses, *Journal of Comparative Neurology*, 238, 286, 1985.

212. Phelps, P. E., Houser, C. R., and Vaughn, J. E., Small cholinergic neurons within field of cholinergic axons characterize olfactory-related regions of rat telencephalon, *Neuroscience,* 48, 121, 1992.

213. Bialowas, J. and Frotscher, M., Choline acetyltransferase-immunoreactive neurons and terminals in the rat septal complex: a combined light and electron microscopic study, *Journal of Comparative Neurology,* 259, 298, 1987.

214. Dimova, R., Vuillet, J., Nieoullon, A., and Kerkerian, L. G. L., Ultrastructural features of the choline acetyltransferase-containing neurons and relationships with nigral dopaminergic and cortical afferent pathways in the rat striatum, *Neuroscience,* 53, 1059, 1993.

215. Ingham, C. A., Bolam, J. P., Wainer, B. H., and Smith, A. D., A correlated light and electron microscopic study of identified cholinergic basal forebrain neurons that project to the cortex in the rat, *Journal of Comparative Neurology,* 239, 176, 1985.

216. Kubota, Y., Leung, E., and Vincent, S. R., Ultrastructure of cholinergic neurons in the laterodorsal tegmental nucleus of the rat: interactions with catecholamine fibers, *Brain Research Bulletin,* 29, 479, 1992.

217. Milner, T. A., Cholinergic neurons in the rat septal complex: ultrastructural characterization and synaptic relations with catecholaminergic terminals, *Journal of Comparative Neurology*, 314, 37, 1991.

218. Naumann, T., Linke, R., and Frotscher, M., Fine structure of rat septo-hippocampal projection neurons by retrograde tracing combined with electron microscopy immunocytochemistry and intracellular staining, *Journal of Comparative Neurology* 325, 207, 1992.

219. Palacios, G., García-Ladona, J., and Codina, M., Ultrastructural study of cholinergic neurons in the medial septal nucleus an vertical limb of the diagonal band of Broca in the basal forebrain of the rat, *Journal of Chemical Neuroanatomy*, 4, 205, 1991.

220. Martínez-Murillo, R., Villalba, R. M., and Rodrigo, J., Immunocytochemical localization of cholinergic terminals in the region of the nucleus basalis magnocellularis of the rat, a correlated light and electron microscopic study, *Neuroscience*, 36, 361, 1990.

221. Beniato, M. and Spencer, R. F., The cholinergic innervation of the rat substantia nigra: a light and electron microscopic immunohistochemical study, *Experimental Brain Research*, 72, 178, 1988.

222. Bolam, J. P., Francis, C. M., and Henderson, Z., Cholinergic input to dopaminergic neurons in the substantia nigra: a double immunocytochemical study, *Neuroscience*, 41, 483, 1991.

223. Carlsen, J. and Heimer, L., A correlated light and electron microscopic immunocytochemical study of cholinergic terminals and neurons in the rat amygdaloid body with special emphasis on the basolateral amygdaloid nucleus, *Journal of Comparative Neurology*, 244, 121, 1986.

224. de Lima, A. D., Montero, V. M., and Singer, W., The cholinergic innervation of the visual thalamus: an EM immunocytochemical study, *Experimental Brain Research*, 59, 206, 1985.

225. Izzo, P. N. and Bolam, J. P., Cholinergic synaptic input to different parts of spiny striatonigral neurons in the rat, *Journal of Comparative Neurology*, 269, 219, 1988.

226. Mesulam, M.-M., Mash, D., Hersh, L., Bothwell, M., and Geula, C., Cholinergic innervation of the human striatum, globus pallidus, subthalamic nucleus, substantia nigra and red nucleus, *Journal of Comparative Neurology*, 323, 252, 1992.

227. Wainer, B. H., Bolam, J. P., Freund, T. F., Henderson, Z., Totterdell, S., and Smith, A. D., Cholinergic synapses in the rat brain: a correlated light and electron microscopic immunohistochemical study employing a monoclonal antibody against choline acetyltransferase, *Brain Research*, 308, 69, 1984.

228. Záborszky, L., Heimer, L., Eckenstein, F., and Léránth, C., GABAergic input to cholinergic forebrain neurons: an ultrastructural study using retrograde tracing of HRP and double immunolabeling, *Journal of Comparative Neurology*, 250, 282, 1986.

229. Span, B. M. and Grofova, I., Cholinergic and non-cholinergic neurons in the rat pedunculopontine tegmental nucleus, *Anatomy and Embryology (Berlin)*, 186, 215, 1992.

230. Walker, L. C., Price, D. L., and Young W. S. III, GABAergic neurons in primate basal forebrain magnocellular complex, *Brain Research*, 499, 188, 1989.

231. Priestley, J. V., Somogyi, P., and Cuello, A. C., Neurotransmitter-specific projection neurons revealed by combining PAP immunohistochemistry with retrograde transport of HRP, *Brain Research*, 220, 231, 1981.

232. Záborszky, L. and Heimer, L., Combinations of tracer techniques, especially HRP and PHA-L, with transmitter identification for correlated light and electron microscopic studies, in *Neuroanatomical Tract-Tracing Methods Vol.2: Recent Progress.*, Heimer, L. and Záborszky, L., Eds., Plenum Press, New York, 1989, 49.

233. Bolam, J. P., Ingham, C. A., Izzo, P. N., Levey, A. I., Rye, D. B., Smith, A. D., and Wainer, B. H., Substance P-containing terminals in synaptic contact with cholinergic neurons in the neostriatum and basal forebrain: a double immunocytochemical study in the rat, *Brain Research*, 397, 279, 1986.

234. Chang, H. T., Penny, G. R., and Kitai, S. T., Enkephalinergic-cholinergic interaction in the rat globus pallidus: a pre-embedding double labeling immunocytochemistry study, *Brain Research*, 426, 197, 1987.

235. Chan-Palay, V., Galanin hiperinnervates surviving neurons of the human basal nucleus of Meynert in dementias of Alzheimer's and Parkinson's disease: a hypothesis for the role of galanin in accentuating cholinergic disfunction in dementia, *Journal of Comparative Neurology*, 273, 543, 1988.

236. Day, J. and Fibiger, H. C., Dopaminergic regulation of cortical acetylcholine release: effects of dopamine receptor agonists, *Neuroscience*, 54, 643, 1993.

237. Ingham, C. A., Bolam, J. P., and Smith, A. D., GABA- immunoreactive synaptic boutons in the rat basal forebrain: comparison of neurons that project to the neocortex with pallidosubthalamic neurons, *Journal of Comparative Neurology*, 273, 263, 1988.

238. Martone, M. E., Armstrong, D. M., Young, S. J., and Groves, P. M., Ultrastructural examination of enkephalin and substance P input to cholinergic neurons within the rat striatum, *Brain Research,* 594, 253, 1992.
239. Nakajima, Y., Stanfield, P. R., Yamaguchi, K., and Nakajima, S., Substance P excites cultured cholinergic neurons in the basal forebrain, *Advances in Experimental Medical Biology,* 295, 157, 1991.
240. Tamiya, R., Hadana, M., Inagaki, S., and Takagi, H., Synaptic relation between neuropeptide Y axons and cholinergic neurons in the rat diagonal band of Broca, *Neuroscience Letters,* 122, 64, 1991.
241. Ulfig, N., Neuronal loss and GABAergic innervation in the basal forebrain, *Trends in Neuroscience,* 11, 209, 1988.
242. Unger, J. W. and Schmidt, Y., Galanin-immunoreactivity in the nucleus basalis of Meynert in the rat: age-related changes and differential response to lesion-induced cholinergic cell loss, *Neuroscience Letters,* 153, 140, 1993.
243. Záborszky, L., Cullinan, W. E., and Braun, A., Afferents to basal forebrain cholinergic projection neurons, an update, *Advances in Experimental Medical Biology,* 295, 43, 1991.
244. Eccles, J. C., *The Physiology of Synapses,* Springer, Berlin, 1964.
245. Uchizono, K., Characteristics of exitatory and inhibitory synapses in the central nervous system of the cat, *Nature (London),* 207, 642, 1965.
246. Lamour, Y. and Blaker, W. D., GABAergic control of the cholinergic projection to the frontal cortex in not tonic, *Brain Research,* 325, 389, 1985.
247. Jones, B. E. and Beaudet, A., Retrograde labeling of neurons in the brainstem following injections of [^{3}H]choline into the forebrain of the rat, *Experimental Brain Research,* 65, 437, 1987.
248. Paxinos, G. and Watson, C., *The Rat Brain in Stereotaxic Coordinates,* Academic Press, Sydney, 1986.
249. Mesulam, M.-M. and Mufson, E. J., Neuronal inputs into the nucleus basalis of the substantia innominata (Ch4) in the Rhesus monkey, *Brain,* 107, 253, 1984.

Chapter 2

Subtypes of Neuronal Muscarinic Receptors: Pharmacological Criteria

Eugenia Monferini

CONTENTS

I. INTRODUCTION

The mammalian brain is enriched in muscarinic cholinergic receptors (mAChR) which constitute a class of related proteins belonging to the family of G-protein-coupled receptors. Initially the postulation of multiple mAChRs was based on pharmacological experiments with compounds such as gallamine or the "atypical" agonist McN-A-343. These compounds presented selective effects on the heart or on the sympathetic ganglia without affecting other muscarinic responses.[1,2] The presence of different mAChR subtypes was definitively established by the development of the tricyclic compound pirenzepine, a muscarinic antagonist able to block gastric acid secretion but without effect on the cardiovascular system.[3] The results of radioligand binding investigations further supported the concept of multiple mAChRs which were classified as M_1, those exibiting high affinity for pirenzepine; and M_2, those showing low affinity for this antagonist.[4] The synthesis of new selective antagonists led to the postulation of further subdivisions of the muscarinic M_2 class (M_2 cardiac, having high affinity for 11[[2-[(diethyl-amino)-methyl]-1-piperidinyl]acetyl]-5,11-dihydro-6H-pyrido[2,3-b][1,4]-benzodiazepine-6-one (AF-DX 116) and methoctramine; and M_2 glandular or M_3, having high affinity for 4-diphenylacetoxy-N-methyl-piperidine, [4-DAMP] and hexahydrosiladiphenidol [HHSiD]).[5-8] The existence of multiple homologous proteins corresponding to different mAChRs was definitively established by the cloning of cDNAs and genes encoding mAChRs.[9-12] To date, five different mAChRs (named m1 to m5) have been cloned and sequenced from the human and rat genome, and mRNAs for all five subtypes have been found in various regions of the brain.[9,13-14] On the basis of the binding properties of the expressed recombinant mAChRs and the distribution of the mRNA, it appears that the receptors designated as m1, m2, and m3 represent the pharmacologically defined M_1, M_2, and M_3 receptors, respectively.[13,15-17] Cloned m4 receptors were available before the corresponding tissue receptors were demonstrated. Recently, *in situ* hybridization and *in vitro* binding studies have provided evidence for M_4 receptors in specific regions of the rat brain.[18,19] As far as the m5 receptor is concerned, there are as yet no reports of pharmacological studies performed in tissues so that cloned m5 receptors are currently the only means to study this receptor subtype.

The presence of five mAChR subtypes in the brain raises questions about the relationship between these subtypes and the classes of binding sites identified by selective muscarinic antagonists. This question is particularly significant considering that none of the antagonists used to classify muscarinic receptors is highly selective for one receptor subtype over all other subtypes. Typically, the currently available selective antagonists display a 10- to 30-fold difference in affinity values.

Numerous investigators have used the radiolabeled and nonradiolabeled forms of AF-DX 116 and pirenzepine and the nonselective muscarinic antagonist [^{3}H]3-quinuclidinyl benzilate (QNB) as tools to classify muscarinic binding sites and have found that various regions of the forebrain contain an abundance of high-affinity pirenzepine sites and a relatively small amount of high-affinity AF-DX 116 sites.[20-22] The actual estimates of the relative proportions of these two types of sites suggest that they might account for all of the muscarinic receptors in the forebrain. These latter results, together with the substantial presence of m3 and m4 mRNA in the forebrain, have led to the suggestion that perhaps the high-affinity pirenzepine site is composed of the m1, m3, and m4 subtypes of the mAChR.[10] A different picture emerges when muscarinic binding sites in the brain are classified using competitive binding assays with the nonselective quaternary ammonium muscarinic antagonist [³H]NMS (N-methylscopolamine). Some investigators have shown that it is necessary to consider at least three types of sites in order to adequately describe the binding properties of pirenzepine and AF-DX 116 in the brain. For example, Giraldo et al.,[23] Ehlert et al.,[24] and Ehlert and Tran[25] have shown that approximately 40 and 66% of the mAChRs labeled by [³H]NMS in the cerebral cortex and corpus striatum, respectively, lack high affinity for both pirenzepine and AF-DX 116. These latter sites have been designated as non-M_1, and non-M_2 by Ehlert et al.[24] and Ehlert and Tran.[25] The considerable abundance of non-M_1 and non-M_2 sites in the corpus striatum and cerebral cortex has led to the suggestion that these sites may be composed of the m3 and m4 subtypes of the mAChR.[24] Detailed kinetic binding studies have also provided evidence for four distinct binding sites in brain homogenate.[26] Thus, the receptor repertoire in different brain regions is likely to be a pharmacologically complex mixture of m1-m4 proteins. Because conventional pharmacological approaches cannot readily distinguish m1-m4, the identities of the receptor proteins in tissues are uncertain. Moreover, conclusions about the molecular identity (i.e., m1-m5) of any particular receptor subclass (i.e., M_1-M_4) involved in ligand binding and/or behavioral and tissue effects of "selective drugs" are only tentative.

Therefore it has become expedient to express the receptor subtypes in isolation in cell lines, by transfection with the encoding genes, and using other molecular methods such as the use of subtype specific antibodies or antisense oligonucleotides to identify the receptor in terms of their genetically defined subtype and to assign a given function.

In this chapter, observations obtained with cells transfected with genes for a type of muscarinic receptor will be compared, when possible, with results obtained in tissues.

II. PHARMACOLOGY OF MUSCARINIC RECEPTOR SUBTYPES

A. MUSCARINIC RECEPTOR AGONISTS

Before the advent of "selective" muscarinic antagonists, a common observation was that the displacement from tissue membranes of a radioactive muscarinic receptor ligand by antagonists followed the law of mass action, whereas displacement by muscarinic agonists did not. At that time, it appeared that antagonists were binding to a uniform population of sites and the heterogeneity in agonist binding was believed to result from the coexistence of more that one binding component of the receptor that had identical antagonist affinities but differing affinities for a given agonist. Later studies demonstrated that the proportion of high and low agonist affinity states could be altered by guanine nucleotides suggesting that heterogeneous agonist binding reflects conformational states of the receptor which arise through interactions with G proteins. Heterogeneity of agonist binding sites has been found in virtually every tissue or cell type examined. Some parameters of muscarinic agonist binding such as the absolute affinity values and the reduction of affinity induced by guanosine triphosphate (GTP) and its analogs are tissue dependent, whereas the ratios of affinity states in a given tissue are agonist dependent.

In tissues in which different mAChR subtypes may be coexpressed, the contribution of subtype heterogeneity to the existence of multiple agonist binding affinity states cannot easily be discerned.

A few reports have been published on the binding affinities of agonists as displacers of specifically bound radiolabeled ligands from muscarinic receptors expressed in different cell lines, and these are summarized in Tables 2-1 and 2-2. In most but not all cell lines, carbachol and oxotremorine displaced specifically bound radioligands in a multiple manner. These observations support the hypothesis that the multiple agonist affinity states found in tissues are due to the recognition by an agonist of different configurations of a particular muscarinic receptor subtype instead of, or in addition to, binding to multiple subclasses of receptors. The finding that agonist displacement curves at a given receptor subtype were multiphasic in some cell lines but monophasic in others indicates that the proportion of affinity states may depend on factors extraneous to the receptor *per se* (receptor density, types, and levels of G proteins).

In the rat forebrain, a region enriched in M_1 receptor subtype, the presence of GTP or 5′-guanylyl-imidodiphosphate (Gpp(NH)p) has little effect on the displacement of specifically bound ligands by muscarinic agonists,[33-36] and this was also observed with cloned m1 receptors in most cell types. On the other hand, GTP or Gpp(NH)p induces significant changes in agonist binding curves at M_2 muscarinic

Table 2-1   **Binding constants of carbachol at cloned muscarinic receptors**

Cell Type		m1	m2	m3	m4	Ref.
HEK	K_H	7.3	0.1	8.1	—	11
	K_L	470	130	560	200	
	% H	30.0	28.4	6.5	—	
CHO	K_H		1.57			27
	K_L		161			
	% H		30.3			
	GTP-shift		Yes			
A9L	IC_{50} (μM)	0.52				28
	nH	0.4				
	GTP-shift	Yes				
RAT-1	K_H	0.1		74		29,30
	K_L	129		1,100		
	% H	25		56		
	GTP-shift	Yes		Yes		
Y1	K_H	3.9				31
	K_L	120				
	% H	26				
B82	K_H	4.3				32
	K_L	45				
	% H	50.3				

Note: K_H and K_L indicate high- and low-affinity values expressed in μM units; %H indicates the percent of total binding sites that display high-affinity for the agonist.

receptors in the cerebellum and brain stem,[33-36] and this was also observed with m2 receptors cloned in Chinese hamster ovary (CHO) cells.

While the proportion of affinity states and the extent of the GTP-induced shift may depend on factors intrinsic to the nature of the cell in which the receptors have been expressed, the ability of an agonist to bind to the receptor, as reflected in its affinity, is a drug receptor related property and as such should be suitable for differentiating receptor subtypes.

Oxotremorine exhibited substantially higher affinity than carbachol for both the high- and the low-affinity states for a given mAChR subtype, consistent with binding studies performed with tissue

Table 2-2   **Binding constants of oxotremorine at cloned muscarinic receptors**

Cell Type		m1	m2	m3	m4	Ref.
HEK	K_H	—	0.008	—	0.022	11
	K_L	4.1	5.3	4.0	4.1	
	% H	—	20	—	5.1	
A9L	IC_{50} (μM)	0.8				28
	nH	0.9				
RAT-1	K_H	0.02		8.3		29,30
	K_L	8.4		47		
	% H	23		46		

Note: K_H and K_L indicate high- and low-affinity values expressed in μM units; %H indicates the percent of total binding sites that display high-affinity for the agonist.

Table 2-3 **Binding constants of muscarinic antagonists at cloned muscarinic receptors**

Compound	Cell Type	m1	m2	m3	m4	Ref.
Atropine	HEK	3.3	16.6	1.1	1.2	11
	CHO-K1	0.21	1.5	0.15	0.29[a]	17
	CHO-K1	0.50	0.90	1.1	0.6	41
QNB	CHO-K1	0.035	0.027	0.088	0.034	41
Scopolamine	CHO-K1	1.1	2.0	0.44	0.8	41
Pirenzepine	HEK	500	12,500	1,200	2,700	11
	CHO-K1	16	906	180	79[a]	17
	CHO-K1	6.31	224	138	37	42
	CHO-K1	8	270	150	28	41
Methoctramine	CHO-K1	16	3.6	118	37[a]	17
	CHO-K1	50	13.2	214	31.6	42
AF-DX 116	HEK	50(K_H)	20(K_H)	300(K_H)	—	11
		6,800(K_L)	800(K_L)	1,100(K_L)	2,600	
	CHO-K1	1,300	186	838	443[a]	17
AF-DX 250	CHO-K1	427	55	692	162	42
HHSiD	CHO-K1	44	249	10	42[a]	17
4-DAMP	CHO-K1	0.58	3.80	0.52	1.17	42

Note: K_D expressed in nM units.

[a] Hill coefficient differs significantly from unity ($p < 0.05$); K_i were calculated by IC_{50} values in Reference 17 (Table 2-1) using the Cheng-Prusoff equation.

preparations.[37-38] The low-affinity binding states for either agonist were comparable for each of the four subtypes when measured in the same cell line, with K_D values ranging from 45 to 1100 μM for carbachol and from 4 to 47 μM for oxotremorine. In the human embryo kidney (HEK) cells, the affinity of carbachol for the high-affinity state was notably higher for m2 than for m1 and m3; while for oxotremorine, the affinity values for the high-affinity binding states of m2 and m4 were comparable. However, in RAT1 cells, binding affinities for both agonists are about ten-fold higher for m1 than for m3 cloned receptors.

When measured in tissues, the M_1 *vs.* M_2 selectivity of these two agonists detemined as the ratio between the K_i values obtained in binding studies performed in the cerebral cortex and cerebellum, respectively, is of 0.06 to 0.17 for carbachol[37-39] and of 0.68 for oxotremorine;[37,38] moreover, in preparations from rabbit hippocampus, the binding data for oxotremorine, but not for carbachol, in displacing [3H]pirenzepine from M_1 receptors were best fitted by a one-binding site model.[40] Because the authors claim that they are probably labeling both m1 and m3 receptor proteins in their conditions,[40] these data are in good agreement with what has been found in clonal HEK cells, where oxotremorine binds in a biphasic manner only to m2 and m4 receptors.

B. MUSCARINIC RECEPTOR ANTAGONISTS

The affinity values of different muscarinic receptor antagonists obtained from binding studies using cells transfected with genes encoding one receptor subtype are summarized in Table 2-3.

The classical muscarinic antagonists atropine, QNB, and scopolamine bound with high affinity to all four subtypes. In contrast to the small regional variations in atropine affinity reported for various tissues, in HEK cells the human m2 subtype bound atropine with significantly lower affinity than the other three mAChR subtypes as well as the natural porcine atrial M_2 mAChR.[43]

Pirenzepine, an antagonist selective for M_1 receptors in tissues,[4] displayed higher affinity for m1 and m4 receptor subtypes (with the exception of HEK cells) suggesting that these mAChR subtypes may have been characterized as M_1 receptors by previous investigations. Methoctramine,[44] AF-DX 116[5] and its (+)-enantiomer AF-DX 250[45] are classified as M_2-selective antagonists, but they possess similarly high affinity for both m2 and m4 receptors. Nevertheless, these compounds represent valuable tools for distinguishing between m2 and m4 *vs.* the m3 subtype. For example, all compounds clearly discriminate between m2 (m4) and m3 receptors with affinity differences ranging from 5- (AF-DX 116) to 16-fold (methoctramine). The ability of these compounds to distinguish between m2 (m4) receptors and m1 receptors is less pronounced.

In tissues, HHSiD had similar affinity at M_1 and M_3 receptors whereas its affinity at M_2 receptors was 20 to 30 times lower.[46] Similar affinity values and selectivity ratios among the different mAChR subtypes were observed in CHO cells. The selective muscarinic antagonist 4-DAMP[8] is widely used in both functional and radioligand binding studies to discriminate between M_1 and M_3 vs. M_2 receptors. In binding studies performed in CHO cells, this compound exhibited similar high affinity for m1, m3, and m4 but up to seven-fold lower affinity for m2 receptors.

Binding curves obtained with some antagonists displayed Hill coefficients of less than unity at one or more of the cloned receptors listed in Table 2-3. Some of these observations may be due to technical problems since a low Hill coefficient reported for pirenzepine at human m4 receptors expressed in CHO-K1 cells[17] was not found in a later study.[47] Possible interpretations of low Hill coefficients include negative cooperativity, multiple noninteracting binding sites, or multiple interconvertible affinity states. Negative cooperativity would be seen if the antagonists bind to a secondary, allosteric site such that the binding kinetic of the radioligand is perturbed; however, most of the compounds which give low Hill coefficients (atropine, pirenzepine, AF-DX 116, HHSiD) behave as competitive antagonists in functional assays. If cells without detectable muscarinic binding sites are transfected with the genetic material for one muscarinic receptor subtype, then the second possibility does not seem to be a likely explanation. Multiple interconvertible affinities are observed with agonists in binding to receptors in the presence of G proteins, but this is usually not the case with antagonists. The observation that low Hill coefficients occur at a given receptor subtype expressed in some but not all cell lines indicates that the expression system may contribute to this parameter. In fact, studies with purified muscarinic receptors from porcine heart and cerebral cortex reconstituted into different lipid environments have demonstrated that the affinities and the proportion of sites that displayed high affinity for pirenzepine were influenced significantly by the lipid composition.[48]

Although none of the tested antagonists showed a marked selectivity for one type over all other subtypes, distinct selectivity profiles are apparent. In general, these profiles are consistent with the known selectivities of antagonists for muscarinic receptors that led to the $M_1/M_2/M_3$ classification. However, the present data also highlight the limitations of this pharmacological scheme, demonstrating the necessity to determine antagonist affinities to all mAChR subtypes. Whereas the antagonist affinities determined for cloned m2 and m3 receptors generally correlated well with those for the pharmacologically defined M_2 and M_3 subtypes, the assignement of the M_1 receptor is more problematic. The antagonist binding properties of both m1 and m4 receptor proteins were similar to those of the putative M_1 receptors. Thus, the identification of muscarinic receptors with M_1-like pharmacological properties in a given tissue may be indicative of either m1 or m4 or a mixture of these two receptor subtypes. In any case, because the limited selectivity of muscarinic antagonists, a combination of them should be used generally to identify a receptor subtype more confidently.

C. MUSCARINIC RECEPTOR SUBTYPES AND SECOND MESSENGERS

Muscarinic receptors are part of a large family of receptors that are linked through guanine-nucleotide binding proteins (G proteins) to effectors that translate activation of a plasma membrane receptor into an intracellular response. In tissues, two major signaling pathways for the muscarinic receptor are the stimulation of phosphoinositide hydrolysis and the inhibition of adenylyl cyclase activity. There is evidence in the brain that the inhibition of adenylyl cyclase is mediated by an M_2 receptor subtype, based on the low potency of pirenzepine in inhibiting this response. By contrast, the stimulation of phosphatidylinositol (PI) metabolism is inhibited with high affinity by pirenzepine, and it has been suggested that this is activated through an M_1 muscarinic receptor subtype.[49] However, a consistent picture of subtype-specific coupling is difficult to obtain in tissues, due to the limited selectivity of the available tools.

Expression of cDNA clones coding for mAChR subtypes in clonal cell lines has allowed an evaluation of the ability of each receptor subtype to mediate cellular responses to mAChR agonists. It is now clear from the work of many groups[10,50-52] that the preferred linkage of m1, m3, and m5 subtypes is to stimulation of phosphoinositide turnover, while that of the m2 and m4 subtypes is to inhibition of adenylyl cyclase. However, this subdivision is not as straightforward. In fact, in HEK cells expressing the m2 and m4 receptor proteins, carbachol efficiently inhibits adenylyl cyclase activity (forskolin-stimulated cyclic adenosine $3',5'$-monophosphate [cAMP] accumulation is decreased by 65 to 75%); and also weakly, but significantly, activates PI hydrolysis.[27] In contrast, carbachol strongly stimulates PI hydrolysis, but does not inhibit adenylyl cyclase activity in cells expressing m1 and m3 receptor proteins;

instead, cAMP levels increase substantially over forskolin-induced levels at high carbachol concentrations.[50] In A9 L cells transformed with m1 and m3 receptors, a modest elevation in cAMP levels and PI hydrolysis was observed with carbachol, which had identical dose-response relationships. Only a pertussis toxin (PTX)-sensitive decrease in cAMP levels was observed in the m4-transformed A9 L cells.[53] Altogether, these findings have suggested that enhanced accumulation of cAMP by m1 and m3 receptors may be due to activation of adenylyl cyclase by a calcium-dependent mechanism secondary to PI metabolism.[54]

Results from a different group[55] suggest no correlation between cAMP levels and PI hydrolysis. When expressed in CHO-K1 cells, m3 and m4 (in the presence of PTX) strongly increase cAMP levels, whereas m5 does not. In spite of having equivalent effects on cAMP accumulation, m3 and m4 differ in their activity on PI turnover, in that m4 has little or no effect on PI hydrolysis. Again, m3 and m5 have similar effects on PI hydrolysis, whereas m5 has much less effect on cAMP levels. Finally, the dose-response relationships of the effects on PI turnover and cAMP accumulation differ by about 100-fold.[55] The increase in cAMP levels elicited by the stimulation of the m4 receptor subtype has a functional correlate in tissues. In the rat olfactory bulb, acetylcholine increases adenylyl cyclase activity by a mechanism which is apparently independent of phospholipid hydrolysis.[56] The agonist profile of the muscarinic stimulation of adenylyl cyclase in the olfactory bulb correlates well with that exhibited by the muscarinic inhibition of the enzyme activity in the striatum, suggesting that the two responses are mediated by a similar receptor subtype.[57] The recent observation that $\beta\gamma$ subunits released from activation of G-protein heterotrimers, can selectively enhance type II and type IV adenylyl cyclase activity, might indicate the mechanism by which these muscarinic receptors are increasing cAMP levels.[58]

Thus, the division of the five muscarinic receptors into two classes (m2 and m4 vs. m1, m3, and m5) is not adequate to explain all of their functional properties.

The ability of mAChR subtypes to interact with multiple G proteins is illustrated also by the observations that m1 and m3 receptors expressed in CHO cells could stimulate PI hydrolysis through both pertussis toxin-sensitive and -insensitive pathways[59] and that m1, but not m3, receptors expressed in RAT1 cells also couple to an inhibition of adenylyl cyclase via a PTX-sensitive G protein.[29,30]

A cellular response typically increases as a function of receptor concentration. In CHO cells transfected with the porcine m2 receptor cDNA, carbachol inhibited adenylyl cyclase with an EC_{50} value of 71 nM, while the EC_{50} for carbachol-stimulated PI hydrolysis was 6000 nM.[27] The stimulation of PI hydrolysis was highly dependent on receptor number. In contrast, the inhibition of adenylyl cyclase activity was similar at each level of receptor expression.[27] Therefore, an agonist might demonstrate an apparent selectivity in activating a response in one cell line vs. another simply because of differences in the number of expressed receptors in each cell line.

These findings indicate that the ability of an individual receptor subtype to recognize various effector systems is differential rather than exclusive, and may be determined by the cellular context in which it is evoked. This emphasizes the importance of comparing agonist potencies at the different mAChR subtypes which have been expressed at similar receptor densities in the same cell line. In one study where this was the case,[60] the two muscarinic agonists pilocarpine and McN-A-343 were notably subtype selective on a functional basis. Pilocarpine was more efficacious in stimulating phosphoinositide hydrolysis linked to m1 as compared to either m3 or m5 muscarinic receptors. McN-A-343, on the other hand, produced marked inhibition of cAMP formation in m4-transfected cells, but only a small response at m2 receptors. These results are qualitatively similar to those reported by other investigators who demonstrated such selectivity of McN-A-343 when its ability to inhibit cAMP formation in the striatum, where M_4 is the predominant receptor subtype,[26] was compared to that observed in m2-CHO cells.[61]

In conclusion, pure populations of mAChRs will provide the means of obtaining pharmacological profiles of each subtype of receptor. This information is necessary to maximize the possibility of synthesizing compounds that are selective for a given receptor subtype although the high degree of amino acid sequence homology among the subtypes of mAChRs may be an obstacle difficult to overcome. The differences observed between receptors in tissues compared to cloned receptors or even among cloned receptors expressed in different cell lines mean that, in the end, animal studies will be conclusive in determining the pharmacological selectivity.

REFERENCES

1. Rathbun, F. J. and Hamilton, J. T., Effect of gallamine on cholinergic receptors, *Can. Anaesth. Soc. J.*, 17, 574, 1970.
2. Eglen, R. M., Kenny, B. H., Michel, A. D., and Whiting, R. L., Muscarinic activity of McN-A-343 and its value in muscarinic receptor classification, *Br. J. Pharmacol.*, 90, 693, 1987.
3. Hirschowitz, B. I. and Molina, E., Classification of muscarinic effects on gastric secretion and heart rate in intact dogs, *Subtypes of muscarinic receptors, Proceedings of the International Symposium on Subtypes of Muscarinic Receptors, Trends in Pharmacological Sciences,* Suppl. January 1984, 69.
4. Hammer, R., Berrie, C. P., Birdsall, N. J. M., Burgen, A. S. V., and Hulme, E. C., Pirenzepine distinguishes between different subclasses of muscarinic receptors, *Nature (London)*, 283, 90, 1980.
5. Hammer, R., Giraldo, E., Schiavi, G. B., Monferini, E., and Ladinsky, H., Binding profile of a novel cardioselective muscarine receptor antagonist, AF-DX 116, to membranes of peripheral tissues and brain in the rat, *Life Sci.*, 38, 1653, 1986.
6. Giraldo, E., Micheletti, R., Montagna, E., Giachetti, A., Viganò, A., Ladinsky, H., and Melchiorre, C., Binding and functional characterization of the cardioselective muscarinic antagonist methoctramine, *J. Pharmacol. Exp. Ther.*, 244, 1016, 1988.
7. Mutschler, E. and Lambrecht, G., Selective muscarinic agonists and antagonists in functional tests, *Subtypes of muscarinic receptors, Proceedings of the International Symposium on Subtypes of Muscarinic Receptors, Trends in Pharmacological Sciences,* Suppl. January 1984, 39.
8. Doods, H. N., Mathy, M.-J., Davidesko, D., Van Charldorp, K. J., De Jonge, A., and Van Zwieten, P. A., Selectivity of muscarinic antagonists in radioligand and *in vivo* experiments for the putative M_1, M_2 and M_3 receptors, *J. Pharmacol. Exp. Ther.*, 242, 257, 1987.
9. Kubo, T., Fukuda, K., Mikami, A., Maeda, A., Takahashi, H., Mishina, M., Haga, T., Haga, K., Ichiyama, A., Kangawa, K., Kojima, M., Matsuo, H., Hirose, T., and Numa, S., Cloning, sequencing and expression of complementary DNA encoding the muscarinic acetylcholine receptor, *Nature (London)*, 323, 411, 1986.
10. Bonner, T. I., Buckley, N. J., Young, A. C., and Brann, M. R., Identification of a family of muscarinic acetylcholine receptor genes, *Science*, 237, 527, 1987.
11. Peralta, E. G., Ashkenazi, A., Winslow, J. W., Smith, D. H., Ramachandran, J., and Capon, D. J., Distinct primary structures, ligand-binding properties and tissue-specific expression of four human muscarinic acetylcholine receptors, *EMBO J.*, 6, 3923, 1987.
12. Buckley, N. J., Bonner, T. I., and Brann, M. R., Localization of a family of muscarinic receptor mRNAs in rat brain, *J. Neurosci.*, 8, 4646, 1988.
13. Maeda, A., Kubo, T., Mishina, M., and Numa, S., Tissue distribution of mRNAs encoding muscarinic acetylcholine receptor subtypes, *FEBS Lett.*, 239, 339, 1988.
14. Weiner, D. M. and Brann, M. R., Distribution of m1-m5 muscarinic receptor mRNAs in rat brain, Subtypes of muscarinic receptors IV, Proceedings of the International Symposium on Subtypes of Muscarinic Receptors, *Trends in Pharmacological Sciences,* Suppl. December 1989, 115.
15. Akiba, I., Kubo, T., Maeda, A., Bujo, H., Nakai, J., Mishina, M., and Numa, S., Primary structure of porcine muscarinic acetylcholine receptor III and antagonist binding studies, *FEBS Lett.*, 235, 257, 1988.
16. Brann, M. R., Buckley, N. J., and Bonner, T. I., The striatum and cerebral cortex express different muscarinic receptor mRNAs, *FEBS Lett.*, 230, 90, 1988.
17. Buckley, N. J., Bonner, T. I., Buckley, C. M., and Brann, M. R., Antagonist binding properties of five cloned muscarinic receptors expressed in CHO-K1 cells, *Mol. Pharmacol.*, 35, 469, 1989.
18. Vilaro, M. T., Wiederhold, K.-H., Palacios, J. M., and Mengod, G., Muscarinic cholinergic receptors in the rat caudate putamen and olfactory tubercle belong predominantly to the m4 class: *in situ* hybridization and receptor autoradiography evidence, *Neuroscience*, 40, 159, 1991.
19. Weiner, D. M., Levey, A. I., and Brann, M. R., Expression of muscarinic acetylcholine and dopamine receptor mRNAs in rat basal ganglia, *Proc. Natl. Acad. Sci. U.S.A.*, 87, 7050, 1990.
20. Fisher, S. K. and Bartus, R. T., Regional differences in the coupling of muscarinic receptors to inositol phospholipid hydrolysis in guinea pig brain, *J. Neurochem.*, 45, 1085, 1985.

21. Watson, M., Roeske, W. R., and Yamamura, H. I., [³H]Pirenzepine and (-)[³H]quinuclidinyl benzilate binding to rat cerebral cortical and cardiac muscarinic cholinergic sites. II. Characterization and regulation of antagonist binding to putative muscarinic subtypes, *J. Pharmacol. Exp. Ther.*, 237, 419, 1986.

22. Wang, J. X., Roeske, W. R., Gulya, K., Wang, W., and Yamamura, H. I., [³H]AF-DX 116 labels subsets of muscarinic cholinergic receptors in rat cerebral cortical and cardiac membranes, *Life Sci.*, 41, 1751, 1987.

23. Giraldo, E., Hammer, R., and Ladinsky, H., Distribution of muscarinic receptor subtypes in rat brain as determined in binding studies with AF-DX 116 and pirenzepine, *Life Sci.*, 40, 833, 1987.

24. Ehlert, F. J., Delen, F. M., Yun, S.H., Friedman, D. J., and Self, D. W., Coupling of subtypes of the muscarinic receptor to adenylate cyclase in the corpus striatum and heart, *J. Pharmacol. Exp. Ther.*, 251, 660, 1989.

25. Ehlert, F. J. and Tran, L. L. P., Regional distribution of M_1, M_2 and non-M_1, non-M_2 subtypes of muscarinic binding sites in rat brain, *J. Pharmacol. Exp. Ther.*, 255, 1148, 1990.

26. Waelbroeck, M., Tastenoy, M., Camus, J., and Christophe, J., Binding of selective antagonists to four muscarinic receptors (M_1-M_4) in rat forebrain, *Mol. Pharmacol.*, 38, 267, 1990.

27. Ashkenazi, A., Winslow, J. W., Peralta, E. G., Peterson, G. L., Schimerlik, M. I., Capon, D. J., and Ramachandran, J., An M_2 muscarinic receptor subtype coupled to both adenylate cyclase and phosphoinositide turnover, *Science*, 238, 672, 1987.

28. Brann, M. R., Buckley, N. J., Jones, S. V. P., and Bonner, T. I., Expression of a cloned muscarinic receptor in A9 L cells, *Mol. Pharmacol.*, 32, 450, 1987.

29. Stein, R., Pinkas-Kramarski, R., and Sokolovsky, M., Cloned M_1 muscarinic receptors mediate both adenylate cyclase inhibition and phosphoinositide turnover, *EMBO J.*, 7, 3031, 1988.

30. Pinkas-Kramarski, R., Stein, R., Zimmer, Y., and Sokolovsky, M., Cloned rat M_3 muscarinic receptors mediate phosphoinositide hydrolysis but not adenylate cyclase inhibition, *FEBS Lett.*, 239, 174, 1988.

31. Shapiro, R. A., Scherer, N. M., Habecker, B. A., Subers, E. M., and Nathanson, N. M., Isolation, sequence, and functional expression of the mouse m1 muscarinic acetylcholine receptor gene, *J. Biol. Chem.*, 263, 18397, 1988.

32. Mei, L., Lai, J., Yamamura, H. I., and Roeske, W. R., The relationship between agonist states of the M_1 muscarinic receptor and the hydrolysis of inositol lipids in transfected murine fibroblast cells (B82) expressing different receptor densities, *J. Pharmacol. Exp. Ther.*, 251, 90, 1989.

33. Gurwitz, D., Kloog, Y., and Sokolovsky, M., High affinity binding of [³H]acetylcholine to muscarinic receptors. Regional distribution and modulation by guanine nucleotides, *Mol. Pharmacol.*, 28, 297, 1985.

34. Korn, S. J., Martin, M. W., and Harden, T. K., *N*-Ethylmaleimide-induced alteration in the interaction of agonists with muscarinic cholinergic receptors of rat brain, *J. Pharmacol. Exp. Ther.*, 224, 118, 1983.

35. Vickroy, T. W., Yamamura, H. I., and Roeske, W. R., Differential regulation of high-affinity agonist binding to muscarinic sites in the rat heart, cerebellum, and cerebral cortex, *Biochem. Biophys. Res. Commun.*, 116, 284, 1983.

36. Watson, M., Yamamura, H. I., and Roeske, W. R., [³H]Pirenzepine and (-)-[³H]quinuclidinyl benzilate binding to rat cerebral cortical and cardiac muscarinic cholinergic sites. I. Characterization and regulation of agonist binding to putative muscarinic subtypes, *J. Pharmacol. Exp. Ther.*, 237, 411, 1986.

37. Schumacher, C., Steinberg, R., Kan, J. P., Michaud, J. C., Bourguignon, J. J., Wermuth, C. G., Feltz, P., Worms, P., and Biziere, K., Pharmacological characterization of the aminopyridazine SR 95639A, a selective M_1 muscarinic agonist, *Eur. J. Pharmacol.*, 166, 139, 1989.

38. Boast, C. A., Leventer, S., Sabb, A., Abelson, M., Bender, R., Giacomo, D., Maurer, S., McArthur, S., Mehta, O., Morris, H., Moyer, J., and Storch, F., Biochemical and behavioral characterization of a novel cholinergic agonist, SR 95639, *Pharmacol. Biochem. Behav.*, 39, 287, 1991.

39. Wanibuchi, F., Konishi, T., Harada, M., Terai, M., Hidaka, K., Tamura, T., Tsukamoto, S., and Usuda, S., Pharmacological studies on novel muscarinic agonists, 1-oxa-8-azaspiro[4.5]decane derivatives, YM796 and YM954, *Eur. J. Pharmacol.*, 187, 479, 1990.

40. Potter, L. T. and Ferrendelli, C. A., Affinities of different cholinergic agonists for the high and low affinity states of hippocampal M_1 muscarine receptors, *J. Pharmacol. Exp. Ther.*, 248, 974, 1989.

41. Bolden, C., Cusack, B., and Richelson, E., Antagonism by antimuscarinic and neuroleptic compounds at the five cloned human muscarinic cholinergic receptors expressed in Chinese hamster ovary cells, *J. Pharmacol. Exp. Ther.*, 260, 576, 1992.

42. Dörje, F., Wess, J., Lambrecht, G., Tacke, R., Mutschler E., and Brann, M. R., Antagonist binding profile of five cloned human muscarinic receptor subtypes, *J. Pharmacol. Exp. Ther.*, 256, 727, 1991.

43. Schimerlik, M. I. and Searles, R. P., Ligand interactions with membrane-bound porcine atrial muscarinic receptor(s), *Biochemistry*, 19, 3407, 1980.

44. Melchiorre, C., Angeli, P., Lambrecht, G., Mutschler, E., Picchio, M. T., and Wess, J., Antimuscarinic action of methoctramine, a new cardioselective M_2 muscarinic antagonist, alone and in combination with atropine and gallamine, *Eur. J. Pharmacol.*, 144, 117, 1987.

45. Engel, W. W., Eberlein, W. G., Mihm, G., Hammer, R., and Trummlitz, G., Tryciclic compounds as selective receptor antagonists. III. Structure-selectivity relationships in a series of cardioselective (M_2) antimuscarinics, *J. Med. Chem.*, 32, 1718, 1989.

46. Lambrecht, G., Fiefel, R., Wagner-Röder. M., Strohmann, C., Zilch, H., Tacke, R., Waelbroeck, M., Christophe, J., Boddeke, H., and Mutschler, E., Affinity profiles of hexahydro-sila-difenidol analogues at muscarinic receptor subtypes, *Eur. J. Pharmacol.*, 168, 71, 1989.

47. Wess, J., Lambrecht, G., Mutschler, E., Brann, M. R., and Dörje, F., Selectivity profile of the novel muscarinic antagonist UH-AH 37 determined by the use of cloned receptors and isolated tissue preparations, *Br. J. Pharmacol.*, 102, 246, 1991.

48. Berstein, G., Haga, T., and Ichiyama, A., Effect of the lipid environment on the differential affinity of purified cerebral and atrial muscarinic acetylcholine receptors for pirenzepine, *Mol. Pharmacol.*, 36, 601, 1989.

49. Gil, D. W. and Wolfe, B. B., Pirenzepine distinguishes between muscarinic receptor-mediated phosphoinositide breakdown and inhibition of adenylate cyclase, *J. Pharmacol. Exp. Ther.*, 232, 608, 1985.

50. Peralta, E. G., Ashkenazi, A., Winslow, J. W., Ramachandran, J., and Capon, D. J., Differential regulation of PI hydrolysis and adenylate cyclase by muscarinic receptor subtypes, *Nature (London)*, 334, 434, 1988.

51. Fukuda, K., Higashida, H., Kubo, T., Maeda, A., Akiba, I., Bujo, H., Mishina, M., and Numa, S., Selective coupling of muscarinic acetylcholine receptor subtypes in NG108-15 cells, *Nature (London)*, 335, 355, 1988.

52. Bonner, T. I., Young, A. C., Brann, M. R., and Buckley, N. J., Cloning and expression of the human and rat m5 muscarinic receptor gene, *Neuron*, 1, 403, 1988.

53. Novotny, E A. and Brann, M. R., Agonist pharmacology of cloned muscarinic receptors, Subtypes of muscarinic receptors IV, Proceedings of the International Symposium on Subtypes of Muscarinic Receptors, *Trends in Pharmacological Sciences,* Suppl. December 1989, 116.

54. Felder, C. C., Kanterman, R. Y., Ma, A. L., and Axelrod, J., A trensfected m1 muscarinic acetylcholine receptor stimulates adenylate cyclase via phosphatidyl inositol hydrolysis, *J. Biol. Chem.*, 264, 20356, 1989.

55. Jones, P. S. V., Heilman, C. J., and Brann, M. R., Functional responses of cloned muscarinic receptors expressed in CHO-K1 cells, *Mol. Pharmacol.*, 40, 242, 1991.

56. Olianas, M. C. and Onali, P., Ca^{2+}-independent stimulation of adenylate cyclase activity by muscarinic receptors in rat olfactory bulb, *J. Neurochem.*, 55, 1083, 1990.

57. Olianas, M. C. and Onali, P., Muscarinic stimulation of adenylate cyclase activity of rat olfactory bulb. I. Analysis of agonist sensitivity, *J. Pharmacol. Exp. Ther.*, 259, 673, 1991.

58. Federman, A. D., Conklin, B. R., Schrader, K. A., Reed, R. R., and Bourne, H. R., Hormonal stimulation of adenylyl cyclase through G_i-protein βy subunits, *Nature (London)*, 356, 159, 1992.

59. Ashkenazi, A., Peralta, E. G., Winslow, J. W., Ramachandran, J., and Capon, D. J., Functionally distinct G proteins selectively couple different receptors to PI hydrolysis in the same cell, *Cell*, 56, 487, 1989.

60. Wang, S. Z. and El-Fakahany, E. E., Application of transfected cell lines in studies of functional receptor subtype selectivity of muscarinic agonists, *J. Pharmacol. Exp. Ther.*, 266, 237, 1993.

61. McKinney, M., Miller, J. H., Gibson, V. A., Nickelson, L., and Aksoy, S., Interactions of agonists with M_2 and M_4 muscarinic receptor subtypes mediating cyclic AMP inhibition, *Mol. Pharmacol.*, 40, 1014, 1991.

Chapter 3

Subtypes of Muscarinic Receptors: Receptor Structure and Molecular Biology

Esam E. El-Fakahany

CONTENTS

I. INTRODUCTION

Using pharmacological tools, several pieces of evidence have been obtained that indicate three distinct subtypes of muscarinic acetylcholine receptors exist.[1,2] Recent advances in technologies related to studies of proteins in general, and neurotransmitter receptors in particular, have resulted in significant progress in gaining knowledge of the structural features of muscarinic acetylcholine receptors. Such knowledge has been derived mainly from studies of the characteristics of solubilized and purified receptors, and more recently as a consequence of the successful molecular cloning of the receptor genes. The latter studies have resulted in the identification of two additional subtypes of muscarinic receptors. The goal of this chapter is to discuss the structural characteristics of muscarinic receptors obtained from both receptor purification and molecular cloning techniques.

II. SOLUBILIZATION AND PURIFICATION
OF MUSCARINIC RECEPTORS

A. SOLUBILIZATION OF MUSCARINIC RECEPTORS
1. General Properties of Solubilized Muscarinic Receptors

Muscarinic receptors have been solubilized with variable efficiencies from different tissues using multiple detergent systems (see Reference 3). Different solubilization procedures vary in their yield of active receptors and in the physical state of the receptor.

Gel electrophoresis of solubilized rat or guinea pig brain muscarinic receptors, which are labeled with the irreversible ligand [³H]propylbenzilylcholine mustard, indicated a molecular weight of ~80,000 Da.[4] Similarly, radiation-inactivation target size analysis of human and rat brain, canine heart, and guinea pig smooth muscle preparations indicated a molecular weight of 78,000 to 82,000 Da.[5] These results suggest the absence of major phylogenic structure diversities in the receptor.

Digitonin-solubilized muscarinic receptors from rat cerebral cortex were separated by fast protein liquid chromatography. Two major distinct receptor peaks were resolved, suggesting that muscarinic receptors in this tissue are formed of multiple entities that differ in electrical charge.[6] Different peaks varied in the ratio of binding of [³H]pirenzepine and [³H]quinuclidinylbenzilate, indicating that they do not belong to the same receptor subtype.[6]

2. Interaction of Agonists and Antagonists with Solubilized Muscarinic Receptors

Gavish and Sokolovsky[7] showed that the affinity of various ligands is maintained for cerebral cortex muscarinic receptors solubilized in 3-[(3-chloramidopropyl)dimethylammonio]-1-propanesulfonate (CHAPS). However, solubilized receptors exhibit a slightly slower rate of ligand dissociation as compared to membrane-bound receptors.[7] In contrast, caudate receptors solubilized with L-α-lysophosphatidylcholine demonstrate a lower binding affinity for both agonists and antagonists as compared to membrane-bound receptors, while the rank order of potency of a series of muscarinic agents remains unchanged.[8]

Rat heart muscarinic receptors solubilized by digitonin exhibit GTP-sensitive high-affinity agonist binding.[9] Thus, muscarinic receptors could be solubilized in a form that is coupled to GTP-binding proteins. This has also been shown by the higher molecular weight of the binding complex of labeled receptor agonists as compared to that of antagonists.[9] These differences in molecular weight might represent different states of coupling of the receptor to GTP-binding proteins.[9] Similar results were obtained by Poyner et al.[10] upon solubilization of cardiac muscarinic receptors by a zwitterionic detergent. In fact, the high- and low-molecular-weight forms of the receptor, which bind agonists with high and low affinity, respectively, have been resolved. Hydrodynamic analysis indicated that the G-protein-coupled high-affinity receptor conformation in fact has a significantly higher molecular weight than the low-affinity receptor species.[10]

However, it should be realized that certain solubilization conditions result in a monomeric form of the receptor, which is uncoupled from GTP-binding proteins in the heart[11] and in the brain.[8] For example, Haga et al.[11] reported that agonists bind to a single population of caudate muscarinic receptors solubilized by L-α-lysophosphatidylcholine.

The M₁ muscarinic receptor antagonist pirenzepine is able to distinguish between multiple receptor subtypes in solubilized preparations of cerebral cortex receptors, although solubilization results in an increase in the proportion of its high-affinity binding sites.[12] In contrast, the low-affinity binding of pirenzepine to cardiac or medulla-pons muscarinic receptors is not maintained upon solubilization, since this manipulation enables pirenzepine to interact with the receptor with high affinity.[13,14] Thus, pirenzepine loses most of its selectivity between brain and heart muscarinic receptors upon extraction of the receptor from the membrane by detergents. It is interesting that there are still significant differences in the kinetics of binding of pirenzepine in solubilized preparations of the cerebral cortex and heart.[15] Solubilization of glandular muscarinic receptors also results in a significant reduction in the affinity of pirenzepine at equilibrium.[13] These findings suggest that the orientation of the receptor into the cell membrane or the interaction between the receptor and signal transduction elements or cytoskeletal components dictates a certain conformation, which is lost upon solubilization. In addition, specific hydrophobic interactions of ligands with cell membrane components such as phospholipids might contribute to the determination of receptor subtype selectivity.

Solubilization of muscarinic receptors also results in the removal of the selectivity of the M_2 receptor antagonist AF-DX 116 between cortical, glandular, and cardiac receptors.[10] In contrast, the allosteric cardioselective muscarinic antagonist gallamine retains its discriminatory properties upon receptor solubilization.[10]

B. PURIFICATION OF MUSCARINIC RECEPTORS

1. General Properties of Purified Muscarinic Receptors

A milestone in the history of purification of muscarinic receptors is represented by the development of affinity ligands such as 3-(2′-aminobenhydroloxyl)-tropane[16] and the azido derivatives of quinuclidinyl benzilate and *N*-methyl-4-piperidyl benzilate,[17] which have been employed successfully in affinity chromatography receptor purification procedures.

Purified cardiac muscarinic receptors exhibit a major band of 78,000 Da and a minor band of 14,800 Da.[18] Since only the major band was labeled by [³H]propylbenzilylcholine mustard, it was suggested that the minor band might simply be a degradation product. Similarly, purified brain receptors have been reported to have a molecular weight of 70,000 Da[19] or 86,000 Da.[17]

The physical properties of muscarinic receptors purified from different tissues have been determined. Purified cardiac muscarinic receptors showed an anomalous electrophoretic migration pattern due to excess charge density and an abnormally large shape parameter.[20] The true molecular weight of the cardiac receptors was estimated to be 50,000 to 60,000 Da.[20] The amino acid composition of purified cardiac[20] and cerebral[19] muscarinic receptors has been determined and shown to be different in the two tissues.

2. Interaction of Agonists and Antagonists with Purified Muscarinic Receptors

The classical muscarinic receptor antagonists atropine, *N*-methylscopolamine, and quinuclidinyl benzilate interact with the same rank order of potency with solubilized and purified brain muscarinic receptors, although they show a general decrease in potency upon purification.[16,21] These findings support a contribution of hydrophobic interactions with components other than the receptor protein in the determination of ligand binding affinity.

Agonists interact with high- and low-affinity conformations of purified cardiac[18] and forebrain[21] muscarinic receptors. However, agonist high-affinity binding to receptors purified from the forebrain is resistant to GTP,[21] suggesting that it might not represent binding to a receptor-G-protein complex. In contrast, Haga and Haga[19] showed that agonists bind to a single affinity state of receptors purified from the brain.

Purification significantly reduces the differences in binding of the subtype-selective muscarinic antagonist pirenzepine in the cerebral cortex and atria.[22] Thus, while receptor solubilization does not modify the heterogeneity of binding of pirenzepine in forebrain, this antagonist binds to the purified receptor with a single intermediate affinity.[21] Tissue selectivity of pirenzepine is fully regained upon reinsertion of purified receptors into cerebral or atrial membranes whose endogenous receptors have been inactivated.[22]

3. Reconstitution of Purified Muscarinic Receptors with G-Proteins

Purified muscarinic receptors could be functionally reconstituted with subunits of GTP-binding proteins in lipid vesicles. Under these conditions the receptors demonstrate high-affinity agonist binding that is sensitive to GTP and its more stable analogs.[23] In addition, activation of the reconstituted receptors results in an increase in the rate of exchange of guanine nucleotides at the GTP-binding proteins.[24–27]

4. Evidence for Dimerization of Muscarinic Receptors Obtained from Purification Studies

Receptor purification studies have provided evidence that cardiac muscarinic receptors might exist in the form of dimers, and that the presence of receptor agonists enhances dimer formation.[20] Receptor polymerization is also supported by the findings of Avissar et al.,[28] who showed that muscarinic receptors in tissues that are rich in the M_2 receptor subtype (e.g., heart, cerebellum, and medulla-pons) are composed of high- (160,000 Da) and low- (86,000 Da) molecular-weight species. Under certain conditions a 40,000-Da species could also be detected.[28] The authors interpreted these data to suggest the formation of receptor dimers (86,000 Da) and tetramers (160,000 Da). However, a more likely interpretation of these data is that the 86,000- and 160,000-Da species represent a monomer and a dimer form of the receptor, respectively. The presence of a 40,000-Da muscarinic receptor has not been substantiated by other studies[29] or by molecular cloning and expression studies (see below), and it might simply be a degradation product. It has also been suggested that the 86,000- and the 160,000-Da forms correspond to the low- and high-affinity agonist binding conformations, respectively.[28] Transition metals that are known to induce agonist high-affinity binding enhance the formation of the 160,000-Da molecular weight species, while guanine nucleotides that promote low-affinity agonist binding induce the dissociation of the receptor into the 86,000-Da variety.[28] There is also supportive pharmacological[30] and molecular biological evidence (see below) for the existence of muscarinic receptor dimers.

III. EFFECTS OF CONSTITUTIVE MEMBRANE COMPONENTS AND CYTOSOLIC FACTORS ON MUSCARINIC RECEPTOR CONFORMATION

There is evidence that membrane phospholipids might play a role in ligand binding to the muscarinic receptor. As discussed earlier, results from receptor solubilization and purification studies indicate significant changes in the pharmacological characteristics of the receptor when it is taken out of its natural environment. Furthermore, treatment of brain membranes with phospholipases A or C resulted in a significant decrease in antagonist binding to the receptor.[31,32] In addition, the affinity of pirenzepine for purified and reconstituted muscarinic receptors could be modified significantly by altering the composition of lipids used for reconstitution.[22] These results suggest that membrane lipids play an important role in the interaction of receptor subtype-selective muscarinic antagonists.

There is also evidence that there are cytosolic factors that modify ligand binding to membrane-bound muscarinic receptors. For example, the presence of cardiac cytosol induces heterogeneity of binding of pirenzepine and the appearance of a high-affinity binding component in heart membranes.[33] Similarly, while pirenzepine binding to brainstem preparations enriched in synaptic membranes displays a single low-affinity state, crude synaptosomal preparations exhibit high- and low-affinity pirenzepine binding sites.[34]

In addition, it has been demonstrated in several laboratories that there are cytosolic factors that influence the affinity of ligand binding to muscarinic receptors or the number of receptors available for binding.[35–38]

IV. IMPORTANT STRUCTURAL MOIETIES INVOLVED IN MUSCARINIC RECEPTOR FUNCTION

A. ROLE OF IONIZABLE GROUPS IN LIGAND BINDING TO MUSCARINIC RECEPTORS

Binding of conventional antagonist ligands to muscarinic receptors is stable over a wide range of pH (6.0 to 10.0).[31,39] However, antagonist binding decreases significantly if pH is lowered below 6.0, and this effect is reversible.[39] In contrast, lowering pH below 4.0 has an irreversible effect in lowering agonist affinity for the receptors.[40] These conditions also result in a permanent loss of the effects of GTP on agonist binding, suggesting a dissociation between the receptors and the G-proteins.[40] Birdsall et al.[41] have demonstrated the existence of three ionizable groups involved in modulation of antagonist binding to cardiac muscarinic receptors, with pK_a values of 5.4, 6.8, and 7.5. The binding affinity of cardioselective antagonists is preferentially affected by the protonation state of the group with pK_a of 6.8.[41] On the other hand, binding of antagonists that are selective for other subtypes of muscarinic receptors is affected more by the protonation state of the residue with a pK_a of 5.4.[41] There is also evidence that the protonated form of scopolamine binds preferentially to the receptor, and that there is competition between ligands and hydrogen ions for an acidic group on the cortical and cardiac muscarinic receptors that has a pK_a of ~5.5.[39]

Cyanogen bromide cleavage experiments suggested that irreversible muscarinic antagonists such as [³H]propylbenzilylcholine mustard attach to an acidic residue in the proposed third transmembrane domain of the muscarinic receptor derived from molecular cloning studies (see below).[42] Uchiyama et al.[43] have also concluded that the binding site of [³H]propylbenzilylcholine mustard in purified brain and cardiac preparations is contained between the N terminus and the second putative intracellular loop. This residue was identified as the aspartic acid in position 105 in the human m1 muscarinic receptor sequence deduced from molecular cloning studies.[44]

B. GLYCOPROTEIN NATURE OF MUSCARINIC RECEPTORS

There is ample evidence that muscarinic receptors are glycoproteins.[1] Thus, muscarinic receptors exhibit characteristics of sialoglycoproteins due to the sensitivity of solubilized receptors to neuraminidase digestion and the avid adsorption of the receptors on lectin affinity resins.[45]

The sequence of the porcine brain m1 muscarinic receptor deduced from the predicted translation product of the cloned gene[46] indicates that the receptor polypeptide contains 54 arginine and lysine residues and 35 aspartate and glutamate residues. Thus, the receptor should be a basic protein with an isoelectric point higher than 7. However, the actual determined isoelectric point of muscarinic receptors in the brain is ~4.5,[47] perhaps due to the contribution of the sialic acid residues that become associated with the receptor during the process of posttranslational modification. In fact, it has been estimated that glucosamine, hexoses, and sialic acid moieties contribute 20 to 30% of the molecular weight of purified cardiac muscarinics receptors.[20]

Various subtypes of muscarinic receptors differ in the number and spacing of their potential glycosylation sites. For example, the number of predicted glycosylation sites on asparagine residues on the N terminus of m1, m2, m3, m4, and m5 receptors is 2, 3, 5, 3, and 3, respectively.[2]

Digestion of sialic acid residues with neuraminidase in lung membranes decreases the number of high-affinity binding sites for agonists, without affecting the binding affinity of antagonists.[48] Similarly, treatment of lung membranes with the parainfluenza virus, which contains neuraminidase activity, does not change the binding of muscarinic antagonist ligands to the receptor. On the other hand, this treatment abolishes the binding of agonists to their high-affinity binding sites.[49]

It is interesting that high-affinity agonist binding to muscarinic receptors in the cerebral cortex[50] and striatum[51] is resistant to the effects of neuraminidase, although these receptors are also known to be glycoproteins.[50]

The state of glycosylation of muscarinic receptors might also be involved in either intracellular receptor transport or incorporation into the cell membrane. Thus, inhibition of protein glycosylation in N1E-115 mouse neuroblastoma cells results in a decrease in the total number of muscarinic receptors.[52] This was mainly due to a reduction in the number of receptors on the cell surface. There was also an attenuation of recovery of receptors after agonist-induced receptor down-regulation.[52]

C. PHOSPHORYLATION SITES ON MUSCARINIC RECEPTORS

Excessive activation of muscarinic receptors generally results in a rapid and reversible decrease in the concentration of cell surface muscarinic receptors, accompanied by an attenuation of receptor function.[53] This is followed by an actual breakdown of the receptor protein (receptor down-regulation).[53] Since it is likely that these regulatory mechanisms are linked to receptor activation, changes in the receptor protein that are induced by products of receptor activation have been sought. One of the biochemical transduction mechanisms linked to activation of muscarinic receptors is the stimulation of phosphoinositide hydrolysis and the possible enhancement of the catalytic activity of protein kinase C.[54] There have been numerous studies that demonstrated a consistent attenuation in the function of muscarinic receptors upon activation of protein kinase C in a variety of tissues by tumor promoter phorbol esters.[55-57] In fact, purified cerebral, but not atrial, receptors are subject to phosphorylation by a specific isozyme of protein kinase C, where four to six phosphate molecules are incorporated per receptor molecule.[58] However, whether this kinase actually plays a physiological role in agonist-induced desensitization of muscarinic receptors remains a matter of debate.[59,60]

On the other hand, mammalian atrial receptors are better substrates for the cyclic AMP-dependent kinase as compared to cerebral receptors.[58] In contrast, protein kinase C readily phosphorylates chick heart muscarinic receptors.[61] These receptors are composed of two subtypes that are similar, but not identical, to the mammalian m2 and m4 receptors.[62] Thus, there is marked receptor subtype specificity for the effects of protein kinases A and C.

Phosphorylation sites of protein kinase C on the cerebral muscarinic receptor have been localized to the C-terminal region, which included a part of the third intracellular loop.[43]

Overall, there is a better similarity in the characteristics of agonist-induced phosphorylation of muscarinic receptors and that induced by a receptor-specific protein kinase similar to the β-adrenergic receptor kinase.[63]

The sequences of the various muscarinic receptor subtypes contain multiple serine and threonine residues, which might represent phosphorylation sites susceptible to different kinases.[2] Most such residues that are localized within the transmembrane domains are highly conserved among the different receptor subtypes and therefore might not be involved in the determination of the subtype-selective susceptibility to various kinases.[2] On the other hand, putative phosphorylation residues within the third cytoplasmic loop might serve such a function.

D. SULFHYDRYL GROUPS AND DISULFIDE BONDS ASSOCIATED WITH MUSCARINIC RECEPTORS

Reductive alkylation of sulfhydryl groups results in conversion of the low-affinity agonist-binding conformation into a high-affinity one in the brain.[64] Several agents that interact with sulfhydryl groups reduce ligand binding to cerebral muscarinic receptors, and the receptor could be protected against such inactivation by both agonists and antagonists.[64]

Treatment of forebrain and cardiac membranes with thiol-modifying agents eliminates the ability of pirenzepine to distinguish between muscarinic receptors in the two tissues, suggesting that key thiol groups in specific oxidation states contribute to the determination of the interaction of subtype-selective ligands.[21]

A model was proposed where there are several sulfhydryl groups present on the muscarinic receptor; the presence of these groups modulates ligand binding to the receptor. The extent and direction of such modulation depend on whether the ligand is an agonist or an antagonist. Furthermore, some of these sulfhydryl groups are not located within the primary ligand binding site on the receptor, but are located adjacent to it.[64] In addition, it has been suggested that a disulfide bond that connects certain receptor sequences plays an important role in the determination of receptor conformation.[65] Proteolytic cleavage of purified brain muscarinic receptors supported the notion of the presence of a disulfide link between the second and third extracellular loops.[42]

Labeling of purified brain muscarinic receptor with [^{3}H]-N-ethylmaleimide resulted in the detection of a conserved cysteine residue that corresponds to amino acid number 98 in the m1 receptor sequence.[44] This residue is located extracellularly and might be involved in the formation of a disulfide bond between two receptor extracellular loops. The existence of such a bond between the first and second extracellular loops has been demonstrated in purified muscarinic receptors.[43]

The predicted sequence of each of the five known muscarinic receptors contains nine conserved cysteine residues. Four of these residues are located extracellularly, four within the transmembrane domains, and one intracellularly.[2]

V. MOLECULAR BIOLOGY OF MUSCARINIC ACETYLCHOLINE RECEPTORS

A. CLONING OF MUSCARINIC RECEPTOR GENES

Numa and colleagues[46] succeeded in obtaining a partial amino acid sequence of purified porcine cerebrum muscarinic receptors. This sequence was utilized to design an oligonucleotide probe that was used to screen a cerebrum cDNA library, resulting in the cloning of a gene encoding a muscarinic receptor with pharmacological characteristics similar to that of M_1 receptors.[46] A similar approach was used to clone the porcine cardiac receptor, which corresponds to the pharmacologically defined M_2 muscarinic receptor.[66] This was followed soon by cloning the genes encoding the rat homologs of M1 and M2 muscarinic receptors by homology screening of cDNA libraries using probes derived from conserved stretches of sequences inherent to the already cloned muscarinic receptors and other G-protein-coupled receptors.[67–69] In addition, two other rat genes that encode muscarinic receptor proteins were also cloned.[68,70] One of these corresponded in pharmacological characteristics to the glandular M_3 muscarinic receptor (termed the m3 receptor), while the other was termed the m4 muscarinic receptor. Meanwhile, the corresponding human genes were cloned by two groups.[68,71] In the following 2 years, another muscarinic receptor gene (encoding the m5 receptor subtype) was cloned by two groups.[72,73] A conventional nomenclature system was adopted whereby the three pharmacologically defined muscarinic receptors are named in upper-case letters (M_1, M_2, and M_3 receptors), while the corresponding cloned receptor genes are assigned lower case letters (m1, m2, and m3 receptor genes). So far, there are no clear pharmacological equivalents to the cloned m4 and m5 muscarinic receptor subtypes. It is also important to note that the numbering assigned to the m3 and the m4 human genes cloned by Peralta et al.[71] corresponds to that of the m4 and m3 rat and human genes cloned by Bonner et al.,[68] respectively. Although there is some evidence that there might be other unknown genes that code for muscarinic receptors or proteins that are highly homologous to them,[68] no such genes have been identified yet.

All cloned muscarinic receptor genes lack introns within their coding sequence or in their 3' untranslated regions.[68,74] This knowledge facilitated cloning some of the muscarinic receptor genes by using genomic rather than cDNA libraries.[72] However, it is possible that there are potential splice sites at the 5' upstream region of the genes encoding some of the receptor subtypes.[68] It is interesting that the 5' domain of the gene encoding m2 muscarinic receptors contains an additional ATG translation initiation codon, which might be responsible for regulating the expression of this particular receptor at the low level found in most tissues.[67]

B. SPECIES DIFFERENCES IN THE PRODUCTS OF CLONED MUSCARINIC RECEPTOR GENES

Thus far, the genes encoding various muscarinic receptors have been cloned from tissues of the following mammalian species: human,[68,71,72] rat,[67,68,70,73,75,76] pig,[46,66,77] and mouse.[78,79] There are some species-related differences in the deduced sequences of cloned muscarinic receptors. For example, the human m4 sequence differs from that of the rat by 24 amino acid substitutions, 1 amino acid deletion, and 1 amino acid insertion.[68] Of these 25 amino acid differences, 15 are in the large third cytoplasmic loop and 6 are in the extracellular amino terminus.[68] Similarly, the human m2 receptor gene differs from the porcine gene by 13 amino acid substitutions, 11 of which also occur in the large third intracellular loop.[68] Evidence has been accumulating for a long time that muscarinic receptors in the chick heart are different from those expressed in mammalian species regarding their pharmacological, biochemical, and immunological properties.[62] The genes of two chick muscarinic receptors that are highly homologous to mammalian m2 and m4 receptors have been cloned.[62,80] However, the few differences in the deduced amino acid sequences between chick and mammalian receptors must be important in terms of the observed pharmacological differences in the two classes of receptors. Furthermore, a *Drosophila* muscarinic receptor was cloned and sequence analysis showed its homology with mammalian muscarinic receptors, particularly the m1, m3, and m5 receptor subtypes.[81] In contrast, this gene has three introns in the segment that encodes the third large cytoplasmic loop.[81] Thus, the muscarinic receptor genes appear to be highly conserved among various species, particularly when it is taken into consideration that in some instances different laboratories have reported differences in the sequences of a given muscarinic receptor cloned from the same species using different cloning strategies.[82]

C. DEDUCED FEATURES OF THE PRODUCTS OF CLONED MUSCARINIC RECEPTOR GENES

Hydropathicity analysis of the sequences of the products of the various muscarinic receptor genes that ranged between 460 and 532 amino acids in length suggested the presence of seven potential transmembrane regions (20 to 30 amino acids each) that are hydrophobic in character, and are connected with extracellular and intracellular loops. The amino terminus of the receptor contains potential glycosylation sites, and the carboxy terminus contains several serine and threonine residues that could serve as sites of phosphorylation by protein kinases. There are also several cytoplasmic consensus sequences that could be phosphorylated by the cyclic AMP-dependent protein kinase.[46]

The various subtypes of muscarinic receptors show a high degree of homology in the putative hydrophobic transmembrane domains. Variable regions are represented by the amino and carboxy terminus and also the third putative cytoplasmic region, which varies in length between 157 and 203 amino acids.[74,83] Excluding the first extracellular loop and the large third cytoplasmic loop, there is more general homology between the m1, m3, and m5 muscarinic receptors, which are collectively different from either the m2 or the m4 receptor.[74,83] The latter two receptors are similar to each other. It should be noted, however, that similarities within each of these two sets of muscarinic receptors also include the first 20 amino acids at the amino terminus of the third cytoplasmic loop.[74,83]

For the most part, there is good correspondence in the pharmacological properties of native muscarinic receptors and the cloned receptors expressed in mammalian cells. It should be noted, however, that there are reports of binding site heterogeneity of some subtype-selective antagonists in cells that express a single receptor subtype.[75,84] However, it is apparent that these antagonists bind to a homogeneous receptor population when the same receptors are expressed in other cells.[85] This confirms the notion that the affinity of ligands for muscarinic receptors is determined not only by the primary sequence of the receptor, but also by the constraints imposed on the receptor conformation, which might vary from one cell type to another. Alternatively, the mechanisms of interaction of subtype-selective antagonists with muscarinic receptors might deviate from simple competition.[86-88] Dependence of receptor properties on the type of cells used for expression is even more marked in the case of coupling of the receptor to signaling pathways. For example, while the chick m4 receptor mediates both an increase in phosphoinositide hydrolysis and an inhibition of adenylate cyclase when expressed in Chinese hamster ovary cells, the same receptor is only coupled to the latter response when expressed in Y1 cells.[80] In this case, the G-protein makeup of the cell and the number of receptors being expressed[2] play an important role in determining the signaling mechanisms.

D. TISSUE DISTRIBUTION OF MUSCARINIC RECEPTOR MESSAGES

The tissue distribution of the messenger RNA encoding the various subtypes of muscarinic receptors has been investigated using Northern hybridization or *in situ* hybridization in tissue sections. It has been shown that exocrine glands contain the message of both m1 and m3 muscarinic receptors, while smooth muscles have the message of both m2 and m3 receptors.[89,90] Mammalian heart exclusively contains the message encoding m2 receptors,[89] and the m4 message is expressed in the adrenal medulla.[91] The brain contains the messages encoding all five muscarinic receptors, albeit with a specific regional distribution. The m1 receptor messenger RNA is prevalent in the pyramidal cell layer of the hippocampus, the granule cell layer of the dentate gyrus, the olfactory bulb, the amygdala, the olfactory tubercle, and the piriform cortex, with moderate levels in the caudate putamen and the cerebral cortex.[92] The m2 receptor message exists at a significantly lower concentration, but is relatively abundant in the medial septum, diagonal band, olfactory bulb, and pontine nuclei.[92] The hindbrain also contains a high concentration of the m2 receptor message.[93,94] In the brain, the m3 and m4 messages are prevalent in the olfactory bulb and the pyramidal cell layer in the hippocampus, but are low in the dentate gyrus.[92] The m3 message is preferentially expressed in the superficial and deep layers of cerebral cortex, in addition to some thalamic and brainstem nuclei, while the m4 message is more abundant in midcortical regions and in the caudate putamen.[92] For a while, the cloned m5 muscarinic receptor remained a mystery since its encoding message could not be detected in any tissue.[72] Improved hybridization strategies have resulted in the ability to detect this message, which is confined to the substantia nigra pars compacta, the ventral tegmental area, lateral habenula, ventromedial hypothalamus, and mammillary body, in addition to the hippocampus (being restricted to the ventral subiculum and the pyramidal cells of the CA1 region).[95,96] There is no evidence for the existence of the m5 receptor message in any peripheral tissues.

E. REGULATION OF MUSCARINIC RECEPTOR MESSAGES

It is widely known that the continued presence of high concentrations of muscarinic agonists results in desensitization of receptor function and also down-regulation of receptor number.[53] It has recently been shown that prolonged incubation with carbachol of Chinese hamster ovary cells that express muscarinic m1 receptors results in a significant decrease in the concentration of messenger RNA encoding the receptor.[97] There is a good correlation in both the time course and the concentration dependence of both receptor and message down-regulation.[97] Thus, receptor down-regulation as a consequence of prolonged receptor activation might be due to a decrease in transcription, which in turn results in a decrease in receptor synthesis. Similar results were obtained in cultured embryonic chick heart cells, where treatment with a muscarinic agonist leads to a decrease in the messages of both m2 and m4 receptors.[98] In the latter study it was shown that the stability of the messages is not altered by agonist treatment, indicating that the level of the message is regulated at the stage of transcription.[98] Exposure of muscarinic receptors in SH-SY5Y cells to agonists resulted in a differential profile of regulation of the messages encoding m2 and m3 receptors.[99] Interestingly, investigation of the time course of message regulation indicated that at certain time points there is a marked increase in the concentration of the message in the presence of the agonist, particularly in the case of m2 receptors.[99] Other *in vitro* studies showed that muscarinic receptor transcripts could also be regulated by microtubule disrupting agents.[100] There is evidence obtained in embryonic chick heart cells that agonist-induced regulation of the muscarinic receptor message involves more than one signal transduction pathway.[101] Recently, evidence has been accumulating that the muscarinic receptor message could be regulated in a heterologous fashion upon the activation of other receptors expressed in the same cell.[102] Thus, it has been demonstrated that the m2 and m4 receptor message is decreased when angiotensin II and A1 adenosine receptors are activated in embryonic chick heart cells, although the decrease is smaller in magnitude compared to that induced by muscarinic agonists.[98] On the other hand, treatment of fetal striatal cultures with nerve growth factor results in a selective increase in the message encoding m2 muscarinic receptors.[103]

In vivo regulation of the messages encoding muscarinic receptors has been investigated in relation to drug treatment, aging and neurological disorders. For example, it has been shown that chronic *in vivo* treatment with muscarinic antagonists[104] or depletion of catecholamines[105] causes a significant increase in the level of the m1 receptor transcript in the brain. Interestingly, depletion of catecholamines actually decreases the concentration of the receptor protein.[105] In contrast, chronic *in vivo* inhibition of acetylcholinesterase results in a marked decrease in the m2 muscarinic receptor message in rat heart, which corresponds to receptor down-regulation.[106] After 1 day of acetylcholinesterase inhibition, however, there is a significant decrease in receptor density without an alteration in the receptor message.[106] These results suggest that various phases of receptor down-regulation might be mechanistically different, and that an alteration in transcription of the receptor message plays an important role only at certain stages of the process of receptor regulation.

Possible changes in the level of muscarinic receptor transcripts in aged rats have been investigated, due to the well-documented decrease in the concentration of muscarinic receptors in various brain regions associated with normal aging.[107] It is rather surprising that no major alterations in the message have been detected when tissues from young and aged rats were compared. Exceptions include the message encoding the m1 muscarinic receptor in the olfactory tubercle[108] and the m2 receptor in the hypothalamus.[109] Similarly, there are no marked changes in the transcripts of m1 through m4 muscarinic receptors measured by solution hybridization in several brain areas of Alzheimer's disease patients, except for a significant decrease in the m1 receptor message in both temporal and occipital cortex.[109] In contrast, it has been reported that there is a marked increase in the m1 receptor message measured by *in situ* hybridization in the temporal cortex in Alzheimer's disease.[110] Whether the use of different methodologies underlies such controversies remains to be determined.

F. INFORMATION DERIVED FROM SITE-DIRECTED MUTAGENESIS STUDIES AND CONSTRUCTION OF RECEPTOR CHIMERAS

A wealth of information related to the potential role of specific amino acid residues or entire regions of the receptor sequence in ligand binding to muscarinic receptors or in the process of receptor activation has been gained through point mutations in the receptor gene or by creating chimeric receptors that represent hybrids between different subtypes of the muscarinic receptors.[2,111–113]

There are several aspartate residues located within the second and third transmembrane regions that are highly conserved among all G-protein-coupled receptors.[2] Mutation of individual aspartate residues in the m1 muscarinic receptor into asparagine resulted in different effects on receptor binding characteristics or function. Results of these studies have indicated that the presence of aspartate 105 is mandatory for the binding of either agonists or antagonists to m1 receptors.[114] Mutation of the aspartate residue at position 71 produces a muscarinic m1 receptor that is capable of binding agonists with a higher affinity as compared to the wild-type receptor, but has a markedly reduced coupling to increased phosphoinositide hydrolysis.[114] Mutation of this particular aspartate residue of the m1 receptor also results in alteration of the interaction of allosteric antagonists with the receptor[115] and in attenuating agonist-induced receptor down-regulation.[116] Similar mutations in the aspartate residues located at positions 99 or 122 of the m1 muscarinic receptor resulted in a reduction in the potency of agonists in stimulating phosphoinositide hydrolysis without a change in the maximal response.[114] Thus, the latter three aspartate moieties (at positions 71, 99, and 122 of the m1 receptor) might be important for receptor-G-protein coupling.

There are also several serine, threonine, and tyrosine residues located in the third, fifth, sixth, and seventh transmembrane regions of muscarinic receptors. These residues are conserved among all five receptor subtypes, but are absent in most other G-protein-coupled receptors.[2] Mutation of a threonine residue in transmembrane 5 or a tyrosine residue in transmembrane 6 of the m3 muscarinic receptor results in a decrease in the affinity of agonist binding to the receptor by 40- to 60-fold and an impairment of receptor coupling to phosphoinositide hydrolysis.[113,117] In contrast, there is no effect of these mutations on the binding of receptor antagonists.[117] Furthermore, mutation of several of the transmembrane serine residues has no effect on ligand binding or receptor function.[117,118] Thus, these threonine and tyrosine residues might be involved in the interaction of the ester group of acetylcholine with the receptor protein by serving as potential donors or acceptors of hydrogen bonds.

Results of three-dimensional modeling of the sequence of the muscarinic receptor suggest that the binding site for receptor agonists is located within a pocket near the extracellular side of the receptor and that the binding process involves a contribution of residues in transmembrane regions 3, 4, 5, 6, and 7.[119] Thus, it has been proposed that the conserved aspartate residue in the third transmembrane domain (residue 105 in the m1 muscarinic receptor sequence) attracts the ammonium head group of acetylcholine, while a hydrophobic pocket containing conserved aromatic residues in transmembranes 4 to 7 stabilizes binding, possibly through formation of hydrogen bonds with the ester moiety of the agonist.[112,119]

As indicated above, receptor purification studies have clearly shown that muscarinic receptors are heavily glycosylated, and potential glycosylation sites have been proposed following cloning of the genes of muscarinic receptors. However, mutation of three candidate asparagine residues located in the extracellular amino terminus of the m2 muscarinic receptor did not result in significant changes in the binding of either agonists or antagonists or in the coupling of this receptor to inhibition of adenylate cyclase.[120] This contrasts with other findings that treatment of heart membranes with agents that remove sialic acid residues decreases the binding affinity of agonists to the receptor.[49] Furthermore, these mutations did not alter the cellular localization of the m2 receptor on the cell surface,[120] in contrast with the reported effects of inhibitors of protein glycosylation in inducing intracellular accumulation of the receptor.[52] However, it should be realized that the latter effects might be due to inhibition of glycosylation of proteins other than muscarinic receptors involved in the process of receptor processing.

It has been suggested that a cysteine residue located in the carboxy terminus of most G-protein-coupled receptors is covalently palmitoylated, a process that results in the formation of an additional cytoplasmic loop.[111] However, mutation of this residue in the m2 muscarinic receptor had no effects on either ligand binding or receptor function.[121]

Savarese et al.[122] have identified two important cysteine residues in the sequence of m1 muscarinic receptors, located in the first and the second putative extracellular loops. Point mutations of either residue results in abolishing both ligand binding and the coupling of the receptor to stimulation of phosphoinositide hydrolysis.[122] Thus, these might be the residues involved in the formation of the disulfide bridge, which is important for the formation of the ligand binding domain or for the determination of the proper folding of the receptor protein.[122] In contrast, mutation of other cysteine residues located in the third extracellular loop or in the seventh transmembrane region reduced receptor function to a lesser extent and had little or no effects on ligand-receptor interactions.[122] In fact, mutation of a certain cysteine residue located in transmembrane 7 results in an increase in the binding affinity of a muscarinic agonist and an enhanced potency in activating phosphoinositide hydrolysis.[122]

In contrast to the absence of a role of serine residues located within the putative transmembrane regions in the process of binding of ligands to the receptor or in agonist-induced receptor activation,[117] site-directed mutagenesis studies identified a series of six serine residues located within an eight amino acid stretch within the third putative cytoplasmic loop of m1 muscarinic receptors, which is crucial for the process of agonist-induced receptor internalization.[123] The same region is also involved in heterologous regulation of muscarinic receptors through activation of β-adrenergic receptors.[123] These residues might be substrates for phosphorylation by protein kinases. In addition, Moro et al.[124] have proposed that a certain arrangement of serine and threonine residues (threonine-serine-serine) in the middle of the large third cytoplasmic loop of the m1 receptor is required for receptor internalization. Mutagenesis of cognate serine and threonine residues in the third cytoplasmic loops of the m2 and m3 muscarinic receptors resulted in attenuation or abolition, respectively, of receptor internalization.[124]

G-Protein-coupled receptors contain several highly conserved proline residues, located within various transmembrane regions, that have been hypothesized to introduce kinks within these important domains. Therefore, it has been proposed that some or all of these residues might be important in the translation of agonist binding to muscarinic receptors into a productive conformational change in the receptor tertiary structure.[2] A point mutation in a proline residue in the seventh transmembrane domain of the m3 muscarinic receptor severely impairs the coupling of the receptor to increased phosphoinositide hydrolysis,[125] supporting the above notion.

There are multiple pieces of evidence supporting the role of the large third cytoplasmic loop in determining the selectivity of coupling of various subtypes of muscarinic receptors to different signal transduction pathways.[83] Kubo et al.[126] first reported on the creation of chimeric receptors by exchanging fragments of the m1 and m2 muscarinic receptors, and their data suggested that replacing the entire large cytoplasmic loop of the m1 receptor with that of the m2 receptor conferred an m2-like profile of electrophysiological changes upon activation of this chimeric m1 receptor. However, there was no evidence for a role of this loop in determining the binding pharmacology of subtype-selective antagonists.[126] More interestingly, a receptor chimera, in which the third cytoplasmic loop of the m1 muscarinic receptor was exchanged with that of the dopamine D_2 receptor, was able to maintain a D_2 receptor pharmacology, but also to respond to dopamine by an increase in calcium mobilization.[127] It was soon realized that most of the third intracellular loop is actually not important for the functionality of muscarinic receptors.[128] In fact, it was shown that only 11 and 8 amino acids located at the amino and the carboxy terminus of this loop are sufficient for full coupling of the m1 muscarinic receptor to enhancement of phosphoinositide hydrolysis.[129] Also, an exchange between m2 and m3 receptors of 9 to 21 amino acids located at the amino terminus of the third intracellular loop resulted in making one receptor behave like the other in a reciprocal fashion, in terms of both the kinetics of the calcium response and the sensitivity to pertussis toxin.[130]

On the other hand, there is evidence that regions other than the third cytoplasmic loop might also contribute to determining the selectivity of coupling to different G-proteins. For example, although transfer of this region from m3 to m2 receptors resulted in coupling of m2 receptors to stimulation of phosphoinositide hydrolysis, this response was significantly less efficient than that of native m3 receptors.[131] Interestingly, although a transfer of the entire large cytoplasmic loop of the m3 receptor to the m2 receptor enabled the latter receptors to be coupled to mobilization of calcium from intracellular stores, these chimeric receptors did not possess the ability of m3 receptors to mediate an enhancement of calcium influx through the cell membrane.[132] In addition, studies of m1/m2 receptor chimeras suggested that additional sequences located upstream from the third cytoplasmic region ensure optimal coupling of m1 receptors to increased phosphoinositide hydrolysis.[133] Further evidence in support of this notion is that the chimeric m2 receptor that has the third intracellular loop of m3 receptors retains its ability to inhibit cyclic AMP formation[131] and that a chimeric m1 receptor that possesses the third cytoplasmic loop of β-adrenergic receptors was still coupled, albeit with a lower efficiency than the wild-type receptor, to stimulation of phosphoinositide hydrolysis.[134]

An interesting feature of most G-protein-coupled receptors is the presence of a triplet sequence (aspartate-arginine-tyrosine), located at the beginning of the second putative cytoplasmic loop.[2] The arginine residue in this triplet is conserved in an invariant manner. Site-directed mutagenesis of this residue in the m1 muscarinic receptor almost abolishes receptor coupling to increased phosphoinositide hydrolysis.[135] This effect might be due to uncoupling of the receptor and the G-protein, since it results in a disappearance of agonist high-affinity binding and also in obliteration of receptor-mediated enhancement of the binding of GTP analogs.[135]

In addition to providing information related to delineation of receptor sequences that are important for coupling to cellular functions, receptor chimeras have also been helpful in studying the receptor domains that are involved in determining the binding specificity of muscarinic receptor subtype-selective antagonists. The results of several such studies support the notion that multiple receptor domains contribute to the discriminatory ability of these selective antagonists. Furthermore, these domains might be different for different ligands, even those that exhibit a similar pharmacological profile in terms of selectivity among receptor subtypes.[118,133,136] In a few cases, however, single short amino acid chains could be the major determinant of ligand affinity.[137]

Recent elegant experiments by Wess and co-workers have illustrated that muscarinic receptors are assembled in the cell membrane as a complex of two receptor subunits.[138,139] Thus, while expression of the N terminus or the C terminus of muscarinic receptors individually results in abolishing ligand binding to the receptor, co-expression of the two receptor halves yields a functional receptor.[138]

VI. CONCLUSIONS

There is general agreement between the results and the predictions of studies of solubilized and purified muscarinic receptors and those deduced from the sequences of the cloned receptor genes. It is obvious that the application of molecular biological techniques to studies of muscarinic receptor structure has rapidly yielded a wealth of knowledge, particularly in terms of assessing the role of certain residues or regions in receptor function through the induction of point mutations or the creation of receptor chimeras. Caution should be exercised, however, in interpreting such data without realizing some of the caveats of such an approach, mainly that some mutations might simply alter the folding pattern of the receptor protein. Future attempts at establishing high-efficiency *in vitro* translation systems to obtain large amounts of the various subtypes of muscarinic receptor proteins should result in an accurate characterization of the three-dimensional conformation of the receptor through the elucidation of its crystal structure.

REFERENCES

1. Nathanson, N. M., Molecular properties of the muscarinic acetylcholine receptor, *Annu. Rev. Neurosci.*, 10, 195–236, 1987.
2. Hulme, E. C., Birdsall, N. J. M., and Buckley, N. J., Muscarinic receptor subtypes, *Annu. Rev. Pharmacol. Toxicol.*, 30, 633–673, 1990.
3. Laduron, P. M. and Ilien, B., Solubilization of brain muscarinic, dopaminergic and serotonergic receptors: a critical analysis, *Biochem. Pharmacol.*, 31, 2145–2151, 1982.
4. Birdsall, N. J. M., Burgen, A. S. V., and Hulme, E. C., A study of the muscarinic receptor by gel electrophoresis, *Br. J. Pharmacol.*, 66, 337–342, 1979.
5. Venter, J. C., Muscarinic cholinergic receptor structure, *J. Biol. Chem.*, 256, 4842–4848, 1983.
6. Pelzer, H. and Raible, A., Separation of muscarinic acetylcholine receptor entities from rat cerebral cortex by ion exchange chromatography, *J. Recept. Res.*, 7, 845–857, 1987.
7. Gavish, M. and Sokolovsky, M., Solubilization of muscarinic acetylcholine receptor by zwitterionic detergent from rat brain cortex, *Biochem. Biophys. Res. Commun.*, 109, 819–824, 1982.
8. Haga, T., Solubilization of muscarinic acetylcholine receptors by L-α-lyso-phosphatidylcholine, *Biomed. Res.*, 1, 265–268, 1980.
9. Berrie, C. P., Birdsall, N. J. M., Hulme, E. C., Keen, M., and Stockton, J. M., Solubilization and characterization of guanine nucleotide-sensitive muscarinic agonist binding sites from rat myocardium, *Br. J. Pharmacol.*, 82, 853–861, 1984.
10. Poyner, D. R., Birdsall, N. J. M., Curtis, C. K., Eveleigh, P., Hulme, E. C., Pedder, E. K., and Wheatley, M., Binding and hydrodynamic properties of muscarinic receptor subtypes solubilized in 3-(3-cholamidopropyl)dimethylammonio-2-hydroxy-1-propanesulfonate, *Mol. Pharmacol.*, 36, 420–429, 1989.
11. Berrie, C. P., Birdsall, N. J. M., Haga, K., Haga, T., and Hulme, E. C., Hydrodynamic properties of muscarinic acetylcholine receptors solubilized from rat forebrain, *Br. J. Pharmacol.*, 82, 839–851, 1984.
12. Berrie, C. P., Birdsall, N. J. M., Hulme, E. C., Keen, M., and Stockton, J. M., Solubilization and characterization of high and low affinity pirenzepine binding sites from rat cerebral cortex, *Br. J. Pharmacol.*, 85, 697–703, 1985.
13. Birdsall, N. J. M., Hulme, E. C., Keen, M., Pedder, E. K., Poyner, D., Stockton, J. M., and Wheatley, M., Soluble and membrane-bound muscarinic acetylcholine receptors, *Biochem. Soc. Symp.*, 52, 23–32, 1987.
14. Baumgold, J., Merril, C., and Gershon, E. S., Loss of pirenzepine regional selectivity following solubilization and partial purification of the putative M_1 and M_2 muscarinic receptor subtypes, *Mol. Brain Res.*, 2, 7–14, 1987.
15. Birdsall, N. J. M., Hulme, E. C., and Keen, M., The binding of pirenzepine to digitonin-solubilized muscarinic acetylcholine receptors from the rat myocardium, *Br. J. Pharmacol.*, 87, 307–316, 1986.
16. Haga, K. and Haga, T., Affinity chromatography of the muscarinic acetylcholine receptor, *J. Biol. Chem.*, 258, 13575–13579, 1983.
17. Amitai, G., Avissar, S., Balderman, D., and Sokolovsky, M., Affinity labeling of muscarinic receptors in rat cerebral cortex with a photolabile antagonist, *Proc. Natl. Acad. Sci. U.S.A.*, 79, 243–247, 1982.
18. Peterson, G. L., Herron, G. S., Yamaki, M., Fullerton, D. S., and Schimerlik, M. I., Purification of the muscarinic acetylcholine receptor from porcine atria, *Proc. Natl. Acad. Sci. U.S.A.*, 81, 4993–4997, 1984.
19. Haga, K. and Haga T., Purification of the muscarinic acetylcholine receptor from porcine brain, *J. Biol. Chem.*, 260, 7927–7935, 1985.
20. Peterson, G. L., Rosenbaum, L. C., Broderick, D. J., and Schimerlik, M. I., Physical properties of the purified cardiac muscarinic acetylcholine receptor, *Biochemistry*, 25, 3189–3202, 1986.
21. Wheatley, M., Birdsall, N. J. M., Curtis, C., Eveleigh, P., Pedder, E. K., Poyner, D., Stockton, J. M., and Hulme, E. C., The structure and properties of the purified muscarinic acetylcholine receptor from rat forebrain, *Biochem. Soc. Trans.*, 15, 113–116, 1987.
22. Berstein, G., Haga, T., and Ichiyama, A., Effect of the lipid environment on the differential affinity of purified cerebral and atrial muscarinic acetylcholine receptors for pirenzepine, *Mol. Pharmacol.*, 36, 601–607, 1989.
23. Kurose, H., Katada, T., Haga, T., Haga, K., Ichiyama, A., and Ui, M., Functional interaction of purified muscarinic receptors with purified inhibitory guanine nucleotide regulatory proteins reconstituted in phospholipid vesicles, *J. Biol. Chem.*, 261, 6423–6428, 1986.

24. Haga, K., Haga, T., and Ichiyama, A., Reconstitution of the muscarinic acetylcholine receptor, *J. Biol. Chem.*, 261, 10133–10140, 1986.

25. Haga, T. and Haga, K., Interaction of the muscarinic acetylcholine receptor and GTP-binding proteins, *Biomed. Res.*, 8, 149–156, 1987.

26. Florio, V. A. and Sternweis, P. C., Mechanisms of muscarinic receptor action on G_0 in reconstituted phospholipid vesicles, *J. Biol. Chem.*, 264, 3909–3915, 1989.

27. Haga, K., Uchiyama, H., Haga, T., Ichiyama, A., Kangawa, K., and Matsuo, H., Cerebral muscarinic acetylcholine receptors interact with three kinds of GTP-binding proteins in a reconstitution system of purified components, *Mol. Pharmacol.*, 35, 286–294, 1989.

28. Avissar, S., Amitai, G., and Sokolovsky, M., Oligomeric structure of muscarinic receptors is shown by photoaffinity labeling: subunit assembly may explain high- and low-affinity agonist states, *Proc. Natl. Acad. Sci. U.S.A.*, 80, 156–159, 1983.

29. Dadi, H. K. and Morris, R. J., Muscarinic cholinergic receptor of rat brain, *Eur. J. Biochem.*, 114, 617–628, 1984.

30. Potter, L. T., Ballesteros, L. A., Biochajian, L. H., Ferrendelli, C. A., Fisher, A., Hanchett, H. E., and Zhang, R., Evidence for paired M_2 muscarinic receptors, *Mol. Pharmacol.*, 39, 211–221, 1991.

31. Aronstam, R. S., Abood, L. G., and Baumgold, J., Role of phospholipids in muscarinic binding by neural membranes, *Biochem. Pharmacol.*, 26, 1689–1995, 1977.

32. El-Fakahany, E. E., Ramkumar, V., and Lai, W. S., Multiple binding affinities of *N*-methylscopolamine to brain muscarinic acetylcholine receptors: differentiation from M_1 and M_2 receptor subtypes, *J. Pharmacol. Exp. Ther.*, 238, 554–563, 1982.

33. Fryer, A. D. and El-Fakahany, E. E., An endogenous factor induces heterogeneity of binding sites of selective muscarinic receptor antagonists in rat heart, *Membr. Biochem.*, 8, 127–132, 1989.

34. van Koppen, C. J. and Sokolovsky, M., Evidence for an endogenous factor involved in maintenance of pirenzepine high-affinity binding in rat brainstem, *Biochem. Biophys. Res. Commun.*, 157, 42–47, 1988.

35. Acton, G., Dailey, J. W., Morris, S. W., and McNatt, L., Evidence for an endogenous factor interfering with antagonist binding at the muscarinic cholinergic receptor, *Eur. J. Pharmacol.*, 58, 343–344, 1979.

36. Diaz-Arrastia, R., Ashizawa, T., and Appel, S. H., Endogenous inhibitor of ligand binding to the muscarinic acetylcholine receptor, *J. Neurochem.*, 44, 622–628, 1985.

37. Creazzo, T. L. and Hartzell, H. C., Reduction of muscarinic acetylcholine receptor number and affinity by an endogenous substance, *J. Neurochem.*, 45, 710–718, 1985.

38. Fang, Y. I., Iijima, M., Ogawa, M., Suzuki, T., and Momose, K., Reduction in the numbers of muscarinic receptors by an endogenous protein, *Biochem. Pharmacol.*, 46, 637–641, 1993.

39. Ehlert, F. J. and Delen, F. M., Influence of pH on the binding of scopolamine and *N*-methylscopolamine to muscarinic receptors in the corpus striatum and heart of rats, *Mol. Pharmacol.*, 38, 143–147, 1990.

40. Anthony, B. L. and Aronstam, R. S., Effect of pH on muscarinic acetylcholine receptors from rat brainstem, *J. Neurochem.*, 46, 556–561, 1986.

41. Birdsall, N. J. M., Chan, S. C., Eveleigh, P., Hulme, E. C., and Miller, K. W., The modes of binding of ligands to cardiac muscarinic receptors, *Trends Pharmacol. Sci.*, Suppl. IV, 31–34, 1990.

42. Curtis, C. A. M., Wheatley, M., Bansal, S., Birdsall, N. J. M., Eveleigh, P., Pedder, E. K., Poyner, D., and Hulme, E. C., Propylbenzilylcholine mustard labels an acidic residue in transmembrane helix 3 of the muscarinic receptor, *J. Biol. Chem.*, 264, 489–495, 1989.

43. Uchiyama, H., Ohara, K., Haga, K., Haga, T., and Ichiyama, A., Location in muscarinic acetylcholine receptors of sites for [^{3}H]propylbenzilylcholine mustard binding and for phosphorylation with protein kinase C, *J. Neurochem.*, 54, 1870–1881, 1990.

44. Kurtenbach, E., Curtis, C. A. M., Pedder, E. K., Aitken, A., Harris, A. C. M., and Hulme, E. C., Peptide sequencing identifies residues involved in antagonist binding and disulfide bond formation, *J. Biol. Chem.*, 265, 13702–13708, 1990.

45. Heron, G. S. and Schimerlik, M. I., Glycoprotein properties of the solubilized atrial muscarinic acetylcholine receptor, *J. Neurochem.*, 41, 1414–1420, 1983.

46. Kubo, T., Fukuda, K., Mikami, A., Maeda, A., Takahashi, H., Mishina, M., Haga, T., Haga, K., Ichiyama, A., Kangawa, K., Kojima, M., Matsuo, H., Hirose, T., and Numa, A., Cloning, sequencing and expression of complementary DNA encoding the muscarinic acetylcholine receptor, *Nature*, 323, 411–416, 1986.

47. Repke, H. and Matthies, H., Biochemical characterization of solubilized muscarinic acetylcholine receptors, *Brain Res. Bull.*, 5, 703–709, 1980.

48. Haddad, E. B. and Gies, J. P., Neuraminidase reduces the number of super-high-affinity [^{3}H]oxotremorine-M binding sites in lung, *Eur. J. Pharmacol.*, 211, 273–276, 1992.

49. Fryer, A. D., El-Fakahany, E. E., and Jacoby, D. B., Parainfluenza virus type 1 reduces the affinity of agonists for muscarinic receptors in guinea-pig lung and heart, *Eur. J. Pharmacol.*, 181, 51–58, 1990.

50. Rauh, J. J., Lambert, M. P., Cho, N. J., Chin, H., and Klein, W. L., Glycoprotein properties of muscarinic acetylcholine receptors from bovine cerebral cortex, *J. Neurochem.*, 46, 23–32, 1986.

51. Gies, J. P. and Landry, Y., Sialic acid is selectively involved in the interaction of agonists with M$_2$ muscarinic acetylcholine receptors, *Biochem. Biophys. Res. Commun.*, 150, 673–680, 1988.

52. Liles, W. C. and Nathanson, N. M., Regulation of neuronal muscarinic acetylcholine receptor number by protein glycosylation, *J. Neurochem.*, 46, 89–95, 1986.

53. El-Fakahany, E. E. and Cioffi, C. L., Molecular mechanisms of regulation of neuronal muscarinic receptor sensitivity, *Membr. Biochem.*, 9, 9–27, 1990.

54. El-Fakahany, E. E., Alger, B. E., Lai, W. S., Pitler, T. A., Worley, P. F., and Baraban, J. M., Neuronal muscarinic responses: role of protein kinase C, *FASEB J.*, 2, 2575–2583, 1988.

55. Lai, W. S. and El-Fakahany, E. E., Phorbol ester-induced inhibition of cyclic GMP formation mediated by muscarinic receptors in murine neuroblastoma cells, *J. Pharmacol. Exp. Ther.*, 241, 366–373, 1987.

56. Lai, W. S. and El-Fakahany, E. E., Regulation of [^{3}H]phorbol-12,13-dibutyrate binding sites in mouse neuroblastoma cells: simultaneous down-regulation by phorbol esters and desensitization of their inhibition of muscarinic receptor function, *J. Pharmacol. Exp. Ther.*, 244, 41–51, 1988.

57. Abdallah, E. A. M., Forray, C., and El-Fakahany, E. E., Relationship between the partial inhibition of muscarinic receptor-mediated phosphoinositide hydrolysis by phorbol esters and tetrodotoxin in rat cerebral cortex, *Mol. Brain Res.*, 8, 1–7, 1990.

58. Haga, T., Haga, K., Kameyama, K., and Nakata, H., Phosphorylation of muscarinic receptors: regulation by G proteins, *Life Sci.*, 52, 421–428, 1993.

59. Liles, W. C., Hunter, D. D., Meier, K. E., and Nathanson, N. M., Activation of protein kinase C induces rapid internalization and subsequent degradation of muscarinic acetylcholine receptors in neuroblastoma cells, *J. Biol. Chem.*, 261, 5307–5313, 1986.

60. Lai, W. S., Rogers, T. B., and El-Fakahany, E. E., Protein kinase C is involved in desensitization of muscarinic receptors induced by phorbol esters but not by receptor agonists, *Biochem. J.*, 267, 23–29, 1990.

61. Richardson, R. M. and Hosey, M. M., Agonist-independent phosphorylation of purified cardiac muscarinic cholinergic receptors by protein kinase C, *Biochemistry*, 29, 8555–8561, 1990.

62. Tietje, K. M. and Nathanson, N. M., Isolation and characterization of a gene encoding a novel m2 muscarinic acetylcholine receptor with high affinity for pirenzepine, *J. Biol. Chem.*, 266, 17382–17387, 1991.

63. Kwatra, M. M., Benovic, J. L., Caron, M. G., Lefkowitz, R. J., and Hosey, M. M., Phosphorylation of chick heart muscarinic cholinergic receptors by the β-adrenergic receptor kinase, *Biochemistry*, 28, 4543–4547, 1989.

64. Aronstam, R. S., Abood, L. G., and Hoss, W., Influence of sulfhydryl reagents and heavy metals on the functional state of the muscarinic acetylcholine receptor in rat brain, *Mol. Pharmacol.*, 14, 575–586, 1978.

65. Hedlund, B. and Bartfai, T., The importance of thiol- and disulfide groups in agonist and antagonist binding to the muscarinic receptor, *Mol. Pharmacol.*, 15, 531–544, 1978.

66. Kubo, T., Maeda, A., Sugimoto, K., Akiba, I., Mikami, A., Takahashi, H., Haga, T., Haga, K., Ichiyama, A., Kangawa, K., Matsuo, H., Hirose, T., and Numa, S., Primary structure of porcine cardiac muscarinic acetylcholine receptor deduced from the cDNA sequence, *FEBS Lett.*, 209, 367–372, 1986.

67. Peralta, E. G., Winslow, J. W., Peterson, G. L., Smith, D. H., Ashkenazi, A., Ramachandran, J., Schimerlik, M. I., and Capon, D. J., Primary structure and biochemical properties of an M$_2$ muscarinic receptor, *Science*, 236, 600–605, 1987.

68. Bonner, T. I., Buckley, N. J., Young, A. C., and Brann, M. R., Identification of a family of muscarinic acetylcholine receptor genes, *Science*, 237, 527–532, 1987.

69. Gocayne, J., Robinson, D. A., Fitzgerald, M. G., Chung, F. Z., Kerlavage, A. R., Lentes, K. U., Lai, J., Wang, C. D., Fraser, C. M., and Venter, J. C., Primary structure of rat cardiac β-adrenergic and muscarinic cholinergic receptor obtained by automated DNA sequence analysis: further evidence for a multigene family, *Proc. Natl. Acad. Sci. U.S.A.*, 84, 8296–8300, 1987.

70. Braun, T., Schofield, P. R., Shivers, B. D., Pritchett, D. B., and Seeburg, P. H., A novel subtype of muscarinic receptor identified by homology screening, *Biochem. Biophys. Res. Commun.*, 149, 125–132, 1987.

71. Peralta, E. G., Ashkenazi, A., Winslow, J. W., Smith, D. H., Ramachandran, J., and Capon, D. J., Distinct primary structures, ligand-binding properties and tissue-specific expression of four human muscarinic acetylcholine receptors, *EMBO J.*, 6, 3923–3929, 1987.

72. Bonner, T. I., Young, A. C., Brann, M. R., and Buckley, N. J., Cloning and expression of the human and rat m5 muscarinic acetylcholine receptor genes, *Neuron*, 1, 403–410, 1988.

73. Liao, D. F., Themmen, A. P. N., Joho, R., Barberis, C., Birnbaumer, M., and Birnbaumer, L., Molecular cloning and expression of a fifth muscarinic acetylcholine receptor, *J. Biol. Chem.*, 264, 7328–7337, 1989.

74. Bonner, T. I., The molecular basis of muscarinic receptor diversity, *Trends Neurosci.*, 12, 148–151, 1989.

75. Stein, R., Pinkas-Kramarski, R., and Sokolovsky, M., Cloned M1 muscarinic receptors mediate both adenylate cyclase inhibition and phosphoinositide turnover, *EMBO J.*, 7, 3031–3035, 1988.

76. Lai, J., Mei, L., Roeske, W. R., Chung, F. Z., Yamamura, H. I., and Venter, J. C., The cloned murine M_1 muscarinic receptor is associated with the hydrolysis of phosphatidylinositols in transfected murine B82 cells, *Life Sci.*, 42, 2489–2502, 1988.

77. Akiba, I., Kubo, T., Maeda, A., Bujo, H., Nakai, J., Mishina, M., and Numa, S., Primary structure of porcine muscarinic acetylcholine receptor III and antagonist binding studies, *FEBS Lett.*, 235, 257–261, 1988.

78. Shapiro, R. A., Scherer, N. M., Habecker, B. A., Subers, E. M., and Nathanson, N. M., Isolation, sequence, and functional expression of the mouse M1 muscarinic acetylcholine receptor gene, *J. Biol. Chem.*, 263, 18397–18403, 1988.

79. van Koppen, C. J., Lenz, W., and Nathanson, N. M., Isolation, sequence and functional expression of the mouse m4 muscarinic acetylcholine receptor gene, *Biochim. Biophys. Acta*, 1173, 342–344, 1993.

80. Tietji, K. M., Goldman, P. S., and Nathanson, N. M., Cloning and functional analysis of a gene encoding a novel muscarinic acetylcholine receptor expressed in chick heart and brain, *J. Biol. Chem.*, 265, 2828–2834, 1990.

81. Shapiro, R. A., Wakimoto, B. T., Subers, E. M., and Nathanson, N. M., Characterization and functional expression in mammalian cells of genomic and cDNA clones encoding a *Drosophila* muscarinic acetylcholine receptor, *Proc. Natl. Acad. Sci. U.S.A.*, 86, 9039–9043, 1989.

82. Kashihara, K., Varga, E. V., Waite, S. L., Roeske, W. R., and Yamamura, H. I., Cloning of the rat M3, M4 and M5 muscarinic acetylcholine receptor genes by the polymerase chain reaction (PCR) and the pharmacological characterization of the expressed genes, *Life Sci.*, 51, 955–971, 1992.

83. Bonner, T. I., Domains of muscarinic acetylcholine receptors that confer specificity of G protein coupling, *Trends Pharmacol. Sci.*, 13, 48–50, 1992.

84. Pinkas-Kramarski, R., Stein, R., Zimmer, Y., and Sokolovsky, M., Cloned rat M3 muscarinic receptors mediate phosphoinositide hydrolysis but not adenylate cyclase inhibition, *FEBS Lett.*, 239, 174–178, 1988.

85. Meng, Y., Hu, J., and El-Fakahany, E. E., *p*-Fluoro-hexahydro-sila-difenidol exhibits poor selectivity between M_3 and M_1 muscarinic receptors, *Membr. Biochem.*, 9, 293–300, 1990.

86. Kenakin, T. and Boselli, C., Pharmacologic discrimination between receptor heterogeneity and allosteric interaction: resultant analysis of gallamine and pirenzepine antagonism of muscarinic responses in rat trachea, *J. Pharmacol. Exp. Ther.*, 250, 944–952, 1989.

87. Lee, N. H. and El-Fakahany, E. E., Allosteric antagonists of the muscarinic acetylcholine receptor, *Biochem. Pharmacol.*, 42, 199–205, 1991.

88. Caulfield, M. P., Muscarinic receptors — characterization, coupling and function, *Pharmacol. Ther.*, 58, 319–379, 1993.

89. Maeda, A., Kubo, T., Mishina, M., and Numa, S., Tissue distribution of mRNAs encoding muscarinic acetylcholine receptor subtypes, *FEBS Lett.*, 239, 339–342, 1988.

90. Zhang, L., Horowitz, B., and Buxton, I. L., Muscarinic receptors in canine colonic circular smooth muscle. I. Coexistence of M_2 and M_3 subtypes, *Mol. Pharmacol.*, 40, 943–951, 1991.

91. Fernando, J. C. R., Abdallah, E. A. M., Evinger, M., Forray, C., and El-Fakahany, E. E., The presence of an M_4 subtype muscarinic receptor in the bovine adrenal medulla revealed by mRNA and receptor binding analyses, *Eur. J. Pharmacol., Mol. Pharmacol. Sec.*, 207, 297–303, 1991.

92. Buckley, N. J., Bonner, T. I., and Brann, M. R., Localization of a family of muscarinic receptor mRNAs in rat brain, *J. Neurosci.*, 8, 4646–4652, 1988.

93. Pinkas-Kramarski, R., Stein, R., and Sokolovsky, M., Postnatal changes in muscarinic receptor subtype mRNAs in rat brain and heart, *J. Mol. Neurosci.*, 1, 209–213, 1989.

94. Wang, S. Z., Zhu, S. Z., Joseph, J. A., and El-Fakahany, E. E., Comparison of the level of MRNA encoding m1 and m2 muscarinic receptors in brains of young and aged rats, *Neurosci. Lett.*, 145, 149–152, 1992.

95. Weiner, D. M., Levey, A. I., and Brann, M. R., Expression of muscarinic acetylcholine and dopamine receptor mRNAs in rat basal ganglia, *Proc. Natl. Acad. Sci. U.S.A.*, 87, 7050–7054, 1990.

96. Vilaro, M. T., Palacios, J. M., and Mengod, G., Localization of m5 muscarinic receptor mRNA in rat brain examined by *in situ* hybridization histochemistry, *Neurosci. Lett.*, 114, 154–159, 1990.

97. Wang, S. Z., Hu, J., Long, R. M., Pou, W. S., Forray, C., and El-Fakahany, E. E., Agonist-induced down-regulation of m1 muscarinic receptors and reduction of their mRNA level in a transfected cell line, *FEBS Lett.*, 276, 185–188, 1990.

98. Habecker, B. A. and Nathanson, N. M., Regulation of muscarinic acetylcholine receptor mRNA expression by activation of homologous and heterologous receptors, *Proc. Natl. Acad. Sci. U.S.A.*, 89, 5035–5038, 1992.

99. Steel, M. D. and Buckley, N. J., Differential regulation of muscarinic receptor mRNA levels in neuroblastoma cells by chronic agonist exposure: a comparative polymerase chain reaction study, *Mol. Pharmacol.*, 43, 694–701, 1993.

100. Fukamauchi, F., Hough, C., and Chuang, D., m2- and m3-muscarinic acetylcholine receptor mRNAs have different responses to microtubule-affecting drugs, *Mol. Cell. Neurosci.*, 2, 315–319, 1991.

101. Habecker, B. A., Wang, H., and Nathanson, N. M., Multiple second-messenger pathways mediate agonist regulation of muscarinic receptor mRNA expression, *Biochemistry*, 32, 4986–4990, 1993.

102. Habecker, B. A., Tietje, K. M., van Koppen, C. J., Creason, S. A., Goldman, P. S., Migeon, J. C., Parenteau, L. A., and Nathanson, N. M., Regulation of expression and function of muscarinic receptors, *Life Sci.*, 52, 429–432, 1993.

103. Ebstein, R. P., Bennett, E. R., Sokoloff, M., and Shoham, S., The effect of nerve growth factor on cholinergic cells in primary fetal striatal cultures: characterization by *in situ* hybridization, *Dev. Brain Res.*, 73, 165–172, 1993.

104. McKinney, M. and Robbins, M., Chronic atropine administration up-regulates rat cortical muscarinic m1 receptor mRNA molecules: assessment with the RT/PCR, *Mol. Brain Res.*, 12, 39–45, 1992.

105. Asanuma, M., Ogawa, N., Haba, K., Hirata, H., and Mori, A., Effects of chronic catecholamine depletions on muscarinic M1-receptor, *J. Neurol. Sci.*, 110, 205–214, 1992.

106. Zhu, S. Z., Wang, S. Z., Abdallah, E. A. M., and El-Fakahany, E. E., DFP-induced regulation of cardiac muscarinic receptor mRNA *in vivo* measured by DNA-excess solution hybridization, *Life Sci.*, 48, 2579–2584, 1991.

107. Surichamorn, W., Kim, O. N., Lee, N. H., Lai, W. S., and El-Fakahany, E. E., Effects of aging on the interaction of quinuclidinyl benzilate, *N*-methylscopolamine, pirenzepine, and gallamine with brain muscarinic receptors, *Neurochem. Res.*, 13, 1183–1191, 1988.

108. Blake, M. J., Appel, N. M., Joseph, J. A., Stagg, C. A., Anson, M., De Souza, E. B., and Roth, G. S., Muscarinic acetylcholine receptor subtype mRNA expression and ligand binding in the aged rat forebrain, *Neurobiol. Aging*, 12, 193–199, 1991.

109. Wang, S. Z., Zhu, S. Z., Mash, D. C., and El-Fakahany, E. E., Comparison of the concentration of messenger RNA encoding four muscarinic receptor subtypes in control and Alzheimer brain, *Mol. Brain Res.*, 16, 64–70, 1992.

110. Harrison, P. J., Barton, A. J. L., Najlerahim, A., McDonald, B., and Pearson, R. C. A., Increased muscarinic receptor messenger RNA in Alzheimer's disease temporal cortex demonstrated by *in situ* hybridization histochemistry, *Mol. Brain Res.*, 9, 15–21, 1991.

111. Savarese, T. M. and Fraser, C. M., *In vitro* mutagenesis and the search for structure-function relationships among G protein-coupled receptors, *Biochem. J.*, 283, 1–19, 1992.

112. Brann, M. R., Klimkowski, V. J., and Ellis, J., Structure/function relationships of muscarinic acetylcholine receptors, *Life Sci.*, 52, 405–412, 1993.

113. Wess, J., Molecular basis of muscarinic acetylcholine receptor function, *Trends Pharmacol. Sci.*, 14, 308–313, 1993.

114. Fraser, C. M., Wang, C. D., Robinson, D. A., Gocayne, J. D., and Venter, J. C., Site-directed mutagenesis of m1 muscarinic acetylcholine receptors: conserved aspartic acids play important roles in receptor function, *Mol. Pharmacol.*, 36, 840–847, 1989.

115. Lee, N. H., Hu, J., and El-Fakahany, E. E., Modulation by certain conserved aspartate residues of the allosteric interaction of gallamine at the m1 muscarinic receptor, *J. Pharmacol. Exp. Ther.*, 262, 312–316, 1992.

116. Hu, J., Wang, S. Z., and El-Fakahany, E. E., Role of conserved aspartate residue 71 in down-regulation of m1 muscarinic receptors, *Pharmacol. Commun.*, 1, 219–225, 1992.

117. Wess, J., Gdula, D., and Brann, M. R., Site-directed mutagenesis of the m3 muscarinic receptor: identification of a series of threonine and tyrosine residues involved in agonist but not antagonist binding, *EMBO J.*, 10, 3729–3734, 1991.

118. Wess. J., Maggio, R., Palmer, J.R., and Vogel, Z., Role of conserved threonine and tyrosine residues in acetylcholine binding and muscarinic receptor activation, *J. Biol. Chem.*, 267, 19313–19319, 1992.

119. Trumpp-Kallmeyer, S., Hoflack, J., Bruinvels, A., and Hibert, M., Modeling of G-protein-coupled receptors: application to dopamine, adrenaline, serotonin, acetylcholine, and mammalian opsin receptors, *J. Med. Chem.*, 35, 3448–3462, 1992.

120. van Koppen, C. J. and Nathanson, N. M., Site-directed mutagenesis of the m2 muscarinic acetylcholine receptor, *J. Biol. Chem.*, 265, 20887–20892, 1990.

121. van Koppen, C. J. and Nathanson, N. M., The cysteine residue in the carboxyl-terminal domain of the m2 muscarinic acetylcholine receptor is not required for receptor-mediated inhibition of adenylate cyclase, *J. Neurochem.*, 57, 1873–1877, 1991.

122. Savarese, T. M., Wang, C. D., and Fraser, C. M., Site-directed mutagenesis of the rat m1 muscarinic acetylcholine receptor, *J. Biol. Chem.*, 267, 11439–11448, 1992.

123. Lee, N. H. and Fraser, C. M., Cross-talk between m1 muscarinic acetylcholine and β_2-adrenergic receptors, *J. Biol. Chem.*, 268, 7949–7957, 1993.

124. Moro, O., Lameh, J., and Sadee, W., Serine- and threonine-rich domain regulates internalization of muscarinic cholinergic receptors, *J. Biol. Chem.*, 268, 6862–6865, 1993.

125. Wess, J., Nanavati, S., Vogel, Z., and Maggio, R., Functional role of proline and tryptophan residues highly conserved among G protein-coupled receptors studied by mutational analysis of the m3 muscarinic receptor, *EMBO J.*, 12, 331–338, 1993.

126. Kubo, T., Bujo, H., Akiba, I., Nakai, J., Mishina, M., and Numa, S., Location of a region of the muscarinic acetylcholine receptor involved in selective effector coupling, *FEBS Lett.*, 241, 119–125, 1988.

127. England, B. P., Ackerman, M. S., and Barrett, R. W., A chimeric D_2 dopamine/m1 muscarinic receptor with D_2 binding specificity mobilizes intracellular calcium in response to dopamine, *FEBS Lett.*, 279, 87–90, 1991.

128. Shapiro, R. A. and Nathanson, N. M., Deletion analysis of the mouse m1 muscarinic acetylcholine receptor: effects on phosphoinositide metabolism and down-regulation, *Biochemistry*, 28, 8946–8950, 1989.

129. Maeda, S., Lameh, J., Mallet, W. G., Philip, M., Ramachandran, J., and Sadee, W., Internalization of the Hm_1 muscarinic cholinergic receptor involves the third cytoplasmic loop, *FEBS Lett.*, 269, 386–388, 1990.

130. Lechleiter, J., Hellmiss, R., Duerson, K., Ennulat, D., David, N., Clapham, D., and Peralta, E., Distinct sequence elements control the specificity of G protein activation by muscarinic acetylcholine receptor subtypes, *EMBO J.*, 9, 4381–4390, 1990.

131. Wess, J., Bonner, T. I., and Brann, M. R., Chimeric m2/m3 muscarinic receptors: role of carboxyl terminal receptor domains in selectivity of ligand binding and coupling to phosphoinositide hydrolysis, *Mol. Pharmacol.*, 38, 872–877, 1990.

132. Felder, C. C., Poulter, M. O., and Wess, J., Muscarinic receptor-operated Ca^{2+} influx in transfected fibroblast cells is independent of inositol phosphates and release of intracellular Ca^{2+}, *Proc. Natl. Acad. Sci. U.S.A.*, 89, 509–513, 1992.

133. Lai, J., Nunan, L., Waite, S. L., Ma, S. W., Bloom, J. W., Roeske, W. R., and Yamamura, H. I., Chimeric M1/M2 muscarinic receptors: correlation of ligand selectivity and functional coupling with structural modifications, *J. Pharmacol. Exp. Ther.*, 262, 173–180, 1992.

134. Wong, S. K. F., Parker, E. M., and Ross, E. M., Chimeric muscarinic cholinergic: β-adrenergic receptors that activate G_s in response to muscarinic agonists, *J. Biol. Chem.*, 265, 6219–6224, 1990.

135. Zhu, S. Z., Wang, S. Z., Hu, J., and El-Fakahany, E. E., An arginine residue conserved in most G-protein-coupled receptors is essential for the function of the m1 muscarinic receptor, *Mol. Pharmacol.*, 45, 517–523, 1994.

136. Wess, J., Bonner, T. I., Dorje, F., and Brann, M. R., Delineation of muscarinic receptor domains conferring selectivity of coupling to guanine nucleotide-binding proteins and second messengers, *Mol. Pharmacol.*, 38, 517–523, 1990.

137. Wess, J., Gdula, D., and Brann, M. R., Structural basis of the subtype selectivity of muscarinic antagonists: a study with chimeric m2/m5 muscarinic receptors, *Mol. Pharmacol.*, 41, 369–374, 1992.

138. Maggio, R., Vogel, Z., and Wess, J., Reconstitution of functional muscarinic receptors by co-expression of amino- and carboxyl-terminal receptor fragments, *FEBS Lett.*, 319, 195–200, 1993.

139. Maggio, R., Vogel, Z., and Wess, J., Coexpression studies with mutant muscarinic/adrenergic receptors provide evidence for intermolecular "cross-talk" between G-protein-linked receptors, *Proc. Natl. Acad. Sci. U.S.A.*, 90, 3103–3107, 1993.

Chapter 4

Distribution of Muscarinic Acetylcholine Receptors in the CNS

Reinhard Schliebs and Steffen Roßner

CONTENTS

I. MUSCARINIC RECEPTOR DIVERSITY

The great majority of acetylcholine effects in the central nervous system are mediated via interaction with muscarinic cholinergic receptors (mAChR). Muscarinic receptors were initially found to be composed of a single population of binding sites when applying antagonist binding. Agonist binding then revealed three classes of sites representing low-, high-, and superhigh-affinity states which were assumed to reflect differences in the coupling of the receptor rather than differences in their structure.[1,2] The nonclassical antagonist pirenzepine provided the basis for the differentiation of muscarinic subtypes:[3] binding sites demonstrating high affinity for pirenzepine were termed M_1 receptors, while those having low affinity for pirenzepine were designated as M_2 sites (see, e.g., Reference 4). In contrast to pirenzepine, the classical muscarinic receptor antagonists such as atropine and *N*-methylscopolamine display similar affinities for the two subtypes.[5]

Muscarinic receptor subtypes were at first defined in the peripheral nervous system; and the distinction between M_1 and M_2 sites was soon extended to the central nervous system, where M_1 and M_2 sites are also present. Both M_1 and M_2 subtypes have been identified in the cerebral cortex by binding studies,[3] by autoradiographic studies,[2] and in functional biochemical assays.[6]

More recently, other studies have indicated that the M_2 receptor class probably does not represent a single entity and can be further subdivided on the basis of the differential affinity of the antagonists 4-diphenylacetoxy-*N*-methyl-piperidine (4-DAMP), and 11[[2-[(diethyl-amino)-methyl]-1-piperidinyl]-acetyl]-5,11-dihydro-6H-pyrido [2,3-b][1,4]-benzodiazepine-6-one (AF-DX 116) in the peripheral nervous system. It has been shown that AF-DX 116 possesses high affinity for M_2 sites in cardiac tissues but is much less active on M_2 receptors found in glandular tissues.[7] The M_2 receptors in the heart, therefore, appear to be different from those in smooth muscle tissues. The M_2 receptors in the heart are denoted as M_{2A} and those in the smooth muscle as M_{2B},[8] respectively, or $M_{2\alpha}$ (heart) and $M_{2\beta}$ (smooth muscle).[9] Later a different nomenclature was introduced designating the M_2 receptors in the heart as M_2 and in glandular tissue as M_3.[7,10] Recently, a further subtype, named M_4, with pharmacological properties different from the M_1 through M_3-mAChRs, has been described in PC12 and NG108-15 cell lines[11,12] and in the brain.[13-16] An even greater diversity of mAChR was suggested by molecular biological approaches which demonstrated the existence of a family of five different mAChR genes expressed in rat brain and which have been designated as m_1, m_2, m_3, m_4, and m_5.[17-25] These five subtypes are distinct proteins having no more than 50% homology in their amino acid sequence.[18,23,24] The mRNAs for the five subtypes are differentially distributed in the brain as shown by *in situ* hybridization and northern blots,[18,19,24,26,27] and have distinct pharmacological[28] and biochemical properties.[23,29]

The identity of each of the five receptors as muscarinic receptors has been established by expression of the clones in cell lines and their detailed characterization. The m_1, m_3, and m_5 receptors transfected into mammalian cells stimulate phosphatidylinositol (PI) metabolism, while m_2 and m_4 receptors inhibit cyclic adenosine 3',5'-monophosphate (cAMP) formation.[17]

From the comparison of the binding properties of individual cloned receptors it was suggested that the pharmacological subtypes M_1, M_2, and M_3 correspond to the cloned m_1, m_2, and m_3 receptors,[23,24,28,30,31] whereas the M_4 subtype most likely represents the cloned m_4 receptor.[15,23,32,33]

Muscarinic receptors are mainly localized postsynaptically on soma and dendrites of noncholinergic cells; however, there exist also some muscarinic receptors on cholinergic nerve terminals (presynaptic autoreceptors; see, e.g., References 34 and 35) as well as on noncholinergic nerve endings (presynaptic heteroreceptors), where they mediate regulation of transmitter release.[36-38]

Most studies localizing muscarinic receptors and their subtypes in the brain have employed receptor autoradiography. These investigations have been complemented by *in situ* hybridization and immunocytochemistry. The aim of this chapter is to review recent data on receptor autoradiography, *in situ* hybridization histochemistry, and immunocytochemistry of muscarinic receptor subtypes at the light and electron microscopical level by trying to combine the information obtained by different experimental techniques. Therefore, in a preceding chapter, the different methodological approaches should be outlined and the advantages and limitations of each technique discussed, thus allowing to assess the different data presented. In order to differentiate between pharmacological and cloned receptors, the pharmacological subtypes are designated with capital M's and the cloned receptors with m's as recommended previously.[39,40]

II. METHODICAL CONSIDERATIONS: RECEPTOR AUTORADIOGRAPHY, *IN SITU* HYBRIDIZATION, IMMUNOCYTOCHEMISTRY

Receptor binding studies are critically limited by the specificity of the radioligands used. With the available radioligands it is not possible to label exclusively one of the mAChR subtypes. Pirenzepine, a ligand specific to M_1 mAChRs, appears to label with high-affinity mAChRs corresponding to the m_1 subtype; but it binds also with moderate affinities to m_3, m_4, and m_5-mAChRs. M_2 radioligands seem to recognize with high affinity both m_2 and m_4 receptors.[17,27,41] These different affinity patterns of radioligands to more than one receptor subtype may result in a certain under- or overestimation of a particular receptor subtype in ligand binding studies (see, e.g., References 41 and 42). Moreover, due to the lack of sufficiently selective radioligands to label M_3 and M_4 receptors indirect estimates have been applied to obtain corresponding data for these mAChR subtypes.[15,43-45] However, regardless of these limitations, receptor autoradiography allows for the anatomic resolution of binding studies, if certain other limitations of the method (e.g., the differential quenching of tritium by gray and white matter when using tritiated radioligands) are taken into account.[46] Due to the poor cellular resolution of the commonly employed autoradiographic procedures, the results do not allow us to determine whether receptors are localized on pre- or postsynaptic elements or to identify the cells which are endowed with a certain receptor subtype.[47] However, this technique has been succesful in characterizing cellular and subcellular receptor localization when using isolated cells.[48]

The availability of molecular data on the mAChR subtypes provided the basis for the synthesis of oligonucleotides which can be used as probes to visualize the cells containing mRNA transcripts of a specific receptor gene by *in situ* hybridization. To detect the hybridized probe both isotopic and nonisotopic methods have been applied. The combination of isotopic and nonisotopic probes for *in situ* hybridization has been shown to be a valuable approach to study the colocalization of various neurotransmitter receptors of the same cell.[49]

In situ hybridization and computer-assisted image analysis of autoradiographs have been used as a technique for measuring mRNA in brain sections, and it has been shown that this method provides a useful tool to quantify *in situ* hybridization experiments.[50-53]

However, it should be pointed out that *in situ* hybridization can give only an estimate of the abundance of the receptor gene transcripts but not of the number of actually expressed receptor molecules themselves.

Due to possible variations in translation rates for different mRNAs as well as differences in turnover rates of the corresponding receptor proteins, it is not possible to correlate the expression pattern of a particular transcript and the density and distribution of the corresponding protein molecule on the basis of these data.[54] The value of *in situ* hybridization, however, lies in the ability to localize cell bodies that express muscarinic receptor subtypes, thus providing complementary information to the receptor autoradiographic measurements.

Application of immunocytochemistry using specific antibodies targeted to various muscarinic receptor subtypes allows a high degree of regional and cellular resolution as compared to receptor autoradiography and can be extended to the ultrastructural level. However, immunocytochemical techniques are generally qualitative rather than quantitative, although immunoprecipitation of prelabeled receptor proteins of a particular tissue region and assaying the amount of radioactivity in the precipitate allow the estimation of the amount of mAChR subtypes.[55,56] Under certain conditions the immunological characterization of muscarinic receptor proteins might, therefore, complement and extend *in situ* hybridization experiments.

III. REGIONAL DISTRIBUTION OF MUSCARINIC ACETYLCHOLINE RECEPTORS

A. RECEPTOR AUTORADIOGRAPHY

Radioligand binding has revealed that there is no brain region that contains exclusively one single muscarinic receptor type.[57,58] Binding and autoradiographic studies have clearly shown a differential distribution of M_1 and M_2 receptor sites in the rat brain.[59-65] It has been suggested that M_1 binding sites are localized mostly on cholinoceptive neurons, whereas M_2 binding sites occur mainly on cholinergic neurons.[2,62-64,66] Indeed, M_1 receptors labeled using various radioligands such as [^{3}H]3-quinuclidinyl benzilate (QNB) (in the presence of carbamylcholine) and [^{3}H]pirenzepine are mainly concentrated in the striatum, hippocampus, and cerebral cortex but not in the brainstem and medulla. On the other hand, M_2 sites visualized using either [^{3}H]QNB (in the presence of pirenzepine to block M_1 sites[61]) or [^{3}H]N-methylscopolamine,[59] [^{3}H]oxotremorine-M,[60,63,67] or the selective antagonists [^{3}H]AF-DX 116[44,68-70] and [^{3}H] (5,11-dihydro-11-{[(2-{2-[(dipropyl- amino)methyl]-1-piperidinyl}ethyl)amino]carbonyl}-6H-pyrido (2,3-b)benzodiazepine-6-one (AF-DX 384),[43,71]) are mostly seen in brain areas enriched with cholinergic cell bodies and terminals such as the pons, medulla, thalamus, medial septum, and nucleus of the diagonal band of Broca.

1. M_1 Receptor Subtype

To label M_1 muscarinic receptor sites radioligands such as [^{3}H]QNB in the presence of carbamylcholine and [^{3}H]pirenzepine were used.[59-61,63,64]

The densities of [^{3}H]pirenzepine binding (preferentially labeling M_1 receptors) were found to be highest in the dentate gyrus, in the CA1 layer of the hippocampus, and in the anterior olfactory nucleus. High binding was also detected in the upper and deeper layers of the cerebral cortex, the caudate putamen, the olfactory tubercle, and the nucleus accumbens, whereas a moderate binding was observed in the external plexiform layer of the olfactory bulb, the lateral amygdala, and the CA3/CA4 subfields of the hippocampal formation. Low densities of M_1 receptors were detected in the lateral and medial septum, vertical limb of the diagonal band, thalamic and hypothalamic nuclei, and superficial gray layers of the superior colliculus and the inferior colliculus.[2,44,59,62-65,72-75] Nearly no labeling for M_1 receptor sites was seen in the cerebellum.[64] In the rabbit hippocampus, M_1 muscarinic binding sites were more abundant in CA1 and CA2 fields than in CA3 and CA4, whereas M_2 receptors appeared more homogeneously distributed.[76]

Although only a low labeling of the cerebellum with the non-subtype-selective [^{3}H]QNB was observed, autoradiographic studies revealed a laminar distribution of the total population of mAChR-binding sites in the cerebellar cortex, cerebellar cortical density in muscarinic receptor sites was highest in the granule and purkinje cell layers, low in the molecular layer, and absent in the white matter.[77-80]

In the basal forebrain region where the neurons of the major central cholinergic projection system are located, a low but significant amount of M_1 receptor was demonstrated with higher M_1 receptor sites in the ventral pallidum and magnocellular preoptic nucleus and lower values in the substantia innominata and basal nucleus of Meynert. The lowest values of M_1 receptors were found in the medial septal nucleus and vertical and horizontal limbs of the diagonal band.[73,75] This regional distribution is completely different from the M_2 pattern in this region (see below).

In the cerebral cortex, M_1 muscarinic receptors are present in all cortical layers but somewhat more concentrated in layers II/III and VI.[59,61,63,64,81] This laminar distribution of M_1 receptor sites is similar to that found in the rat visual cortex using [^{3}H]pirenzepine binding.[81-84]

In Figure 4-1 representative examples of M_1 receptor autoradiographs using [^{3}H]pirenzepine binding in serial rat brain sections are shown.

2. M_2 Receptor Subtype

The M_2 receptor subtype has been localized in brain areas enriched in cholinergic cell bodies such as the interpeduncular nucleus, facial and trigeminal nerve nuclei, and basal forebrain.[59,61,63] M_2 receptor sites have been visualized using either [^{3}H]QNB (in the presence of pirenzepine to block M_1 sites[61]), [^{3}H]acetylcholine (in the presence of nicotine[62]), and [^{3}H]N-methylscopolamine,[59] [^{3}H]oxotremorine-M;[60,63,67] or the selective antagonists [^{3}H]AF-DX 116[44,68-70] and [^{3}H]AF-DX384.[43,71]

Autoradiographic studies using [^{3}H]AF-DX 116 as a specific radioligand to label M_2 mAChR in the rat brain revealed a heterogeneous regional distribution:[68-70] areas with high densities include some layers of parietal cortex, certain thalamic nuclei, superior colliculus, interpeduncular nucleus, dorsomedial nucleus of the hypothalamus, and septal nuclei. Intermediate densities have been found in the hippocampus, dentate gyrus, caudate-putamen, frontal cortex, and nucleus accumbens. Areas of low densities include the substantia nigra, globus pallidus, and cerebellum. High densities of M_2 receptor sites as visualized by [^{3}H]AF-DX 116 autoradiography were also found in the external plexiform layer of the olfactory bulb, the anteroventral nucleus of the thalamus, and the motor trigeminal nucleus. The pyramidal layer of the hippocampus demonstrates high M_2 binding sites followed by the oriens layers of the CA1 and CA3 fields. In the dentate gyrus the granular layer is more enriched than the molecular layer.[44,68,69,71]

Using the novel radioligand [^{3}H]AF-DX 384 as selective for M_2 sites, it was revealed that most binding sites were found in the external plexiform layer of the olfactory bulb, the lateroposterior nucleus of the thalamus, the superficial gray layer of the superior colliculus, the pontine nucleus, and the hypoglossal nucleus; high densities are located in the olfactory tubercule, the nucleus accumbens, the caudate putamen, the anteroventral nucleus of the thalamus, and the motor trigeminalis nucleus whereas low densities of M_2 binding sites were observed in the ventral posteriormedial nucleus of the thalamus and in the medial part of the geniculate nucleus; and very low labeling was seen in the substantia nigra and in the cerebellum.[71] In Figure 4-3 representative examples of M_2 receptor autoradiographs using [^{3}H]AF-DX 384 binding in serial rat brain sections are shown.

However, it has also to be taken into account that [^{3}H]AF-DX 384 labels to a certain extent also M_4 receptor sites.[43,85] This is especially important for those brain regions enriched with M_4 subtype (caudate-putamen and olfactory tubercule, see also below).

Currently, a novel most potent M_2-selective muscarinic antagonist, (5-({4-4-[(diethylamino)butyl]-1-piperidinyl}acetyl)-10,11-dihydrobenzo(b,e)(1,4)diazepine-11-one) (DIBA), was reported, which appears to be one order more potent than AF-DX 384 and should become a useful probe for M_2 receptor detection.[86]

Within the basal forebrain region, the highest M_2 receptor sites were observed in the medial septal nucleus and vertical and horizontal limbs of the diagonal band, and the lowest M_2 receptor site was seen in the substantia innominata and nucleus basalis using [^{3}H]oxotremorine-M as a specific radioligand.[75] As most of the cholinergic afferents originating in the brainstem terminate in those areas of the basal forebrain region in which the highest M_2 receptor density was found, a postsynaptic localization of M_2 receptors and/or presynaptic localization on cholinergic terminals was suggested.[75]

Concerning the cortical distribution M_2 receptors are more concentrated in layers IV and V of the parietal cortex than in other cortical layers.[70] On the other hand, Regenold et al.[69] and Quirion et al.[62] found the highest densities of M_2 receptors in cortical layers III, V, and VI. Using a new M_2-selective muscarinic antagonist, Gitler et al.[86] found a similar pattern: intermediate layers > deep layers > superficial layers.

In the primate brain, the cortical laminar distribution of M_2 receptors is controversial: Miyoshi et al.[87] found a diffuse distribution of AF-DX 116 binding throughout all layers of the parietal, insular, and temporal cortex, whereas Wagster et al.[88] revealed highest binding in layers IV and V and lowest in layer I using [^{3}H]oxotremorine-M as a radioligand.

In the rat visual cortex, M_2 receptor sites were detected in upper layers IV and VI, a laminar distribution which seems to be complementary to the laminar distribution of M_1 receptor sites[82,84] (see also Figures 4-1 and 4-3).

3. M_3 Receptor Subtype

There are no ligands sufficiently selective for labeling the M_3 receptor subtype. Although 4-DAMP and hexahydro-sila-difenidol (HHSD) were found to have a high affinity for the M_3 subtype,[89,90] they do not sufficiently discriminate between M_1 and M_3 receptors.[33] However, in certain regions of the rat brain which are enriched with that subtype, [^{3}H]4-DAMP appears to be a useful ligand to label muscarinic M_3 subtype receptors.[91] Similarly, [^{3}H]N-methyl-scopolamine binding in the presence of unlabeled pirenzepine to block M_1 sites was suggested to characterize the distribution of the M_3 subtype in the rat forebrain.[92]

However, recently the different affinity patterns of muscarinic antagonists such as pirenzepine (high affinity for M_1, low affinity for M_2 and M_3), AF-DX (high affinity for M_2, low affinity for M_1 and M_3), and 4-DAMP (high affinity for M_1 and M_3, low affinity for M_2) have been exploited for a more distinct pharmacological differentiation of the muscarinic receptor subtypes M_1 through M_3. Quantitative auto-radiography of [^{3}H]4-DAMP in the presence of unlabeled pirenzepine and AF-DX 116 to block M_1 and M_2 sites was found to permit a direct estimation of M_3 receptors and has been used for mapping the M_3 receptor in the rat brain.[44,45] The M_3 receptors are widely distributed throughout the brain at a relatively low prevalence (less than 20 to 30% of the total density of muscarinic receptors). The highest M_3 receptor binding was found in the olfactory tubercle, striatum, nucleus accumbens, and CA1 subfield of the hippocampus; moderate binding remained in neocortical areas, the CA3 subfield and dentate gyrus of the hippocampus, thalamus, posterior hypothalamus, amygdala, and pons.[44,45] Low amounts of M_3 receptors were found in the medial septum, vertical limb of the diagonal band, ventral thalamic nucleus, medial geniculate nucleus, hypothalamic nuclei, and inferior colliculus.[44,45] In the cerebral cortex, M_3 receptor densities are highest in layers I-II and lowest in layer VI.[44] When compared with the other muscarinic subtypes, M_3 receptors are enriched in several diencephalic and brainstem regions including thalamic and hypothalamic nuclei, the substantia nigra, the superior colliculus, the periaqueductal gray, and the pons.[45] This distinct distribution of M_3 receptors is in agreement with observations on the regional distribution of m_3 receptor mRNA expressed in the cerebral cortex, hippocampus, thalamic nuclei, superior colliculus, periaqueductal gray, and pons.[27] However, in the striatum the m_3 mRNA level has been shown to be very low.[26] Immunocytochemistry revealed a similar distribution (see below).

4. M_4 Receptor Subtype

Radioligand binding studies in the rat striatum revealed that about 40% of the muscarinic receptors showed pharmacological properties which are different from those of the known M_1 to M_3 receptors but very similar to the cloned m_4 receptor.[16] Therefore, these binding sites have been named M_4 receptor sites.[16] This was confirmed by other studies demonstrating that in the rat caudate putamen and olfactory tubercle, a major population of muscarinic cholinergic receptors appears to belong to the M_4 type.[15,43] Furthermore, kinetic experiments demonstrated that the M_4 receptors predominate in homogenates from rat striatum but they are also present to a lower proportion in the cortex and hippocampus.[16]

Recently it was found that several muscarinic M_2-selective ligands including [^{3}H]oxotremorine-M, [^{3}H]AF-DX 116, and [^{3}H]AF-DX 384 label both M_2 and M_4 receptors sites,[43,85] a fact which should be considered when estimating M_2 receptor autoradiography.[85]

5. M_5 Receptor Subtype

Although mRNA for m_5 receptors has been located in discrete regions of the rat brain,[25,93] there are as yet no reports of pharmacological studies of this receptor in brain tissue due to a lack of sufficiently selective radioligands.

B. RECEPTOR IMMUNOCYTOCHEMISTRY

The availability of monoclonal antibodies specific for mAChRs and their subtypes has made it possible to visualize individual receptor-bearing structures in the central nervous system at the light and electron microscopic level.

Several different monoclonal and polyclonal antibodies have been raised against mAChRs[94-98] (for review, see Reference 99) and their subtypes.[56,100-104]

To visualize mAChRs in the rat brain the monoclonal antibody M35 has been used, which does not differentiate between M_1 and M_2 subtypes.[105-108] In the neocortex, M35 immunoreactivity was mainly seen in layer V pyramidal neurons and some layer III pyramidal and layer II stellate neurons including their apical and basal dendrites.[106] Ultrastructurally, immunoreactive sites were associated with the intradendritic microtubular system and dendritic postsynaptic membranes.[105] In the hippocampal formation, CA2 to

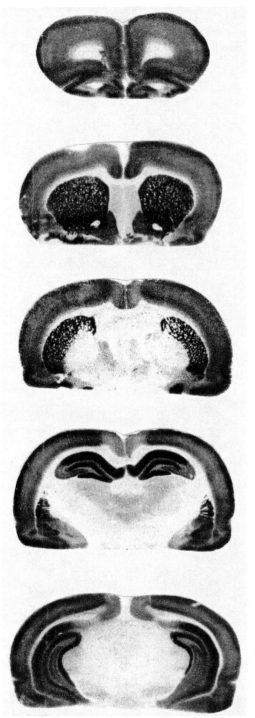

Figure 4-1 Autoradiographic distribution of M_1 muscarinic acetylcholine receptor subtype in the rat brain using [³H]pirenzepine as a radioligand. Autoradiography of [³H]pirenzepine binding at a ligand concentration of 10 nM on thaw-mounted cryo-cut sections of the rat brain was performed as described elsewhere.[82,83]

CA4 pyramidal neurons and granular cells in the dentate gyrus showed immunoreactivity for the mAChR.[106] It is interesting to note that approximately 90% of cholinoceptive neurons is endowed with both muscarinic and nicotinic AChRs using combined immunocytochemistry.[108] This high degree of colocalization suggests that the interaction between muscarinic and nicotinic cholinergic input occurs at individual neurons.[109] In the adult cat visual cortex a predominant immunolabeling of neurons and dendrites in layers I to III, V, and VI has been observed which was developmentally regulated.[110]

m1

m3

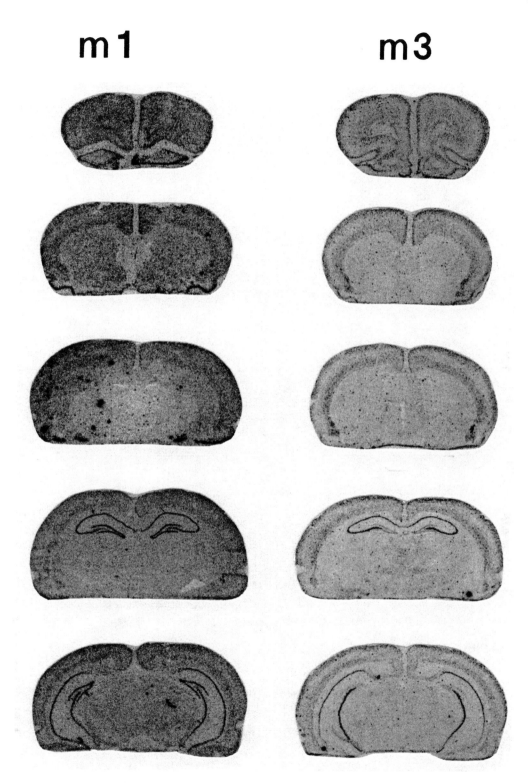

Figure 4-2 Regional distribution of gene transcripts of the m_1 and m_3 muscarinic acetylcholine receptor in the rat brain using ^{35}S-labeled oligonucleotides as specific probes. *In situ* hybridization experiments were performed as described previously.[112]

M35 immunoreactivity is also expressed in the thalamus, demonstrating a selective distribution of muscarinic receptors in thalamic neuronal subsets.[111]

The monoclonal antibody M35 does not differentiate between muscarinic receptor subtypes. However, by cloning of the distinct muscarinic receptors and identification of regions of divergence, it has been possible to target antibodies to all of the individual receptor subtypes[55,102-104] (for review, see Reference 102). The antibodies to m_1-m_5 have allowed research to identify and quantify the native receptor subtypes in the brain.[103] Using immunoprecipitation experiments, the m_1, m_2, and m_4 receptor proteins represented the vast majority of the total solubilized muscarinic receptors in rat brain.[103] The immunocytochemical studies in rat brain slices demonstrated that m_1, m_2, and m_4 are the most abundant receptor subtypes with marked regional differences. The m_1 protein is abundant in the cerebral cortex, hippocampus, and striatum; and is localized in cell bodies and neurites, suggesting its role as a postsynaptic receptor. The m_2 receptor protein is the predominant subtype in the basal forebrain, mesopontine tegmentum, and cranial motor nuclei pointing to its role as an autoreceptor,[55,103,104] whereas the m_4 subtype was detected with high levels, particularly in the neostriatum and olfactory tubercle.[103] When expressed as percentage of total muscarinic receptors, the densities of m_2 receptors varied considerably between brain regions with high enrichment in the hindbrain, brain stem, and cerebellum (about 70% of total amount of muscarinic receptors); and lower expression in the cortex, hippocampus, and striatum (about 15%).[55] The absolute density of the m_2 receptors, however, appears relatively uniformly distributed throughout the brain.[55]

The m_3 receptor accounts for only a small proportion of total receptor sites (less than 12% of total muscarinic receptors expressed) throughout the brain with somewhat higher levels in the cortex and hippocampus followed by olfactory tubercle and striatum.[56] Fewer m_3 receptors were found in the thalamus, pons/medulla, and cerebellum. The m_5 receptor has yet to be detected by the immunocytochemical approach.[102]

In the cerebral cortex, the m_1 immunoreactivity was particularly dense in layers II, III and VI,[103] which is consistent to the laminar distribution of the m_1 mRNA.[27,112,113] The m_2 receptor protein was found to be dense in layer IV and the borders of layers V and VI[103] which is in good correspondence to the laminar distribution of m_2 mRNA.[113,114] The m_4 immunolabel, considerably less than the other subtypes, was localized in layers II, III, and IV, being consistent with the laminar pattern of m_4 mRNA distribution.[112,113]

Cellularly, the cortical m_2 receptor protein was mostly associated with fibers and terminals and only occasionally with perikarya, whereas the m_1 and m_4 immunoreactivity was mainly localized in the neuropil and in scattered perikarya.[103]

When comparing receptor protein, mRNA and ligand binding sites, the regional distributions of m_1-m_4 receptor proteins are in good agreement with the corresponding mRNA localization and recent estimates of the proportions of M_1-M_4 binding sites.[102] However, there are also some differences in the ratios of immunocytochemically detected m_2/m_4 and the pharmacologically classified M_2/M_4 receptors suggesting that different populations of receptors might be assayed by radioligand binding due to a lack of selective ligands as already indicated above.

Although all four muscarinic receptor proteins are present in the cerebral cortex, hippocampus, striatum, thalamus, amygdala, and hypothalamus, the subtypes appear to have a highly complementary distribution within each region, in particular in the laminar pattern of the cerebral cortex.[102]

C. DISTRIBUTION OF MUSCARINIC RECEPTOR GENES

In situ hybridization experiments have demonstrated that brain tissues express a mixture of muscarinic receptor genes.[26,27] In Figures 4-2 and 4-4 representative examples of autoradiograms of serial rat brain sections hybridized with ^{35}S-labeled oligonucleotide probes specific for m_1 through m_4 mAChR mRNA are shown clearly demonstrating striking regional changes in the appearance of each label.

In situ hybridization indicated that m_1, m_3, and m_4 mRNAs are abundant in the cerebral cortex, with m_1 uniformly distributed but m_3 localized primary in the inner and outer layers and m_4 being more concentrated in layer IV (see Figures 4-2 and 4-4). Despite of the low level of m_2 mRNA in the cerebral cortex, a bimodal distribution can be observed with m_2 mRNA more concentrated in layer IV and to a lower extent in upper layer VI (Figure 4-4). A similar cortical distribution of mAChR gene transcripts has been observed in the rat visual cortex. The m_1 receptor mRNA is almost homogeneously distributed throughout the visual cortex, whereas the m_2 mAChR mRNA predominates in layer IV with lower levels in upper layer VI. The m_3 mRNA is localized primarily in layers II and III and to a lower extent in layer VI, while the m_4 mRNA is slightly more labeled in layer IV as compared to the other layers[84,112,114] (see also Figure 4-4).

The m_1 and m_3 mRNAs are highly abundant in the hippocampus but m_1 is much more concentrated than m_3 in the dentate gyrus. m_4 mRNA is only moderately present in the hippocampus, is low in abundance in the CA2 field, and is very low in the dentate gyrus (Figures 4-2 and 4-4). m_1 receptors are highly expressed in pyramidal and granule cells of the hippocampus, in contrast to m_3 and m_4 subtypes which are predominantly expressed in pyramidal cells and only at a very low level in granule cells.[26,27]

In addition to the hippocampus, m_1 mRNA is prevalent in the olfactory nuclei and plexiform layers of the olfactory bulb, the olfactory tubercle, the amygdala, and the piriform cortex, and to a lesser amount in the caudate putamen.[26,27] m_2 mRNA was observed in the pontine nuclei, plexiform layers of the olfactory bulb, medial septal nuclei, and diagonal band[27] (Figure 4-4).

m_3 mRNA was heavily labeled in the piriform cortex and olfactory tubercle and was also present in the plexiform layers of the olfactory bulb and in a number of thalamic nuclei including habenula, anteroventral, ventrolateral, and midline thalamic nuclei; only very low levels of m_3 mRNA were found in the striatum[26,27] (see Figure 2). m_4 mRNA predominated in the rat caudate putamen, olfactory bulb, and piriform cortex, whereas very low levels were detected in the thalamus and cerebellum[26,27,112] (see also Figure 4-4).

In the cerebellum, a significant amount of m_1 and m_3 mRNA is present mainly localized in the granular layer, whereas m_4 mRNA could also be observed but to a very low extent.[112] There was no labeling of m_2 mRNA in the cerebellum.[27]

The m_5 receptor mRNA is present in the rat brain at low abundance and is particularly enriched in the hippocampus where it is restricted to the ventral subiculum, and the pyramidal cells of the CA1, and, in lower amounts to the CA2 subfields.[25] Hybridization signals were also seen in the substantia nigra pars compacta, ventral tegmental area, lateral habenula, ventromedial hypothalamic nucleus, and mammillary bodies, suggesting an involvement of the m_5 receptor in the regulation of the dopaminergic nigrostriatal pathway.[25]

The mRNA for the m_2 receptor subtype is very poorly expressed in the basal ganglia including the caudate putamen, nucleus accumbens, and olfactory tubercle as opposed to the detected labeling in these areas by M_2 ligands (see Figures 4-3 and 4-4). In contrast, the m_4 receptor mRNA is heavily expressed in these areas[15,26,27,93] as can also be seen in Figure 4-4, suggesting that in these regions [^{3}H]AF-DX 384 recognizes preferentially M_4 receptors.

However, in most regions there is a very good correlation between the distribution of M_2 receptor binding sites and and m_2 mRNA[67] (compare also Figures 4-3 and 4-4). Cholinergic cell groups such as the medial septum-diagonal band complex, nucleus basalis, pedunculopontine and laterodorsal tegmental nuclei, nucleus parabigeminalis, several motor nuclei of the brain stem, and motor neurons of the spinal cord contain m_2 mRNA which is colocalized with ChAT mRNA, suggesting that at least a fraction of these receptors act as presynaptic autoreceptors.[67] However, in other cholinergic cell groups such as the medial habenula and some cranial nerve nuclei no hybridization with the m_2 probe was detected.[67]

In general, the regional distribution of muscarinic receptors as obtained by pharmacological analysis is in good agreement with the distribution of transcripts for the different receptor subtype genes.[54] However, there are also exceptions of this correlation (see, e.g., Reference 54).

IV. CELLULAR AND SUBCELLULAR LOCALIZATION OF MUSCARINIC RECEPTORS

In contrast to receptor autoradiography, immunocytochemistry at the ultrastructural level allows for a cell-type and compartment-specific visualization of receptor proteins. The data on the subcellular localization of receptor proteins appear to additionally give estimates on receptor processing: labeling of perikaryal compartments indicates receptor synthesis and processing, whereas labeling of dendrites and postsynaptic densities points to transport and membrane incorporation, respectively.[109]

Immunoreactivity for muscarinic receptor subtypes m_1, m_2, and m_4 was found to be concentrated in the surface membranes of cells or neuritic processes. Cytoplasmic staining was also occasionally evident and adjacent to the Golgi apparatus.[103] It is interesting to note that nuclei or glia cells could not be stained with any of the antibodies used,[103] whereas the presence of muscarinic receptors on glia cells in culture was demonstrated.[115] Immunocytochemistry might also be helpful to provide answers regarding the synaptic localization of muscarinic receptor subtypes. While it is generally accepted that the location of m_1 receptors is predominantly postsynaptic, there are conflicting reports concerning the location of m_2 receptors. Some authors proposed that M_2 receptors are mostly presynaptically localized,[61,116-118] whereas

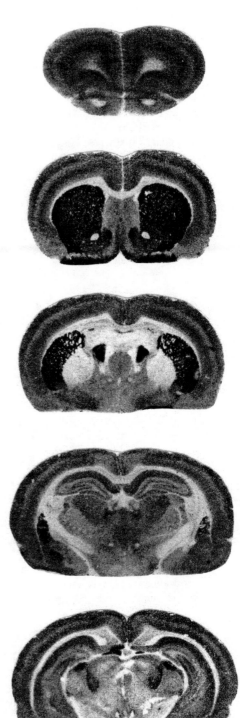

Figure 4-3 Autoradiographic distribution of M_2 muscarinic acetylcholine receptor subtype in the rat brain using [^{3}H]AF-DX 384 as a radioligand. Autoradiography of [^{3}H]AF-DX 384 binding at a ligand concentration of 30 nM on thaw-mounted cryo-cut sections of the rat brain was performed as described elsewhere.[84,124]

other groups suggest a mainly postsynaptic localization of the M_2 subtype.[119-121] Immunostaining of the olfactory bulb for the m_2 receptor revealed that at least in this region the location of the m_2 subtype is predominantly postsynaptic.[101]

V. SUMMARY AND CONCLUSIONS

There are at least five distinct muscarinic receptor subtypes in the brain which can be detected by receptor autoradiography, *in situ* hybridization, and immunocytochemistry. Receptor autoradiography allows for

m2   m4

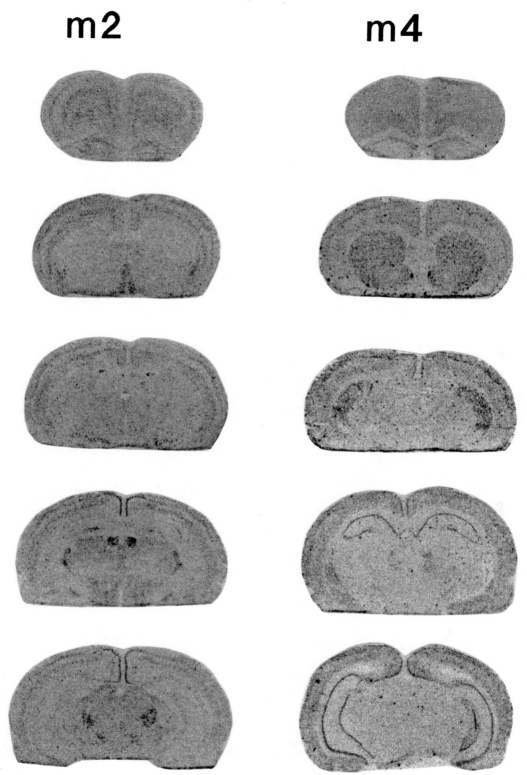

Figure 4-4 Regional distribution of gene transcripts of the m_2 and m_4 muscarinic acetylcholine receptor in the rat brain using ^{35}S-labeled oligonucleotides as specific probes. *In situ* hybridization experiments were performed as described previously.[112,114]

the anatomic resolution of binding studies but is limited by the specificity of the ligands available. *In situ* hybridization can provide data on the distribution of cells which express a distinct receptor subtype but cannot give any information on the abundance or subcellular distribution of the receptor itself. Immuno-cytochemistry allows for a high degree of cellular resolution including the ultrastructural level, but is limited by a lack of quantification. However, when correlating all three experimental approaches, a tremendous amount of information on the nature of muscarinic receptors in particular areas of the brain has been accumulated since the first definition of muscarinic receptors in 1914 by Dale[122] (see also Reference 123).

M_1 receptor sites appear to be mostly present in cholinoceptive target regions including the cerebral cortex, some hippocampal areas, and the striatum, whereas M_2 sites are predominant in brain areas enriched with cholinergic cell bodies such as nuclei of the basal forebrain, thalamus, and pons. M_3 receptors which are distributed throughout the brain but represent only 20 to 30% of the total amount of mAChRs are concentrated in the olfactory tubercle, striatum, nucleus accumbens, and CA1 subfield of hippocampus and to a lesser extent in the neocortical areas. The M_4 receptor type predominates in the striatum and olfactory tubercle. There are as yet no reports on a pharmacological characterization of the M_5 subtype in brain tissue. However, the m_5 receptor mRNA is present in the rat brain at low abundance and particularly enriched in the hippocampus.

In addition to the cholinergic innervation pattern, the exact knowledge of the regional and cellular distribution of muscarinic receptors might provide further information to detect functional neuronal circuitries with respect to a selective and specific transmission of cholinergic signals.

REFERENCES

1. Birdsall, N. J. M., Burgen, A. S. V., and Hulme E. C., The binding of agonists to brain muscarinic receptors, *Mol. Pharmacol.*, 14, 723, 1978.
2. Potter, L. T., Flynn, D. D., Hanchett, H. E., Kalinoski, D. L., Luber-Narod, J., and Mash, D. C., Independent M1 and M2 receptors: ligands, autoradiography and functions, *Trends Pharmacol. Sci.*, Suppl., 22, 1984.
3. Hammer, R., Berrie, C. P., Birdsall, N. J. M., Burgen A. S. V., and Hulme E. C., Pirenzepine distinguishes between different subclasses of muscarinic receptors, *Nature (London)*, 283, 90, 1980.
4. Hirschowitz, B. I., Hammer, R., Giachetti, A., Keirns, J. J., and Levine R. R., *Subtypes of Muscarinic Receptors*, Elsevier, Amsterdam, 1984.
5. Watson, M., Vickroy, T. W., Roeske, W. R., and Yamamura, H. I., Subclassification of muscarinic receptors based upon the selective antagonist pirenzepine, *Trends Pharmacol. Sci.*, Febr. Suppl., 9, 1984.
6. Marchi, M. and Raiteri, M., On the presence in the cerebral cortex of muscarinic receptor subtypes which differ in neuronal localization, function and pharmacological properties, *J. Pharmacol. Exp. Ther.* 235, 230, 1985.
7. Hammer, R., Giraldo E., Schiavi, G. B., Monferini, E., and Ladinsky, H., Binding profile of a novel cardioselective receptor antagonist AFDX-116 to membranes of peripheral tissues and brain in the rat, *Life Sci.*, 38, 1653, 1986.
8. Eglen, R. M. and Whiting, R. L., Muscarinic receptor subtypes: a critique of the current classification and a proposal for a working nomenclature, *J. Autonom. Pharmacol.*, 5, 323, 1986.
9. Lambrecht, G., Mutschler, E., Moser, U., Riotte, J., Wagner, M., and Wess J., Heterogeneity in muscarinic receptors: evidence from pharmacological and electrophysiological studies with selective agonists, in *International Symposium on Muscarinic Cholinergic Mechanisms*, Cohen, S. and Sokolovsky, M., Eds., Freund Publishing House, London, 1987, 245.
10. De Jonge, A., Dodds, H. N., Riezebos, J., and Van Zwieten, P. A., Heterogeneity of muscarinic binding sites in rat brain, submandibular gland and atrium, *Br. J. Pharmacol.*, 89, 551, 1986.
11. Michel, A. D., Stefanich, E., and Whiting, R. L., PC12 phaeochromocytoma cells contain an atypical muscarinic receptor binding site, *Br. J. Pharmacol.*, 97, 914, 1989.
12. Michel, A. D., Delmendo, R., Stefanich, E., and Whiting, R. L., Binding characteristics of the muscarinic receptor subtype of the NG108-15 cell line, *Naunyn-Schmiedeberg's Arch. Pharmacol.*, 340, 62, 1989.
13. Ladinsky, H., Schiavai, G. B., Monferini, E., and Giraldo, E., Pharmacological muscarinic receptor subtypes, *Prog. Brain Res.*, 84, 193, 1990.

14. McKinney, M., Anderson, D., Forray, C., and El-Fakahany, E. E., Characterization of the striatal M2 muscarinic receptor mediating inhibition of cyclic AMP using selective antagonists: a comparison with the brainstem M2 receptor, *J. Pharmacol. Exp. Ther.*, 250, 565, 1989.

15. Vilaró, M. T., Wiederhold, K.-H., Palacios, J. M., and Mengod, G., Muscarinic cholinergic receptors in the rat caudate-putamen and olfactory tubercle belong predominantly to the m_4 class: *in situ* hybridization and receptor autoradiography evidence, *Neuroscience*, 40, 159, 1991.

16. Waelbroek, M., Tastenoy, M., Camus, J., and Christophe, J., Binding of selective antagonists to four muscarinic receptors (M_1 to M_4) in rat forebrain, *Mol. Pharmacol.*, 38, 267, 1990.

17. Bonner, T. I., The molecular basis of receptor diversity, *Trends Neurosci.*, 12, 148, 1989.

18. Bonner, T. I., Buckley, N. J., Young, A. C., and Brann, M. R., Identification of a family of muscarinic acetylcholine receptor genes, *Science*, 237, 527, 1987.

19. Bonner, T. I., Young, A. C., Brann, M. R., and Buckley, N. J., Cloning and expression of the human and rat m_5 muscarinic acetylcholine receptor genes, *Neuron*, 1, 403, 1988.

20. Kerlavage, A. R., Fraser, C. M., and Venter, J. C., Muscarinic cholinergic receptor structure: molecular biological support for subtypes, *Trends Pharmacol. Sci.*, 8, 426, 1987.

21. Kubo, T., Fukuda, K., Mikami, A., Maeda, A., Takahashi, H., Mishina, M., Haga, T., Haga, K., Ichiyama, A., Kangawa, K., Kojima, M., Matsuo, H., Hirose, T., and Numa, S., Cloning, sequencing and expression of complementary DNA encoding the muscarinic acetylcholine receptor, *Nature (London)*, 323, 411, 1986.

22. Maeda, A., Kubo, T., Mishina, M., and Numa, S., Tissue distribution of mRNAs encoding muscarinic acetylcholine receptor subtypes, *FEBS Lett.*, 239, 339, 1988.

23. Peralta, E. G., Winslow, J. W., Peterson, G.-L., Smith, D. H., Ashkenazi, A., Ramachandran, J., Schimerlik, M. I., and Capon, D. J., Primary structure and biochemical properties of an M_2 muscarinic receptor, *Science*, 236, 600, 1987.

24. Peralta, F. G., Ashkenazi, A., Winslow, J.W., Smith, D. H., Ramachandran, J., and Capon, D. J., Distinct primary structures, ligand binding properties and tissue-specific expression of four human muscarinic acetylcholine receptors, *EMBO J.*, 6, 3923, 1987.

25. Vilaró, M. T., Palacios, J. M., and Mengod, G., Localization of m_5 muscarinic receptor mRNA in rat brain examined by *in situ* hybridization histochemistry, *Neurosci. Lett.*, 114, 154, 1990.

26. Brann, M. R., Buckley, N. J., and Bonner, T. I., The striatum and cerebral cortex express different muscarinic receptor RNAs, *FEBS Lett.*, 230, 90, 1988.

27. Buckley, N. J., Bonner, T. I., and Brann, M. R., Localization of a family of muscarinic receptor mRNAs in rat brain, *J. Neurosci.*, 8, 4646, 1988.

28. Buckley, N. J., Bonner, T. I., Buckley, C. M., and Brann, M. R., Antagonist binding properties of five cloned muscarinic receptors expressed in CHO-K1 cells, *Mol. Pharmacol.*, 35, 469, 1989.

29. Peralta, F. G., Ashkenazi, A., Winslow, J. W., Ramachandran, J., and Capon, D. J., Differential regulation of PI hydrolysis and adenyl cyclase by muscarinic receptor subtypes, *Nature (London)*, 334, 434, 1988.

30. Akiba, I., Kubo, T., Maeda, A., Bujo, H., Nakai, J., Mishina, M., and Numa, S., Primary structure of porcine muscarinic acetylcholine receptor III and antagonist binding studies, *FEBS Lett.*, 235, 257, 1988.

31. Fukuda, K., Kubo, T., Akiba, I., Maeda, A., Mishina, M., and Numa, S., Molecular distinction between muscarinic acetylcholine receptor subtypes, *Nature (London)*, 327, 623, 1987.

32. Fukuda, K., Higashida, H., Kubo, T., Maeda, A., Akiba, I., Bujo, H., Mishina, M., and Numa, S., Selective coupling with K^+ currents of muscarinic acetylcholine receptor subtypes in NG108-15 cells, *Nature (London)*, 355, 1988.

33. Michel, A. D., Stefanich, E., and Whiting, R. L., Direct labeling of rat M_3-muscarinic receptors by ^3H-4-DAMP, *Eur. J. Pharmacol.*, 166, 459, 1989.

34. McKinney, M., Miller, J. H., and Aagaard, P. J., Pharmacological characterization of the rat hippocampal muscarinic autoreceptor, *J. Pharmacol. Exp. Ther.*, 264, 74, 1993.

35. Ono, S., Saito, Y., Ohgane, N., Kawanishi, G., and Mizobe, F., Heterogeneity of muscarinic autoreceptors and heteroreceptors in the rat brain: effect of a novel M1 agonist, AF102B, *Eur. J. Pharmacol.*, 155, 77, 1988.

36. Kilbinger, H., Presynaptic muscarine receptors modulating acetylcholine release, *Trends Pharmacol. Sci.*, 5, 103, 1984.

37. Meyer, E. M. and Otero, D. H., Pharmacological and ionic characterizations of the muscarinic receptors modulating [^3H]acetylcholine release from rat cortical synaptosomes, *J. Neurosci.*, 5, 1202, 1985.

38. Raiteri, M., Leardi, R., and Marchi, M., Heterogeneity of presynaptic muscarinic receptors regulating transmitter release in the rat brain, *J. Pharmacol. Exp. Ther.* 228, 209, 1984.

39. Levine, R. R. and Birdsall, N. J. M., Eds. Subtypes of muscarinic receptors IV, *Trends Pharmacol. Sci.*, Suppl., 1989.

40. Watson, S. and Abbott, A., Eds., Receptor nomenclature, *Trends Pharmacol. Sci.*, Suppl., 1992.

41. Richards, M. H., Pharmacology and second messenger interactions of cloned muscarinic receptors, *Biochem. Pharmacol.*, 42, 1645, 1991.

42. Mei, L., Roeske, W. R., and Yamamura, H. I., Molecular pharmacology of muscarinic receptor heterogeneity, *Life Sci.*, 45, 1831, 1989.

43. Miller, J. H., Gibson, V. A., and McKinney, M., Binding of [^{3}H]AF-DX 384 to cloned and native muscarinic receptors, *J. Pharmacol. Exp. Ther.*, 259, 601, 1991.

44. Smith, T. D., Annis, S. J., Ehlert, F. J., and Leslie, F. M., N-[^{3}H]methylscopolamine labeling of non-M$_1$, non-M$_2$ muscarinic receptor binding sites in rat brain, *J. Pharmacol. Exp. Ther.*, 256, 1173, 1991.

45. Zubieta, J. K. and Frey, K. A., Autoradiographic mapping of M$_3$ muscarinic receptors in the rat brain, *J. Pharmacol. Exp. Ther.*, 264, 415, 1993.

46. Whitehouse, P. I., Receptor autoradiography: applications in neuropathology, *Trends Neurosci.*, 8, 434, 1985.

47. Kuhar, M. J., Recent progress in receptor mapping: which neurons contain the receptors?, *Trends Neurosci.*, 10, 308, 1987.

48. James, W. M. and Klein, W. L., Localization of acetylcholine receptors on isolated CNS neurons: cellular and subcellular differentiation, *J. Neurosci.*, 8, 4225, 1988.

49. Palacios, J. M., Mengod, G., Vilaró, M. T., and Ramm, P., Recent trends in receptor analysis techniques and instrumentation, *J. Chem. Neuroanat.*, 4, 343, 1991.

50. Fuxe, K., Agnati, L. F., Rosén, L., Bjelke, B., Cintra, A., Bortolotti, F., Tinner, B., Andersson, C., Hasselroth, U., Steinbusch, H., Gustafsson, J.-A., and Benfenati, F., Computer-assisted image analysis techniques allow a characterization of the compartments within the basal ganglia. Focus on functional compartments produced by *d*-amphetamine activation of the c-fos gene and its relationship to the glucocorticoid receptor, *J. Chem. Neuroanat.*, 4, 355, 1991.

51. Lewis, M. E., Krause, II, R. G., and Roberts-Lewis, J. M., Recent developments in the use of synthetic oligonucleotides for *in situ* hybridization histochemistry, *Synapse*, 2, 308, 1988.

52. Nunez, D. J., Davenport, A. P., Emson, P. C., and Brown, M. J., A quantitative '*in-situ*' hybridization method using computer-assisted image analysis, *Biochem. J.*, 263, 121, 1989.

53. Rogers, W. T., Schwaber, J. S., and Lewis, M. E., Quantitation of cellular resolution *in situ* hybridization histochemistry in brain image analysis, *Neurosci. Lett.*, 82, 315, 1987.

54. Palacios, J. M., Mengod, G., Sarasa, M., Vilaró, M. T., Pompeiano, M., and Martinez-Mir, M. I., The use of *in situ* hybridization histochemistry for the analysis of neurotransmitter receptor expression at the microscopic level, *J. Recept. Res.*, 11, 459, 1991.

55. Li, M., Yasuda, R. P., Wall, S. J., Wellstein, A., and Wolfe, B. B., Distribution of m$_2$ muscarinic receptors in rat brain using antisera selective for m$_2$ receptors, *Mol. Pharmacol.*, 40, 28, 1991.

56. Wall, S. J., Yasuda, R. P., Li, M., and Wolfe, B. B., Development of an antiserum against m$_3$ muscarinic receptors: distribution of m$_3$ receptors in rat tissues and clonal cell lines, *Mol. Pharmacol.*, 40, 783, 1991.

57. Ehlert, F. J. and Tran, L. L. P., Regional distribution of M1, M2 and non-M1, non-M2 subtypes of muscarinic binding sites in rat brain, *J. Pharmacol. Exp. Ther.*, 256, 1148, 1990.

58. Giraldo, E., Hammer, R., and Ladinsky, H., Distribution of muscarinic receptor subtypes in rat brain as determined in binding studies with AF-DX 116 and pirenzepine, *Life Sci.*, 40, 833, 1987.

59. Cortes, R. and Palacios, J. M., Muscarinic cholinergic receptor subtypes in the rat brain. I. Quantitative autoradiographic studies, *Brain Res.*, 362, 227, 1986.

60. Horvath, E., Van Rooijen, L. A. A., Traber, J., and Spencer, D. G., Jr., Effects of N-methylmaleimide on muscarinic acetylcholine receptor subtype autoradiography and inositide response in rat brain, *Life Sci.*, 39, 1129, 1986.

61. Mash, D. C. and Potter L. T., Autoradiographic localization of M1 and M2 muscarine receptors in the rat brain, *Neuroscience*, 19, 551, 1986.

62. Quirion, R., Araujo, D., Regenold, W., and Boksa, P., Characterization and quantitative autoradiographic distribution of (3H)acetylcholine muscarinic receptors in mammalian brain. Apparent labeling of an M2-like receptor sub-type, *Neuroscience*, 29, 271, 1989.

63. Spencer, D. G., Jr., Horvath, E., and Traber, J., Direct autoradiographic determination of M1 and M2 muscarinic acetylcholine receptor distribution in the rat brain: relation to cholinergic nuclei and projections, *Brain Res.,* 380, 59, 1986.

64. Tonnaer, J. A. D. M., Ernste, B. H. W., Wester, J., and Kelder, K., Cholinergic innervation and topographical organization of muscarinic binding sites in rat brain: a comparative autoradiographic study, *J. Chem. Neuroanat.,* 1, 95, 1988.

65. Frey, K. A. and Howland, M. M., Quantitative autoradiography of muscarinic cholinergic receptor binding in the rat brain: distinction of receptor subtypes in antagonist competition assays, *J. Pharmacol. Exp. Ther.,* 263, 1391, 1992.

66. Palacios, J. M., Mengod, G., Vilaró, M. T., Wiederhold, K. H., Bodekke, H., Alvarez, F. J., Chinaglia, G., and Probst, A., Cholinergic receptors in the rat and human brain: microscopic visualization, *Prog. Brain Res.,* 84, 243, 1990.

67. Vilaró, M. T., Wiederhold, K.-K., Palacios, J. M., and Mengod G., Muscarinic M_2 receptor mRNA expression in cholinergic and non-cholinergic cells in the rat brain: a correlative study using *in situ* hybridization histochemistry and receptor autoradiography, *Neuroscience,* 47, 367, 1992.

68. Regenold, W., Araujo, D., and Quirion, R., Direct visualization of brain M_2 muscarinic receptors using the selective antagonist [^{3}H]AF-DX 116, *Eur. J. Pharmacol.,* 144, 417, 1987.

69. Regenold, W., Araujo, D., and Quirion, R., Quantitative autoradiographic distribution of [^{3}H]AF-DX 116 muscarinic-M_2 receptor binding sites in rat brain, *Synapse,* 4, 115, 1989.

70. Wang, J. X., Roeske, W. R., Hawkins, K. N., Gehlert, D. R., and Yamamura, H. I., Quantitative autoradiography of M_2 muscarinic receptors in the rat brain identified by using a selective radioligand [^{3}H]AF-DX 116, *Brain Res.,* 477, 322, 1989.

71. Aubert, I., Cécyre, D., Gauthier, S., and Quirion, R., Characterization and autoradiographic distribution of [^{3}H]AF-DX 384 binding to putative muscarinic M_2 receptors in the rat brain, *Eur. J. Pharmacol.,* 217, 173, 1992.

72. Wamsley, J. K., Gehlert, D. R., Roeske, W. R., and Yamamura, H. I., Muscarinic antagonist binding site heterogeneity as evidenced by autoradiography after direct labeling with [^{3}H]-QNB and [^{3}H]-pirenzepine, *Life Sci.,* 34, 1395, 1984.

73. Schliebs, R. and Stewart, M. G., Unilateral decortication affects muscarinic cholinergic receptor binding in rat basal forebrain, *Neurochem. Int.,* 16, 81, 1990.

74. Schwab, C., Brückner, G., Rothe, T., Castellano, C., and Oliverio, A., Autoradiography of muscarinic cholinergic receptors in cortical and subcortical brain regions of C57Bl/6 and DBA/2 mice, *Neurochem. Res.,* 17, 1057, 1992.

75. Zilles, K., Werner, L., Qü, M., Schleicher, A., and Gross G., Quantitative autoradiography of 11 different transmitter binding sites in the basal forebrain region of the rat — evidence of heterogeneity in distribution patterns, *Neuroscience,* 42, 473, 1991.

76. Nio, J., Besson, M. J., and Breton, P., Ontogenic distribution of muscarinic receptors and acetylcholinesterase in the rabbit hippocampus, *Brain Res. Bull.,* 31, 723, 1993.

77. Russo-Neustadt, A., Rotter, A., and Frostholm, A., Distribution of muscarinic receptors in the developing rodent cerebellum, *Brain Res.,* 548, 179, 1991.

78. Miyoshi, R., Kito, S., Shimizu, M., and Matsubayashi H., Ontogeny of muscarinic receptors in the rat brain with emphasis on the differentiation of M_1- and M_2-subtypes — semiquantitative *in vitro* autoradiography, *Brain Res.,* 420, 302, 1987.

79. Neustadt, A., Frostholm, A., and Rotter, A., On the cellular localization of cerebellar muscarinic receptors: an autoradiographic analysis in weaver, reeler, Pukinje cell degeneration and staggerer mice, *Brain Res. Bull.,* 20, 163, 1988.

80. Neustadt, A., Frostholm, A., and Rotter, A., Topographical visualization of muscarinic cholinergic receptors in the cerebellar cortex of mouse, rat guinea pig, and rabbit: a species comparison, *J. Comp. Neurol.,* 272, 317, 1988.

81. Schliebs, R., Walch, C., and Stewart, M. G., Laminar pattern of cholinergic and adrenergic receptors in rat visual cortex using quantitative receptor autoradiography, *J. Hirnforsch.,* 30, 303, 1989.

82. Kumar, A. and Schliebs, R., Postnatal development of cholinergic receptors, protein kinase C and dihydropyridine-sensitive calcium antagonist binding in rat visual cortex. Effect of visual deprivation, *Int. J. Dev. Neurosci.,* 10, 491, 1992.

83. Schliebs, R. and Stewart, M. G., Laminar postnatal development of muscarinic cholinergic receptors in rat visual cortex and the effect of monocular deprivation, *Neurochem. Int.,* 19, 143, 1991.

84. Schliebs, R., Roßner, S., Kumar, A., and Bigl, V., Muscarinic acetylcholine receptor subtypes in rat visual cortex — a comparative study using quantitative receptor autoradiography and *in situ* hybridization, *Ind. J. Exp. Biol.*, 32, 25, 1994.

85. Vilaró, M. T., Wiederhold, K.-H., Palacios, J. M., and Mengod, G., Muscarinic M_2-selective ligands also recognize M_4 receptors in the rat brain: evidence from combined *in situ* hybridization and receptor autoradiography, *Synapse*, 171, 1992.

86. Gitler, M. S., Reba, R. C., Cohen, V.I., Rzeszotarski, W. J., and Baumgold, J., A novel m_2-selective muscarinic antagonist: binding characteristics and autoradiographic distribution in rat brain, *Brain Res.*, 582, 253, 1992.

87. Miyoshi, R., Kito, S., and Shimoyama, M., Quantitative autoradiographic localization of the m_1 and m_2 subtypes of muscarinic acetylcholine receptors in the monkey brain, *Jpn. J. Pharmacol.*, 51, 247, 1989.

88. Wagster, M. V., Whitehouse, P. J., Walker, L. C., Kellar, K. J., and Price, D. L., Laminar organization and age-related loss of cholinergic receptors in temporal cortex of Rhesus monkey, *J. Neurosci.*, 9, 2879, 1990.

89. Doods, H. N., Mathy, M. J., Davidesko, D., Van Charldorp, K. J., De Jonge, A., and Van Zwieten, P. A., Selectivity of muscarinic antagonists in radioligand and *in vivo* experiments for the putative M1, M2 and M3 receptors, *J. Pharmacol. Exp. Ther.*, 242, 257, 1987.

90. Lambrecht, G., Feifel, R., Moser, U., Wagner-Röder, M., Choo, L. K., Camus, J., Tastenoy, M., Waelbroeck, M., Strohmann, C., Tacke, R., Rodrigues de Miranda, J. F., Christophe, J., and Mutschler, E., Pharmacology of hexahydro-difenidol, hexahydro-sila-difenidol and related selective muscarinic antagonists, *Trends Pharmacol. Sci.*, 10 (Supplement: Subtypes of Muscarinic Receptors IV), 60, 1989.

91. Araujo, D. M., Lapchak, P. A., and Quirion, R., Heterogeneous binding of [^{3}H]4-DAMP to muscarinic cholinergic sites in the rat brain: evidence from membrane binding and autoradiographic studies, *Synapse*, 9, 165, 1991.

92. Pavia, J., Marquez, E., Laukkonen, S., Martos, F., Gómez, A., and Sánchez de la Cuesta, F., M_1 and M_3 muscarinic receptor subtypes in rat forebrain, *Meth. Find. Exp. Clin. Pharmacol.*, 13, 653, 1991.

93. Weiner, D. M., Levey, A. I., and Brann, M. R., Expression of muscarinic acetylcholine and dopamine receptor mRNAs in rat basal ganglia, *Proc. Natl. Acad. Sci. U.S.A.*, 87, 7050, 1990.

94. Andre, C., Guillet, J. A., DeBacker, J. P., Vanderheyden, P., Hoebeke, J., and Strosberg, A. D., Monoclonal antibodies against the native or denatured forms of muscarinic acetylcholine receptors, *EMBO J.*, 3, 17, 1984.

95. Luetje, C. W., Brumwell, C., Gainer-Norman, M., Peterson, G. L., Schimerlik, M. I., and Nathanson, N. M., Isolation and characterization of monoclonal antibodies specific for the cardiac muscarinic acetylcholine receptor, *Biochemistry*, 26, 6892, 1987.

96. Levey, A. I., Stormann, T. M., and Brann, M. R., Bacterial expression of human muscarinic receptor fusion proteins and generation of subtype-specific antisera, *FEBS Lett.*, 275, 65, 1990.

97. Luthin, G. R., Harkness, J., Artymashan, R. P., and Wolfe B. B., Antibodies to a synthetic peptide can be used to distinguish between muscarinic acetylcholine receptor binding sites in brain and heart, *Mol. Pharmacol.*, 34, 327, 1988.

98. Venter, J. C., Eddy, B., Hall, L. M., and Fraser, C. M., Monoclonal antibodies detect the conservation of muscarinic cholinergic receptor structure from *Drosophila* to human brain and detect possible structural homology with $\alpha 1$-adrenergic receptors, *Proc. Natl. Acad. Sci. U.S.A.*, 81, 272, 1984.

99. Schröder, H., Immunohistochemistry of cholinergic receptors, *Anat. Embryol. (Berl.)*, 186, 407, 1992.

100. Dörje, F., Levey, A. I., and Brann, M. R., Immunological detection of muscarinic subtype proteins (m_1-m_5) in rabbit peripheral tissues, *Mol. Pharmacol.*, 40, 459, 1991.

101. Fonseca, M. I., Aguilar, J. S., Skopura, A. F., and Klein, W. L., Cellular mapping of m_2 muscarinic receptors in rat olfactory bulb using antiserum raised against a cytoplasmic loop peptide, *Brain Res.*, 563, 163, 1991.

102. Levey, A. I., Immunological localization of m_1-m_5 muscarinic acetylcholine receptors in peripheral tissues and brain, *Life Sci.*, 52, 441, 1993.

103. Levey, A. I., Kitt, C. A., Simonds, W. F., Price, D. L., and Brann, M. R., Identification and localization of muscarinic acetylcholine receptor proteins in brain with subtype-specific antibodies, *J. Neurosci.*, 11, 3218, 1991.

104. Wall, S. J., Yasuda, R. P., Hory, F., Flagg, S., Martin, B. M., Ginns, E. I., and Wolfe, B. B., Production of antisera selective for m_1 muscarinic receptors using fusion proteins: distribution of m_1 receptors in rat brain, *Mol. Pharmacol.*, 39, 643, 1991.

105. Matsuyama, T., Luiten, P. G. M., Spencer, Jr., D. G., and Strosberg, A. D., Ultrastructural localization of immunoreactive sites for muscarinic acetylcholine receptor proteins in the rat cerebral cortex, *Neurosci. Res. Commun.*, 2, 69, 1988.

106. van der Zee, E. A., Matsuyama, T., Strosberg, A. D., Traber, J., and Luiten, P. G. M., Demonstration of muscarinic acetylcholine receptor-like immunoreactivity in the rat forebrain and upper brain stem, *Histochemistry*, 92, 475, 1989.

107. van der Zee, E. A., Streefland, C., Strosberg, A. D., Schröder, H., and Luiten, P. G. M., Colocalization of muscarinic and nicotinic receptors in cholinoceptive neurons of the suprachiasmatic region in young and aged rats, *Brain Res.*, 542, 348, 1991.

108. van der Zee, E. A., Streefland, C., Strosberg, A. D., Schröder, H., and Luiten, P. G. M., Visualization of cholinoceptive neurons in the rat neocortex: colocalization of muscarinic and nicotinic acetylcholine receptors, *Mol. Brain Res.*, 14, 326, 1992.

109. Schröder, H., Monoclonal antibodies reveal the cellular localization and expression of cortical nicotinic and muscarinic cholinoceptors in human cerebral cortex, *Prog. Histo-Cytochem.*, 26, 266, 1992.

110. van Huizen, F., Strosberg, A. D., and Cynader, M. S., Cellular and subcellular localisation of muscarinic acetylcholine receptors during postnatal development in cat visual cortex using immunocytochemical procedures, *Dev. Brain Res.*, 44, 296, 1988.

111. Bertini, G., Everts, H., and Bentivoglio, M., The distribution of muscarinic receptors in the thalamus: an immunocytochemical study in the rat, *Eur. J. Neurosci.*, Suppl. 5, 57, 1235, 1992.

112. Roßner, S., Kues, W., Witzemann, V., and Schliebs, R., Laminar expression of m_1-, m_3, and m_4-muscarinic cholinergic receptor genes in the developing rat visual cortex using *in situ* hibridization histochemistry. Effect of monocular deprivation, *Int. J. Dev. Neurosci.*, 11, 369, 1993.

113. Weiner, D. M. and Brann, M. R., Distribution of m_1-m_5 muscarinic receptor mRNAs in rat brain, *Trends Pharmacol. Sci.*, Suppl., 10, 115, 1989.

114. Roßner, S., Kumar, A., Witzemann, V., and Schliebs, R., Development of laminar expression of the m_2 muscarinic cholinergic receptor gene in rat visual cortex and the effect of monocular visual deprivation, *Dev. Brain Res.*, 77, 55, 1994.

115. Hösli, L., Hösli, E., Maelicke, A., and Schröder, H., Peptidergic and cholinergic receptors on cultured astrocytes of different regions of the rat CNS, *Prog. Brain Res.*, 94, 317, 1992.

116. Lapchak, P. A., Araujo, D., Quirion, R., and Collier, B., Binding sites for [^{3}H]AF-DX 116 and effect of AF-DX 116 on endogenous acetylcholine release from rat brain slices, *Brain Res.*, 496, 285, 1989.

117. Mash, D. C., Flynn, D. D., and Potter, L. T., Loss of M2 muscarine receptors in the cerebral cortex in Alzheimer's disease and experimental cholinergic denervation, *Science*, 228, 1115, 1985.

118. Vizi, E. S., Kobayashi, O., Torocsik, A., Kinjo, M., Nagashima, H., Manabe, N., Goldiner, P. L., Potter, P. E., and Foldes, F. F., Heterogeneity of presynaptic muscarinic receptors involved in modulation of transmitter release, *Neuroscience*, 31, 259, 1989.

119. Dawson, V. L., Hunt, M. E., and Wamsley, J. K., Alterations in cortical muscarinic receptors following cholinotoxin (AF64A) lesion of the rat nucleus basalis magnocellularis, *Neurobiol. Aging*, 13, 25, 1991.

120. Joyce, J. N., Gibbs, R. B., Cotman, C. W., and Marshall, J. F., Regulation of muscarinic receptors in hippocampus following cholinergic denervation and reinnervation by septal and striatal transplants, *J. Neurosci.*, 9, 2776, 1989.

121. Smith, C. J., Perry, E. K., Perry, R. H., Candy, J. M., Johnson, M., Bonham, J. R., Dick, D. J., Fairbairn, A., Blessed G., and Birdsall, N. J. M., Muscarinic cholinergic receptor subtypes in hippocampus in human cognitive disorders, *J. Neurochem.*, 50, 847, 1988.

122. Dale, H. H., The action of certain esters and ethers of choline, and their relation to muscarine, *J. Pharmacol. Exp. Ther.*, 68, 147, 1914.

123. Hulme, E. C., Birdsall, N. J. M., and Buckley, N. J., Muscarinic receptor subtypes, *Annu. Rev. Pharmacol. Toxicol.*, 30, 633, 1990.

124. Schliebs, R., Feist, T., Roßner, S., and Bigl, V., Receptor function in cortical rat brain regions after lesion of nucleus basalis, *J. Neural Transm.*, Suppl. 44, 1994, in press.

Chapter 5

Distribution of Nicotinic Receptors in the CNS

Jennifer A. Court and Elaine K. Perry

CONTENTS

I. INTRODUCTION

The original classification of cholinergic receptors into two classes (muscarinic and nicotinic), based on drug interactions, indicated the key importance of nicotinic receptors (nAchRs) at the muscle end plate and a preponderance of muscarinic receptors (mAchRs) in the central nervous system (CNS). However, recent findings suggest that although brain nAchRs are less numerous than mAchRs they are likely to have an important role in behavior and cognition,[1-6] are affected by aging and dementia (see below), and may have neurotrophic and neuroprotective functions.[7-10] CNS nAchRs are a family of ligand-gated cation channels, with a pentameric structure generally comprised of two alpha and three beta subunits.[11-14] Pharmacological studies have suggested heterogeneity of this class of cholinergic receptor in the brain (more than one binding site for nicotine[15] and subpopulations with varying agonist and antagonist sensitivity[16-20]), and this has been confirmed by more recent molecular biological techniques, largely conducted on brain from rodent (see below) and chick.[21] In the rat at least six alpha subunits (α_2, α_3, α_4, α_5, α_6, α_7) and three beta subunits (β_2, β_3, β_4) have been identified, and different subunit composition appears to result in distinct, although not exclusive, receptor pharmacological profiles (for reviews see References 11, 12, 19, and 22). Although it is the α subunit that possesses the agonist binding site, it has become clear that the β composition of the receptor may also affect the ligand specificity and rate of desensitization of receptor subtypes.[11,23-25]

The distribution of nAchRs in the CNS was first investigated using α-bungarotoxin (αBT),[26,27] which had been employed to identify the muscle nicotinic acetylcholine receptor, and for which it has a high affinity. It subsequently became clear that this ligand does not generally bind to the high-affinity nAchRs in brain, except possibly at selected sites, for example on cerebellar interneurons[16] and in fetal rat hippocampus.[28] The kinetics, pharmacological profile, and distribution of high-affinity [3H]nicotine binding resemble those of [3H]acetylcholine receptors (in the presence of atropine to block mAchRs), whereas the binding of [125I]αBT does not.[29-32] Recent molecular biological studies have helped to clarify the identity of nicotinic and αBT binding sites. From the coincidence of cDNA signals for specific receptor subtype mRNAs,[33] the use of ligand affinity chromatography, amino acid sequencing and specific antibodies,[34-37] and the transfection of subunits into single cell systems,[38,39] it has been deduced that the major nAchR in rat and chick brain, which binds nicotine (and other agonists, e.g., cytisine and methyl carbamylcholine) with high affinity (in the nanomolar range), is likely to be composed of α_4 and β_2 subunits. However, α_2, α_3, and β_4, although not as prevalent as α_4 and β_2, may also contribute to the brain nicotinic cholinergic activity.[11,23,29]

αBT binds to a separate subset of receptors (αBTRs), which contain α_7 subunits and are apparently as numerous as nicotine binding sites.[14] These receptors have a low affinity for nicotine (micromolar),[15] and their role and possible endogenous ligands are subjects of considerable interest. It has been recently demonstrated that both the endogenous polypeptide thymopoietin[40] and a toxin isolated from delphinium (*Delphinium brownii*) seeds, methyllycaconitine,[41] are potent displacers of [^{125}I]α-bungarotoxin in rat brain. The α_7 subunits are able to form active cation channels, blocked by α-bungarotoxin, in the absence of β subunits.[42-44] A second neurotoxin (neuronal bungarotoxin, also termed kappa-bungarotoxin, toxin F, or bungarotoxin 3.1) exhibits some selectivity for α_3-containing (the ganglionic-type) nicotinic receptors, but since it also binds to αBTRs and other nAchRs (e.g., α_4-containing), albeit with lower affinity,[12,20,45] it has not proved to be a useful ligand to map subsets of receptors.

II. DISTRIBUTION OF NICOTINIC RECEPTORS IN RODENT BRAIN

The distribution of high-affinity nicotine and αBT binding has been characterized throughout rodent brain using autoradiography (Table 5-1). In the rat, [^{3}H]nicotine binding is highest in the interpeduncular nucleus, all thalamic nuclei (with the exception of the posterior and intralaminar nuclei), the superior colliculus, and the medial habenula. Binding is also prominent in the substantia nigra (pars compacta), the ventral tegmental area, selected areas of the hippocampal formation (the dentate molecular layer, the subiculum, and the presubiculum), the cerebral cortex (laminae I, III, and IV), and the septum (bed nucleus stria terminalis). Moderate binding was found in the neostriatum, the ventral striatum, the dorsal tegmental nucleus, and the cerebellum.[31] This distribution in the rat has been confirmed using [^{3}H]acetylcholine (in the presence of atropine) and [^{3}H]methylcarbamylcholine as nicotinic ligand for autoradiography. The highest binding was again apparent in thalamic regions and the medial habenula nucleus, with moderate binding in the zona compacta of the substantia nigra, the ventral tegmental area, the dorsal tegmental nucleus, the striatum, and laminae III and IV of the cerebral cortex.[50-52]

The distribution in mice appears somewhat different[47,48] (see Table 5-1), with the highest binding being in the interpeduncular nuclei and relatively less binding in the thalamus, medial habenula, the subiculum, and the septum. Lower levels were observed in the cerebral cortex, dorsal raphé, and pontine nucleus, and little or no binding was present in hippocampus, entorhinal cortex, hypothalamus, caudate/ putamen, locus ceruleus, and cerebellum. If these disparities between such closely related species as mice and rats are genuine and not a reflection of variation in techniques, it is untenable to extrapolate details of nicotinic distribution from rodents to other species, notably the human.

The distribution of mRNAs coding for α_2, α_3, α_4, and β_2 receptor in rodent brain has been evaluated by Wada et al.[33] and Marks et al.[49] (see Table 5-1). β_2 mRNA appears to be present in most regions of the CNS, with highest levels in the thalamus, the ventral tegmental area and pars compacta of the substantia nigra, the piriform cortex, layer II of the entorhinal cortex, and in the somatosensory and motor areas of the brainstem in rats and in the medial habenula in mice. The α subunits were expressed in distinctive distributions, but with some overlap. In rats the α_4 gene is the most widespread, and the highest concentration of α_4 subunit mRNAs was found in the thalamus, with the next highest levels in the substantia nigra (pars compacta and ventral tegmental area), the piriform cortex, the endopiriform nucleus, the amygdala (cortical nucleus and the basolateral nucleus), the subicular complex, the septum, and the brainstem (interpeduncular nuclei, somatosensory areas, and the caudal linear raphé; Table 5-1). Moderate concentrations were found in the cerebral cortex, some hypothalamic nuclei, and other brainstem nuclei. In mice the highest signals for α_4 were found in the medial habenula and substantia nigra (Table 5-1).

The α_3 gene was most intensely expressed in the medial habenula, layer II of the entorhinal cortex, lamina IV of the cerebral cortex, the anterior, ventral, and geniculate thalamic nuclei, the supramammillary nucleus, the posterior area of the lateral zone of the hypothalamus, and localized areas of the brainstem (trigeminal ganglion, locus ceruleus, and motor areas) in the rat and the medial habenula in the mouse (Table 5-1). α_3 mRNA appears to be expressed in some of the same brain regions as α_4 (Table 5-1), suggesting that different neurons in the same population may express different receptor subtypes or that some neurons may express more than one α subtype, even possibly as part of the same ion channel. α_2 is concentrated in only a few brain regions (Table 5-1), certain interpeduncular nuclei, the dorsal tegmental nucleus, the inferior colliculus, the parabrachial nucleus, the pyramidal layer and stratum radiatum of the subiculum, and the stratum oriens of the CA1 region of the hippocampus in rats and in the interpeduncular and dorsal tegmental nuclei in mice.

A comparison of nicotine autoradiography and *in situ* hybridization studies (Table 5-1) indicates a considerable degree of concordance of high-affinity nicotine binding with expression of α_4 and β_2

subunits, for example in most areas of the thalamus, the medial habenula, and the substantia nigra, and in some areas of the cerebral cortex, the subiculum, the septum, and the brainstem. However, in other regions correlation is not apparent and nicotinic binding appears to be present in the absence of subunit expression or vice versa. In the striatum in rats, nicotinic binding is present, but there is little corresponding mRNA. Conversely, in a number of other brain areas [e.g., the posterior and intralaminar nuclei of the thalamus (rat), laminae II and V of the cerebral cortex (rat), the claustrum (rat and mouse), the hypothalamus (rat and mouse), and the somatosensory and vestibular areas of the brainstem (rat)] there are apparently moderate to high levels of α_4 and β_2 mRNAs but little or no detectable nicotinic binding. This may simply reflect variations in signal detection, the presence of other nicotine binding receptor subunits, not as yet recognized, or particularly rapid mRNA or receptor turnover. Alternatively, the discrepancy may be an indication of the presynaptic location of some nicotinic receptors such that the terminals on which the receptors are situated are some distance from the perikarya.

The $[^{125}I]\alpha BT$ binding exhibited a different pattern of distribution from the $[^3H]$nicotine binding in rats[31] (Table 5-1). It was present in the cerebral cortex (highest intensity in laminae I and IV), the endopiriform nucleus, the hypothalamus (particularly in the mamillary body and in the suprachiasmatic, supraoptic, paraventricular, and posterior nuclei), the hippocampus (dense labeling in the CA4 region), certain brainstem nuclei (the superior and inferior colliculi, the locus ceruleus, the dorsal tegmental nucleus, the interpeduncular nucleus, the dorsal and ventral parabrachial nuclei, the inferior and lateral superior olive, the medial vestibular nucleus, the dorsal cochlear nucleus, the gigantocellular reticular nucleus and the nucleus of the spinal tract of the trigeminal nerve, and the medial septal nucleus), the amygdala (the basolateral, lateral, and medial nuclei), and the ventral lateral geniculate nucleus. A more detailed study of $[^{125}I]\alpha BT$ binding in the hippocampus indicates that the activity is located in interneurons (which may or may not be cholinergic) in the dentate situated between the granular and molecular layers, in the CA1 in the striatum radiatum and striatum lacunosum moleculare, and in the CA3 in the stratum radiatum and the stratum lucidum.[53] An intense *in situ* hybridization signal for α_7 was also observed in the interneurons of CA1-4 in the stratum oriens and stratum moleculare.[44] There was little or no binding in other thalamic nuclei, striatum, substantia nigra, and ventral tegmental nucleus.[31] As for nicotine binding there were also species differences between $[^{125}I]\alpha BT$ binding in mice and rats (see Table 5-1); generally, mice expressed relatively more $[^{125}I]\alpha BT$ binding in the brainstem and less in the cerebral cortex.[47]

The distribution of mRNA for the α_7 subunit in rats[44] appears to closely parallel that of $[^{125}I]\alpha BT$ binding (Table 5-1). Moderate to high transcription levels were found in the olfactory areas, the cerebral cortex (superficial and deep layers), the endopiriform nucleus, the claustrum, the hippocampus (the granule cell layer of the dentate gyrus and the pyramidal cell layer of the CA fields 1 to 4), the hypothalamus (suprachiasmatic, arcuate, supraoptic, dorso- and ventromedial nuclei, the mammillary complex, and the preoptic area), the medial lateral septum, the amygdala and associated cortical and hippocampal areas, and in the brainstem (superior and inferior colliculi, the central gray, the dorsal and median raphé nuclei, the tegmental nuclei, dorsal and ventral nuclei of the lateral lemnicus, and the vestibular nuclei). α_7 transcripts were also detectable in the zona incerta, the lateral geniculate nucleus, the subthalamic nucleus, the medial habenula, the interpeduncular nuclei, and a weak signal was observed in the entorhinal cortex, the subiculum, and the Purkinje cell layer of the cerebellum. There was little or no expression in the other thalamic nuclei, substantia nigra, or striatum.

The β_4 subunit, present in diverse loci in rat brain,[54] is able to form functional nicotinic acetylcholine receptors with α_2, α_3, and α_4 subunits.[55,56] Its transcription signal is reported to be greatest in the medial habenula, but is also distributed throughout the isocortex, the forebrain olfactory regions, and the hippocampal formation (pre- and parasubiculum, subiculum, dentate gyrus, CA1 and CA3). There was isolated but intense hybridization in the interpeduncular nucleus and the trigeminal motor V nucleus, and there were more moderate signals in the amygdalo-hippocampal fissure, the locus ceruleus, the pontine nucleus, the supramammillary nucleus of the hypothalamus, the septal nucleus, and the Purkinje cell layer of the cerebellum. There was no apparent signal in the thalamic nuclei or other areas of the brainstem, septum, amygdala, hypothalamus, or cerebellum.[54] β_4 was colocalized with α_4, throughout its distribution and in certain locations with α_3 and α_2. β_3, in common with β_4, is not as widespread as β_2, has its highest expression in the medial habenula, and is able to form functional nicotinic channels with α_2, α_3, and α_4 subunits.[12]

α_5 Has been detected in a small number of localized sites in the rat brain,[46] notably in the pyramidal layer of the subiculum, layers IV and VI of the pre- and parasubiculum, the substantia nigra (pars compacta and

Table 5-1. Distribution of $[^{3}H]$nicotine and $[^{125}I]$bungarotoxin binding and mRNAs for nicotinic receptor subunits in rodent brain

	$[^{3}H]$Nicotine binding	In situ hybridization					$[^{125}I]$-α-Bungarotoxin binding	In situ hybridization (α_7)
		α_2	α_3	α_4	α_5	β_2		
Cerebral cortex								
Laminae I	++	−	−	+		(+)	+++ (*)	++
II	−	−	−	++	(+)	++	+	
III	++	−	−	++ *(*)	(+)	(+)	+ (*)	
IV	++	−	+++	+		(+)	+	
V	−	+	(+)	++		+	+	
VI				++	++ *	+	+++	
Olfactory cortex						+++	+++	++
Claustrum	(*)			+++ *(*)		+ [−]		+++
Hippocampal formation								
Dentate	(*)	+ (*)	+	++		*(*)		+++
Dentate granular	++							
Dentate molecular	[−]			[−]				
CA1/CA2	[−]	+ [−]	+ [−]	++		+(+) *(*) *(*)	+	+++
CA3						*(*)	+	
CA4							+++	
Para/presubiculum	++ *	−	−	+++ **	+++	+(+) [−]		+++
Subiculum	*(*)	++/−	+(+)	+++ *(*)	+++ *(*)	++ *	*	(+)
Entorhinal cortex								
Generally	(*)	−	−	+		(+) *		(+)
Medial area II (lateral)		−	++++	++	++ (also layers IV and VI)	++ (lateral)		(+)
Amygdala	[−]	−	−	++		+(+) *	+ *	++
Septum	*(*)	−/++ (+)	− −	−/++ **		++ *(*)	−/++	++

Thalamus (nuclei except those below)	+++ **	–	+++ *	+++ **	+++	++	– [–]
Posterior nucleus	– –*	+++ *	+++ **	+++ *(*)	*(*)		
Intra laminar nucleus		(+)	+	++ *	+++ (*)	+++	
Parafascicular nucleus							**
Medial habenula	+++ **	+++ ****	+++ ***	++ ***	*	++ ***	++
Hypothalamus	(*)	+++ *	–/++ *(*)	+ **		+ **	+++ *
Striatum	+ (*)	–(*)	– (*)	(+) (*)			– *
Substantia nigra pars compacta	++ *	–	+ *	+++ *****	+++ **	+++ *(*)	–
Ventral tegmental area	++ *	++	++ ****	++ *****	+++	+++	
Brainstem							
Somatosensory areas			++++	++++	++	++++	
Motor areas		–	–	+++	++	++	
Vestibular areas		+ *	++	++ *	++	+	+
Dorsal raphé	*	–(*)	+ *(*)	+ *(*) **	+	*(*)	*(*)
Interpeduncular nuclei	+++ ***	+++ ***	+++ **	(+) *(*)	+++ ***	(+) *(*)	+ *(*)
Superior colliculus	+++	*					
Inferior colliculus				++			++
Tegmental nuclei	+ *	***	**	+++ *(*)	+++	+++ *(*)	+
Locus ceruleus	(*)	*(*)	*(*)	++ ****	*(*)	++ ****	
Pontine nuclei	(*)	*(*)	*(*)	++ *	(*)	++ *	
Cerebellum	+ [–]	–	(+) [–]	+ ** } granule	–	+ **	(+) (p cells)

Note: + Refers to rats,[31,33,46] * and [–] refer to mice,[47–49] –/++ indicates high concentrations in localized areas, –, (+), +(+), ++, +++, ++++ indicate no signal → very intense signal (similarly for *).

ventral tegmental area), the interpeduncular nucleus, and the dorsal motor nucleus of the vagus nerve. Lower levels were observed in a small number of sites including the isocortex and the anterior olfactory cortex. No hybridization signal above background was observed in the amygdala, septum, thalamus, hypothalamus, or cerebellum. The significance of this putative acetylcholine binding subunit has yet to be elucidated. It does not appear to form a functional acetylcholine-gated ion channel nor to bind $[^{125}I]\alpha BT$ in combination with β_2, β_3, or β_4.

III. DISTRIBUTION OF NICOTINIC RECEPTORS IN HUMAN (AND OTHER PRIMATE) BRAIN

A complete topographical analysis of the distribution of nAchRs and the transcripts of nAchR gene products in the human brain has not been achieved to date. This is in part due to the daunting size and complexity of the human brain, but is also compounded by the need for species-specific *in situ* hybridization probes and the effect of perimortem factors such as mode of death, which can influence RNA autolysis. Thus far it appears that the distribution pattern of nAchRs in primate brain bears some similarity to that in the rodent, but important distinctions are apparent. Table 5-2 collates data from both autoradiographic studies and binding to isolated membrane fractions.

In the human the highest levels of $[^3H]$nicotine binding were present in the thalamus (Table 5-2 and Figures 5-1 and 5-2), in particular the lateral geniculate nucleus; the principal anterior, dorsomedial, and medial geniculate nuclei showed a slightly lower intensity and the pulvinar, dorsolateral, reticular, and ventral groups of nuclei demonstrated moderate binding[61,64] (Figure 5-1A to D). Binding in the human striatum (in contrast to rat and mouse) and in substantia nigra was almost as high as that in the dorsomedial thalamus. In the substantia nigra binding was concentrated in the pars compacta (Figure 5-1E and H), where it exactly parallels the distribution and density of pigmented dopaminergic neurons and is lost in conjunction with their disappearance in Parkinson's disease (PD).[63] It is higher in the medial (including A9 subgroup) compared with lateral portions, the reverse of the relative neuronal vulnerability in PD, but indicative of a more active role for the nicotine receptor in cortical, including limbic, compared with dopaminergic projections. Nicotinic binding was moderately dense in the dorsal tegmental area adjacent to the substantia nigra, which is at the rostal end of the Ch5/6 pedunculopontine-dorsolateral tegmental cholinergic nuclei. Periaqueductal grey matter was also high in nicotine binding, including the dorsal raphé nucleus.[57,63] An early report that the substantia innominata region of the human brain is high in binding measured in isolated membranes[65] has not been confirmed.[58,61] In the cynomolgus monkey high-affinity nicotine binding was also most predominant in the thalamus together with the medial habenula.[66]

Although lower levels of nicotinic binding are generally observed in human neo- and archicortex (Table 5-2 and Figures 5-1D and 5-2), a distinct pattern of distribution and some areas of high intensity have been reported.[62] Binding was greatest in the subicular complex, particularly the presubiculum, where it was prominent in the parvo- and magnocellular islands, and in the middle layers of the entorhinal cortex. This relatively high density, almost reaching that of the thalamus, may be related to the extensive phylogenetic development of these regions,[67] which has occurred in conjunction with the neocortical association areas. Such multimodal cortical association areas are among those projecting to the subicular complex and entorhinal cortex.[68,69] In the somatosensory (Brodman areas 3, 1, and 2) (Figure 5-2B) and occipital visual cortex (area 17) (Figure 5-2C) binding was highest in the upper and lower layers, and relatively sparse in the sensory input, layer IV. In the primary motor (area 4) (Figure 5-2A and B) and temporal (area 21) cortex (Figure 5-2D), binding in the outer half of the cortical ribbon was denser than that in the inner half, and a distinct band was apparent in the temporal and cingulate cortex (area 32) in the lower portion of layer III. In the prefrontal association cortex the pattern of binding was less distinct, although slightly higher in the lower architectonic areas. There was generally little binding in the hippocampus (areas CA1-4) and the dentate gyrus, with the exception of the stratum lacunosum moleculare in CA2-3 and, to a lesser extent, the supra- and subgranular zones of the dentate (Figures 5-1D and 5-2E). In the amygdaloid complex, as in the hippocampus, there was little binding, although the adjacent entorhinal cortex was strikingly and relatively selectively labeled, compared with other temporal cortical areas (Figure 5-2D).

In the cerebellum, high-affinity nicotine binding is present at moderate levels in the dentate nucleus and granule cell layer of the cerebellar cortex (Figure 5-1G).

The distribution of α_3 expression has been studied in primate brain in the cynomolgus monkey.[66] The mRNA coding for this subunit was greatest in the hippocampal dentate gyrus, the dorsal and ventral

Table 5-2. **Distribution of nicotinic binding in human brain**

	Adem et al.[57]	Araujo et al.[58]	Nordberg et al.[59]	Perry et al.[60-62]
Cerebral cortex				
Frontal, occipital, and temporal	++	(+)	+(+)	+
Olfactory tubicle			+(+)	
Claustrum	+			
Hippocampal formation				
Dentate gyrus	++			
Dentate granular	+(+)		+	+
Dentate molecular	+			
Pyramidal layer	+			(+)
Stratum lacunosum molecular				+(+)
Para/presubiculum				++
Subiculum				+(+)
Entorhinal cortex				+(+)
Amygdala				(+)
Thalamus		+++	+++	+++
Lateral geniculate nuclei				++++
Periaqueductal grey	+++			
Hypothalamus			+(+)	
Striatum				
Caudate	+	++	++	++(+)
Putamen	+++		++	++
Globus pallidus		+	(+)	
Substantia nigra				
Pars compacta	+++		++	++(+)
Pars reticulata	++			+
Ventral tegmental area				+(+)
Pons			+	
Medulla oblongata			+	
Cerebellum	++			+
Spinal cord	(+)			(+)

Note: Data are derived from References 57 to 63; symbols are as for Table 5-1.

geniculate nuclei, and the medial habenula; it was also evident in the cerebral cortex, other areas of the hippocampus (e.g., CA1-3), the subicular complex, the entorhinal cortex, the laterodorsal and parafascicular thalamic nuclei, the subcortical nuclei, and the granule cell layer of the cerebellar cortex.

Recently the distribution of the transcription products of human α_3 and β_2 genes has been studied in thalamic and hippocampal areas, from cases obtained at autopsy after sudden death.[70] In common with studies in rats and the cynomolgus monkey there was an intense α_3 hybridization signal in certain thalamic nuclei,[33,66] in particular in the dorsomedial, lateral posterior, ventro-posteromedial, reticular, and lateral geniculate nuclei. There was little or no α_3 signal in the caudate and putamen despite high levels of nicotine binding. At variance with rat and mouse, in humans the β_2 signal appeared low in the thalamus, suggesting the possibility of species variation in the major β subunit contribution to nicotinic binding in this brain area. In the hippocampal formation there was a closer correlation between rat and human, with a generally higher signal for β_2 than α_3 and with the greatest intensity in the dentate granule layer, the pyramidal layer of the CA2-3, and the subiculum. The distribution of high-affinity nicotine binding, measured in contiguous sections, did not show concordance with α_3 or β_2 *in situ* hybridization activity in the hippocampus, but reflected α_3 activity in the thalamus.

There have been few studies to date exploring the distribution of αBT sites in the human brain. Recent data from this laboratory[70,71] indicate differences when compared with nAchR distribution in human brain, but similarities when compared with αBT binding in rat brain. In the hippocampus, αBTR binding is concentrated in the dentate and CA1/2 regions with little binding in the subicular complex and

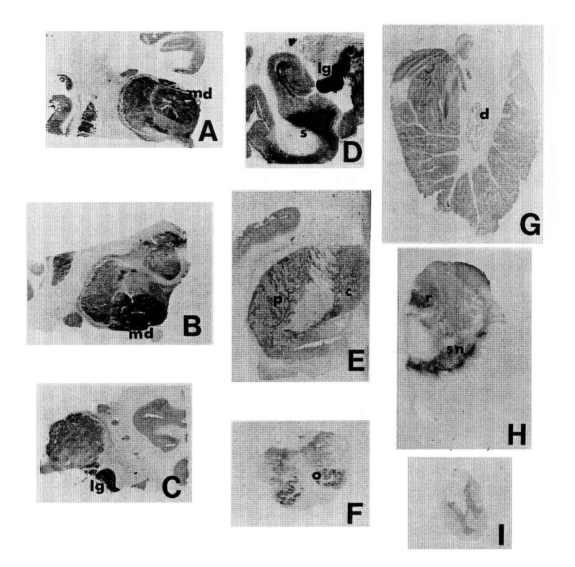

Figure 5-1 [³H]Nicotine binding in different regions of the human brain. A to D, thalamus at different coronal levels, illustrating high binding in dorsomedial nucleus (dm) and intense binding in lateral geniculate (lg) [note also high binding in subicular complex (s) in C as in E]; E, striatum enlarged to demonstrate lattice-like binding in putamen (p) and caudate (c); F, brainstem binding in olivary nucleus (o); G, cerebellum, binding in both cortex and dentate nucleus (d); H, binding in midbrain is concentrated in substantia nigra pars compacta (sn) and also raphé (r); I, low level of binding in ventral horn of lumbar spinal cord.

entorhinal cortex. A strong *in situ* hybridization signal for α_7 was also detected in the hippocampal dentate (granule cell layer), and binding was evident in the CA3 region of the hippocampus.[70] In the human cerebellum, αBT binding, present at relatively low levels in the molecular layer of the cortex, was absent from the dentate nucleus.[71]

IV. CELLULAR LOCATION OF NICOTINIC RECEPTORS

A presynaptic location for CNS nAchRs is indicated in the striatum, the cerebral cortex, and the hippocampus, since stimulation with nicotinic agonists of nerve ending particles (synaptosomes) or tissue slices from these brain regions (in rat or mouse), evokes release of neurotransmitters.[72-76] This is

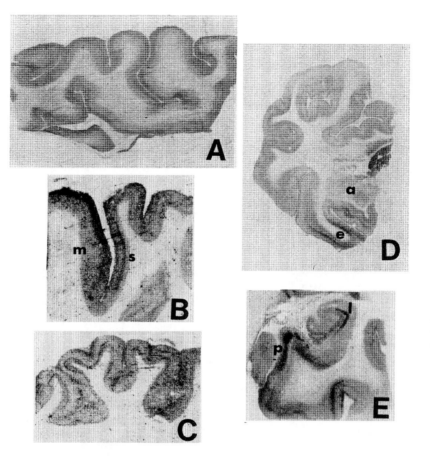

Figure 5-2 [3H]nicotine binding in different regions of the human cortex. A, motor cortex (saggital section); B, enlarged motor cortex (m) and adjacent somatosensory cortex (s); C, visual cortex, Brodmann area 17; D, temporal cortex (coronal section), illustrating high binding in entorhinal cortex (e), but little or no binding in the amygdala (a); E, hippocampus and adjacent parahippocampal gyrus (coronal section), illustrating relatively high binding in the striatum lacunosum-moleculare (l) and presubiculum (p).

supported by the data discussed above that indicate the presence of receptors in the absence of mRNA for the nAchR subunits in the striatum in both rodent and human. In addition, the loss of cortical nAchRs together with cholineacetyltransferase (ChAT) activity in Alzheimer's disease (AD) and other dementing conditions in the elderly (see below) gave rise to the concept that nAchRs in the forebrain were sited presynaptically on cholinergic neurons. However, lesioning studies in experimental animals have largely not supported this view. Although injection of colchicine into the nucleus basalis magnocellularis of the rat (the major forebrain site of cortical cholinergic innervation) resulted in a slight reduction of cortical nAchRs (17%, measured by [³H]acetylcholine binding in the presence of atropine),[77] electrolytic or excitotoxic lesions in the septum (the major source of hippocampal innervation) and the nucleus basalis magnocellularis failed to reduce nicotinic binding in the hippocampus and cortex, respectively, despite dramatically reducing ChAT activity in these areas.[78,79] Presumably only a minority of nAchRs exist presynaptically on basal forebrain cholinergic inputs to the hippocampus and cortex, others being situated on other afferents (noncholinergic or nonbasal forebrain inputs). Since a significant proportion of receptors may in addition be postsynaptic, up-regulation of these after presynaptic lesioning may compensate for presynaptic receptor losses. Indeed, expression of the mRNA coding for the α_4 nicotinic subunit was up-regulated in the rat cerebral cortex by 82% and that for the β_2 subunit by 19% one week after a unilateral nucleus basalis magnocellularis excitotoxic (ibotenic acid) lesion, while no up-regulation of the α_3 subunit was observed.[79]

The possibility that nAchRs in the cerebral cortex exist on afferents other than those from the basal forebrain is substantiated by a recent study in which injections of the excitotoxin kainic acid or electrolytic lesions were employed to destroy thalamocortical fibers projecting to the sensory cortex of rat.[80] A loss of nicotinic binding occurred in the known projection zone of the cortex, together with a loss of μ-opiate receptors, but with no loss of muscarinic binding. There is also evidence that nAchRs are located on axonal terminals of the lateral geniculate nucleus input in the cat visual cortex. Lesions of the lateral geniculate nucleus abolish nicotinic binding concentrated in layer IV of cortical areas 17 and 18, and binding can also be reduced by undercutting the visual cortex.[81]

NAchRs may be sited "preterminally" on the axons of intrinsic GABAergic neurons of the interpeduncular nucleus.[82] Nicotine, and nicotinic agonists at low micromolar concentrations, increased the frequency of postsynaptic GABAergic currents, measured by whole-cell recordings in thin tissue slices. The response was blocked by neuronal nicotinic inhibitors (dihydro-β-erythroidine, hexamethonium, and mecamylamine) and suppressed by tetrodotoxin. There is also some suggestion that in the cerebral cortex of rat and human *in vitro* presynaptic (or even preterminal) nAchRs can control the release of γ-aminobutyric acid (GABA).[83]

Immunohistochemistry employing monoclonal antibodies raised against torpedo nAchR,[84-86] immunoaffinity-purified chick neuronal receptors,[87] and fusion protein constructs for the β_2 subunit[88] indicates that nicotinic receptor-like activity is essentially neuronal and is situated both pre- and postsynaptically. In rat and human neocortex, nAchR-like immunoactivity was associated with cerebral cortical pyramidal neurons and their apical dendrites. Nonpyramidal neurons were also positive, including fusiform cells in layer VI and small, round or oval cells in layers I and II to V. No immunostained axons or presynaptic sites were observed in the cerebral cortex,[84] but numerous postsynaptic sites were present on dendrites (in rat and human).[84,88] Using fluorescent retrograde tracers, Bravo and Karten[87] demonstrated that in the rat the nAchR-bearing pyramidal neurons in layer III project mainly to subcortical targets such as the caudate-putamen, superior colliculus, and pontine nuclei, while very few project to other areas of the cerebral cortex. In addition, Schröder et al.[85] and van Zee[86] suggest that a high proportion of nAchR-like positive neurons in the rat cortex are also positive for mAchR-like activity. Outside the cerebral cortex, immunostaining of dendrites was also observed on cerebellar Purkinje cells, and in the striatum, axon terminals were positive.[88] Thus immunohistochemical studies confirm a presynaptic location for nAchRs in the striatum, whereas in the cerebral cortex receptors are likely to be in part postsynaptic (possibly colocalized with muscarinic receptors).

V. RECEPTOR DISTRIBUTION IN RELATION TO CENTRAL CHOLINERGIC PATHWAYS AND BRAIN FUNCTIONS

Cholinergic neurons and their axonal projections have been most reliably mapped in rodent and primate (including human) brain using antibodies to purified choline acetyltransferase, the enzyme synthesizing acetylcholine, which is (unlike acetylcholinesterase) exclusively localized presynaptically, often in combination with lesioning or retrograde labeling in nonhumans. There are a number of distinct projection systems that appear to be common to all mammalian species.[89-91] These include neuronal groups (Ch1-Ch4) situated in the basal forebrain extending from the septal region rostrally to the subthalamic nucleus caudally. These neuronal groups project topographically to the interpeduncular nucleus, medial habenular nucleus, and ventral tegmentum (Ch1-Ch3); hippocampus (Ch1 and Ch2); olfactory bulb (Ch3); amygdaloid complex (anterolateral portion of Ch4); entire neocortex (remainder of Ch4); and select thalamic nuclei, especially the nucleus reticularis (Ch4). There are, in addition, more circumspect nuclei in the midbrain and brainstem: a small cell group in the habenula projects to the interpeduncular nucleus and a larger complex in the lateral tegmental area [including parts of the medial and lateral parabrachial nuclei and the pedunculopontine tegmental nucleus (designated Ch5 and Ch6)] projects to the superior colliculus, pretectal area, and several thalamic nuclei, especially the medial and dorsolateral geniculate bodies, the intralaminar nuclei, and the anterior and lateral nuclear groups (Ch5 is said to project to all thalamic nuclei). Other descending cholinergic projection target areas from this group include the caudal pontine reticular formation, the raphé nuclei, and the locus ceruleus. Among other areas innervated by ascending projections from Ch5/6 are the substantia nigra and, to a lesser extent, basal forebrain, dorsal and lateral septum, prefrontal cortex, amygdala (medial, basolateral, and central), and lateral hypothalamus. Apart from Ch1-Ch6, further projection systems include cells in the periolivary nuclei innervating the cochlear auditory receptor cells in the organ of Corti, which are probably cholinergic, and the skeletal α and γ motoneurons together with preganglionic autonomic neurons. Local circuit cholinergic neurons occur in the neostriatum (caudate, putamen, and nucleus accumbens), and also in the neo- and archicortex in rodents, but not primates. These various cholinergic systems are distinguishable not only anatomically but also neurochemically by different metabolic enzymes (e.g., intense NADP-H diaphorase immunostaining in Ch5/6), coexisting peptides (e.g., galanin in Ch1, substance P in Ch5/6), and growth factor dependence (e.g., NGF for Ch1-4, but not Ch5-6).

The regional distributions of neither specific nicotinic radioligand binding nor receptor subunit mRNA exactly parallel any one or all of the above pathways. However, the pattern for nicotine binding, concentrated in such areas as thalamus, superior colliculus, and substantia nigra pars compacta, which are also high in α_4 subunit mRNA, is indicative of a particular concentration of nAchRs in neurons immediately postsynaptic to the brainstem Ch5 and Ch6 cholinergic projections. These cholinergic projections, as part of the reticular activating system, are likely to play a major role in arousal and, together with the thalamic control of sensory input, also in selective attention. It is perhaps at this anatomical level that administration of nicotine increases performance in tasks dependent on attention in experimental animals and in normal and demented humans.[2,3,5] The concentration of nicotine binding in neurons of thalamic nuclei receiving input from the brainstem, together with evidence that destruction of thalamic nuclei is followed by a reduction in cortical nicotine binding (especially in visual cortex following lateral geniculate lesions), suggests that nicotinic receptors may be localized in cholinoceptive thalamic neurons, which project to the neocortex. Thus information regarding sensory modalities reaching the cortex via the thalamus is likely to be regulated by the brainstem cholinergic reticular activating system. It is intriguing to note that the onset of REM sleep (associated with a high degree of dreaming) is accompanied by rapid bursts of firing from certain pedunculopontine cholinergic neurons[92] and that the characteristic ponto-geniculo-occipital (PGO) spikes evident in REM sleep may be associated with nAchR activation. Perhaps this provides an explanation for why smokers do not tend to have their sleep interrupted by the urge to smoke!

The high density of nicotine binding and α_4 subunits in substantia nigra pars compacta is compatible with a role for the nAchR in controlling dopaminergic projections to the striatum, which in turn govern the extrapyramidal motor system. Interestingly, nicotine binding in the human midbrain is denser in the A10 than in the A9 subgroup, indicative of the greater responsiveness of the mesocortical compared with mesostriatal dopaminergic systems. This is consistent with the behavioral effects of nicotine associated with reward (e.g., self-adminstration by animals and humans for "pleasure") and the role of the mesolimbic dopaminergic system in reward pathways.

With regard to the cerebral cortex, although the concentration of cholinergic axonal processes in various archicortical (e.g., hippocampus, basolateral amygdaloid nucleus) compared with neocortical areas is not reflected in high nicotine binding density, it is notable that cholinergic axons are generally more dense in the superficial compared with deeper layers of the human neocortex,[91] similar to the general pattern for nicotinic binding, especially in frontal and temporal cortex. In cynomolgus monkeys, frontal cortex[93] choline acetyltransferase is higher in motor compared to premotor and cingulate and lowest in prefrontal cortex, similar to nicotinic autoradiographic distribution in humans (Figure 5-2). The high density of nicotine binding in entorhinal cortex of human and other species (and diminution of this binding in AD[63]) suggests this receptor may be associated with the perforant pathway. Effects of nicotine on learning may be mediated at this locus.

The high density of α_3 subunits in the habenula raises the question of whether a specific nicotinic subtype with distinct pharmacological profile may be associated with the habenular-interpeduncular cholinergic pathway.

In contrast to distribution of nicotine binding and the α_4 subunit, αBT binding and the α_7 subunit are clearly not generally associated with the Ch5/6 group and may by virtue of their concentration in cortex, particularly hippocampus, be primarily associated with the forebrain (Ch1-Ch4, particularly Ch1) cholinergic system. The relatively intense αBT binding in CA2 and CA3 of Ammon's horn and the low level of binding in the subicular complex observed in the human hippocampus parallels the pattern of acetylcholinesterase reactivity.[62] The localization of αBT binding and α_7 in both rat and human suggests this receptor type is not associated with intrinsic cholinergic neurons (identified in rodent but not human), but instead is likely to be postsynaptic to the afferent cholinergic input. There is clearly a need for more detailed and extensive mapping of αBTRs in human brain.

Although comparisons between presynaptic transmitter activities and receptors inevitably reveal mismatch on account of differential pre- and postsynaptic receptor localization, it is of some interest to compare nicotinic and muscarinic subtypes in relation to cholinergic anatomy. Identified mAchR subtypes include the different molecular forms m_1 to m_5, although pharmacological specificity for CNS receptors has only been clearly established for M_1 (highest affinity for pirenzepine). This subtype generally exhibits a rostral-caudal (decreasing) gradient in density with higher densities in the cerebral cortex — a pattern that bears a closer resemblance to αBT than nicotine binding and a predominant association with Ch1-4. However, pirenzepine binding (M_1) is also high in the striatum. The distribution of nicotine binding is more similar to that of non-M_1 sites (previously referred to as M_2 with low affinity to pirenzepine and higher agonist affinity), which are elevated in some subcortical areas such as striatum and brainstem (although not thalamus or substantia nigra) compared with cortex.[60,94]

VI. CHANGES IN NICOTINIC RECEPTOR BINDING IN RESPONSE TO AGONISTS

Chronic exposure to some nicotinic agonists (nicotine, anabasine, and acetylcholine — via acetylcholinesterase blockade) and tobacco smoke causes up-regulation of brain nAchRs in experimental animals and humans.[48,95-100] The mechanism of this response is unknown; it is not produced by all agonists (e.g., lobeline[99]) and does not involve increased levels of mRNAs for α_2, α_3, α_4, α_5, or β_2 subunits.[49] In rats this up-regulation is associated with an increase in α_4, β_2-containing receptors[37] and does not occur uniformly throughout the brain in either rodents or humans. In human brain obtained at autopsy from individuals who had smoked tobacco, the greatest increases in nicotinic binding compared with nonsmokers were observed in the entorhinal cortex and hippocampus. Increases were also evident in the cerebral cortex and cerebellar cortex, but little or no change occurred in the raphé nuclei or medulla oblongata.[95] In mice the subiculum, hippocampal dentate gyrus, frontal cortex, claustrum, hypothalamus, raphé nuclei, and certain nuclei in the pons (including the dorsal tegmental nucleus) appeared the most, and the cerebellum, the septum, and thalamus the least, sensitive to nicotine administration.[47] In view of the species specificity in the distribution of receptor binding and mRNA, similar differences in responsivity to agonists are, as these initial studies suggest, highly likely.

VII. ALTERATIONS IN HUMAN BRAIN RECEPTOR DISTRIBUTION IN PATHOLOGICAL CONDITIONS

The finding that cortical nicotinic (nicotine, acetylcholine, or methylcarbamylcholine) binding is diminished in AD ranks with the loss of choline acetyltransferase as one of the most consistently observed neurotransmitter-related changes in this disease (Table 5-3). Subcortical regions such as striatum and thalamus are generally normal in AD with respect to both the receptor and presynaptic cholinergic activity. This anatomical correlation between the two parameters originally raised the question of whether the receptor loss reflects degeneration of the cortical cholinergic input with a presynaptic receptor localization. However, as mentioned above, lesioning experiments in rodents do not support a major presynaptic cholinergic localization. In contrast, immunocytochemical evidence based on reactivity to the monoclonal antibody WF6 indicates a loss of intrinsic neuronal immunoreactivity in the cortex.[111] Interestingly, one study[107] indicates a more extensive loss in occipital compared to other cortical areas such as temporal, despite greater pathology in the latter. This may reflect the localization of the receptor to thalamic input in primary sensory cortex, and it is likely that alterations in cortical nAchR binding in AD are associated with a variety of pathological processes affecting both cholinergic and noncholinergic input pathways together with intrinsic cholinoceptive neurons. In contrast to high-affinity nicotinic binding, binding of the antagonist αBT is unchanged in AD.[110]

Fewer studies have so far been conducted in PD and Lewy body dementia (LBD) (Table 5-3). These suggest the loss of nicotinic high-affinity agonist binding may be more widespread than in AD, affecting not only the cortex (which in PD is generally free of neuropathology), but also the striatum, the thalamus, the substantia nigra (pars compacta), and the dorsolateral tegmental region adjacent to the substantia nigra, which contains a proportion of the pedunculopontine cholinergic neurons. The loss of nicotine binding in the pars compacta is greater than that of the pigmented dopaminergic neurons in both PD and LBD and equal in both diseases, despite the more extensive neuronal loss in the latter disease — raising the intriguing possibility that receptor attenuation may precede degeneration of the cells. If confirmed, this finding has important implications for the pathogenetic role of the receptor in neurodegeneration of this cell group.

Table 5-3. **Alterations in the distribution of nicotinic receptor (high-affinity site for acetylcholine/nicotine/methylcarbamylcholine) in Alzheimer's, Parkinson's, and Lewy body disease**

Brain region	Alzheimer's disease	Parkinson's disease	Lewy body dementia
Frontal cortex	↓20,61,101,103,104,108,109	↓101,108	
Temporal cortex			
Neocortex	↓61,101-103,105,108	↓101,108	↓63
Hippocampus	↓63,65,101,103,106,108	↓101,106,108	
Entorhinal cortex	↓63		→63
Parietal cortex	↓58,103,61	↓61	↓107
Occipital cortex	↓58,61,103		
Striatum/caudate nucleus	→61,101↓108	↓101,108→61	
Thalamus	→64,101	↓101→64	
Meynert nucleus	→58↓65		
Substantia nigra			
Pars compacta		↓63	↓63
Pars reticulata		→63	→63
Dorsolateral tegmentum		↓63	↓63
Dorsal raphé		→63	→63

Note: Numbers indicate report references. Where available, reports indicate reductions in cortical choline acetyltransferase in AD, PD, and LBD but no alterations in enzyme activity in striatum, thalamus, or globus pallidus in AD and no alteration in striatum in PD and LBD. With respect to other nicotinic ligands, αBT binding is reported normal in AD cortex[20,110] and there is one report that the low-affinity κBT binding site is reduced.[20] Nicotine binding is also reported decreased in hippocampus and parietal cortex in middle-aged Down's syndrome (but not Huntington's disease, motor neuron disease, or alcoholic dementia).

The changes in nAchR distribution in the diseases so far investigated (Table 3) can be interpreted in relation to involvement of the various cholinergic systems in the brain. Thus, in AD there is a selective involvement of the Ch1-Ch4 cholinergic neurons projecting primarily to the cortex (and sparing of cholinergic striatal interneurons and Ch5-6). In contrast, in PD and LBD both the forebrain Ch1-4 and brainstem Ch5-6 are affected — a pathological pattern that is likely to contribute to the differing clinical profiles, especially extrapyramidal movement disorder and also perhaps some of the cognitive symptoms in these disorders compared to AD. The question of whether nicotinic binding is affected in the neostriatum in AD and PD is not clearly answered, given the inconsistency among the few published reports (Table 5-3). One possible explanation for the inconsistency in AD may be the inadvertent inclusion of cases of clinically diagnosed AD who have Lewy bodies in the brainstem (and other areas) and may more properly belong to the LBD category. The latter is distinct from classical AD, where numerous cortical plaques and tangles occur in the absence of Lewy bodies. In PD more detailed investigation of the striatum, including putamen, is required, since the putamen is more affected (by dopamine loss) than the caudate. If nicotinic receptor loss in this area is confirmed — together with the evidence of extensive loss of the receptor in the substantia nigra, which projects to the striatum — the selective localization of striatal nicotinic receptor to the dopaminergic input would be established. If the receptor is unchanged, then the contribution of striatal interneurons will need to be considered — an issue that might further be resolved by examining tissue from patients with Huntington's chorea, which primarily affects striatal interneurons.

The possibility that loss of cortical nAchR binding might be of diagnostic significance in dementia has been explored by Nordberg et al.[112] using positron emission tomography (PET). Following intravenous injections of (*S*)(−) and *R*(+) [^{11}C]nicotine markers, reductions in both enantiomers have been observed in the frontal cortex of patients with AD. It is curious that the more biologically active of the enantiomers is not more affected, and the possibility that the alteration in uptake may reflect changes in blood flow needs to be eliminated. Moreover, in view of parallel changes in the cortical receptor seen in autopsy tissue in AD, PD, and LBD, PET scanning of the nicotinic receptor is unlikely to differentiate among these diseases, although it could be of great value in selecting cases for and analyzing the results of cholinergic therapy. Preliminary PET data indicate that tacrine administration in a patient with AD restored nicotinic binding to normal.[113] If verified, this important finding suggests the nAchR loss in the cortex is reversible in AD and in the disease (and possibly in normal aging — see below) may primarily reflect down-regulation in the absence of normal levels of synaptic transmitter.

VIII. NICOTINIC RECEPTOR IN AGING AND DEVELOPMENT

In normal human aging, loss of brain high-affinity nicotine binding has been observed in the frontal cortex[102,114] and hippocampus.[106] In hippocampus, decreased nicotine binding was strongly correlated with age in 11 normal individuals between 40 and 90 years of age.[106] In the frontal cortex a loss of approximately 50% from 20 to 100 years of age was observed in the absence of any change in ChAT activity over the same age range.[114] NAchRs lost during normal aging are thus not apparently sited on cholinergic afferents, but may be postsynaptic. This is supported by the reported 50% loss of neurons positive for the nicotinic monoclonal antibody WF6 in the human frontal cortex from middle to old age.[111] Alternatively, the receptors lost in aging may be sited on noncholinergic terminals. In experimental animals only minimal loss of basal forebrain cholinergic innervation of the cerebral cortex or hippocampus occurs during aging.[115] However, a reduction of both synthesis and stimulated release of acetylcholine does occur with age,[115] and since exposure to agonists up-regulates brain nAchRs, the dearth of nAchRs in the elderly (and possibly in dementia) may be in response to reduced acetylcholine release causing a failure to maintain receptor numbers.

Alterations in receptor distribution in aging postmaturity are likely to be relevant in the context of neurodegenerative changes and predisposition to diseases (such as AD and PD) in senescence. Of equal importance is the subject of alterations in the pattern of nAchRs during development where transient innervation and neuronal elimination occur. In contrast to the apparent beneficial effects of nicotine in certain degenerative diseases,[2,3,9,10] prenatal exposure to tobacco smoking in humans and to nicotine in experimental animals is considered to be detrimental to normal development of the fetal brain. Navarro et al.[116] have demonstrated that prenatal nicotine adversely affects transient peaks of choline acetyltransferase and high-affinity choline uptake in rat cerebral cortex (but not in brainstem and cerebellum). In fetal human brain the distribution of nicotine binding at 15 weeks gestation is, unlike that of the adult, concentrated in the nucleus of Meynert, globus pallidus, caudate nucleus, and parietal cortex and is less dense in thalamus.[117] In frontal cortex, binding was only measurable at 14 weeks,[118] and a perinatal peak was followed by a steady decline to old age.[114] At 25 to 27 weeks gestation nicotine binding is evident at low levels in the hippocampus and subiculum, is intense in the cerebellar dentate nucleus, and is very intense in the brainstem.[71] A similarly high level of αBT binding was not observed in fetal brainstem or dentate nucleus.[71] Neuronal vulnerability to pre- as opposed to postnatal nicotine may then relate to relatively high nAchR binding in the prenatal period in certain brain areas.

IX. CONCLUSIONS

Nicotine and nicotinic agonists such as acetylcholine, methylcarbamyl choline, and cytisine at nanomolar concentrations bind to a separate group of CNS receptors than α-bungarotoxin, and recent studies have confirmed different molecular structures for these two groups of receptors. High-affinity nicotine receptors are composed of two α and three β subunits, and a variety of α and β subunits occur in mammalian CNS. The various subunits are expressed in distinct distribution patterns, and different subunit combinations result in a range of receptor pharmacological profiles. By contrast, α-bungarotoxin-sensitive ion channels can be formed solely of homomeric α_7 receptors. The levels of high-affinity nicotinic binding and expression of corresponding mRNAs are highest in the thalamus and other midbrain and brainstem structures (e.g., lateral geniculate nucleus, medial habenula, substantia nigra, striatum, and a number of brainstem nuclei) and may be associated with brainstem Ch5 and Ch6 cholinergic projections. Although at generally lower levels, nicotinic binding in the cerebral cortex displays distinctive patterns of distribution, with the greatest intensity in human brain in the subicular complex and entorhinal cortex. Nicotinic receptors in the striatum are likely to be presynaptic, whereas in the cerebral cortex there is evidence for both presynaptic (though few on cholinergic basal forebrain terminals) and postsynaptic locations. α-Bungarotoxin binding is more concentrated in the forebrain, with highest levels in the cerebral cortex, the hippocampus, the hypothalamus, and the inferior colliculus, with little or no binding in the thalamus. There is species variation in the distribution of both nicotine and α-bungarotoxin receptors not only between rodents and primates but also apparently between rats and mice, indicating that it is inappropriate to extrapolate between species.

CNS nicotinic receptors are reduced in neuropathological conditions such as Alzheimer's disease, Parkinson's disease, and Lewy body dementia as well as in normal aging. However, the number of these receptors is up-regulated by exposure to some (but not all) nicotinic agonists and tobacco smoke. Both the loss of nicotinic receptors in pathological conditions and the increase in receptor number in response to certain nicotinic compounds do not occur universally in all brain areas, the neo- and archicortex appearing the most susceptible to these changes. Since nicotinic receptors have been shown to have a role in plasticity and memory tasks involving attention, they may represent a possible route for therapy in dementing conditions of the elderly, especially if such treatment can be directed toward specific receptor subtypes and brain areas.

ACKNOWLEDGMENTS

We are most grateful for the expert secretarial assistance of Maureen Middlemist, to Margaret Piggott for her help in the preparation of the text, and to *Molecular Geriatrics* for their financial support.

REFERENCES

1. Clarke, P. B. S., Nicotine and smoking: a perspective from animal studies, *Psychopharmacology*, 92, 135-143, 1987.
2. Sahakian, B., Jones, G., Levy, R., Gray, J., and Warburton, D., The effects of nicotine on attention, information processing and short term memory in patients with dementia of the Alzheimer type, *Br. J. Psych.*, 154, 797-800, 1989.
3. Jones, G. M. M., Sahakian, B. J., Levy, R., Warburton, D. M., and Gray, J. A., Effects of acute subcutaneous nicotine on attention, information processing and short term memory in Alzheimer's disease, *Psychopharmacology*, 108, 485-494, 1992.
4. Decker, M. W., Majchrzak, M. J., and Anderson, D. J., Effects of nicotine on spacial memoral deficits in rats with septal lesions, *Brain Res.*, 572, 281-285, 1992.
5. Levin, E. D., Nicotinic systems and cognitive function, *Psychopharmacology*, 108, 417-431, 1992.
6. Newhouse, P. A., Potter, A., Corwin, J., and Lenox, R., Acute nicotine blockade produced cognitive impairment in normal humans, *Psychopharmacology*, 108, 480-484, 1992.
7. Janson, A. M., Fuxe, K., Agnati, L. F., Jansson, A., Bjelke, B., Sundström, E., Andersson, K., Häfstrand, A., Goldstein, M., and Owman, C., Protective effects of chronic nicotine treatment on lesioned nigrostriatal dopamine neurons in the male rat, *Prog. Brain Res.*, 79, 257-265, 1989.
8. Lipton, S. A. and Kater, S. B., Neurotransmitter regulation of neuronal outgrowth, plasticity and survival, *Trends Neurosci.*, 12, 265-270, 1989.
9. van Duijn, C. M. and Hofman, A., Relation between nicotine intake and Alzheimer's disease, *Br. Med. J.*, 302, 1491-1494, 1991.
10. Brenner, D. E., Kukull, W. A., van Belle, G., Bowen, J. D., McCormick, W. C., Teri, M. P. H., and Larson, E. B., Relationship between cigarette smoking and Alzheimer's disease in a population-based case-control study, *Neurology*, 43, 293-300, 1993.
11. Heinemann, S., Boulter, J., Connolly, J., Deneris, E., Duvoisin, R., Hartley, M., Hermans-Borgmeyer, I., Hollman, M., O'Shea-Greenfield, A., Papke, R., and Patrick, J., The nicotinic receptor genes, *Clin. Neuropharmacol.*, 14, S45-S61, 1991.
12. Deneris, E. S., Connolly, J., Rogers, S. W., and Duvoisin, R., Pharmacological and functional diversity of neuronal nicotinic acetylcholine receptors, *Trends Pharmacol. Sci.*, 12, 34-40, 1991.
13. Anand, R., Conroy, W. G., Schoepfer, R., Whiting, P., and Lindstrom, J., Neuronal nicotinic acetylcholine receptors expressed in Xenopus oocytes have a pentameric quaternary structure, *J. Biol. Chem.*, 266, 11192-11198, 1991.
14. Clarke, P. B. S., The fall and rise of neuronal α-bungarotoxin binding proteins, *Trends Pharmacol. Sci.*, 13, 407-413, 1992.
15. Wonnacott, S., α-bungarotoxin binds to low-affinity nicotine binding sites in rat brain, *J. Neurochem.*, 47, 1706-1712, 1986.

16. de La Garza, R., Bickford-Wimer, P. C., Hoffer, B. J., and Freedman, R., Heterogeneity of nicotine actions in the rat cerebellum: an *in vivo* electrophysiologic study, *J. Pharmacol. Exp. Ther.*, 240, 689-695, 1987.

17. de la Garza, R., Freedman, R., and Hoffer, B. J., κ-Bungarotoxin blockade of nicotine electrophysiological actions in cerebellar Purkinje neurons, *Neurosci. Lett.*, 99, 95-100, 1989.

18. Mulle, C., Vidal, C., Benoit, P., and Changeux, J.-P., Existence of different subtypes of nicotinic acetylcholine receptors in the rat habenulo-interpeduncular system, *J. Neurosci.*, 11, 2588-2597, 1991.

19. Sargent, P. B., The diversity of neuronal nicotinic acetylcholine receptors, *Annu. Rev. Neurosci.*, 16, 403-443, 1993.

20. Sugaya, K., Giacobini, E., and Chiappinelli, V. A., Nicotinic acetylcholine receptor subtypes in frontal cortex: changes in Alzheimer's disease, *J. Neurosci. Res.*, 27, 349-359, 1990.

21. Morris, B. J., Hicks, A. A., Wisden, W., Darlison, M. G., Hunt, S. P., and Barnard, E. A., Distinct regional expression of nicotinic acetylcholine receptor genes in chick brain, *Mol. Brain Res.*, 7, 305-315, 1990.

22. Role, L. W., Diversity in primary structure and function of neuronal nicotinic acetylcholine receptor channels, *Curr. Opin. Neurobiol.*, 2, 254-264, 1992.

23. Luetje, C. W. and Patrick, J., Both α- and β-subunits contribute to the agonist sensitivity of neuronal nicotine acetylcholine receptors, *J. Neurosci.*, 11, 837-845, 1991.

24. Figl, A., Cohen, B. N., Quick, M. W., Davidson, N., and Lester, H. A., Regions of beta 4, beta 2 subunit chimeras that contribute to the agonist selectivity of neuronal nicotinic receptors, *FEBS Lett.*, 308, 245-248, 1992.

25. Cachelin, A. B. and Jaggi, R., β subunits determine the time course of desensitization in rat alpha 3 neuronal nicotinic acetylcholine receptors, *Pfluegers Arch.*, 419, 579-582, 1991.

26. Polz-Tejera, G., Schmidt, J., and Karten, H. J., Autoradiographic localisation of α-bungarotoxin-binding sites in the central nervous system, *Nature*, 258, 349-351, 1975.

27. Morley, B. J., Lorden, J. F., Brown, G. B., Kemp, G. E., and Bradley, R. J., Regional distribution of nicotinic acetylcholine receptor in rat brain, *Brain Res.*, 134, 161-166, 1977.

28. Alkondon, M. and Albuquerque, E. X., Initial characterization of the nicotinic acetylcholine receptors in rat hippocampal neurons, *J. Receptor Res.*, 11, 1001-1021, 1991.

29. Martino-Barrows, A. M. and Kellar, K. J., [^{3}H]Acetylcholine and [^{3}H](–)nicotine label the same recognition site in rat brain, *Mol. Pharmacol.*, 31, 169-174, 1987.

30. Marks, M. J., Stitzel, J. A., Romm, E., Wehner, J. M., and Collins, A. C., Nicotinic binding sites in rat and mouse brain: comparison of acetylcholine, nicotine and alpha-bungarotoxin, *Mol. Pharmacol.*, 30, 427-436, 1986.

31. Clarke, P. B. S., Schwartz, R. D., Paul, S. M., Pert, C. B., and Pert, A., Nicotinic binding in rat brain: autoradiographic comparison of [^{3}H]acetylcholine, [^{3}H]nicotine, and [^{125}I]αbungarotoxin, *J. Neurosci.*, 5, 1307-1315, 1985.

32. Härfstrand, A., Adem, A., Fuxe, K., Agnati, L., Andersson, K., and Nordberg, A., Distribution of nicotinic cholinergic receptors in the rat tel- and diencephalon: a quantitative receptor autoradiographical study using [^{3}H]-acetylcholine, [α^{125}I]bungarotoxin and [^{3}H]nicotine, *Acta Physiol. Scand.*, 132, 1-14, 1988.

33. Wada, E., Wada, K., Boulter, J., Deneris, E., Heinemann, S., Patrick, J., and Swanson, L. W., Distribution of alpha 2, alpha 3, alpha 4, and beta 2 neuronal nicotinic receptor subunit mRNAs in the central nervous system: a hybridization histochemical study in the rat, *J. Comp. Neurol.*, 284, 314-335, 1989.

34. Schoepfer, R., Whiting, P., Esch, F., Blacher, R., Shimasaki, S., and Lindstrom, J., cDNA clones coding for the structural subunit of a chicken brain nicotinic acetylcholine receptor, *Neuron*, 1, 241-248, 1988.

35. Whiting, P. J., Esch, F., Shimasaki, S., and Lindstrom, J., Neuronal nicotinic acetylcholine receptor β subunit is coded by the cDNA $α_4$, *FEBS Lett.*, 219, 459-463, 1987.

36. Nakayama, H., Nakashima, T., and Kurogochi, Y., $α_4$ is a major acetylcholine binding subunit of cholinergic ligand affinity — purified nicotinic acetylcholine receptor from rat brains, *Neurosci. Lett.*, 121, 122-124, 1991.

37. Flores, C. M., Rogers, S. W., Pabreza, L. A., Wolfe, B. B., and Kellar, K. J., A subtype of nicotinic cholinergic receptor in rat brain is composed of $α_4$ $β_2$ subunits and is upregulated by chronic nicotine treatment, *Mol. Pharmacol.*, 41, 31-37, 1992.

38. Whiting, P., Schoepfer, R., Lindstrom, J., and Priestly, T., Structural and pharmacological character- ization of the major brain nicotinic acetylcholine receptor subtype stably expressed in mouse fibroblasts, *Mol. Pharmacol.*, 40, 463-472, 1991.

39. Connolly, J., Boulter, J., and Heinemann, S. F., $\alpha_4 b_2$ and other nicotinic acetylcholine receptor subtypes as targets of psychoactive and addictive drugs, *Br. J. Pharmacol.*, 105, 657-666, 1992.

40. Afar, R., Clarke, P. B. S., Goldstein, G., and Qurk, M., Thymopoietin, a polypeptide ligand for the α- bungartoxin binding site in brain: an autoradiographic study, *Neuroscience*, 48, 641-653, 1992.

41. Ward, J. M., Cockcroft, V. B., Lunt, G. G., Smillie, F. S., and Wonnacott, S., Methyllycaconitine: a selective probe for neuronal α-bungarotoxin binding sites, *FEBS Lett.*, 270, 45-58, 1990.

42. Couturier, S., Bertrand, D., Matter, J.-M., Hernandez, M.-C., Bertrand, S., Millar, N., Valera, S., Barkas, T., and Ballivet, M., A neuronal nicotinic acetylcholine receptor subunit (α7) is developmen- tally regulated and forms a homo-oligomeric channel blocked by αBT, *Neuron*, 5, 847-856, 1990.

43. Bertrand, D., Bertrand, S., and Ballivet, M., Pharmacological properties of the homomeric α_7 receptor, *Neurosci. Lett.*, 146, 87-90, 1992.

44. Séguéla, P., Wadiche, J., Dineley-Miller, K., Dani, J. A., and Patrick, J. W., Molecular cloning, functional properties, and distribution of rat brain α_7: a nicotinic cation channel highly permeable to calcium, *J. Neurosci.*, 13, 596-604, 1993.

45. Schulz, D. W., Loring, R. H., Aizenman, E., and Zigmond, R. E., Autoradiographic localization of putative nicotinic receptors in the rat brain using ^{125}I-neuronal bungarotoxin, *J. Neurosci.*, 11, 287-297, 1991.

46. Wada, E., McKinnon, D., Heinemann, S., Patrick, J., and Swanson, L. W., The distribution of mRNA encoded by a new family member of the neuronal nicotinic acetylcholine receptor gene family (α_5) in the rat central nervous system, *Brain Res.*, 526, 45-53, 1990.

47. Pauly, J. R., Marks, M. J., Gross, S. D., and Collins, A. C., An autoradiographic analysis of cholinergic receptors in mouse brain after chronic nicotine treatment, *J. Pharmacol. Exp. Ther.*, 258, 1127-1136, 1991.

48. Marks, M. J., Campbell, S. M., Romm, E., and Collins, A. C., Genotype influences the development of tolerance to nicotine in the mouse, *J. Pharmacol. Exp. Ther.*, 259, 392-402, 1991.

49. Marks, M. J., Pauly, J. R., Gross, S. D., Deneris, E. S., Hermans-Borgmeyer, I., Heineman, S. F., and Collins, A. C., Nicotine binding and nicotinic receptor subunit RNA after chronic nicotine treatment, *J. Neurosci.*, 12, 2765-2784, 1992.

50. Schwartz, R. D., Autoradiographic distribution of high affinity muscarinic and nicotinic cholinergic receptors labeled with [3H] acetylcholine in rat brain, *Life Sci.*, 38, 2111-2119, 1986.

51. Yamada, S., Gehlert, D. R., Hawkins, K. N., Nakayama, K., Roeske, W. R., and Yamamura, H. I., Autoradiographic localization of nicotinic receptor binding in rat brain using (3H)methylcarbamylcholine, a novel radioligand, *Life Sci.*, 41, 2851-2861, 1987.

52. Araujo, D. M., Lapchak, P. A., Collier, B., and Quirion, R., N[3H]methylcarbamylcholine binding sites in the rat and human brain: relationship to functional nicotinic autoreceptors and alterations in Alzheimer's disease, *Prog. Brain Res.*, 79, 345-352, 1989.

53. Freedman, R., Wetmore, C., Strömberg, I., Leonard, S., and Olson, L., α-Bungarotoxin binding to hippocampal interneurons: immunocytochemical characterization and effects on growth factor expres- sion, *J. Neurosci.*, 13, 1965-1975, 1993.

54. Dineley-Miller, K. and Patrick, J., Gene transcripts for the nicotinic acetylcholine receptor subunit, beta4 are distributed in multiple areas of the rat central nervous system, *Mol. Brain Res.*, 16, 339-344, 1992.

55. Papke, R. L. and Heinemann, S. F., The role of the β_4-subunit in determining the kinetic properties of rat neuronal nicotinic acetylcholine α_3-receptors, *J. Physiol.*, 440, 95-112, 1991.

56. Duvoisin, R. M., Deneris, E. S., Patrick, J., and Heineman, S., Functional diversity of the neuronal nicotinic acetylcholine receptors is increased by a novel subunit: β_4, *Neuron*, 3, 487-496, 1989.

57. Adem, A., Nordberg, A., Jossan, S. S., Sara, V., and Gillberg, P.-G., Quantitative autoradiography of nicotinic receptors in large cryosections of human brain hemispheres, *Neurosci. Lett.*, 101, 247-252, 1989.

58. Araujo, D. M., Lapchak, P. A., Robitaille, Y., Gauthier, S., and Quirion, R., Differential alteration of various cholinergic markers in cortical and subcortical regions of human brain in Alzheimer's disease, *J. Neurochem.*, 50, 1914-1920, 1988.

59. Nordberg, A., Nilsson-Hakansson, L., Adem, A., Hardy, J., Alufuzoff, I., Lai, Z., Herrera-Marschitz, M., and Winblad, B., The role of nicotinic receptors in the pathophysiology of Alzheimer's disease, *Prog. Brain Res.*, 79, 353-361, 1989.

60. Perry, E. K., Smith, C. J., Perry, R. H., Whitford, C., Johnson, M., and Birdsall, N. J., Regional distribution of muscarinic and nicotinic cholinergic receptor binding in the human brain, *J. Chem. Neuroanat.*, 2, 189-199, 1989.

61. Perry, E. K., Smith, C. J., Perry, R. H., Johnson, M., and Fairbairn, A. F., Nicotinic (^{3}H-nicotine) receptor binding in human brain: characterization and involvement in cholinergic neuropathology, *Neurosci. Res. Commun.*, 5, 117-124, 1989.

62. Perry, E. K., Court, J. A., Johnson, M., Piggott, M. A., and Perry, R. H., Autoradiographic distribution of [^{3}H]nicotine binding in human cortex: relative abundance in subicular complex, *J. Clin. Anat.*, 5, 399-405, 1992.

63. Perry, E. K., Morris, C. M., Court, J. A., Cheng, A., Fairbairn, A. G., McKeith, I. G., Irving, D., Brown, A., and Perry, R. H., Alteration in nicotine binding sites in Parkinson's disease, Lewy body dementia and Alzheimer's disease: an index of early neuropathology, *Neuroscience*, in press.

64. Xuereb, J. H., Perry, E. K., Candy, J. M., Bonham, J. R., Perry, R. H., and Marshall, E., Parameters of cholinergic neurotransmission in the thalamus in Parkinson's disease and Alzheimer's disease, *J. Neurol. Sci.*, 99, 185-197, 1990.

65. Shimohama, S., Tamiguchi, T., Fujiwara, M., and Kameyama, M., Changes in nicotinic and muscarinic cholinergic receptors in Alzheimer-type dementia, *J. Neurochem.*, 46, 288-293, 1986.

66. Cimino, M., Marini, P., Fornasari, D., Cattabeni, F., and Clementi, F., Distribution of nicotinic receptors in cynomolgus monkey brain and ganglia: localization of α_3 subunit mRNA, α-bungarotoxin and nicotine binding sites, *Neuroscience*, 51, 77-86, 1992.

67. Amaral, D. G. and Insausti, R., Hippocampal formation, in *The Human Nervous System*, Paxinos, G., Ed., Academic Press, New York, 1990, 751-756.

68. Insausti, R., Amaral, D. G., and Cowan, W. M., The entorhinal cortex of the monkey. II. Cortical afferents, *J. Comp. Neurol.*, 264, 396-408, 1987.

69. Swanson, L. W., Köhler, C., and Björklund, A., The limbic region. The septohippocampal system, in *Handbook of Chemical Neuroanatomy*, Vol. 3, Part 1, Björklund, A., Hökfelt, T., and Swanson, L. W., Eds., Elsevier, Amsterdam, 1987, 125-277.

70. Rubboli, F., Court, J. A., Sala, C., Morris, C., Chini, B., Perry, E., and Clementi, F., Distribution of nicotinic receptor subtypes in the human brain, *Eur. J. Neurosci.*, in press.

71. Court, J. A., unpublished data, 1993.

72. Rowell, P. P. and Winkler, D. L., Nicotinic stimulation of [^{3}H]acetylcholine release from mouse cerebral cortical synaptosomes, *J. Neurochem.*, 43, 1593-1598, 1984.

73. Araujo, D. M., Lapchak, P. A., Collier, B., and Quirion, R., Characterization of N-[3H]methylcarbamylcholine binding sites and effect of N-methylcarbamylcholine on acetylcholine release in rat brain, *J. Neurochem.*, 51, 292-299, 1988.

74. Rapier, C., Lunt, G. G., and Wonnacott, S., Stereoselective nicotine-induced release of dopamine from striatal synaptosomes: concentration dependence and repetitive stimulation, *J. Neurochem.*, 50, 1123-1130, 1988.

75. Rapier, C., Lunt, G. G., and Wonnacott, S., Nicotinic modulation of [^{3}H]dopamine release from striatal synaptosomes: pharmacological characterisation, *J. Neurochem.*, 54, 937-945, 1990.

76. Grady, S., Marks, M. M., Wonnacott, S., and Collins, A. C., Characterization of nicotinic receptor-mediated [^{3}H]dopamine release from synaptosomes prepared from mouse striatum, *J. Neurochem.*, 59, 848-856, 1992.

77. Tilson, H. A., Schwartz, R. D., Ali, S. F., and McLamb, R. L., Colchicine administration into the area of the nucleus basalis decreases cortical nicotinic cholinergic receptors labeled by [^{3}H]-acetylcholine, *Neuropharmacology*, 28, 855-861, 1989.

78. Smith, C. J., Court, J. A., Keith, A. B., and Perry, E. K., Increases in muscarinic stimulated hydrolysis of inositol phospholipids in rat hippocampus following cholinergic deafferentation are not paralleled by alterations in cholinergic receptor density, *Brain Res.*, 485, 317-324, 1989.

79. Miyai, I., Veno, S., Yorifuiji, S., Fujimura, H., and Tarvi, S., Alterations in neocortical expression of nicotinic acetylcholine receptor in mRNAs following unilateral lesions of the rat nucleus basalis magnocellularis, *J. Neural Transm.*, 82, 79-91, 1990.

80. Sahin, M., Bowen, W. D., and Donoghue, J. P., Location of nicotinic and muscarinic cholinergic and μ-opiate receptors in rat cerebral neocortex: evidence from thalamic and cortical lesions, *Brain Res.*, 579, 135-147, 1992.

81. Prusky, G. T., Shaw, C., and Cynader, M. S., Nicotine receptors are located on lateral geniculate nucleus terminals in cat visual cortex, *Brain Res.*, 412, 131-138, 1987.

82. Léna, C., Changeux, J.-P., and Mulle, C., Evidence of "preterminal" nicotinic receptors on GABAergic axons in the rat interpeduncular nucleus, *J. Neurosci.*, 13, 2680-2688, 1993.

83. Court, J. A., Perry, E. K., and Perry, R. H., Nicotine inhibits the release of [^{3}H]GABA from rat and human cerebral cortex, *in vitro, Neurosci. Res. Commun.*, 7, 9-15, 1990.

84. Schröder, H., Zilles, K., Maelicke, A., and Hajós, F., Immunohisto- and cytochemical localization of cortical nicotinic cholinoceptors in rat and man, *Brain Res.*, 502, 287-295, 1989.

85. Schröder, H., Zilles, K., Luiten, P. G. M., Strosberg, A. D., and Aghchi, A., Human cortical neurons contain both nicotinic and muscarinic acetylcholine receptors: an immunocytochemical double-labeling study, *Synapse*, 4, 319-326, 1989.

86. van der Zee, E. A., Streefland, C., Strosberg, A. D., Schröder, H., and Luiten, P. G. M., Visualization of cholinoceptive neurons in the rat neocortex: colocalization of muscarinic and nicotinic acetylcholine receptors, *Mol. Brain Res.*, 14, 326-336, 1992.

87. Bravo, H. and Karten, H. J., Pyramidal neurons of the rat cerebral cortex, immunoreactive to nicotinic acetylcholine receptors, project mainly to subcortical targets, *J. Comp. Neurol.*, 320, 62-68, 1992.

88. Hill, J. A., Zoli, M., Bourgeois, J.-P., and Changeux, J.-P., Immunocytochemical localization of a neuronal nicotinic receptor: β$_2$-subunit, *J. Neurosci.*, 13, 1551-1568, 1993.

89. Fibiger, H. C. and Vicent, S. R., Anatomy of central cholinergic neurons, in *Psychopharmacology: The Third Generation of Progress*, Meltzer, H. Y., Ed., Raven Press, New York, 1987, 211-218.

90. Wainer, B. H. and Mesulam, M.-M., Ascending cholinergic pathways in the rat brain, in *Brain Cholinergic Systems*, Steriade, M. and Biesold, D., Eds., Oxford University Press, Oxford, 1990, 65-119.

91. Mesulam, M.-M., Hersh, L. B., Mash, D. C., and Geula, C., Differential cholinergic innervation within functional subdivisions of the human cerebral cortex: a choline acetyltransferase study, *J. Comp. Neurol.*, 318, 316-328, 1992.

92. Jones, B. E., Paradoxical sleep and its chemical/structural substrates in the brain, *Neuroscience*, 40, 637-656, 1991.

93. Lewis, D. A., Distribution of choline acetyltransferase-immunoreactive axons in monkey frontal cortex, *Neuroscience*, 40, 363-374, 1991.

94. Cortes, R., Probst, A., and Palacios, J. M., Quantitative light microscopic autoradiographic localization of cholinergic muscarinic receptors in the human brain: forebrain, *Neuroscience*, 20, 65-107, 1987.

95. Benwell, M. E. M., Balfour, D. J. K., and Anderson, J. M., Evidence that tobacco smoking increases the density of (–)-[^{3}H]nicotine binding sites in human brain, *J. Neurochem.*, 50, 1243-1247, 1988.

96. Wonnacott, S., The paradox of nicotinic acetylcholine receptor upregulation by nicotine, *Trends Pharmacol. Sci.*, 11, 216-219, 1990.

97. Collins, A. C., Romm, E., and Wehner, J. M., Dissociation of the apparent relationship between nicotine tolerance and up-regulation of nicotinic receptors, *Brain Res. Bull.*, 25, 373-379, 1990.

98. Bhat, R. V., Turner, S. L., Marks, M. J., and Collins, A. C., Selective changes in sensitivity to cholinergic agonists and receptor changes elicited by continuous physostigmine infusion, *J. Pharmacol. Exp. Ther.*, 255, 187-196, 1990.

99. Bhat, R. V., Turner, S. L., Selvaag, S. R., Marks, M. J., and Collins, A. C., Regulation of brain nicotinic receptors by chronic agonist infusion, *J. Neurochem.*, 56, 1932-1939, 1991.

100. Sanderson, E. M., Drasdo, A. L., McCrea, K., and Wonnacott, S., Upregulation of nicotinic receptors following continuous infusion of nicotine is brain-region-specific, *Brain Res.*, 617, 349-352, 1993.

101. Aubert, I., Araujo, D. M., Cécyre, D., Robitaille, Y., Gauthier, S., and Quirion, R., Comparative alterations of nicotine and muscarinic binding sites in Alzheimer's disease and Parkinson's disease, *J. Neurochem.*, 58, 529-541, 1992.

102. Flynn, D. D. and Mash, D. C., Characterization of L-(^{3}H)nicotine binding in human cerebral cortex: comparison between Alzheimer's disease and the normal, *J. Neurochem.*, 47, 1948-1954, 1986.

103. London, E. D., Ball, M. J., and Waller, S. B., Nicotinic binding sites in cerebral cortex and hippocampus in Alzheimer's dementia, *Neurochem. Res.*, 14, 745-750, 1989.

104. Nordberg, A. and Winblad, B., Reduced number of ^{3}H-nicotine and ^{3}H-acetylcholine binding sites in the frontal cortex of Alzheimer brains, *Neurosci. Lett.*, 72, 115-119, 1986.

105. Nordberg, A., Adem, A., Hardy, J., and Winblad, B., Change in nicotinic receptor subtypes in temporal cortex of Alzheimer brains, *Neurosci. Lett.*, 86, 317-321, 1988.

106. Perry, E. K., Perry, R. H., Smith, C. J., Purohit, D., Bonham, J., Dick, D. J., Candy, J. M., Edwardson, J. A., and Fairbairn, A., Cholinergic receptors in cognitive disorders, *Can. J. Neurol. Sci.*, 13, 521-527, 1986.

107. Perry, E. K., Smith, C. J., Court, J. A., and Perry, R. H., Cholinergic nicotinic and muscarinic receptors in dementia of Alzheimer, Parkinson and Lewy body types, *J. Neural. Transm. (P-D Sect).*, 2, 149-158, 1990.

108. Rinne, J. O., Myllykyla, T., Lonnberg, P., and Marjamaki, P., A postmortem study of brain nicotinic receptors in Parkinson's and Alzheimer's disease, *Brain Res.*, 547, 167-170, 1991.

109. Whitehouse, P. J., Martino, A. M., Antuono, P. G., Lowenstein, P. R., Coyle, J. T., Price, D. L., and Kellar, K. J., Nicotinic acetylcholine binding sites in Alzheimer's disease, *Brain Res.*, 371, 146-151, 1986.

110. Davis, P. and Feisullin, S., Postmortem stability of alpha-bungarotoxin binding sites in mouse and human brain, *Brain Res.*, 216, 449-454, 1981.

111. Schroder, H., Giacobini, E., Struble, R. G., Zilles, K., and Maelicke, A., Nicotinic cholinoceptive neurons in the frontal cortex are reduced in Alzheimer's disease, *Neurobiol. Aging*, 12, 259-262, 1991.

112. Nordberg, A., Hartvig, P., Lilja, A., Viitanen, M., Amberla, K., Lundqvist, H., Ulin, J., Andersson, Y., Langstrom, B., and Winblad, B., Brain nicotinic receptor deficits in Alzheimer patients as studied by positron emission tomography techniques, in *Alzhiemer's Disease: Basic Mechanisms, Diagnosis and Therapeutic Strategies*, Iqbal, K., McLachlan, D. R. C., Winblad, B., and Wisniewski, H., Eds., John Wiley & Sons, New York, 1991, 517-523.

113. Nordberg, A., *In vivo* detection of neurotransmitter changes in Alzheimer's disease, *Ann. N.Y. Acad. Sci.*, 695, 27-33, 1993.

114. Court, J. A., Piggott, M. A., Perry, E. K., Barlow, R. B., and Perry, R. H., Age associated decline in high affinity nicotinic binding in human brain frontal cortex does not correlate with changes in choline acetyltransferase activity, *Neurosci. Res. Commun.*, 10, 125-133, 1992.

115. Decker, M. W., The effects of aging on hippocampal and cortical projections of the forebrain cholinergic system, *Brain Res. Rev.*, 12, 423-438, 1987.

116. Navarro, H. A., Seidler, F. J., Eylers, J. P., Baker, F. E., Dobbins, S. S., Lappi, S. E., and Slotkin, T. A., Effects of prenatal exposure on development of central and peripheral cholinergic neurotransmitter systems. Evidence for cholinergic trophic influences in developing brain, *J. Pharmacol. Exp. Ther.*, 251, 894-900, 1989.

117. Cairns, N. J., and Wonnacott, S., (^{3}H)-nicotine binding sites in fetal human brain, *Brain Res.*, 475, 1-7, 1988.

118. Perry, E. K., Smith, C. J., Atack, J. R., Candy, J. M., Johnson, M., and Perry, R. H., Neocortical cholinergic enzyme and receptor activities in the human fetal brain, *J. Neurochem.*, 47, 1262-1269, 1986.

Chapter 6

Function of Nicotinic Receptors in the CNS

Lynn Wecker and Z. Jian Yu

CONTENTS

I. INTRODUCTION

Nicotine is a highly active pharmacological compound with numerous actions affecting both the peripheral and central nervous systems (CNS). Through interactions within the CNS, nicotine initiates or modulates numerous biochemical and physiological processes including: (1) the release of acetylcholine (ACh) and transduction of its signal; (2) the release of other neurotransmitters including serotonin (5-HT), dopamine (DA), and norepinephrine (NE); (3) neuroendocrine systems and the release of prolactin, vasopressin, adrenocorticotrophic hormone, and corticosterone; and (4) peptide processing and release. (For recent reviews on the neurochemical actions of nicotine, the reader is referred to Nordberg et al.[1]) Nicotine also has been shown to increase cerebral glucose metabolism[2-5] and more recently, the amount and activities of specific isoforms of cytochrome P450 in various brain regions.[6-7] In addition to these neurochemical actions, nicotine affects animal behaviors including: (1) locomotor activity; (2) schedule-controlled behaviors; and (3) attention, information processing, and short-term memory (the reader is referred to recent reviews[8-11] and Chapters 12 and 13 of this volume). Many, if not all, of these actions of nicotine in the CNS are thought to be initiated by the interaction of nicotine with specific receptive sites, collectively termed nicotinic receptors.

In the early 1970s it was thought that two types of nicotinic receptors existed, namely, those present at the skeletal neuromuscular junction with actions that were blocked by the snake venom component α-bungarotoxin (αBTX), and those present at ganglia with actions that were blocked by the classical ganglionic antagonists mecamylamine and hexamethonium. By the mid to late 1980s, it was evident that at least two major families of nicotinic receptors existed in the brain. Studies indicated that the agonist ligands [^{3}H]nicotine and [^{3}H]ACh — and more recently [^{3}H]cytisine and [^{3}H]*N*-methylcarbamylcholine (MCC) — bound with high affinity to one class of sites with regional distribution, kinetic characteristics, and chemical reactivity that differed from the sites that bound [^{125}I]αBTX.[12-18]

During the past decade, it has become evident that there is a great diversity of cholinergic nicotinic receptors. Within the CNS, the term nicotinic receptor now refers to a population of membrane associated proteins composed of α and β subunits. To date, 10 genes have been cloned from neural tissue coding for 7 α (α2-α8) and 3 β (β2-β4) subunits.[19] These receptor subtypes differ with respect to: (1) the subunit association necessary for the reconstitution of functional receptors, i.e., combinations of α and β are necessary for reconstituting functional receptors with α2-α6, whereas α7 and α8 seem to form homooligomeric channels when expressed alone; (2) sensitivity to antagonists and agonists, with receptors containing α subunits 2–6 insensitive to αBTX and those containing α subunits 7 and/or 8 sensitive to blockade by αBTX; and (3) channel and kinetic properties. Studies on the structural and molecular distinctions and diversities among these subtypes, and elucidation of specific channel properties of these

0-8493-7630-0/95/$0.00+$.50

receptors are highly active areas of research, and will not be discussed at length in this chapter. Instead, for excellent treatises on these subjects, the reader is referred to recent reviews dedicated to these topics.[19-21]

This chapter focuses on two perhaps related aspects of nicotinic receptor function in the CNS, namely, neurotransmitter release and regulation of cell survival. Wherever possible the relationships between these functional aspects are elucidated, and attempts are made to correlate these functions with specific receptor subtypes.

II. NEUROTRANSMITTER RELEASE

A. ACUTE STUDIES

Nicotine-binding sites in the brain are located on cholinergic, dopaminergic, noradrenergic, and serotonergic nerve terminals where they function to enhance neurotransmitter release. Initial studies in the CNS arose primarily from findings in the periphery indicating that nicotine evoked powerful sympathomimetic responses through actions at ganglia, at chromaffin cells, and on nerve terminals. Indeed, more than 30 years ago, Burn[22] suggested that "...the pleasure of smoking is derived in part from the release of noradrenaline from its store in the brain." Since that time, numerous studies have focused on elucidating the effects of nicotine on neurotransmitter release.

In the 1970s, studies demonstrated that nicotine increased the release of [^{3}H]NE from brain slices and synaptosomes[23-25] and that this effect was sensitive to blockade by the ganglionic antagonist hexamethonium;[26] the calcium dependency of the response was equivocal.[24,26] Similarly, many studies demonstrated that nicotine increased [^{3}H]DA release from brain preparations, and that this response was blocked by hexamethonium and *d*-tubocurarine, but not tetrodotoxin.[25-27] At about the same time, nicotine was shown to increase the release of ACh[28] as well as 5-HT.[29] During the past decade, there has been an explosion of studies characterizing the effects of nicotine on neurotransmitter release. Evidence supports unequivocally that nicotine and nicotinic agonists interact with presynaptic sites to modulate the release of DA[30-37] and ACh,[33,38-44] with modest evidence supporting modulation of the release of 5-HT[45-47] and equivocal results obtained from studies investigating the release of NE.[47,48]

B. CHRONIC STUDIES

Until recently, studies could not distinguish between the receptors mediating ACh release and those mediating DA release. In general, both responses were sensitive to blockade by ganglionic antagonists such as hexamethonium, mecamylamine, and pempidine, as well as by the nicotinic antagonist dihydro-β-erythroidine, but insensitive to blockade by αBTX.[35-38,41,43] Much insight into possible differences between the receptors mediating ACh release and those mediating DA release has been gained from studies investigating the consequences of chronic agonist administration.

Although is was well known that the chronic administration of nicotine or nicotine agonists increased the density of nicotine binding sites in the brain from both laboratory animals and humans,[49-52] until recently it was unclear whether this increased receptor number was associated with an increased receptor-mediated release of ACh or DA. Studies investigating the agonist-induced release of neurotransmitters, following the chronic administration of nicotine or the nicotinic agonists dimethylphenylpiperazinium (DMPP) and anatoxin-a yielded inconclusive results; evidence for increased, as well as decreased, nicotinic receptor-mediated neurotransmitter release was apparent (Table 6-1). Because these studies differed with respect to agonist administered, species and brain regions studied, and route, dose, and treatment regimen, our laboratory used a treatment regimen that increases the density of nicotine receptors in the brain[52] and determined whether the receptor-mediated release of ACh and DA were differentially affected by the chronic administration of nicotine. Results from these studies (Table 6-2) indicated that chronic nicotine administration consistently and significantly increased the ability of nicotine to stimulate [^{3}H]DA release, but significantly decreased the agonist-induced release of [^{3}H]ACh.[47] These results were in agreement with the findings of Rowell and Wonnacott[56] and Lapchak et al.,[51] respectively, and demonstrated differential agonist responses following the chronic administration of nicotine. These findings underscore the great diversity of both the regulation of function and the subtypes of cholinergic nicotinic receptors in the brain, as is now evident from molecular studies.

Pharmacological and molecular evidence indicates that nicotine-induced DA release is blocked by the snake venom component neuronal bungarotoxin (nBTX) at concentrations that do not affect the binding

Table 6-1 **Effects of chronic nicotinic agonist administration on the stimulus-induced release of ACh and DA**

Neurotransmitter	Stimulus	Treatment Regimen	Preparation	Change	Ref.
DA	DMPP[a]	DMPP 17 mg/kg/d, minipump 14 d	Rat striatal slices	↓	53
	Nicotine	Nicotine 4 mg/kg/h, i.v. 10 d	Mouse striatal P_2	↓	54
	Nicotine	Nicotine 0.35 mg/kg, s.c. 12–15 d	Rat nucleus accumbens	↔	55
	Nicotine	Anatoxin-a 96 nmol/d, minipump 7 d	Rat striatal P_2	↑	56
	Nicotine	Nicotine 0.4 mg/kg, s. c. 2 ×/d, 14 d	Mouse striatum	↔	57
ACh	Nicotine	Nicotine 0.45 mg/kg, s.c. 2 ×/d, 16 d	Rat parietal cortex	↔	58
	MCC[b]	Nicotine 3.6 μmol/kg, s.c., 2 ×/d, 10 d	Rat hippocampal and cortical slices	↓	51

[a] DMPP, dimethylphenylpiperazinium; [b] MCC, *N*-methylcarbamylcholine.

of high-affinity ligands such as [³H]nicotine[35,37] which binds to the $\alpha4\beta2$ subtype.[59-61] These results, in concert with evidence that nBTX blocks the $\alpha3\beta2$ receptor,[62] suggest that the receptor regulating DA release may be of the $\alpha3\beta2$ subtype. Furthermore, a recent study, comparing the potencies of agonists for their ability to stimulate ACh release from hippocampal slices and to induce Rb^+ flux into M10 fibroblasts expressing chicken $\alpha4\beta2$ receptor subunits, suggests that receptors mediating ACh release may be of the $\alpha4\beta2$ subtype.[44] In light of these findings, chronic studies may be interpreted to suggest that persistent agonist administration increases the function and/or expression of $\alpha3\beta2$ receptors, and decreases the function and/or expression of $\alpha4\beta2$ receptors. It is somewhat difficult to reconcile these findings with those indicating that the chronic administration of nicotine increases the density of $\alpha4\beta2$ receptors.[61] Although this apparent inconsistency remains to be resolved, it is evident that the chronic administration of nicotine has differential effects, depending on the specific receptor subtype examined. Furthermore, results from chronic studies suggest that the $\alpha4\beta2$ receptor may have a greater propensity to exhibit desensitization than the $\alpha3\beta2$ receptor. Indeed, distinct alterations in the expression and function of specific receptor subtypes may be induced by the chronic administration of nicotine.

Table 6-2 **Effects of chronic nicotine administration on the nicotine-induced release of [³H]DA and [³H]ACh from rat brain slices**

Neurotransmitter	Brain Region	Change
[³H]DA	Striatum	Concentration-dependent 14–90% increase
[³H]ACh	Striatum	Concentration-dependent 36–77% decrease
	Hippocampus	Concentration-dependent 42–47% decrease

Note: Rats received injections of saline or nicotine bitartrate (1.76 mg/kg, s.c.,) twice daily for 10 d. Brain slices were prepared, preincubated with [³H]DA or [³H]choline, and superfused or incubated with 1–100 μ*M* nicotine. Adapted from Reference 47.

III. REGULATION OF CELL SURVIVAL

A. *IN VIVO* STUDIES

While the function of nicotinic receptors in modulating neurotransmitter release has been studied for nearly 20 years, a relatively new area of research has begun to focus on the possible role of nicotinic receptors in regulating cell survival, growth, and proliferation. During the past several years, studies have sought to determine whether nicotine, through an action at CNS nicotinic receptors, affects neuronal viability. The rationale for these studies was based on the reproducible, but somewhat controversial epidemiological evidence that cigarette smoking exerts a protective effect against the development of Parkinson's disease, and that this effect may be attributed to the ability of nicotine to alter dopaminergic neurotransmission in the brain.[63] Beginning in the mid 1980s, studies investigated whether nicotine administration could prevent the degeneration of nigrostriatal dopaminergic neurons following the administration of the neurotoxin 1-methyl-4-phenyl-1,2,3,6-tetrahydropyridine (MPTP). Unfortunately, results from these studies have been inconclusive, i.e., evidence has supported no effect by nicotine,[64-66] protection against the neurotoxicity of MPTP,[67,68] as well as an exacerbation[69] of the effect of the neurotoxin. These differential results may be attributed to the different doses of neurotoxin used, as well as the different doses and duration of the nicotine treatment regimens. Thus, our laboratory determined whether nicotine, at a dose regimen that increases the density of nicotine receptors in the brain[52] and alters the activity of selective isoforms of cytochrome P450 in brain,[6,7] could protect against the striatal depletion of DA following the unilateral intranigral injection of MPTP. For comparative purposes, we also investigated: (1) the effects of deprenyl pretreatment, which have been shown to protect nigrostriatal neurons from MPTP-induced DA depletion;[70,71] and (2) whether nicotine altered the depletion of striatal DA following the intranigral injection of 6-hydroxydopamine (6-OHDA). Figure 6-1A depicts the effects of pretreatment with nicotine or deprenyl on striatal DA 1 week after the intranigral injection of MPTP. Results indicate that both nicotine and deprenyl significantly attenuated, but did not totally prevent MPTP-induced striatal DA depletion. Furthermore, nicotine was ineffective in attenuating the 6-OHDA-induced depletion of striatal DA (Figure 6-1B). Thus, results indicate that nicotine does have a partial protective effect against MPTP-induced degeneration of nigrostriatal neurons, and that this effect is specific for this neurotoxin.

In addition to studies with MPTP neurotoxicity, evidence supports the idea that nicotine protects nigrostriatal neurons from the consequences of partial mesodiencephalic hemitransections, as assessed by morphological, biochemical, and stereological techniques.[72-74] Although the mechanisms mediating this effect have not been elucidated, Janson and Moller[74] suggested that desensitization of nicotinic receptors on DA neurons as a consequence of chronic agonist exposure may be involved. While this possibility certainly exists, it is not compatible with evidence indicating that chronic nicotine administration increases the responsiveness of striatal DA neurons to an agonist.[47] Furthermore, recent studies have indicated that the neuroprotective effects of nicotine are not selective for dopaminergic neurons, but extend to cholinergic neurons as well. Studies have shown that the chronic administration of nicotine prevents the decreased neuronal density in cerebral cortical layers II, III, and IV following ibotenic acid lesions of the nucleus basalis.[75-76]

B. *IN VITRO* STUDIES

Although the mechanisms mediating the aforementioned effects are not readily apparent, there is direct evidence to suggest that nicotinic receptor stimulation may alter cell survival, growth, and proliferation. A nicotinic receptor-mediated neurotrophic action is supported by studies indicating that carbachol and choline (at concentrations of choline that stimulate nicotinic receptors) prevent the death of sympathetic neurons that ensues when these cells are deprived of nerve growth factor (NGF), an effect that was blocked by mecamylamine, but not by the muscarinic antagonist atropine.[77] However, other studies suggest that stimulation of nicotinic receptors inhibits cell growth. Based on findings that the antagonists *d*-tubocurarine and mecamylamine increased neurite outgrowth from retinal ganglion cells, Lipton et al.[78] suggested that the normal stimulation of nicotinic receptors by ACh exerts an inhibitory effect on the growth of these cells, and that antagonists have a growth-enhancing effect. These studies have been supported and extended by Chan and Quik[79] who demonstrated that nicotine decreased neurite outgrowth of PC12 cells, and that this effect was prevented completely by αBTX at a concentration that totally blocks labeled toxin binding to the cells. Furthermore, because the concentration of nicotine required for

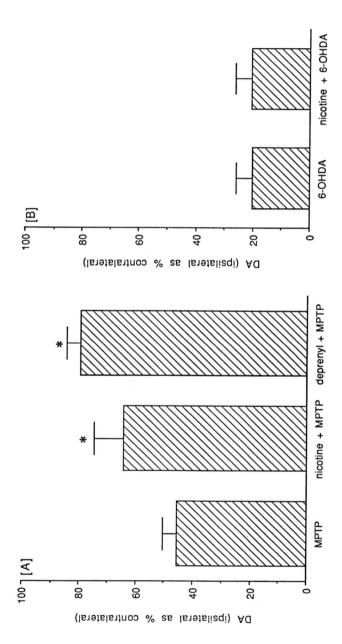

Figure 6-1 Effects of chronic nicotine administration on the neurotoxin-induced depletion of DA in rat striatum. (A) Rats received injections of saline or nicotine bitartrate (1.76 mg/kg, s.c.) twice daily for 10 d prior to and 7 d after the unilateral intranigral injection of MPTP (150 μg). For deprenyl studies, rats received an acute injection of saline or deprenyl (5 mg/kg, i.p.) 1 h prior to the unilateral intranigral injection of MPTP (150 μg). (B) Animals received injections of saline or nicotine bitartrate (1.76 mg/kg, s.c.) twice daily for 10 d prior to and 7 d after the unilateral intranigral injection of 6-OHDA (4 μg). Animals were killed 7d following the administration of the neurotoxins, the striata were isolated, and DA levels were quantitated by high-pressure liquid chromatography (HPLC). Ipsilateral DA levels are expressed relative to contralateral values; contralateral DA levels in all groups did not differ from DA levels determined in naive animals and were 11.4 ± 0.39 μg/g wet weight. Each value is the mean ± SEM of determinations from preparations from 6 to 14 rats per group. *Significantly different from MPTP alone, *p* < 0.05.

inhibition of neurite outgrowth correlated with the affinity of nicotine for the αBTX sensitive receptor, studies suggest that these sites (receptors containing subunits α7 or α8) may inhibit neurite outgrowth.

In sum, studies to date suggest that nicotinic receptors may modulate cell survival, growth, and proliferation. Although the specific effects and receptor subtypes involved have not been unequivocally determined, indirect and direct evidence suggests that both αBTX-sensitive and αBTX-insensitive receptor subtypes may be involved.

C. ROLE OF IMMEDIATE EARLY GENES

While the cellular and molecular mechanisms mediating the ability of nicotine to exert neuroprotective actions are equivocal, evidence has suggested that alterations in the expression of immediate early genes (IEGs) may be involved. In 1986, Greenberg et al.[80] demonstrated that nicotine induced the transcription of both c-*fos* and actin, and increased the level of c-*fos* mRNA in PC12 cells. This effect was selective for these two genes because the expression of other genes including α-tubulin and c-*raf* was unaffected. Furthermore, the effects of nicotine differed from those of both a depolarizing concentration of KCl and NGF, i.e., KCl did not induce actin, and the effect of NGF was not calcium dependent, whereas the nicotine effect required an influx of extracellular calcium through voltage-sensitive calcium channels. The ability of nicotine to increase c-*fos* was blocked by the nicotinic antagonist mecamylamine, suggesting that this effect of nicotine was receptor mediated.

Since these initial findings, numerous investigators have provided evidence supporting the nicotinic receptor-mediated induction of c-*fos* expression in the CNS. Following the acute administration of nicotine to rats, intense *fos*-like immunoreactivity was observed in the midbrain, an effect that was blocked by the administration of mecamylamine.[81] Similarly, nicotine induced the expression of c-*fos*, *jun*-B, c-*jun*, and *jun*-D in the superior cervical ganglion, again responses completely inhibited by both hexamethonium and mecamylamine.[82] Insight into the population of receptors mediating this effect was provided Ren and Sagar[83] who determined that the distribution of c-*fos* positive neurons in central visual pathways following the administration of nicotine corresponded generally to the distribution of nBTX-binding sites that were displaceable by nicotine, but not by αBTX; areas included the medial habenula, the dorsal lateral geniculate nucleus, and the superior colliculus. However, dopaminergic-rich regions of the brain that are known to contain a high density of nicotine binding sites, including the substantia nigra and the ventral tegmental area, failed to exhibit increased staining.

Recently, Sharp et al.[84] demonstrated that the expression of c-*fos* following the administration of nicotine differed markedly among brain regions with respect to dose effect, duration of the maximal response, and desensitization to repeated administration. Furthermore, pretreatment with mecamylamine totally prevented or significantly attenuated the c-*fos* response in all brain regions examined (dentate gyrus, hippocampus, medial habenula, pyriform cortex, and cingulate gyrus) with the exception of the cerebellar cortex. Although the authors could not correlate regional differences in sensitivity with differences in the density of either high-affinity nicotine or αBTX-binding sites, they did determine that the repeated administration of nicotine induced a desensitization in all regions except the medial habenula and locus ceruleus. If the α4β2 receptor has a greater propensity to exhibit desensitization than the α3β2 receptor, findings would suggest that the receptor subtype in the medial habenula and locus ceruleus mediating the nicotine-induced expression of c-*fos* may contain α4 rather than α3 subunits. Thus, although nicotine appears to simulate the expression of IEGs through a receptor-mediated action, the subtype of receptor and possible neurotransmitter involved remains unknown.

IV. CONCLUSIONS

In sum, nicotinic receptors appear to be associated with many types of neurons in the brain where they subserve numerous neurochemical and molecular functions. During the past several years, with advances in molecular techniques and the cloning of receptor subunits, we have learned much about this voltage-gated ion channel family of receptors. It is likely that further studies coupling molecular approaches with functional measures will lead to a greater understanding of the role of each of the receptor subtypes in modulating neuronal function within the CNS.

ACKNOWLEDGMENT

The authors gratefully acknowledge the support of grant no. 0411 from the STRC, Inc.

REFERENCES

1. Nordberg, A., Fuxe, K., Holmstedt, B., and Sundwall, A., Eds., *Nicotinic Receptors in the CNS — Their Role in Synaptic Transmission, Prog. Brain Research*, Vol. 79, Elsevier, Amsterdam, 1989.
2. Grunwald, F., Schrock, H., Theilen, H., Biber, A., and Kuschinsky, W., Local cerebral glucose utilization of the awake rat during chronic administration of nicotine, *Brain Res.*, 456, 350, 1988.
3. London, E. D., Connolly, R. J., Szikszay, M., Wamsley, J. K., and Dam, M., Effects of nicotine on local cerebral glucose utilization in the rat, *J. Neurosci.*, 8, 3920, 1988.
4. London, E. D., Dam, M., and Fanelli, R. J., Nicotine enhances cerebral glucose utilization in central components of the rat visual system, *Brain Res. Bull.*, 20, 381, 1988.
5. London, E. D., Fanelli, R. J., Kimes, A. S., and Moses, R. L., Effects of chronic nicotine on cerebral glucose utilization in the rat, *Brain Res.*, 520, 208, 1990.
6. Anandatheerthavarada, H. K., Williams, J. F., and Wecker, L., The chronic administration of nicotine increases cytochrome P450 in rat brain, *J. Neurochem.*, 60, 1941, 1993.
7. Anandatheerthavarada, H. K., Williams, J. F., and Wecker, L., Differential effect of chronic nicotine administration on brain cytochrome P4501A1/2 and P4502E1, *Biochem. Biophys. Res. Commun.*, 194, 312, 1993.
8. Stolerman, I. P. and Reavill, C., Primary cholinergic and indirect dopaminergic mediation of behavioral effects of nicotine, in *Nicotinic Receptors in the CNS — Their Role in Synaptic Transmission, Prog. Brain Research*, Vol. 79, Nordberg, A., Fuxe, K., Holmstedt, B., and Sundwall, A., Eds., Elsevier, Amsterdam, 1989, 227.
9. Rosecrans, J. A., Stimler, C. A., Hendry, J. S., and Meltzer, L. T., Nicotine-induced tolerance and dependence in rats and mice: studies involving schedule-controlled behavior, in *Nicotinic Receptors in the CNS — Their Role in Synaptic Transmission, Prog. Brain Research*, Vol. 79, Nordberg, A., Fuxe, K., Holmstedt, B., and Sundwall, A., Eds., Elsevier, Amsterdam, 1989, 239.
10. Stolerman, I. P., Behavioral pharmacology of nicotine: implications for multiple brain nicotinic receptors, in *The Biology of Nicotine Dependence*, Bock, G. and Marsh, J., Eds., John Wiley & Sons, Chichester, 1990, 3.
11. Clarke, P. B. S., Mesolimbic dopamine activation — the key to nicotine reinforcement?, in *The Biology of Nicotine Dependence*, Bock, G. and Marsh, J., Eds., John Wiley & Sons, Chichester, 1990, 153.
12. Schwartz, R. D., McGee, R., and Kellar, K. J., Nicotinic cholinergic receptors labeled by [^{3}H]acetylcholine in rat brain, *Mol. Pharmacol.*, 22, 56, 1982.
13. Marks, M. J. and Collins, A. C., Characterization of nicotine binding in mouse brain and comparison with the binding of α-bungarotoxin and quinuclidinyl benzilate, *Mol. Pharmacol.*, 22, 554, 1982.
14. Clarke, P. B. S., Schwartz, R. D., Paul, S. M., Pert, C. B., and Pert, A., Nicotinic binding in rat brain: autoradiographic comparison of [^{3}H]acetylcholine, [^{3}H]nicotine, and [^{125}I]-α-bungarotoxin, *J. Neurosci.*, 5, 1307, 1985.
15. Marks, M. J., Stitzel, J. A., Romm, E., Wehner, J. M., and Collins, A. C., Nicotinic binding sites in rat and mouse brain: comparison of acetylcholine, nicotine, and α-bungarotoxin, *Mol. Pharmacol.*, 30, 427, 1986.
16. Abood, L. G. and Grassi, S., [^{3}H]Methylcarbamylcholine, a new radioligand for studying brain nicotinic receptors, *Biochem. Pharmacol.*, 35, 4199, 1986.
17. Boksa, P. and Quirion, R., [^{3}H]*N*-methyl-carbamylcholine, a new radioligand specific for nicotinic acetylcholine receptors in brain, *Eur. J. Pharmacol.*, 139, 323, 1987.
18. Harfstrand, A., Adem, A., Fuxe, K., Agnati, L., Andersson, K., and Nordberg, A., Distribution of nicotinic cholinergic receptors in the rat tel- and diencephalon: a quantitative receptor autoradiographical study using [^{3}H]-acetylcholine, [α-^{125}I]bungarotoxin and [^{3}H]nicotine, *Acta Physiol. Scand.*, 132, 1, 1988.
19. Heinemann, S., Boulter, J., Connolly, J., Deneris, E., Duvoisin, R., Hartley, M., Hermans-Borgmeyer, I., Hollmann, M., O'Shea-Greenfield, A., Papke, R., Rogers, S., and Patrick, J., The nicotinic receptor genes, *Clin. Neuropharmacol.*, 14 (Suppl. 1), S45, 1991.
20. Papke, R. L., The kinetic properties of neuronal nicotinic receptors: genetic basis of functional diversity, *Prog. Neurobiol.*, 41, 509, 1993.
21. Sargent, P. B., The Diversity of neuronal nicotinic acetylcholine receptors, *Annu. Rev. Neurosci.*, 16, 403, 1993.
22. Burn, J. H., The action of nicotine on the peripheral circulation, *Ann. N.Y. Acad. Sci.*, 90, 81, 1960.

23. Hall, G. H. and Turner, D. M., Effects of nicotine on the release of [^{3}H]-noradrenaline from the hypothalamus, *Biochem. Pharmacol.*, 21, 1829, 1972.

24. Goodman, F. R., Effects of nicotine on distribution and release of [^{14}C]-norepinephrine and [^{14}C]-dopamine in rat brain striatum and hypothalamus slices, *Neuropharmacology*, 13, 1025, 1974.

25. Arqueros, L., Naquira, D., and Zunino, E., Nicotine-induced release of catecholamines from rat hippocampus and striatum, *Biochem. Pharmacol.*, 27, 2667, 1978.

26. Westfall, T. C., Effect of nicotine and other drugs on the release of [^{3}H]-norepinephrine and [^{3}H]-dopamine from rat brain slices, *Neuropharmacology*, 13, 693, 1974.

27. Giorguieff-Chesselet, M. F., Kemel, M. L., Wandscheer, D., and Glowinski, J., Regulation of dopamine release by presynaptic nicotinic receptors in rat striatal slices: effect of nicotine in a low concentration, *Life Sci.*, 25, 1257, 1979.

28. Chiou, C. Y., Long, J. P., Potrepka, R., and Spratt, J. L., The ability of various nicotinic agents to release acetylcholine from synaptic vesicles, *Arch. Int. Pharmacodyn.*, 187, 88, 1970.

29. Hery, F., Bourgoin, S., Hamon, M., Ternaux, J. P., and Glowinski, J., Control of the release of newly synthesized ^{3}H-5-hydroxytryptamine by nicotinic and muscarinic receptors in rat hypothalamic slices, *Naunyn-Schmiedeberg's Arch. Pharmacol.*, 296, 91, 1977.

30. Westfall, T. C., Grant, H., and Perry, H., Release of dopamine and 5-hydroxytryptamine from rat striatal slices following activation of nicotinic cholinergic receptors, *Gen. Pharmacol.*, 14, 321, 1983.

31. Rapier, C., Harrison, R., Lunt, G. G., and Wonnacott, S., Neosurugatoxin blocks nicotinic acetylcholine receptors in the brain, *Neurochem. Int.*, 7, 389, 1985.

32. Rapier, C. M., Harrison, R., Lunt, G. G., and Wonnacott, S., Comparison of the effects of neosurugatoxin and α-bungarotoxin on nicotinic acetylcholine receptors in rat brain, *Biochem. Soc. Trans.*, 13, 1212, 1985.

33. Rapier, C., Wonnacott, S., Lunt, G. G., and Albuquerque, E. X., The neurotoxin histrionicotoxin interacts with the putative ion channel of the nicotinic acetylcholine receptors in the central nervous system, *FEBS Lett.*, 212, 292, 1987.

34. Rapier, C., Lunt, G. G., and Wonnacott, S., Stereoselective nicotine-induced release of dopamine from striatal synaptosomes: concentration dependence and repetitive stimulation, *J. Neurochem.*, 50, 1123, 1988.

35. Schulz, D. W. and Zigmond, R. E., Neuronal bungarotoxin blocks the nicotinic stimulation of endogenous dopamine release from rat striatum, *Neurosci. Lett.*, 98, 310, 1989.

36. Rapier, C., Lunt, G. G., and Wonnacott, S., Nicotinic modulation of [^{3}H]dopamine release from striatal synaptosomes: pharmacological characterisation, *J. Neurochem.*, 54, 937, 1990.

37. Grady, S., Marks, M. J., Wonnacott, S., and Collins, A. C., Characterization of nicotinic receptor-mediated [^{3}H]dopamine release from synaptosomes prepared from mouse striatum, *J. Neurochem.*, 59, 848, 1992.

38. Rowell, P. P. and Winkler, D. L., Nicotinic stimulation of [^{3}H]acetylcholine release from mouse cerebral cortical synaptosomes, *J. Neurochem.*, 43, 1593, 1984.

39. Moss, S. J. and Wonnacott, S., Presynaptic nicotinic autoreceptors in rat hippocampus, *Biochem. Soc. Trans.*, 13, 1985, 1985.

40. Beani, L., Bianchi, C., Nilsson, L., Nordberg, A., Romanelli, L., and Sivilotti, L., The effect of nicotine and cytisine on [^{3}H]-acetylcholine release from cortical slices of guinea-pig brain, *Naunyn-Schmiedeberg's Arch. Pharmacol.*, 331, 293, 1985.

41. Araujo, D. M., Lapchak, P. A., Collier, B., and Quirion, R., Characterization of *N*-[^{3}H]methylcarbamylcholine binding sites and effect of *N*-methylcarbamylcholine on acetylcholine release in rat brain, *J. Neurochem.*, 51, 292, 1988.

42. Toide, K. and Arima, T., Effects of cholinergic drugs on extracellular levels of acetylcholine and choline in rat cortex, hippocampus and striatum studied by brain dialysis, *Eur. J. Pharmacol.*, 173, 133, 1989.

43. Lapchak, P. A., Araujo, D. M., Quirion, R., and Collier, B., Presynaptic cholinergic mechanisms in the rat cerebellum: evidence for nicotinic, but not muscarinic autoreceptors, *J. Neurochem.*, 53, 1843, 1989.

44. Wilkie, G. I., Hutson, P. H., Stephens, M. W., Whiting, P., and Wonnacott, S., Hippocampal nicotinic autoreceptors modulate acetylcholine release, *Biochem. Soc. Trans.*, 21, 429, 1993.

45. Westfall, T. C., Grant, H., Naes, L., and Meldrum, M., The effect of opioid drugs on the release of dopamine and 5-hydroxytryptamine from rat striatum following activation of nicotinic-cholinergic receptors, *Eur. J. Pharmacol.*, 92, 35, 1983.

46. Ribeiro, E. B., Bettiker, R. L., Bogdanov, M., and Wurtman, R. J., Effects of systemic nicotine on serotonin release in rat brain, *Brain Res.*, 621, 311, 1993.
47. Yu, Z. J. and Wecker, L., Chronic nicotine administration differentially affects neurotransmitter release from rat striatal slices, *J. Neurochem.*, 63, 186, 1994.
48. Yoshida, K., Kato, Y., and Imura, H., Nicotine-induced release of noradrenaline from hypothalamic synaptosomes, *Brain Res.*, 182, 361, 1980.
49. Schwartz, R. D. and Kellar, K. J., *In vivo* regulation of [³H]acetylcholine recognition sites in brain by nicotinic cholinergic drugs, *J. Neurochem.*, 45, 427, 1985.
50. Benwell, M. E. M., Balfour, D. J. K., and Anderson, J. M., Evidence that tobacco smoking increases the density of (-)-[³H]nicotine binding sites in human brain, *J. Neurochem.*, 50, 1243, 1988.
51. Lapchak, P. A., Araujo, D. M., Quirion, R., and Collier, B., Effect of chronic nicotine treatment on nicotinic autoreceptor function and N-[³H]methylcarbamylcholine binding sites in the rat brain, *J. Neurochem.*, 52, 483, 1989.
52. Coutcher, J. B., Cawley, G., and Wecker, L., Dietary choline supplementation increases the density of nicotine binding sites in rat brain, *J. Pharmacol. Exp. Ther.*, 262, 1128, 1992.
53. Westfall, T. C. and Perry, H., The nicotinic-induced release of endogenous dopamine from rat striatal slices from animals chronically exposed to dimethylphenylpiperazinium (DMPP), *Neurosci. Lett.*, 71, 340, 1986.
54. Marks, M. J., Grady, S. R., and Collins, A. C., Downregulation of nicotinic receptor function after chronic nicotine infusion, *J. Pharmacol. Exp. Ther.*, 266, 1268, 1993.
55. Damsma, G., Day, J., and Fibiger, H. C., Lack of tolerance to nicotine-induced dopamine release in the nucleus accumbens, *Eur. J. Pharmacol.*, 168, 363, 1989.
56. Rowell, P. P. and Wonnacott, S., Evidence for functional activity of up-regulated nicotine binding sites in rat striatal synaptosomes, *J. Neurochem.*, 55, 2105, 1990.
57. Harsing, L. G., Jr., Sershen, H., and Lajtha, A., Dopamine efflux from striatum after chronic nicotine: evidence for autoreceptor desensitization, *J. Neurochem.*, 59, 48, 1992.
58. Nordberg, A., Romanelli, L., Sundwall, A., Bianchi, C., and Beani, L., Effect of acute and subchronic nicotine treatment on cortical acetylcholine release and on nicotinic receptors in rats and guinea-pigs, *Br. J. Pharmacol.*, 98, 71, 1989.
59. Whiting, P. J., Schoepfer, R., Conroy, W. G., Gore, M. J., Keyser, K. T., Shimasaki, S., Esch, F., and Lindstrom, J. M., Expression of nicotinic acetylcholine receptor subtypes in brain and retina, *Mol. Brain Res.*, 10, 61, 1991.
60. Nakayama, H., Nakashima, T., and Kurogochi, Y., α-4 is a major acetylcholine binding subunit of cholinergic ligand affinity-purified nicotinic acetylcholine receptor from rat brains, *Neurosci. Lett.*, 121, 122, 1991.
61. Flores, C. M., Rogers, S. W., Pabreza, L. A., Wolfe, B. B., and Kellar, K. J., A subtype of nicotinic cholinergic receptor in rat brain is composed of α4 and β2 subunits and is upregulated by chronic nicotine treatment, *Mol. Pharmacol.*, 41, 31, 1992.
62. Luetje, C. W. and Patrick, J., Both α- and β-subunits contribute to the agonist sensitivity of neuronal nicotinic acetylcholine receptors, *J. Neurosci.*, 11, 837, 1991.
63. Baron, J. A., Cigarette smoking and Parkinson's disease, *Neurology*, 36, 1490, 1986.
64. Sershen, H., Hashim, A., and Lajtha, A., Behavioral and biochemical effects of nicotine in an MPTP-induced mouse model of Parkinson's disease, *Pharmacol. Biochem. Behav.*, 28, 299, 1987.
65. Sershen, H., Hashim, A., Wiener, H. L., and Lajtha, A., Effect of chronic oral nicotine on dopaminergic function in the MPTP-treated mouse, *Neurosci. Lett.*, 93, 270, 1988.
66. Carr, L. A. and Basham, J. K., Effects of tobacco smoke constituents on MPTP-induced toxicity and monoamine oxidase activity in the mouse brain, *Life Sci.*, 48, 1173, 1991.
67. Janson, A. M., Fuxe, K., and Goldstein, M., Differential effects of acute and chronic nicotine treatment on MPTP-(1-methyl-4-phenyl-1,2,3,6-tetrahydropyridine) induced degeneration of nigrostriatal dopamine neurons in the black mouse, *Clin. Invest.*, 70, 232, 1992.
68. Janson, A. M., Fuxe, K., Agnati, L. F., Jansson, A., Bjelke, B., Sundstrom, E., Andersson, K., Harfstrand, A., Goldstein, M., and Owman, C., Protective effects of chronic nicotine treatment on lesioned nigrostriatal dopamine neurons in the male rat, in *Nicotinic Receptors in the CNS — Their Role in Synaptic Transmission*, Prog. Brain Research, Vol 79, Nordberg, A., Fuxe, K., Holmstedt, B., and Sundwall, A., Eds., Elsevier, Amsterdam, 1989, 257.
69. Behmand, R. A. and Harik, S. I., Nicotine enhances 1-methyl-4-phenyl-1,2,3,6-tetrahydropyridine neurotoxicity, *J. Neurochem.*, 58, 776, 1992.

70. Fuller, R. W. and Hemrick-Luecke, S. K., Deprenyl protection against striatal dopamine depletion by 1-methyl-4-phenyl-1,2,3,6-tetrahydropyridine in mice, *Res. Commun. Subst. Abuse*, 5, 241, 1984.

71. Heinonen, E. H. and Lammintausta, R., A review of the pharmacology of selegilene, *Acta Neurol. Scand.*, 84 (Supp 136), 44, 1991.

72. Janson, A. M., Fuxe, K., Agnati, L. F., Kitayama, I., Harfstrand, A., Andersson, K., and Goldstein, M., Chronic nicotine treatment counteracts the disappearance of tyrosine-hydroxylase-immunoreactive nerve cell bodies, dendrites and terminals in the mesostriatal dopamine system of the male rat after partial hemitransection, *Brain Res.*, 455, 332, 1988.

73. Fuxe, K., Janson, A. M., Jansson, A., Andersson, K., Eneroth, P., and Agnati, L. F., Chronic nicotine treatment increases dopamine levels and reduces dopamine utilization in substantia nigra and in surviving forebrain dopamine nerve terminals after a partial di-mesencephalic hemitransection, *Naunyn-Schmiedeberg's Arch. Pharmacol.*, 341, 171, 1990.

74. Janson, A. M. and Moller, A., Chronic nicotine treatment counteracts nigral cell loss induced by a partial mesodiencephalic hemitransection: an analysis of the total number and mean volume of neurons and glia in substantia nigra of the male rat, *Neuroscience*, 57, 931, 1993.

75. Sjak-Shie, N. N. and Meyer, E. M., Effects of chronic nicotine and pilocarpine administration on neocortical neuronal density and [^{3}H]GABA uptake in nucleus basalis lesioned rats, *Brain Res.*, 624, 295, 1993.

76. Socci, D. J. and Arendash, G. W., Chronic nicotine prevents neocortical neuronal loss resulting from nucleus basalis lesions, *Soc. Neurosci. Abstr.*, 19, 913, 1993.

77. Koike, T., Martin, D. P., and Johnson, E. M., Jr., Role of Ca^{2+} channels in the ability of membrane depolarization to prevent neuronal death induced by trophic-factor deprivation: evidence that levels of internal Ca^{2+} determine nerve growth factor dependence of sympathetic ganglion cells, *Proc. Natl. Acad. Sci.*, 86, 6421, 1989.

78. Lipton, S. A., Frosch, M. P., Phillips, M. D., Tauck, D. L., and Aizenman, E., Nicotinic antagonists enhance process outgrowth by rat retinal ganglion cells in culture, *Science*, 239, 1293, 1988.

79. Chan, J. and Quik, M., A role for the nicotinic α-bungarotoxin receptor in neurite outgrowth in PC12 cells, *Neuroscience*, 56, 441, 1993.

80. Greenberg, M. E., Ziff, E. B., and Greene, L. A., Stimulation of neuronal acetylcholine receptors induces rapid gene transcription, *Science*, 234, 80, 1986.

81. Pang, Y., Kiba, H., and Jayaraman, A., Acute nicotine injections induce c-*fos* mostly in non-dopaminergic neurons of the midbrain of the rat, *Mol. Brain Res.*, 20, 162, 1993.

82. Koistinaho, J., Pelto-Huikko, M., Sagar, S. M., Dagerlind, A., Roivainen, R., and Hokfelt, T., Differential expression of immediate early genes in the superior cervical ganglion after nicotine treatment, *Neuroscience*, 56, 729, 1993.

83. Ren, T. and Sagar, S. M., Induction of c-*fos* immunostaining in the rat brain after the systemic administration of nicotine, *Brain Res. Bull.*, 29, 589, 1992.

84. Sharp, B. M., Beyer, H. S., McAllen, K. M., Hart, D., and Matta, S. G., Induction and desensitization of the c-Fos mRNA response to nicotine in rat brain, *Mol. Cell. Neurosci.*, 4, 199, 1993.

Chapter 7

Electrophysiological Effects of Acetylcholine on Central Neurons (Whole Cells and Interactions)

James H. Pirch

CONTENTS

I. INTRODUCTION

Electrophysiological effects of acetylcholine on brain neurons have been the subject of a multitude of investigations for more than three decades. Because it is impossible to review the complete electrophysiology of acetylcholine (ACh), this chapter will be limited to recordings of whole neuron responses. Electroencephalogram (EEG) or field potential effects will not be covered unless specifically related to the neuronal responses being discussed. Several reports have considered the role of cholinergic systems in neocortical EEG activation, arousal mechanisms, and the sleep-waking cycle.[1-7] Acetylcholine has been the subject of many *in vitro* studies to characterize its effects on membrane currents in neurons from several brain areas. In addition to Chapter 8 in this volume, there are several excellent reviews and studies regarding channels affected by ACh.[3,8-16] Details of transduction processes and long-term potentiation (LTP) mechanisms involving ACh are subjects of Chapters 9 and 11.

Acetylcholine influences neuronal activity via nicotinic or muscarinic receptors and the result can be (1) enhancement of excitability to other transmitters, (2) depolarization and initiation of action potentials, or (3) inhibition. Neurons in the thalamic relay nuclei, reticular nucleus of the thalamus, somatosensory cortex, and visual cortex have received most attention concerning the actions of ACh. Acetylcholine, acting on the same receptor, may induce facilitation or inhibition, depending on the type of channel associated with the receptor in a particular neuron. An example of this is the production of opposite effects on excitability by ACh acting via muscarinic receptors in thalamic relay and reticular nucleus neurons. ACh depolarizes membrane potentials of relay neurons in the dorsal lateral geniculate nucleus and promotes a shift from bursting to tonic action potential generation by closing a non-G-protein K^+ channel, whereas in reticular nucleus neurons, ACh hyperpolarizes membrane potentials and induces burst firing by opening a G-protein-mediated K^+ channel.[3,17] Although ACh exerts inhibitory actions on certain neurons, the predominant action of ACh in the CNS is enhancement of excitability.[16]

This review will be organized primarily around the *in vivo* effects of ACh on neurons in specific brain areas; *in vitro* studies will be examined where appropriate to ascribe mechanism of action. Detailed descriptions of selected studies will be used to provide an overview of ACh effects on neurons in the area under consideration. The approach will focus on responses to ACh and cholinergic agonists applied directly to neurons via microiontophoresis, responses to drugs applied to *in vitro* slice preparations, and effects of antagonists used to elucidate the receptor(s) involved.

II. THALAMIC AND NUCLEUS RETICULARIS NEURONS

A. THALAMIC RELAY NEURONS

Iontophoretic application of acetylcholine to medial and/or lateral geniculate neurons *in vivo* can result in either increases or decreases in firing rate.[18-22] The excitation may be rapid in onset and offset or delayed in both onset and offset. Both nicotinic and muscarinic receptors are involved in these responses. Neurons in the nucleus reticularis are predominantly inhibited by acetylcholine. Evidence from several laboratories suggests that activation of cholinergic receptors leads to a shift from bursting to single spike firing patterns in thalamic relay neurons.[3]

On relay neurons in dorsal lateral geniculate nucleus (LGNd) of anesthetized cats, the predominant effect of iontophoretically applied acetylcholine is enhancement of the spontaneous discharge, facilitation of excitatory responses to light, suppression of evoked inhibitory episodes, and reduction of postinhibitory rebound bursts.[23] Electrical stimulation of the mesencephalic reticular formation (MRF) produced similar effects. ACh and MRF effects increased during desynchronization of the cortical EEG, suggesting an influence of arousal state on the cholinergic responses. Iontophoretic application of a muscarinic antagonist, scopolamine, consistently blocked the effects of ACh and either blocked or reduced the effects of MRF stimulation. *In vivo* intracellular recordings from cat LGNd relay neurons revealed that iontophoretic ACh suppressed the hyperpolarizing potentials evoked by light or optic chiasm stimulation, produced a slowly developing depolarization shift, and reduced spike afterhyperpolarization.[23]

Marks and Roffwarg[24] examined the influence of cholinergic agonists on spontaneous activity of LGNd relay neurons during waking, slow-wave sleep, and REM sleep in unanesthetized, head-restrained rats. ACh and carbamylcholine facilitated neuronal activity during waking or REM sleep in all neurons.[24] In contrast, no response (13/21), inhibition (5/21), or inhibition followed by facilitation (3/21) were observed during slow-wave sleep. Both facilitatory and inhibitory effects were antagonized by scopolamine. Although the authors state that in the vast majority of relay cells, ACh facilitates activity only in waking and REM sleep, it should be remembered that such facilitatory effects have also been repeatedly observed in anesthetized animals by other investigators.

Hartveit and Heggelund[25] examined the effect of ACh on the visual response of lagged and nonlagged cells in the LGNd of anesthetized cats. Lagged cells are initially suppressed by a flashing light spot, and have a longer latency to the visual response than nonlagged cells. Iontophoretic ACh enhanced the excitatory visual response of both lagged and nonlagged neurons, and generally had little effect on the initial suppression in lagged cells.

Davidowa et al.[26] recorded responses of "slow" and "fast" lateral geniculate neurons to diffuse light stimuli in rats, and examined the effect of iontophoretically applied ACh on these responses. ACh primarily enhanced excitatory responses to light, and also augmented postinhibitory off discharge. These effects were antagonized by iontophoretic atropine, indicating a muscarinic receptor involvement. In the rat, cholinergic fibers are concentrated in regions that contain slow responding neurons, and the influence of ACh was most pronounced on these cells.

Stimulation of the mesopontine peribrachial and laterodorsal cholinergic nuclei induced two types of cholinergically mediated depolarizing responses in relay neurons in the cat anterior thalamic, ventroanterior-ventrolateral, and rostral intralaminar centrolateral nuclei.[27] A short-lasting depolarizing response (latency of 150 msec and duration of 1.3 s) was associated with an increase in membrane conductance and was blocked by intravenous mecamylamine, a nicotinic antagonist. The long-lasting depolarizing response (latency of 1.2 s and duration of 20 s) was associated with an increase in apparent input resistance and was abolished by intravenous scopolamine but not mecamylamine. The prolonged response was also observed in another study in which stimulation of brain stem peribrachial cholinergic afferents produced a slow depolarization and tonic firing in cat ventral lateral nucleus neurons.[28] The depolarization was associated with an increase in membrane resistance. Reactivity to a corticothalamic volley was also increased during the slow depolarization. Thus, acetylcholine appears to exert excitatory influences on a variety of thalamic relay neurons primarily via muscarinic receptors but also via nicotinic receptors.

B. PERIGENICULATE/NUCLEUS RETICULARIS NEURONS

The perigeniculate nucleus (PGN) in the cat represents the visual segment of the thalamic reticular nucleus. Neurons in this nucleus utilize γ-aminobutyric acid (GABA) as a neurotransmitter to inhibit LGNd relay cells. In anesthetized cats or in unanesthetized cats with midpontine pretrigeminal transection, iontophoretically applied ACh consistently depressed spontaneous discharge of nearly all perigeniculate neurons.[23,29,30,31] MRF stimulation also consistently inhibited PGN neurons.[23,30] It appears that ACh enhances excitability of relay cells but reduces excitability of inhibitory interneurons and GABAergic PGN neurons, producing an overall facilitation of excitatory transmission.[23] The inhibitions induced by ACh were antagonized by atropine but not by mecamylamine, bicuculline, or strychnine.[29,30,31] MRF stimulation invoked an early and late phase of inhibition; atropine did not affect the early phase, but antagonized the late phase in 9 of 14 neurons.[30] ACh application and MRF stimulation increased burst firing of nucleus reticularis neurons, whereas iontophoretic atropine tended to suppress burst firing.[30,31]

C. *IN VITRO* STUDIES

In vitro studies have revealed that there are both depolarizing and hyperpolarizing influences of acetylcholine on thalamic neurons. Generally, nicotinic receptors mediate rapid excitatory responses, whereas muscarinic receptors mediate slow excitatory and inhibitory responses.[32] Application of ACh to guinea pig slices *in vitro* caused hyperpolarization in all lateral and medial geniculate neurons followed by slow depolarization in approximately 50% of cells.[32] Methacholine and muscarine mimicked the ACh-induced effect, and scopolamine blocked the responses. The ACh-induced hyperpolarization was mediated by an increase K^+ conductance, and the slow depolarization was associated with a decrease in K^+ conductance. In contrast to the initial hyperpolarization in guinea pig geniculate neurons, application of ACh to cat lateral and medical geniculate neurons caused a rapid depolarization in nearly all cells, followed in some cells by hyperpolarization and/or slow depolarization. The hyperpolarization and slow depolarization in cat geniculate neurons were mediated by muscarinic receptors, whereas the rapid depolarization was proposed to be nicotinic in nature. The rapid excitatory response was associated with an increase in membrane conductance with a reversal potential of –49 to –4 mV, and may have been due to activation of cation conductance. ACh-induced hyperpolarization in thalamic reticular nucleus neurons is blocked by scopolamine but not pirenzepine, indicating the effect is mediated by muscarinic receptors which are not of the M_1 type.[33]

Application of ACh to medial habenula neurons *in vitro* resulted in rapid excitation followed by inhibition; both effects were not abolished by blockade of synaptic transmission.[34] Dimethylphenylpiperazinium (DMPP), a nicotinic agonist, mimicked the ACh effects; and the nicotinic antagonist, hexamethonium, blocked both the excitatory and the inhibitory responses induced by ACh. The responses to ACh were not altered by scopolamine or atropine. ACh or nicotine produced an increase in membrane conductance associated with depolarization, with a reversal potential that indicated that the excitation was due to an increase in membrane cation conductance. The inhibitory response that followed the repetitive firing induced by ACh was associated with hyperpolarization and an increase in membrane conductance. A similar effect could be induced by direct depolarization or glutamate, suggesting that the postexcitatory hyperpolarizing response to ACh may be an endogenous potential not directly coupled to nicotinic receptors. Thus, cholinergic transmission in the medial habenula nucleus may be primarily nicotinic, and this area may be significantly involved in the central actions of nicotine.[34]

D. INTERACTIONS

Among the few studies regarding interactions between ACh and other transmitters on thalamic neurons was that of Funke and Eysel[35] who observed that 5-hydroxytryptamine (5-HT, serotonin) exerted an opposing interaction with ACh on perigeniculate and thalamic reticular neurons in the cat. Iontophoretic application of 5-HT replaced the usual ACh-induced inhibition of tonic firing and production of burst activity with regular tonic activity. It was proposed that the balance between ACh and 5-HT influences on PGN neurons controls the degree of tonic inhibition of LGN relay neurons, thereby inhibiting or facilitating retino-geniculo-cortical transmission. McCormick[3] and Nicoll et al.[8,9] have provided detailed reviews concerning the effects of several neurotransmitters on thalamic relay and reticular nucleus neurons.[3,8]

III. VISUAL CORTEX NEURONS

Several investigators have recorded facilitatory effects of ACh on spontaneous and/or evoked activity of visual cortex neurons.[36–40,42] Sillito and Kemp[38] observed facilitation of visual responses in 61% of cat striate cortex neurons during iontophoretic application of ACh, and depression of responses in 31%. There was a large increase in stimulus-specific responses without any loss in direction or orientation selectivity in 90% of the cases (i.e., ACh increased visual responses to optimal stimuli but not those to nonoptimal stimuli). The ACh-induced facilitation was antagonized with iontophoretic atropine.

In a different study, Sato et al.[39] observed that iontophoretic ACh facilitated responses to visual input and electrical stimulation of lateral geniculate nucleus in 74% of cat striate cortex neurons, whereas suppression of responses was seen in 16% of neurons. ACh effects were antagonized by atropine but not hexamethonium. Atropine suppressed visual responses of cells facilitated by ACh and enhanced responses of cells inhibited by ACh, suggesting a tonic modulation of responsivity by ACh. Facilitation was produced by ACh even in the absence of an increase in spontaneous activity (i.e., improved signal-to-noise ratio). In many facilitated neurons, ACh also increased responses to nonoptimal stimuli, resulting in worsening of orientation selectivity or direction selectivity (54% of facilitated neurons). This observation was contrary to the observation of Sillito and Kemp[38] that ACh tended to produce an increase in orientation and direction selectivity. A similar observation was made by Muller and Singer,[40] who reported that the predominant effect of iontophoretically applied ACh was slow facilitation of neuronal responses to moving light bars, which was sometimes paralleled by a decrease of the cell selectivity to the direction of movement (approximately 29% of cells). An inhibitory effect of ACh on responses to moving light bars was found in 26% of the neurons examined; this inhibitory action was antagonized by either scopolamine or bicuculline (GABA antagonist), indicating that the inhibitory responses were due to stimulation of a GABA interneuron by ACh acting on muscarinic receptors. This is consistent with subsequent studies by McCormick and Prince[41] regarding GABA interneuron mediation of inhibitory responses to ACh in the neocortex. Subsequently, Murphy and Sillito[42] reexamined the responses of cat visual cortex neurons, focusing on the effect of iontophoretic ACh on directional bias. In agreement with previous work from that laboratory, the majority of cells demonstrated either unchanged (47%) or increased (35%) directional selectivity. Of the neurons that were excited by ACh, 33% demonstrated an increase in directional bias and 9% had a decrease. Of neurons that were inhibited by ACh, directional selectivity was increased in 43% and decreased in 24%. The reasons for the conflicting observations of Sillito and Kemp,[38] Murphy and Sillito,[42] Müller and Singer,[40] and Sato et al.[39] were unresolved. ACh may facilitate sending of processed information to other visual centers and may improve the signal-to-noise ratio of information processing. Facilitation of visual cortex neuron responses by ACh appears to be related to alterations in K$^+$ conductance similar to effects observed in other cortical areas, whereas the inhibitory effects may be mediated by GABA interneurons.[41]

Unilateral kainic acid-induced lesions of the nucleus basal magnocellularis in cats resulted in a marked reduction of visual responsivity of neurons in the ipsilateral primary visual cortex.[43] Iontophoretic ACh facilitated responses, and cells that had been unresponsive to visual stimuli became clearly responsive during ACh application. It was suggested that exogenous ACh reversed the impairment in responsiveness caused by interruption of cholinergic innervation to the visual cortex.

A. *IN VITRO* STUDIES

Application of ACh to burst-generating cells in layer V of the guinea pig visual or somatosensory cortex *in vitro* resulted in depolarization and shift in firing pattern from spontaneously bursting to tonic single-spike activity.[44] This ACh effect could be mimicked by acetyl-β-methylcholine and was blocked by scopolamine, indicating that the effect was mediated via a muscarinic receptor. Intracellular recordings revealed that the depolarizing response was associated with a decrease in potassium conductance that consisted of a voltage-independent component and a voltage- and Ca^{2+}-sensitive component, but the currents were not classified.

B. INTERACTIONS

Iontophoretic application of ACh enhanced visually evoked and spontaneous activity of cat striate cortex neurons, and this excitation was reduced or blocked by iontophoretic norepinephrine or dopamine.[45] The catecholamine-induced blockade was of long duration, lasting for several minutes. Additionally, the catecholamines alone inhibited visually evoked responses. These observations would appear to be inconsistent with the observation of Wang and McCormick[44] that norepinephrine and ACh produced similar effects on layer V visual cortex neurons, and that increased activity in ascending noradrenergic and cholinergic pathways during periods of increased attentiveness and arousal may increase the responsiveness of these cells to fast synaptic inputs.[44]

IV. SOMATOSENSORY CORTEX NEURONS

A large number of studies have been conducted concerning the electrophysiological effects of ACh on neurons in the somatosensory cortex. As early as 1963, it was known that iontophoretic application of ACh excited a significant number of these neurons, and that receptors for the excitatory action were muscarinic in nature.[36,46,47] The excitatory effect of iontophoretic ACh on somatosensory neurons is consistently observed, and several studies have indicated that sensitivity to the excitatory effect varies according to depth.[16,36,48–53] Following electrolytic lesions of the nucleus basalis in rats, the percentage of somatosensory cortex neurons excited by iontophoretic ACh and carbachol and their individual sensitivity was higher in lesioned animals.[54]

The excitatory effect of ACh on somatosensory neurons is antagonized by muscarinic antagonists.[47,49,53,55] The M_1 muscarinic receptor antagonist, pirenzepine, significantly attenuated excitatory responses of rat parietal cortex neurons to ACh, carbachol, and McN-A-343 (a selective M_1 receptor agonist).[56] Thus, it appears that M_1 receptors are involved in the muscarinic action of ACh on at least some neocortical neurons.

In 1971, Krnjevic and co-workers[16] reexamined the actions of ACh on pericruciate and suprasylvian neurons in the cat, using extracellular and intracellular recordings. ACh induced a muscarinic receptor-mediated depolarizing effect associated with an increase in membrane resistance, and promoted repetitive firing by slowing repolarization after spikes. The authors concluded that ACh probably acted by reducing the resting K^+ conductance and also the delayed K^+ current of the action potential. They also speculated that the liberation of ACh by ascending cholinergic pathways may change both the general responsiveness and the pattern of firing of small or large groups of cortical neurons.

Several investigators have since examined the effect of ACh on responsiveness of somatosensory cortex neurons. The most common effect of iontophoretic ACh is facilitation of responses to somatic sensory stimuli, although suppression of responses is occasionally observed.[57–60] This action of ACh is consistently blocked by muscarinic antagonists. Certain neurons responded to sensory input only in the presence of ACh.[57,58] In some studies, it was observed that ACh tended to enhance one attribute selectively rather than to act as a general excitant.[58] In many neurons the ACh-induced enhancement of responses to somatic stimuli lasted more than 5 min and sometimes up to an hour.[59] In a related study, cat somatosensory cortex neurons were tested for their responses to skin stimuli before and after pairing of the cutaneous stimulation with stimulation of the basal forebrain cholinergic area.[60] Of the neurons studied, 60% demonstrated enhanced responses to the somatic stimulus after pairing; the enhancement sometimes lasted more than an hour. If atropine was administered iontophoretically while the basal forebrain stimulus was paired with the skin stimulus, the pairing produced no enhancement. On the other hand, atropine was unable to reverse the enhancement once it had developed.

A. *IN VITRO* STUDIES

Krnjevic observed early on that ACh applied to somatosensory neurons induced a muscarinic receptor-mediated depolarizing effect associated with an increase in membrane resistance and slowed repolarization after spikes.[16] It was concluded that ACh reduced resting K^+ conductance and the delayed K^+ current of the action potential. Since then, several investigations have confirmed those observations and have further characterized the ionic responses to ACh.

Consistent with the extracellular iontophoretic studies of Bradshaw et al.,[56] McCormick and Prince[33] found that pirenzepine, an M_1 muscarinic receptor antagonist, blocked ACh-induced slow excitation of pyramidal neurons in the guinea pig sensorimotor, visual, and anterior cingulate cortex. The slow excitation is due to suppression by ACh of the voltage dependent K^+ current (M current) and the Ca^{2+}-activated K^+ current associated with the slow afterhyperpolarization (sAHP). On the other hand, pirenzepine was not effective in blocking the rapid scopolamine-sensitive excitatory response of GABA interneurons to ACh which produced a rapid inhibition of pyramidal neurons. Neither was pirenzepine effective in blocking the scopolamine-sensitive, ACh-induced hyperpolarization of neurons in the reticular nucleus of the thalamus.[33]

In a series of investigations on cat sensorimotor cortex slices, Schwindt et al.[15,61,62] characterized ionic conductances in large layer V neurons and examined the influence of muscarinic agents on the slow conductances. The sAHP had a mean duration of 13.5 s following repetitive 100-Hz spikes, and consisted of two components with time constants of several hundred milliseconds (the early sAHP) and several seconds (the late sAHP).[15,61] The early component was associated with a Ca^{2+}-mediated K^+ conductance, whereas the late component was Ca^{2+}-independent and was subsequently shown to reflect a Na^+-dependent K^+ current.[62] Muscarine reduced or abolished both the early and late sAHPs, and the entire sAHP was replaced with a Ca^{2+}-mediated sADP.[61] Thus, ACh reduced the sAHP directly by reducing the underlying outward K^+ currents and indirectly by causing the sADP. Excitability could be influenced by the currents associated with the sAHP (slower firing rate) and the sADP (continued repetitive firing). The results of these studies were consistent with studies regarding ionic mechanisms associated with ACh-induced excitability changes in hippocampal pyramidal neurons and neocortex.[14,41,63]

B. INTERACTIONS

In the frontoparietal cortex of urethane anesthetized rats, the predominant interaction of norepinephrine, dopamine, and serotonin with ACh was depression of the ACh-induced excitation.[64] An increase in ACh excitation was seen in only 5 of 102 neurons studied. Different results were observed in the somatosensory cortex of halothane anesthetized rats where norepinephrine primarily facilitated excitatory responses to ACh and dopamine suppressed responses.[65] Phenylephrine, an α-receptor agonist, also enhanced ACh excitation, whereas the β receptor agonist, isoproterenol, had no effect or suppressed the ACh response.[66] The α receptor antagonist, phentolamine, blocked the facilitating effect of norepinephrine. In unanesthetized rats, the excitatory response of somatosensory cortex neurons to ACh was depressed by norepinephrine (55%), dopamine (33%), and serotonin (59%), effects more like those seen in the urethane anesthetized animals.[67] Depressant effects of monoamines on excitatory responses to ACh have also been observed in the visual and cingulate cortex, hypothalamus, and thalamus.[45,68,69,81,84]

Various peptides may also modify ACh-induced excitation of somatosensory neurons. Even though cholecystokinin, vasoactive intestinal polypeptide (VIP), angiotensin II, and substance P could all excite somatosensory neurons, the excitation produced by ACh could be depressed by simultaneous administration of the peptide.[70] Substance P was the most potent in this respect, whereas VIP enhanced ACh excitatory responses in some neurons.

A number of studies have revealed an interaction between ACh and glutamate. Krnjevic et al.[16] noted that responses to glutamate were strikingly enhanced in intensity and duration following iontophoretic release of ACh onto pericruciate or suprasylvian neurons in the cat. In other studies, application of ACh to cat somatosensory cortex neurons produced a prolonged enhancement of excitatory responses to glutamate that sometimes lasted as long as an hour.[59] Stimulation of the basal forebrain cholinergic area predominantly facilitated the response of somatosensory neurons to glutamate in a manner similar to the facilitation produced by iontophoretic ACh.[71] Atropine given during the stimulation prevented facilitation of the glutamate response.

V. AUDITORY CORTEX NEURONS

Few laboratories have addressed the actions of ACh in the auditory cortex, and the effects of ACh on auditory cortex neurons appear to be more complex than effects in other cortical areas. In unanesthetized squirrel monkeys, ACh increased spontaneous activity of auditory cortex neurons but did not produce a differential effect on responses to recorded squirrel monkey vocalizations.[72] In other studies, responses of some cat auditory cortex neurons were facilitated and others were inhibited by ACh.[73–75] Atropine was able to inhibit tone-evoked responses, but facilitated responses of a number of auditory neurons, suggesting that endogenous acetylcholine may exert an inhibitory influence on some auditory cortex neurons.[73–75] Auditory cortex neuron responses exhibited long-term facilitation or depression following pairing of iontophoretic acetylcholine with a tone; the facilitatory effect was antagonized by atropine or pirenzepine, suggesting that this effect of ACh was mediated via M_1 receptors.[75] In unanesthetized rats, basal forebrain (BF) stimulation facilitated auditory cortex neuron responses to tones that were paired with stimulation but failed to enhance responses to unpaired tones. The facilitating action of BF stimulation was prevented by pretreatment with atropine, indicating that the effect of BF stimulation was mediated by acetylcholine acting on muscarinic receptors.[76] In a related study in the rat auditory cortex, basal forebrain stimulation produced an atropine-sensitive facilitation of field potentials, single neuron discharges, and monosynaptic excitatory postsynaptic potentials (EPSPs) elicited by stimulation of the medial geniculate.[77] *In vivo* intracellular recordings revealed that BF stimulation depolarized rat auditory cortex neurons, increased the frequency of subthreshold membrane potential fluctuations, and resulted in a shift of spike discharge pattern from phasic to tonic.[78] It appears, then, that the predominant effect of ACh in the auditory cortex is similar to that in the visual and somatosensory cortex where facilitation of responsiveness is observed.

VI. FRONTAL, CINGULATE, AND OLFACTORY CORTEX NEURONS

Iontophoretic application of ACh predominantly enhances the firing rate of neurons in the cingulate cortex, frontal cortex, and olfactory cortex.[12,79,80–82] These excitatory responses are invariably abolished by muscarinic receptor antagonists. Intracellular studies in guinea pig cingulate cortex slices revealed that ACh induced a short latency hyperpolarization followed by a long-lasting depolarization and action potential generation.[41] The inhibitory response was abolished by muscarinic antagonists, by GABA antagonists, or by blockade of synaptic transmission, indicating that the inhibition was caused by muscarinic activation of GABA interneurons. Some nonpyramidal neurons displayed a short-latency excitation of a time course similar to the short-latency inhibition displayed in pyramidal neurons. The long-lasting depolarization was blocked by muscarinic antagonists including the M_1 receptor selective antagonist, pirenzepine, which is poorly effective in blocking the inhibitory response.[83] The long-duration excitatory response was associated with ACh blockade of the voltage-dependent M current (I_M) and the Ca^{2+}-activated K^+ conductance underlying the slow afterhyperpolarization (I_{AHP}). In guinea pig olfactory cortex neurons *in vitro*, muscarinic agonists inhibit I_{AHP} at lower concentrations than are required to inhibit I_M; and it was suggested that cholinergic modulation of I_{AHP} may be highly significant in controlling cell excitability in cortical neurons.[12]

Iontophoretically applied norepinephrine can increase or decrease the excitatory response of rat cingulate cortex neurons to ACh.[84] Enhancement was concurrent with a norepinephrine-induced increase in spontaneous activity, whereas decreases in ACh response were associated with norepinephrine inhibition of firing rate. Methionine-enkephalin inhibited ACh-evoked excitation in nine of ten rat frontal cortex neurons in which the interaction was tested.[80] Dopaminergic modulation of cholinergic responses in the rat medial prefrontal cortex has been studied in detail.[81] Application of dopamine (DA) at low iontophoretic current decreased baseline spontaneous firing in prefrontal neurons and suppressed ACh-evoked activity in many cells. However, at these currents spontaneous activity was suppressed more than the ACh-evoked activity, resulting in an increase in the ratio of the ACh-evoked response (signal) to baseline spontaneous discharges (noise). Viewing the data in terms of signal-to-noise ratio is controversial and may be misleading. Ten units had an increase in ACh-evoked response accompanied by a suppression of spontaneous activity by DA (called potentiation). In 13 units spontaneous activity was suppressed more than ACh-evoked activity (called enhancement). DA suppressed the ACh-evoked response in 10 units more than it suppressed spontaneous activity. Viewed in a different manner, the ACh-evoked response was increased in 10 neurons and decreased in 23 neurons. The suppression of ACh-evoked responses by higher DA currents was frequently of long duration (>15 min after termination of DA application). Additional analysis indicated that activation of D_1 receptors suppressed ACh excitatory responses, whereas D_2 receptors enhanced ACh responses.[81]

VII. HIPPOCAMPAL AND SEPTAL NEURONS

Application of ACh to hippocampal pyramidal neurons elicits a slow excitation mediated by muscarinic receptors.[63,85–89] Experiments to characterize the slow cholinergic postsynaptic potential and the pharmacology of cholinergic excitatory responses were performed *in vitro* on hippocampal slices by Cole and Nicoll.[85,86] ACh depolarized hippocampal CA1 pyramidal cells with an associated increased input resistance, blocked a calcium-activated potassium conductance (I_{AHP}), and blocked accommodation of action potential discharge; these effects were enhanced by physostigmine and antagonized by atropine, but were unaffected by nicotinic receptor antagonists. Electrical stimulation of the stratum oriens induced an initial fast EPSP, a subsequent inhibitory postsynaptic potential (IPSP), and a slow EPSP that lasted for 20 to 30 s. The slow EPSP, which resembled the EPSP produced by iontophoretic ACh, was enhanced by physostigmine and blocked by atropine. Voltage clamp analysis of cholinergic action in the hippocampus revealed that the calcium-activated potassium current associated with the afterhyperpolarization,

I_{AHP}, was tenfold more sensitive to the muscarinic agonist, carbachol, than the voltage-dependent K^+ current, I_M.[14] Furthermore, the slow EPSP appeared to be generated by blockade of a voltage-independent, resting potassium current rather than I_{AHP} or I_M. It was proposed that ACh blockade of I_{AHP} is involved in facilitation of action potential discharge to depolarizing stimuli, whereas the slow EPSP directly causes action potential discharge.[14] The muscarinic receptor subtypes involved in these responses is unclear.[9]

Acetylcholine facilitated slow depolarizing responses to *N*-methyl-D-aspartate in rat CA1 hippocampal cells.[90] ACh-induced excitation of hippocampal CA1 and CA3 pyramidal cells was facilitated by iontophoretic application of somatostatin-14 or by intraperitoneal injection of ethanol.[91,92] The mechanisms of these potentiating effects are unknown.

Identified septo-hippocampal neurons in the medial septum and nucleus of the diagonal band of Broca were excited by iontophoretic application of ACh or cholinergic agonists.[93] The ACh-induced excitation of these presumably cholinergic neurons was readily antagonized by atropine. If these were cholinergic neurons, then it is unlikely that the effect of ACh represents feedback regulation of their activity.[93] Instead, ACh released by terminals of cholinergic neurons in the pedunculopontine nucleus and/or dorsolateral tegmental nucleus may serve to regulate excitability of septo-hippocampal neurons.[94,95]

VIII. ACETYLCHOLINE AND LEARNING-RELATED NEURONAL ACTIVITY

It seems clear that ACh released from terminals of BF cholinergic neurons plays an important role in memory and cognition.[96,97] Neurons in the globus pallidus, basal nucleus of Meynert, or substantia innominata of primates demonstrate altered discharge rates during certain portions of behavioral tasks.[98–103] A large proportion of these neurons have altered firing rates during phases of the task or stimuli that precede reward or reinforcement.[103] Changes in activity of rat BF neurons in response to tone or light stimuli preceding MFB stimulation have been observed in our laboratory.[104,105] Some of our experiments involved visual discrimination in which light applied to one eye was followed by MFB stimulation while light to the other was not. BF neurons exhibited discriminatory responses between light to the eye associated with MFB stimulation and light to the other eye.[105] The responses of BF neurons were not modality specific and did not depend on physical properties of the sensory stimulus. On the other hand, the responses did reflect learning of the association between sensory stimuli and availability or delivery of reinforcement. It has been suggested that activity of BF cholinergic neurons under these conditions may facilitate the neural changes in other brain areas that underlie conditioned responses[101] or associative learning.[100] BF lesions alter electrophysiological correlates of conditioning in the neocortex.[104,106–107] The observations above suggest that learning-related neuronal responses in areas that receive BF cholinergic innervation are mediated or influenced by ACh.

Considerable evidence reviewed above supports a role for ACh in modulation of neuronal responses to sensory input or other transmitters such as glutamate. Such ACh-induced alterations of responsiveness to synaptic input may be involved in fundamental mechanisms of plasticity associated with learning and memory. Basically, two types of effect of iontophoretic application of ACh have been observed: a transient alteration of response which subsides rapidly following termination of application and a long-term enhancement, the development of which is induced by simultaneous application of ACh with stimulation of input or application of glutamate. The prolonged enhancement sometimes outlasts the application of ACh for periods of an hour or more. Evoked responses to primary sensory input were predominantly facilitated by ACh in the visual[38,39] and somatosensory cortex.[57–60] Glutamate-induced excitation of somatosensory cortex neurons was also enhanced by ACh.[59] The effects of ACh on auditory cortex neurons is more complex, with some responses being facilitated and others inhibited.[73–75] Atropine was able to inhibit visually evoked responses in visual cortex[39] and tone-evoked responses in auditory cortex.[74] Lesions of basal forebrain cholinergic neurons reduced responsiveness of visual cortex neurons to visual stimuli.[43] The results with atropine and basal forebrain lesions suggest that endogenous ACh normally may be directly involved in cortical neuron excitatory responses. Acetylcholine promoted long-term facilitation of responsiveness of visual and somatosensory cortex neurons to sensory input or to iontophoretic glutamate.[38,59,60,108] Auditory cortex neurons exhibited predominantly facilitation of tone-evoked responses following pairing of a tone with iontophoretic ACh or basal forebrain stimulation.[75] Additionally, basal forebrain stimulation paired with a somatic stimulus produced long-term enhancement of the response of somatosensory cortex neurons to the somatic stimulus.[60] Early evidence that cholinergic mechanisms may be involved in cortical neuron plasticity associated with learning was provided by the studies of Woody[109] who demonstrated that eye-blink conditioning and iontophoretically-applied ACh induced similar increases in excitability of pyramidal tract neurons in the cat.

Several investigators have examined the effect of ACh and/or muscarinic receptor antagonists on learning-related neuronal responses. The predominant role of ACh is facilitation of excitatory learning-related neuronal responses. In the primate orbitofrontal and dorsolateral prefrontal cortex, ACh-sensitive neurons exhibited excitatory responses associated with various phases of a complex bar-press feeding task that were altered by iontophoretic administration of ACh or atropine.[110,111] Continuous application of ACh enhanced excitatory responses to task events and to cue lights or tones, whereas atropine diminished such responses. Stimulation of the basal nucleus of Meynert produced an excitatory response in the same cortical neurons, and this excitation was blocked by iontophoretic application of atropine. Application of ACh to monkey amygdala neurons during cue signaled conditioned bar-press feeding enhanced task-related excitatory responses, or attenuated inhibitory responses.[112] Atropine evoked or enhanced inhibitory response patterns. In a series of experiments in awake rats that employed sensory-sensory conditioning procedures, pairing of two vibrissal stimulations produced an excitatory response of barrel field cortex neurons to a stimulus that was ineffective before pairing and modified preexisting responses by eliciting the appearance of long-latency excitatory components.[113] Iontophoretic application of atropine abolished these "conditioned" responses, whereas ACh facilitated an enlargement of the receptive field and induced a more sustained discharge to vibrissal stimulation. Rat lateral hypothalamus neurons that exhibited excitatory responses to tone cues preceding electric shock or tail pinch were also excited by iontophoretic ACh.[114] Atropine blocked the excitatory neuronal responses to the tone cue as well as excitatory responses to iontophoretic ACh. Hypothalamic neurons may receive their cholinergic input partly from the pedunculopontine nucleus as well as from basal forebrain cholinergic neurons.[94,95]

In our laboratory, rat frontal cortex neurons displayed discriminative responses to tones during a classical associative conditioning procedure in which medial forebrain bundle (MFB) stimulation served as the unconditioned stimulus (UCS) and tones of different frequencies were utilized as conditioned stimuli (CS).[82,115] One tone (CS+) was paired with the UCS, whereas the other tone was unpaired. When a cortical neuron that exhibited a discriminative response to the paired tone was isolated, the effects of iontophoretically applied ACh on the rate of spontaneous discharge and the effect of a muscarinic receptor antagonist applied iontophoretically on the conditioning-related response were examined. If the neurons were classified according to responses to CS+ and ACh as well as the effect of the antagonist drugs on these responses, the largest and most clearly defined group of neurons constituted 46% of the population. These were cells with an excitatory response to CS+ and an excitatory response to ACh, both of which were significantly attenuated during iontophoretic application of the muscarinic antagonist (atropine or tropicamide).[82] The simplest interpretation is that ACh, acting on muscarinic receptors, was directly involved in the conditioned response of the frontal cortex neurons. It could not be determined definitively whether the source of ACh is extrinsic from neurons with cell bodies in the basal forebrain or neurons intrinsic to the frontal cortex. However, the observations that many nucleus basalis and substantia innominata neurons in the rat exhibited excitatory responses to tone- or light-conditioned stimuli would be consistent with a significant contribution from basal forebrain cholinergic neurons.[104,105] Furthermore, the fact that suppression of basal forebrain activity by microinjection of GABA or procaine attenuated the conditioned response of cortical units that were excited by a tone CS provides strong evidence of an extrinsic influence from the basal forebrain which is very likely to involve cholinergic neurons.[116] That a unilateral kainic acid-induced lesion of the rat nucleus basalis area reduced the percentage of ipsilateral cortical units which demonstrated an excitatory response to a tone CS to approximately one third that found in the contralateral cortex, provides additional evidence consistent with a role of extrinsic cholinergic input in neocortical neuron excitatory responses to conditioned stimuli.[104,107] In the presence of iontophoretically applied acetylcholine, frontal cortex neurons that were previously unresponsive exhibited selective excitatory responses to CS+; the responses subsided when application of acetylcholine was terminated.[117] The fact that iontophoretically applied atropine attenuated conditioned responses that were already established indicates that acetylcholine is significantly involved in *maintenance* of the conditioning-related responses of neurons in the cortex, amygdala, and hypothalamus.[82,110–114] The suppression of already established responses of frontal cortex neurons to conditioned tone stimuli by microinjection of GABA or procaine into the nucleus basalis area provides additional indirect evidence of an *active* role for cholinergic basal forebrain neurons in maintaining cortical responses to conditioned stimuli.[107,116]

REFERENCES

1. Shute, C. C. D. and Lewis, P.R., The ascending cholinergic reticular system: neocortical, olfactory and subcortical projections, *Brain*, 90, 497, 1967.
2. McCormick, D. A., Cellular mechanisms of cholinergic control of neocortical and thalamic neuronal excitability, in *Brain Cholinergic Systems*, Steriade, M. and Biesold, D., Eds., Oxford University Press, New York, 1990, 236.
3. McCormick, D. A., Neurotransmitter actions in the thalamus and cerebral cortex and their role in neuromodulation of thalamocortical activity, *Prog. Neurobiol.*, 39, 337, 1992.
4. Steriade, M., Datta, S., Paré, D., Oakson, G., and Curró Dossi, R., Neuronal activities in brain-stem cholinergic nuclei related to tonic activation processes in thalamocortical systems, *J. Neurosci.*, 10, 2541, 1990.
5. Stewart, D. J., MacFabe, D. F., and Vanderwolf, C.H., Cholinergic activation of the electrocorticogram: role of substantia innominata and effects of atropine and quinuclidinyl benzilate, *Brain Res.*, 322, 219, 1984.
6. Buzsaki, G., Bickford, R. G., Ponomareff, G., Thal, L. J., Mandel, R., and Gage, F.H., Nucleus basalis and thalamic control of neocortical activity in the freely moving rat, *J. Neurosci.*, 8, 4007, 1988.
7. Szymusiak, R. and McGinty, D., Sleep-related neuronal discharge in the basal forebrain of cats, *Brain Res.*, 370, 82, 1986.
8. Nicoll, R. A., The coupling of neurotransmitter receptors to ion channels in the brain, *Science*, 241, 545, 1988.
9. Nicoll, R. A., Malenka, R. C., and Kauer, J. A., Functional comparison of neurotransmitter receptor subtypes in mammalian central nervous system, *Physiol. Rev.*, 70, 513, 1990.
10. Brown, D. A., Slow cholinergic excitation — a mechanism for increasing neuronal excitability, *TINS*, 6, 302, 1983.
11. Constanti, A. and Galvan, M., M-current in voltage-clamped olfactory cortex neurones, *Neurosci. Lett.*, 39, 65, 1983.
12. Constanti, A. and Sim, J. A., Calcium-dependent potassium conductance in guinea-pig olfactory cortex neurones *in vitro*, *J. Physiol.*, 387, 173, 1987.
13. Halliwell, J. V., M-current in human neocortical neurones, *Neurosci. Lett.*, 67, 1, 1986.
14. Madison, D. V., Lancaster, B., and Nicoll, R. A., Voltage clamp analysis of cholinergic action in the hippocampus, *J. Neurosci.*, 7, 733, 1987.
15. Schwindt, P. D., Spain, W. J., Foehring, R. C., Chubb, M. C., and Crill, W. E., Slow conductance changes in neurons from cat sensorimotor cortex *in vitro* and their role in slow excitability changes, *J. Neurophysiol.*, 59, 450, 1988.
16. Krnjevic, K., Pumain, R., and Renaud, L., The mechanism of excitation by acetylcholine in the cerebral cortex, *J. Physiol.*, 215, 447, 1971.
17. McCormick, D. A. and Prince, D. A., Acetylcholine induces burst firing in thalamic reticular neurones by activating a potassium conductance, *Nature (London)*, 319, 402, 1986.
18. Phillis, J. W., The pharmacology of thalamic and geniculate neurons, *Int. Rev. Neurobiol.*, 14, 1, 1971.
19. Tebecis, A. K., Cholinergic and non-cholinergic transmission in the medial geniculate nucleus of the cat, *J. Physiol.*, 226, 153, 1972.
20. Krnjevic, K., Chemical nature of synaptic transmission in vertebrates, *Physiol. Rev.*, 54, 418, 1974.
21. Sillito, A. M., Kemp, J. A., and Berardi, N., The cholinergic influence on the function of the cat dorsal lateral geniculate nucleus (dLGN), *Brain Res.*, 280, 299, 1983.
22. Eysel, U. T., Pape, H.-C., and Van Schayck, R., Excitatory and differential actions of acetylcholine in the lateral geniculate nucleus of the cat, *J. Physiol.*, 370, 233, 1986.
23. Francesconi, W., Muller, C.M., and Singer, W., Cholinergic mechanisms in the reticular control of transmission in the cat lateral geniculate nucleus, *J. Neurophysiol.*, 59, 1690, 1988.
24. Marks, G. A. and Roffwarg, H. P., The cholinergic influence upon rat dorsal lateral geniculate nucleus is dependent on state of arousal, *Brain Res.*, 494, 294, 1989.
25. Hartveit, E. and Heggelund, P., The effect of acetylcholine on the visual response of lagged cells in the cat dorsal lateral geniculate nucleus, *Exp. Brain Res.*, 95, 443, 1993.
26. Davidowa, H., Albrecht, D., Gabriel, H.-J., and Zippel, U., SLOW and FAST lateral geniculate neurons are differently influenced by acetylcholine, *Brain Res. Bull.*, 31, 455, 1993.

27. Curró Dossi, R., Paré, D., and Steriade, M., Short-lasting nicotinic and long-lasting muscarinic depolarizing responses of thalamocortical neurons to stimulation of mesopontine cholinergic nuclei, *J. Neurophysiol.*, 65, 393, 1991.

28. Deschênes, M. and Hu, B., Membrane resistance increase induced in thalamic neurons by stimulation of brainstem cholinergic afferents, *Brain Res.*, 513, 339, 1990.

29. Godfraind, J. M., Acetylcholine and somatically evoked inhibition on perigeniculate neurones in the cat, *Br. J. Pharmacol.*, 63, 295, 1978.

30. Dingledine, R. and Kelly, J. S., Brain stem stimulation and the acetylcholine-evoked inhibition of neurones in the feline nucleus reticularis thalami, *J. Physiol.*, 271, 135, 1977.

31. Ben Ari, Y., Dingledine, R., Kanazawa, I., and Kelly, J.S., Inhibitory effects of acetylcholine on neurones in the feline nucleus reticularis thalami, *J. Physiol.*, 261, 647, 1976.

32. McCormick, D. A. and Prince, D. A., Actions of acetylcholine in the guinea-pig and cat medial and lateral geniculate nuclei, *in vitro, J. Physiol.*, 392, 147, 1987.

33. McCormick, D. A. and Prince, D.A., Pirenzepine discriminates among ionic responses to acetylcholine in guinea-pig cerebral cortex and reticular nucleus of thalamus, *TIPS*, Feb. Suppl., 72, 1986.

34. McCormick, D. A. and Prince, D. A., Acetylcholine causes rapid nicotinic excitation in the medial habenular nucleus of guinea pig, *in vitro, J. Neurosci.*, 7, 742, 1987.

35. Funke, K. and Eysel, U. T., Modulatory effects of acetylcholine, serotonin and noradrenaline on the activity of cat perigeniculate neurons, *Exp. Brain Res.*, 95, 409, 1993.

36. Krnjevic, K. and Phillis, J. W., Acetylcholine sensitive cells in the cerebral cortex, *J. Physiol.*, 166, 296, 1963.

37. Spehlman, R., Daniels, J. C., and Smathers, C. C., Acetylcholine and the synaptic transmission of specific impulses to the visual cortex, *Brain Res.*, 94, 125, 1971.

38. Sillito, A. M. and Kemp, J. A., Cholinergic modulation of the functional organization of the cat visual cortex, *Brain Res.*, 289, 143, 1983.

39. Sato, H., Hata, Y., and Tsumoto, T., A functional role of cholinergic innervation to neurons in the cat visual cortex, *J. Neurophysiol.*, 58, 765, 1987.

40. Müller, C. M. and Singer, W., Acetylcholine-induced inhibition in the cat visual cortex is mediated by a GABAergic mechanism, *Brain Res.*, 487, 335, 1989.

41. McCormick, D. A. and Prince, D. A., Mechanisms of action of acetylcholine in the guinea-pig cerebral cortex *in vitro, J. Physiol.*, 375, 169, 1986.

42. Murphy, P. C. and Sillito, A.M., Cholinergic enhancement of direction selectivity in the visual cortex of the cat, *Neuroscience*, 40, 13, 1991.

43. Sato, H., Hata, Y., Hagihara, K., and Tsumoto, T. A., Effects of cholinergic depletion on neuron activities in the cat visual cortex, *J. Neurophysiol.*, 58, 765, 1987.

44. Wang, Z. and McCormick, D.A., Control of firing mode of corticotectal and corticopontine layer V burst-generating neurons by norepinephrine, acetylcholine, and 1*S*,3*R*-ACPD, *J. Neurosci.*, 13, 2199, 1993.

45. Reader, T. A., De Champlain, J., and Jasper, H. H., Participation of presynaptic and postsynaptic receptors in acetylcholine-catecholamine interactions in cerebral cortex, *Adv. Biosci.*, 18, 363, 1979.

46. Krnjevic, K. and Phillis, J. W., Iontophoretic studies of neurons in the mammalian cerebral cortex, *J. Physiol.*, 165, 274, 1963.

47. Krnjevic, K. and Phillis, J. W., Pharmacological properties of acetylcholine sensitive cells in the cerebral cortex, *J. Physiol.*, 166, 328, 1963.

48. Crawford, J. M. and Curtis, D. R., Pharmacological studies on feline Betz cells, *J. Physiol.*, 186, 121, 1966.

49. Stone, T.W., Cholinergic mechanisms in the rat somatosensory cerebral cortex, *J. Physiol.*, 225, 485, 1972.

50. Bassant, M. H., Baleyte, J. M., and Lamour, Y., Effects of acetylcholine on single cortical somatosensory neurons in the unanesthetized rat, *Neuroscience*, 39, 189, 1990.

51. Crawford, J. M., The sensitivity of cortical neurones to acidic amino acids and acetylcholine, *Brain Res.*, 17, 287, 1970.

52. Lamour, Y., Dutar, P., and Jobert, A., Excitatory effect of acetylcholine on different types of neurons in the first somatosensory neocortex of the rat: laminar distribution and pharmacological properties, *Neuroscience*, 7, 1483, 1982.

53. Lamour, Y., Dutar, P., and Jobert, A., A comparative study of two populations of acetylcholine-sensitive neurons in rat somatosensory cortex, *Brain Res.*, 289, 157, 1983.

54. Lamour, Y., Dutar, P., and Jobert, A., Spread of acetylcholine sensitivity in the neocortex following lesion of the nucleus basalis, *Brain Res.*, 252, 377, 1982.

55. Bevan, P., Bradshaw, C. M., and Szabadi, E., The antagonism of neuronal responses to acetylcholine by atropine: a quantitative study, *Brain Res.*, 88, 568, 1975.

56. Bradshaw, C. M., Sheridan, R. D., and Szabadi, E., Involvement of M1-muscarinic receptors in the excitation of neocortical neurones by acetylcholine, *Neuropharmacology*, 26, 1195, 1987.

57. Donoghue, J. P. and Carroll, K. L., Cholinergic modulation of sensory responses in rat primary somatic sensory cortex, *Brain Res.*, 408, 367, 1987.

58. Metherate, R., Tremblay, N., and Dykes, R. W., The effects of acetylcholine on response properties of cat somatosensory cortical neurons, *J. Neurophysiol.*, 59, 1231, 1988.

59. Metherate, R., Tremblay, N., and Dykes, R. W., Transient and prolonged effects of acetylcholine on responsiveness of cat somatosensory cortical neurons, *J. Neurophysiol.*, 59, 1253, 1988.

60. Tremblay, N., Warren, R. A., and Dykes, R. W., Electrophysiological studies of acetylcholine and the role of the basal forebrain in the somatosensory cortex of the cat. II. Cortical neurons excited by somatic stimuli, *J. Neurophysiol.*, 64, 1212, 1990.

61. Schwindt, P. D., Spain, W. J., Foehring, R. C., Stafstrom, C. E., and Crill, W. E., Multiple potassium conductances and their functions in neurons from cat sensorimotor cortex *in vitro*, *J. Neurophysiol.*, 59, 424, 1988.

62. Schwindt, P. D., Spain, W. J., and Crill, W.E., Long-lasting reduction of excitability by a sodium-dependent potassium current in cat neocortical neurons, *J. Neurophysiol.*, 61, 233, 1989.

63. Benardo, L. S. and Prince, D. A., Ionic mechanisms of cholinergic excitation in mammalian hippocampal pyramidal cells, *Brain Res.*, 249, 333, 1982.

64. Reader, T. A., Ferron, A., Descarries, L., and Jasper, H. H., Modulatory role for biogenic amines in the cerebral cortex. Microiontophoretic studies, *Brain Res.*, 160, 217, 1979.

65. Waterhouse, B. D., Moises, H. C., and Woodward, D. J., Noradrenergic modulation of somatosensory cortical neuronal responses to iontophoretically applied putative neurotransmitters, *Exp. Neurol.*, 69, 30, 1980.

66. Waterhouse, B. D., Moises, H. C., and Woodward, D. J., α-receptor-mediated facilitation of somatosensory cortical neuronal responses to excitatory synaptic inputs and iontophoretically applied acetylcholine, *Neuropharmacology*, 20, 907, 1981.

67. Bassant, M. H., Ennouri, K., and Lamour, Y., Effects of iontophoretically applied monoamines on somatosensory cortical neurons of unanesthetized rats, *Neuroscience*, 39, 431, 1990.

68. Bloom, F. E., Oliver, A. P., and Salmoiraghi, G. C., The responsiveness of individual hypothalamic neurons to microiontophoretically administered endogenous amines, *Int. J. Neuropharmacol.*, 2, 181, 1963.

69. Phillis, J. W. and Tebecis, A. K., The responses of thalamic neurons to iontophoretically applied monoamines, *J. Physiol.*, 193, 715, 1969.

70. Lamour, Y., Dutar, P., and Jobert, A., Effects of neuropeptides on rat cortical neurons: laminar distribution and interaction with the effect of acetylcholine, *Neuroscience*, 10, 107, 1983.

71. Tremblay, N., Warren, R. A., and Dykes, R. W., Electrophysiological studies of acetylcholine and the role of the basal forebrain in the somatosensory cortex of the cat. I. Cortical neurons excited by glutamate, *J. Neurophysiol.*, 64, 1199, 1990.

72. Foote, S. L., Freedman, R., and Oliver, A.P., Effects of putative neurotransmitters on neuronal activity in monkey auditory cortex, *Brain Res.*, 86, 229, 1975.

73. McKenna, T. M, Ashe, J. H., Hui, G. K., and Weinberger, N. M., Muscarinic agonists modulate spontaneous and evoked unit discharge in auditory cortex of cat, *Synapse*, 2, 54, 1988.

74. McKenna, T. M., Ashe, J. H., and Weinberger, N. M., Cholinergic modulation of frequency receptive fields in auditory cortex. I. Frequency-specific effects of muscarinic agonists, *Synapse*, 4, 30, 1989.

75. Metherate, R. and Weinberger, N. M., Cholinergic modulation of responses to single tones produces tone-specific receptive field alterations in cat auditory cortex, *Synapse*, 6, 133, 1990.

76. Hars, B., Maho, C., Edeline, J.-M., and Hennevin, E., Basal forebrain stimulation facilitates tone-evoked responses in the auditory cortex of awake rat, *Neuroscience*, 56, 61, 1993.

77. Metherate, R. and Ashe, J. H., Nucleus basalis stimulation facilitates thalamocortical synaptic transmission in rat auditory cortex, *Synapse*, 14, 132, 1993.

78. Metherate, R., Cox, C. L., and Ashe, J.H., Cellular bases of neocortical activation: modulation of neural oscillations by the nucleus basalis and endogenous acetylcholine, *J. Neurosci.*, 12, 4701, 1992.

79. Jones, R. S. G. and Olpe, H.-R., On the role of the baseline firing rate in determining the responsiveness of cingulate cortical neurons to iontophoretically applied substance P and acetylcholine, *J. Pharm. Pharmacol.*, 36, 623, 1984.

80. Palmer, M. R., Morris, D. H., Taylor, D. A., Stewart, J. M., and Hoffer, B.J., Electrophysiological effects of enkephalin analogs in rat cortex, *Life Sci.*, 23, 851, 1978.

81. Yang, C. R. and Mogenson, G. J., Dopaminergic modulation of cholinergic responses in rat medial prefrontal cortex: an electrophysiological study, *Brain Res.*, 524, 271, 1990.

82. Pirch, J. H., Turco, K., and Rucker, H. K., A role for acetylcholine in conditioning-related responses of rat frontal cortex neurons: microiontophoretic evidence, *Brain Res.*, 586, 19, 1992.

83. McCormick, D. A. and Prince, D.A., Two types of muscarinic response to acetylcholine in mammalian cortical neurons, *Proc. Natl. Acad. Sci. U.S.A.*, 82, 6344, 1985.

84. Jones, R. S. G. and Olpe, H.-R., Monoaminergic modulation of the sensitivity of neurones in the cingulate cortex to iontophoretically applied substance P, *Brain Res.*, 311, 297, 1984.

85. Cole, A. E. and Nicoll, R. A., Characterization of a slow cholinergic post-synaptic potential recorded *in vitro* from rat hippocampal pyramidal cells, *J. Physiol.*, 352, 173, 1984.

86. Cole, A. E. and Nicoll, R. A., The pharmacology of cholinergic excitatory responses in hippocampal pyramidal cells, *Brain Res.*, 305, 283, 1984.

87. Ben-Ari, Y., Krnjevic, K., Reinhardt, W., and Ropert, N., Intracellular observations on the disinhibitory action of acetylcholine in hippocampus, *Neuroscience*, 6, 2445, 1981.

88. Dodd, J., Dingledine, R., and Kelly, J. S., The excitatory action of acetylcholine on hippocampal neurones of the guinea-pig and rat maintained *in vitro*, *Brain Res.*, 207, 109, 1981.

89. Halliwell, J. V. and Adams, P. R., Voltage-clamp analysis of muscarinic excitation in hippocampal neurons, *Brain Res.*, 250, 71, 1982.

90. Markram, H. and Segal, M., Long-lasting facilitation of excitatory postsynaptic potentials in the rat hippocampus by acetylcholine, *J. Physiol.*, 427, 381, 1990.

91. Mancillas, J. R., Siggins, G. R., and Bloom, F. E., Somatostatin selectively enhances acetylcholine-induced excitations in rat hippocampus and cortex, *Proc. Natl. Acad. Sci. U.S.A.*, 83, 7518, 1986.

92. Mancillas, J. R., Siggins, G. R., and Bloom, F. E., Systemic ethanol: selective enhancement of responses to acetylcholine and somatostatin in hippocampus, *Science*, 231, 161, 1986.

93. Lamour, Y., Dutar, P., and Jobert, A., Septo-hippocampal and other medial septum-diagonal band neurons: electrophysiological and pharmacological properties, *Brain Res.*, 309, 227, 1984.

94. Garcia-Rill, E., The pedunculopontine nucleus, *Prog. Neurobiol.*, 36, 363, 1991.

95. Woolf, N. J. and Butcher, L. L., Cholinergic systems in the rat brain. III. Projections from the pontomesencephalic tegmentum to the thalamus, tectum, basal ganglia, and basal forebrain, *Brain Res. Bull.*, 16, 603, 1986.

96. Olton, D. S., Dementia: animal models of the cognitive impairments following damage to the basal forebrain cholinergic system, *Brain Res. Bull.*, 25, 499, 1990.

97. Dekker, A. J., Connor, D. J., and Thal, L. J., The role of cholinergic projections from the nucleus basalis in memory, *Neurosci. Biobehav. Rev.*, 15, 299, 1991.

98. Delong, M. R., Activity of pallidal neurons during movement, *J. Neurophysiol.*, 34, 414, 1971.

99. Mitchell, S. J., Richardson, R. T., Baker, F. H., and Delong, M. R., The primate nucleus basalis of Meynert: neuronal activity related to a visuomotor tracking task, *Exp. Brain Res.*, 68, 506, 1987.

100. Richardson, R. T. and DeLong, M. R., A reappraisal of the functions of the nucleus basalis of Meynert, *TINS*, 11, 264, 1988.

101. Richardson, R. T., Mitchell, S. J., Baker, F. H., and DeLong, M. R., Responses of nucleus basalis of Meynert neurons in behaving monkeys, in *Cellular Mechanisms of Conditioning and Behavioral Plasticity*, Woody, C. D., Alcon, D. L., and McGaugh, J. L., Eds, Plenum, New York, 1988, 161.

102. Rolls, E. T., Sanghera, M. K., and Roper-Hall, A., The latency of activation of neurones in the lateral hypothalamus and substantia innominata during feeding in the monkey, *Brain Res.*, 164, 121, 1979.

103. Wilson, F. A. W. and Rolls, E. T., Learning and memory is reflected in the responses of reinforcement-related neurons in the primate basal forebrain, *J. Neurosci.*, 10, 1254, 1990.

104. Rigdon, G. C. and Pirch, J. H., Nucleus basalis involvement in conditioned neuronal responses in rat frontal cortex, *J. Neurosci.*, 6, 2535, 1986.

105. Pirch, J. H., Basal forebrain and frontal cortex neuron responses during visual discrimination in the rat, *Brain Res. Bull.*, 31, 73, 1993.

106. Pirch, J. H., Corbus, M. J., Rigdon, G. C., and Lyness, W. H., Generation of cortical event-related slow potentials in the rat involves nucleus basalis cholinergic innervation, *Electroencephalogr., Clin. Neurophysiol.*, 63, 464, 1986.

107. Pirch, J., Rigdon, G., Rucker, H., and Turco, K., Basal forebrain modulation of cortical cell activity during conditioning, in *The Basal Forebrain: Anatomy to Function*, Napier, T. C., Kalivas, P. W., and Hanin, I., Eds., Plenum Press, New York, 1991, 219.

108. Greuel, J. M., Luhmann, H. J., and Singer, W., Pharmacological induction of use-dependent receptive field modifications in the visual cortex, *Science*, 242, 74, 1988.

109. Woody, C. D., Acquisition of conditioned facial reflexes in the cat: cortical control of different facial movements, *Fed. Proc.*, 41, 2160, 1982.

110. Aou, S., Oomura, Y., and Nishino, H., Influence of acetylcholine on neuronal activity in monkey orbitofrontal cortex during bar press feeding task, *Brain Res.*, 275, 178, 1983.

111. Inoue, M., Oomura, Y., Nishino, H., Aou, S., Sikdar, S.K., Hynes, M., Mizuno, Y., and Katabuchi, T., Cholinergic role in monkey dorsolateral prefrontal cortex during bar-press feeding behavior, *Brain Res.*, 278, 185, 1983.

112. Lénárd, L., Oomura, Y., Nakano, Y., Aou, S., and Nishino, H., Influence of acetylcholine on neuronal activity of monkey amygdala during bar press feeding behavior, *Brain Res.*, 500, 359, 1989.

113. Delacour, J., Houcine, O., and Costa, J. C., Evidence for a cholinergic mechanism of "learned" changes in the responses of barrel field neurons of the awake and undrugged rat, *Neuroscience*, 34, 1, 1990.

114. Ono, T., Nakamura, K., Fukuda, M., and Kobayashi, T., Catecholamine and acetylcholine sensitivity of rat lateral hypothalamic neurons related to learning, *J. Neurophysiol.*, 67, 265, 1992.

115. Rucker, H. K., Corbus, M. J., and Pirch, J. H., Discriminative conditioning-related slow potential and single-unit responses in the frontal cortex of urethane-anesthetized rats, *Brain Res.*, 376, 368, 1986.

116. Rigdon, G. C. and Pirch, J. H., Microinjection of procaine of GABA into the nucleus basalis magnocellularis affects cue-elicited unit responses in the rat frontal cortex, *Exp. Neurol.*, 85, 283, 1984.

117. Pirch, J. H., unpublished data, 1992.

Chapter 8

Acetylcholine-Operated Ionic Conductances in Central Neurons

Hylan C. Moises and Mark D. Womble

CONTENTS

I. INTRODUCTION

A prominent role for acetylcholine (ACh) as a neurotransmitter in the peripheral nervous system has been recognized since the pioneering efforts of Loewi[1] and Dale.[2] As a result of work by Eccles and colleagues[3] some 20 years later, ACh was the first compound clearly demonstrated to function as a neurotransmitter in the mammalian central nervous system (CNS). Subsequently, numerous studies have provided evidence for ACh acting as a transmitter at a variety of central synapses (see recent reviews by North,[4] Halliwell,[5] Nicoll et al.[6]). In this review, we focus on the multiple responses elicited by ACh in neurons of the mammalian CNS.

The central neuronal effects of ACh can be partitioned into two broad pharmacological categories, designated nicotinic and muscarinic, based on the abilities of the compounds nicotine and muscarine to selectively mimic ACh action at these receptor sites. Nicotinic cholinergic responses in the CNS generally yield rapid depolarization of the postsynaptic neuron and thus are excitatory. This is due to the opening of a nonspecific cation channel, which is part of the receptor complex. In this regard, central nicotinic synaptic transmission is similar to that seen at peripheral ganglionic synapses and the neuromuscular junction.

Activation of muscarinic receptors elicits a wide variety of postsynaptic responses in central neurons. These may involve a direct cholinergic action on the neuron, producing excitation or inhibition, or may indirectly alter neuronal activity by modulating presynaptic transmitter release. These effects result from the inhibition, potentiation, or modulation of a variety of ionic conductances, including a large family of potassium currents, one or more mixed cationic currents, and voltage-dependent calcium currents.

One of the most important muscarinic actions of ACh in the CNS is the enhancement of neuronal excitability. This response has been observed in a number of cortical brain areas and is generated by several different mechanisms. One of these is associated with a prolonged depolarization of the neuron, an effect produced by the muscarinic inhibition of several different potassium conductances, including the M-current, a K^+ leak current, and an inwardly rectifying K^+ current. Neuronal excitability is also enhanced by the muscarinic inhibition of the slowly decaying afterhyperpolarization (AHP) that normally follows a burst of action potentials. Loss of the AHP is attributable to blockade of the slowly decaying Ca^{2+}- or Na^+-activated K^+ currents underlying its production. An additional mechanism for increasing excitability that accompanies the muscarinic inhibition of the slow AHP in some central neurons is the generation of a prolonged afterdepolarization. This may result from the muscarinic activation of a Ca^{2+}-activated mixed cation current or the Ca^{2+}-dependent blockade of a K^+ current.

0-8493-7630-0/95/$0.00+$.50
© 1995 by CRC Press, Inc.

In contrast to the cholinergic excitation predominant in cortical neurons, inhibitory muscarinic responses have most often been observed in subcortical and brainstem neurons. In these areas, the neuron is hyperpolarized by the muscarinic activation of a K^+ conductance. Similarly, potentiation of a delayed rectifier K^+ current may represent an additional muscarinic mechanism for reducing neuronal excitability. Muscarinic receptor activation can also alter neuronal firing characteristics by modulating the voltage dependence for activation and inactivation of the transient K^+ current, I_A. Depending on the resting state of the neuron, the modulation of I_A can produce different neuronal responses to excitatory synaptic inputs. Finally, muscarinic receptor activation has been shown to modulate the activity of voltage-gated Ca^{2+} channels. Inhibition of the high voltage-activated calcium current may be responsible for the decrease in presynaptic neurotransmitter release that underlies the muscarinic-induced reductions in excitatory and inhibitory postsynaptic potentials observed in many CNS neurons.

Unlike the rapid, depolarizing nicotinic response, the postsynaptic actions of muscarinic receptor stimulation are generally slow to develop and may be very long lasting. These responses appear to be mediated by G-proteins and the subsequent activation of a second messenger transduction system.[7-9] The large variety of muscarinic responses observed in central neurons may be explained in part by recent pharmacological and molecular cloning evidence demonstrating the presence of a number of G-protein classes and subtypes[10,11] and the presence of at least five different subtypes of muscarinic receptors (M_1 to M_5).[12-14] These variations allow several possible pathways for mediation of muscarinic actions (see Chapters 2 and 9 in this volume). The muscarinic receptor-linked intracellular transduction pathway that has received the most attention is the phosphoinositide (PI) hydrolysis system. Phosphoinositide turnover leads to the production of diacylglycerol, an activator of protein kinase C (PKC), and inositol trisphosphate (IP_3), which generates release of Ca^{2+} from intracellular stores.[15-17] However, the precise identification of the G-proteins, muscarinic receptor subtypes, and second messenger transduction pathways involved in mediating the variety of muscarinic actions in neurons of the mammalian CNS is still far from complete.

II. NICOTINIC CHOLINERGIC RESPONSES

Nicotinic responses to ACh play an important role in excitatory neurotransmission in several areas of the mammalian CNS. Intracellular recordings from neurons of the rat locus ceruleus, the guinea pig medial habenula nucleus, the lateral geniculate nucleus of the cat or guinea pig, and the rat vestibular nucleus have demonstrated that application of ACh or nicotinic agonists produces a rapid depolarization.[18-21] The depolarization was associated with an increase in membrane conductance and was blocked by nicotinic, but not muscarinic, antagonists (Figure 8-1). Voltage-clamp analysis revealed that ACh induced the appearance of an inward current. This current had a reversal potential of approximately –15 mV, suggesting that the open channel was permeable to both Na^+ and K^+ ions. These findings demonstrate that the rapid nicotinic excitation observed in CNS neurons results from the opening of a mixed cationic channel, presumably one directly associated with the receptor complex. Thus, nicotinic receptors of the mammalian CNS appear to be similar to the nicotinic ACh receptor found at peripheral ganglionic synapses and at the neuromuscular junction. In medial habenula neurons, recordings of membrane current together with fura-2 fluorescence measurements demonstrated that the ion channel associated with the nicotinic receptor also exhibited significant permeability to calcium (Ca^{2+}) ions.[22] Activation of these receptors resulted in a significant rise in cytosolic Ca^{2+} concentration, even in the absence of Ca^{2+} influx through voltage-activated Ca^{2+} channels. The Ca^{2+} influx through the nicotinic receptor was sufficient to activate a Ca^{2+}-dependent Cl^- conductance and decrease the cellular response to γ-aminobutyric acid (GABA), suggesting that this influx could play an important physiological role.

Recently, an unusual cholinergic inhibitory response has been described in neurons of the rat dorso-lateral septal nucleus.[23,24] In the presence of the muscarinic antagonist atropine, nicotinic agonists produced a slowly developing, long-lasting membrane hyperpolarization. This response was associated with a decrease in membrane resistance, reversed near the equilibrium potential for K^+ (E_K), and was sensitive to changes in extracellular K^+ concentration. The hyperpolarizing response was decreased by the removal of extracellular Ca^{2+}, injection of the Ca^{2+} chelator ethyleneglycol-*bis*-(aminoethyl ether)-tetraacetic acid (EGTA) into the cell, or during application of K^+ channels blockers such as tetraethylammonium (TEA) or apamin. These findings suggest that nicotinic receptor activation in septal neurons produces membrane hyperpolarization and inhibition via a Ca^{2+}-dependent increase in K^+ conductance. The transduction mechanism for activation of this current is unknown, but the slow time course implies the involvement of a second messenger cascade.

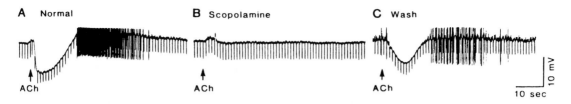

A Normal   B Scopolamine   C Wash

ACh   ACh   ACh

                                                                        10 mV

                                                          10 sec

Figure 8-1 Acetylcholine causes a rapid depolarization, followed by a hyperpolarization and slow excitation, in a guinea pig lateral geniculate nucleus neuron through the activation of nicotinic and muscarinic receptors. (A) Application of ACh at upward arrow to neuron depolarized with intracellular injection of direct current to near firing threshold (–55 mV). (B) A second application of ACh after a brief local application of the muscarinic antagonist scopolamine elicits only a fast, nicotinic depolarization. (C) Partial recovery of the hyperpolarizing and slow depolarizing responses after wash-out of the scopolamine. Action potential amplitudes have been truncated. Hyperpolarizing conductance test pulses were delivered at a rate of 1 Hz. (From McCormick, D. A. and Prince, D. A., *J. Physiol. (Lond.)*, 392, 147, 1987. With permission.)

III. MUSCARINIC CHOLINERGIC RESPONSES

Muscarinic receptor activation elicits a wide variety of postsynaptic responses in neurons of the mammalian CNS. The responses may be broadly divided into those resulting from the inhibition of active membrane currents, the direct activation of a current by muscarinic receptors, and the modulation of existing voltage-gated currents.

A. MUSCARINIC CURRENT INHIBITION AND THE ENHANCEMENT OF NEURONAL EXCITABILITY

One of the first muscarinic responses to be demonstrated in the mammalian CNS was the acceleration in firing of cat hippocampal neurons following iontophoretic application of ACh *in vivo*, an effect that was blocked by the muscarinic antagonist atropine.[25] Subsequently, intracellular recordings from a variety of cortical neurons *in vitro* have shown that the enhancement in neuronal excitability and changes in spike discharge patterns produced by muscarinic receptor stimulation involved several mechanisms, including a prolonged membrane depolarization, an increase in membrane resistance, blockade of spike frequency accommodation, and reduction of the slow AHP.[20,26-37]

1. Muscarinic Depolarization

The most noticeable change that contributes to the muscarinic increase in excitability of cortical neurons is a slowly developing, long-lasting depolarization (Figure 8-2). This change in membrane potential may last for tens of seconds and can initiate action potential firing either directly or indirectly by bringing the membrane potential closer to the threshold for action potential production and increasing the likelihood that subsequent excitatory inputs will initiate firing. Associated with the muscarinic depolarization is an increase in membrane input resistance, a change that effectively increases the neuronal length constant and enhances the effectiveness of synaptic inputs, especially those distal from the cell body. These effects were originally observed following ionophoresis of ACh onto cat neocortical neurons *in vivo* by Krnjevic et al.,[39] who suggested that the changes were due to a decrease in K+ conductance. Subsequently, recordings from neurons in brain slice preparations of the hippocampus, neocortex, geniculate nucleus, and basolateral amygdala, among others, have confirmed that ACh or its stable analogue carbachol strongly depolarized central neurons and increased input resistance in an atropine-sensitive manner.[20,21,27,29,30,32,33,35-37,40-44] The magnitude of the depolarization and resistance change was voltage sensitive, being larger when the cell was depolarized to just below the threshold for action potential production and disappearing at hyperpolarized membrane potentials.[26,29,35,36,40] Current-clamp analysis also revealed that the depolarizing response to carbachol had a reversal potential negative to –80 mV and was sensitive to changes in extracellular K+ concentration, suggesting the involvement of a K+ conductance.[20,35-36]

A similar depolarizing response can be elicited upon activation of cholinergic afferent pathways (Figure 8-2B). Several groups have demonstrated that the repetitive stimulation of these pathways in brain slice preparations produces a series of noncholinergic excitatory and inhibitory postsynaptic potentials (EPSPs and IPSPs) followed by a long-lasting cholinergic depolarization, the slow EPSP.[27,29,35,36,41,45,46] The slow

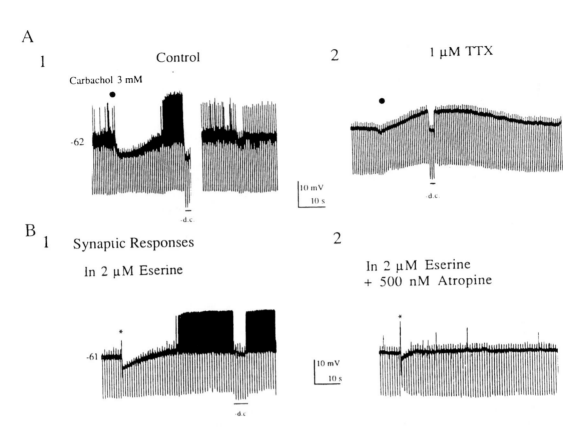

Figure 8-2 Slow muscarinic depolarizations elicited in pyramidal neurons by carbachol or cholinergic afferent pathway stimulation in a slice preparation of the rat basolateral amygdala. (A1) When recordings were made in normal medium, pressure ejection of carbachol (filled circle) evoked an initial hyperpolarization (mediated indirectly via the muscarinic excitation of GABAergic interneurons[35]), followed by a slow, sustained depolarization and action potential firing. Constant-current hyperpolarizing pulses were passed through the recording electrode to monitor membrane input resistance. The carbachol-induced slow depolarization was accompanied by a marked increase in input resistance. (A2) Recordings in the presence of tetrodotoxin demonstrate that the slow depolarization was associated with an increase in input resistance and resulted from a direct action on the neuron. (B1) In another pyramidal neuron, stimulation of cholinergic afferent fibers in the slice (asterisk) evoked a series of noncholinergic fast EPSPs, followed by a GABAergic IPSP and a long-lasting slow EPSP that was associated with an increase in input resistance. (B2) Addition of atropine (500 n*M*) to the bathing medium blocked the production of the slow EPSP. (Part B modified from Washburn, M. S. and Moises, H. C., *J. Physiol. (Lond.)*, 449, 121, 1992.)

EPSP sometimes lasted for up to several minutes and, if of sufficient amplitude, could elicit action potentials. It was associated with an increase in membrane resistance and showed voltage-sensitive changes in amplitude, increasing in magnitude as the cell was depolarized from rest prior to synaptic stimulation. The slow EPSP was enhanced in the presence of cholinesterase inhibitors such as eserine and was blocked by prior application of atropine, tetrodotoxin, or cadmium, indicating that it was produced by synaptically released ACh acting on postsynaptic muscarinic receptors.[27,29,35,36,41,46] Overall, these results suggest that ACh acts on muscarinic receptors in a variety of cortical neurons to inhibit a tonically active, voltage-dependent K[+] current. The resulting decrease in net outward current produces membrane depolarization.

A muscarinic-induced slow depolarization has also been described in bullfrog sympathetic neurons. Recordings from these cells indicated that the change in membrane potential was due to the inhibition of a persistently active, voltage-dependent K[+] current, the muscarinic-sensitive M-current (I_M).[47] Subsequent studies of mammalian CNS neurons revealed that M-current inhibition also contributed to the production

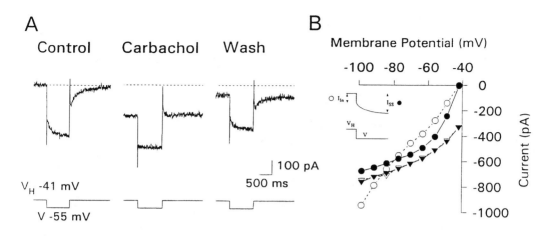

Figure 8-3 Muscarinic blockade of the M-current and Leak current in a pyramidal neuron of the rat basolateral amygdala. (A) In control medium, a hyperpolarizing voltage step to –55 mV from a holding potential of –41 mV evoked a slow, inward-going current relaxation due to deactivation of I_M. Bath application of carbachol (40 μM) produced an inward shift in the level of holding current and eliminated the M-current relaxation, effects that were partially reversed after drug washout. All bathing media contained TTX (1 μM). (B) Current-voltage (I-V) relationships determined from the cell in (A), showing control instantaneous (I_{In}, open circles) and steady-state (I_{SS}, filled circles) current responses and instantaneous (open triangles) and steady-state (filled triangles) responses in the presence of carbachol, plotted against step command potential. The muscarinic inhibition of I_M is indicated by the reduction of outward rectification in the steady-state I-V relationship observed between the command potentials of –40 and –70 mV. At more negative membrane potentials, carbachol produced a uniform inward shift in steady-state current, indicating the inhibition of a voltage-insensitive K+ leak conductance. (Modified from Womble, M. D. and Moises, H. C., *J. Physiol. (Lond.)*, 457, 93, 1992. With permission.)

of muscarinic depolarizations in a number of cortical neurons, including hippocampal pyramidal neurons,[48] olfactory cortical neurons,[49] neocortical neurons,[30,43,50] and amygdaloid neurons.[51] The M-current is a sustained outward current that is activated in central neurons by membrane depolarization above an activation threshold of approximately –65 mV.[48,51] To identify this current, neurons were initially voltage-clamped at a relatively depolarized membrane potential to activate I_M. Application of a 1-s hyperpolarizing voltage-step produced the time-dependent deactivation of I_M, identified as a slow, inward-going current relaxation during the hyperpolarizing step (Figure 8-3). Measurements of the reversal potential indicated that I_M was largely a K+ conductance.[48,51] Carbachol inhibited I_M and eliminated the inward current relaxation normally observed during the hyperpolarizing voltage step, leaving only a flat, ohmic current response.[48-54] The inhibitory effect of carbachol on I_M was dose dependent and was prevented in the presence of atropine, indicating mediation by muscarinic receptor activation. The loss of I_M was accompanied by a decrease in membrane conductance and a sharp drop (inward shift) in the level of steady-state holding current recorded at –40 mV.

Since the threshold for activation of the M-current is close to the normal resting potential of the neuron, this conductance can function to stabilize the membrane potential at the resting level by resisting the actions of brief depolarizing influences. With prolonged depolarization, the M-current gradually activates over a period of about 100 ms, allowing it to contribute in part to the process of spike frequency accommodation.[55] Once activated by depolarization, I_M is noninactivating and decreases neuronal excitability by adding a sustained outward component to the steady-state current at membrane potentials positive to its activation threshold. The muscarinic inhibition of a persistently active M-current produces an inward shift in holding current level, reflected in unclamped neurons as a membrane depolarization. The voltage-sensitive nature of M-current activation is responsible for the observed voltage-dependent behavior of the slow muscarinic depolarization and for the larger inward current shifts produced by muscarinic agonists when tested at more depolarized holding potentials. Even at membrane potentials where muscarinic receptor activation does not produce overt membrane depolarization, the muscarinic-induced loss of I_M can contribute to a reduction in the accommodation response and the enhancement of neuronal responses to depolarizing synaptic events.

However, blockade of the M-current alone cannot fully account for production of the muscarinic depolarization and slow EPSP in all central neurons. For example, current-clamp recordings from neurons of the hippocampus, neocortex, geniculate nucleus, and amygdala have demonstrated that the muscarinic depolarization and increased input resistance were not always voltage dependent and could be observed at hyperpolarized membrane potentials where I_M was inactive.[20,33,35,37,40] These observations suggested that the muscarinic inhibition of a second voltage-insensitive current also contributed to production of the muscarinic depolarizing response. Observations obtained under voltage-clamp from a variety of central neurons have provided direct evidence for this possibility (Figure 8-3).[33,44,51,53,54,56] First, measurements of the magnitude of the inward current shift produced by carbachol at depolarized holding potentials demonstrated that even complete inhibition of I_M was insufficient to fully account for the loss of sustained outward current. Second, carbachol was found to reduce membrane conductance and produce an inward current shift at membrane potentials negative to –70 mV, where I_M was inactive. Third, the instantaneous current-voltage (I-V) relationship at these negative membrane potentials was linear and in the presence of carbachol showed a uniform decrease in conductance. Fourth, the response was sensitive to changes in extracellular K^+ concentration and prevented by atropine. Finally, the muscarinic-induced depolarization or inward current shift could be observed in the absence of M-current inhibition or in cells, such as olfactory cortex neurons, which lacked I_M. These findings indicate that the muscarinic inhibition of a tonically active, voltage-insensitive K^+ leak current (I_{Leak}) also contributed to production of the slow muscarinic depolarization and the increase in input resistance. Since I_M is largely inactive at the normal resting potential of most neurons, inhibition of I_{Leak} seems likely to be the major factor underlying production of the slow muscarinic depolarization. However, any concurrent inhibition of I_M would also aid in the development of the slow depolarization by removing the current primarily responsible for opposing a depolarizing change in membrane potential. In thalamic relay neurons, the muscarinic inhibition of I_{Leak} induces a change in neuronal firing pattern, from a rhythmically phasic or bursting firing pattern to a tonic mode, in which the rapid production of single action potentials predominates.[33]

Investigations into the identity of the muscarinic receptor subtypes responsible for inhibition of the M-current and production of the slow depolarization initially attempted to classify them into M_1 and non-M_1 receptor subtypes, based on the ability of pirenzepine (M_1 selective antagonist) and gallamine (non-M_1 antagonist) to block the responses. Using this approach, several studies found that gallamine, but not pirenzepine, was effective at preventing the muscarinic inhibition of I_M, suggesting that blockade of this current was mediated by a non-M_1 receptor subtype.[57,58] In contrast, the muscarinic slow depolarization resulting from inhibition of I_{Leak} was sensitive to blockade by low concentrations of pirenzepine, with a dissociation coefficient of approximately 16 nM, suggesting that this muscarinic response was mediated by the M_1 receptor subtype.[32,56,58,59] More recent studies utilizing a broader range of newly developed muscarinic antagonists have concluded that the block of I_{Leak} is mediated by the M_3 receptor subtype.[44,60] Since the M_3 receptor subtype has a moderately high affinity for pirenzepine, this finding is not necessarily incompatible with earlier studies demonstrating that pirenzepine antagonized production of the muscarinic slow depolarization.

The possible role of G-proteins and intracellular second messenger cascades in the mediation of the muscarinic depolarizing response has also been examined in several types of central neurons. In hippocampal and geniculate neurons, intracellular dialysis with the non-hydrolyzable GTP analogue GTP-γ-S rendered the muscarinic inhibition of I_{Leak} and production of the slow depolarization largely irreversible, indicating the involvement of a G-protein.[33,61] However, production of the muscarinic depolarization and blockade of I_M were not prevented by pretreatment with pertussis toxin (PTX), which inactivates the G_i and G_o types of G-proteins.[33,58,61,62] To test for involvement of the PI transduction system, several groups applied phorbol esters, which directly activate PKC to hippocampal pyramidal neurons, in an attempt to mimic the muscarinic inhibitions of I_M and I_{Leak}. These compounds produced membrane depolarization and an increase in membrane resistance, suggesting blockade of I_{Leak}.[58,62-63] The M-current was unaffected by phorbol ester treatment, but was inhibited by the direct intracellular application of IP_3.[58,62] The finding that the muscarinic inhibition of I_M could not be prevented by preloading the cells with Ca^{2+} chelators suggested a possible direct inhibitory action of IP_3 on the M-channel, independent of any IP_3-mediated release of internal calcium. Taken together, these results indicate that the muscarinic inhibitions of I_M and I_{Leak} may be mediated by a pertussis toxin-insensitive G-protein and activation of phosphoinositide turnover.

2. Muscarinic Reductions in Accommodation and the Slow Afterhyperpolarization

Many central neurons, including pyramidal neurons from the olfactory and cerebral cortex, amygdala, and hippocampus, respond to a prolonged depolarizing stimulus with a series of action potentials whose rate of firing decreases or accommodates with time, thus limiting the number of action potentials produced during the depolarization (Figure 8-4A). A prolonged AHP is normally observed immediately following termination of such a burst of action potentials (Figure 8-4B). The AHP consists of a rapidly decaying medium AHP with a time course of tens to a few hundred milliseconds, followed by a long-lasting slow AHP that decays over a period of several seconds. The slow AHP functions to hyperpolarize the membrane, decreasing the ability of excitatory inputs to bring the cell to threshold and slowing the rate of firing during repetitive stimulation. Muscarinic receptor activation enhances neuronal firing rates by decreasing both spike frequency accommodation and the amplitude and duration of the slow AHP. These effects, together with the slow muscarinic depolarization, enable the neuron to fire more readily and robustly to subsequent excitatory inputs. The muscarinic-induced enhancement of neuronal excitability and changes in firing pattern have been postulated to play a role in the processes of associative learning and memory formation, as well as controlling attention and arousal.[33-34,38]

Several studies have demonstrated that the slow AHP resulted from the action of a slowly decaying, Ca^{2+}-activated K^+ current (I_{AHP}) (Figure 8-4C).[31,53,64-67] This current was activated by Ca^{2+} entry during the preceding depolarization, had a reversal potential near E_K, was voltage insensitive, and was resistant to blockade by TEA. Activation of I_{AHP} during the period of membrane depolarization was primarily responsible for spike frequency accommodation.[55] The application of ACh or muscarinic agonists resulted in an increased rate of neuronal firing to subsequent depolarizations because of reductions in both accommodation and the slow AHP.[26,27,29,30,35,40] Voltage-clamp recordings from cortical neurons in brain slice preparations have shown that these changes were associated with the muscarinic inhibition of I_{AHP}.[31,43,53,54,56,65-67] Stimulation of presynaptic cholinergic fibers in the slice produced similar reductions in accommodation, the slow AHP, and I_{AHP} (Figure 8-4C). All of these actions were enhanced by eserine and blocked by atropine, indicating mediation via the activation of muscarinic receptors.[27,29,35,36,67] The I_{AHP} was not active at membrane potentials negative to –50 mV and therefore did not contribute to the steady-state resting membrane conductance nor to production of the slow cholinergic depolarizing response.[5,54] However, the reductions of spike frequency accommodation and the slow AHP that accompanied the muscarinic suppression of I_{AHP} allowed neurons to greatly increase the rate and prolong the duration of action potential firing.

Several studies have attempted to determine the muscarinic receptor subtype that mediates inhibition of the slow AHP. Müller and Misgeld[59] reported in dentate granule cells and CA3 hippocampal pyramidal neurons that inhibition of the slow AHP by carbachol or stimulation of cholinergic afferents was not prevented by 1 μM pirenzepine, suggesting activation of a pirenzepine-insensitive (non-M_1) receptor subtype. In olfactory cortex neurons, the carbachol-induced inhibitions of the slow AHP and I_{AHP} were not prevented by pirenzepine (300 nM), but were antagonized by gallamine, also suggesting the involvement of a non-M_1 receptor subtype.[53] In contrast, Dutar and Nicoll[58] found that similar concentrations of pirenzepine (0.3 to 1.0 μM) could fully block the inhibitory action of carbachol on the slow AHP in CA1 hippocampal neurons, suggesting the involvement of an M_1 (pirenzepine-sensitive) receptor. Similarly, pirenzepine (0.1 to 1.0 μM) produces a dose-dependent competitive antagonism of the carbachol-induced inhibition of the slow AHP in pyramidal neurons of the basolateral amygdala, indicating mediation by an M_1 muscarinic receptor subtype.[35]

The inhibition of I_{AHP} and the slow AHP in hippocampal pyramidal neurons by muscarinic agonists was mimicked by direct activation of PKC by phorbol esters but not prevented by pretreatment with PTX.[63,68] These data suggest the involvement of a pertussis toxin-insensitive G-protein and activation of protein kinase C. However, more recent studies in CA1 and CA3 hippocampal neurons have shown that inhibition of the slow AHP induced by low concentrations of carbachol (0.1 to 0.5 μM) was specifically prevented by intracellular injection of an inhibitory peptide for Ca^{2+}/calmodulin-dependent protein kinase II (CaMKII), and mimicked by blockers of protein phosphatases but not by blockers of PKC.[69] These findings suggest that the muscarinic inhibition of the slow AHP was mediated by activation of CaMKII with a subsequent phosphorylation step. Interestingly, the inhibitory actions of higher carbachol concentrations could not be prevented by CaMKII inhibition. Thus, the presence of a second, lower affinity class of muscarinic receptors that block I_{AHP} through another intracellular pathway, possibly involving the activation of PKC, remains a distinct possibility. A possible role for cyclic-GMP in mediating the

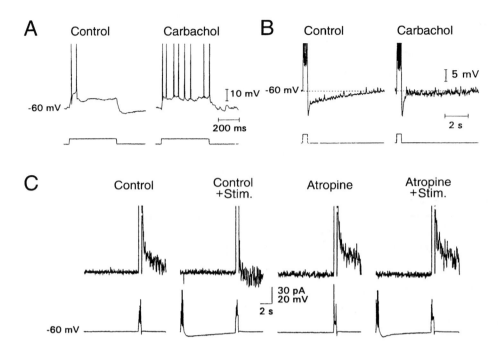

Figure 8-4 Muscarinic blockade of accommodation, the slow AHP and underlying I_{AHP} in pyramidal neurons of the rat basolateral amygdala. (A) Injection of depolarizing current into the neuron evoked only an initial rapid burst of action potentials. During bath application of carbachol (10 μ*M*), a similar current pulse elicited action potentials throughout the depolarization. (B) In the same neuron, a larger depolarizing current pulse evoked a burst of action potentials followed by a prolonged, biphasic AHP, consisting of a rapidly decaying medium AHP and a long-lasting slow AHP. Carbachol (10 μ*M*) selectively blocked the slow AHP. (C) Hybrid-clamp records obtained from a different neuron in the presence of TTX, showing inhibition of the slowly decaying portion of the AHP tail current following stimulation of the cholinergic afferent fibers and its blockade by atropine. (Part C from Womble, M. D. and Moises, H. C., *Brain Res.*, 621, 87, 1993. With permission.)

muscarinic inhibition of the slow AHP has also been suggested,[28] but evidence from a later study did not support this idea.[70]

Although the inhibitions of I_M, $I_{Leak,}$ and I_{AHP} all contribute to the increase in neuronal excitability and firing observed following muscarinic receptor activation, they do not appear to do so equally. Comparisons of the ability of bath-applied carbachol to block each of these currents indicated that I_{AHP} was the most sensitive current ($IC_{50} = 0.3$ to 0.5 μ*M*), I_{Leak} was intermediate ($IC_{50} = 1$ to 2 μ*M*), and I_M was the least sensitive ($IC_{50} = 2$ to 5 μ*M*) to muscarinic inhibition.[51,53,54,56,58,67] Similarly, stimulation of cholinergic afferent pathways in brain slice preparations of the hippocampus or basolateral amygdala could lead to reductions in spike frequency accommodation and blockade of the slow AHP and I_{AHP} in the absence of membrane depolarization.[29,35,54,67] Likewise, voltage-clamp recordings from these neurons indicated that the inward current shift that accompanies the inhibition of I_{Leak} and underlies production of the slow muscarinic depolarization could be produced by stimulation of cholinergic afferents independently of any M-current blockade (Womble and Moises, unpublished observations).[54] Taken together, these data suggest that inhibition of I_{Leak} was largely responsible for production of the slow muscarinic depolarization. An exception to this finding has been seen in hippocampal neurons cocultured with septal explants, wherein the inhibition of I_M clearly contributed to production of the slow muscarinic EPSP following stimulation of cholinergic septal afferents.[45] However, this may have been due to hyperinnervation of the cultured hippocampal neurons by cholinergic afferent fibers, resulting in a higher than normal concentration of ACh release following septal stimulation.

These findings suggest that the predominant effect arising from cholinergic synaptic transmission in cortical neurons is the muscarinic inhibition of I_{AHP}. Thus, ACh can increase neuronal excitability by producing reductions in spike frequency accommodation and the slow AHP, enabling the neuron to fire action potentials at faster rates and for longer periods of time. The muscarinic inhibition of I_{AHP} also serves to uncouple Ca^{2+} influx from activation of I_{AHP}, producing a large increase in intradendritic free Ca^{2+} accumulation.[71] At the same time, muscarinic receptor activation stimulates Ca^{2+}/calmodulin protein kinase activity[69] and potentiates N-methyl-D-aspartic acid (NMDA) receptor-mediated synaptic responses.[72] It is interesting to note that the induction of long-term potentiation (LTP) requires elevation of postsynaptic Ca^{2+} levels,[73] activation of CaMKII,[74-76] and stimulation of NMDA receptors.[77] Since LTP is thought to represent a putative cellular model for memory formation,[78] these findings suggest a possible link between the known importance of ACh to the processes of learning and memory formation and the cellular mechanisms of synaptic plasticity within the mammalian CNS.

3. Muscarinic Inhibition of Other Potassium Conductances

As in other cortical neurons, prolonged depolarization of cat sensorimotor cortex neurons elicited a long-lasting AHP. In these cells, the slowly decaying portion of the AHP could be divided into an early and a late component, both of which were reduced by muscarinic receptor activation.[31,66,79] The early slow AHP was sensitive to blockade of Ca^{2+} influx and was attributed to the activation of the Ca^{2+}-activated K^+ conductance, I_{AHP}. The late slow AHP also resulted from activation of a K^+ conductance, as evidenced by its reversal potential near E_K and dependence upon extracellular K^+ concentration, but was abolished by application of the Na^+ channel blocker tetrodotoxin (TTX). The current underlying production of the late slow AHP was found to be a slowly decaying, Na^+-activated K^+ conductance ($I_{K(Na)}$). Muscarinic agonists reduced spike frequency accommodation and the early and late slow AHPs in these cells by inhibiting both I_{AHP} and $I_{K(Na)}$, a functionally appropriate response since these currents have complementary functions.

In rat nucleus accumbens and locus ceruleus neurons, muscarinic receptor activation produces a slow depolarization, which in voltage-clamp recordings was associated with an inward current shift.[80-81] In these neurons, the depolarization was produced by the inhibition of an inwardly rectifying K^+ current that normally contributed to the resting membrane conductance. This muscarinic action was mediated by the M_1 receptor subtype and may involve activation of protein kinase C, since phorbol esters produced a similar membrane depolarization and prevented additional depolarization during subsequent application of muscarine.[80] In basolateral amygdaloid neurons, a rapidly developing inward rectifier current (I_{IR}) could be activated by hyperpolarizations from rest, and this current was similarly inhibited by muscarinic receptor activation.[82] However, in these neurons the inward rectifier did not contribute to the resting conductance, and thus inhibition of this current did not contribute to production of the slow muscarinic depolarization.

B. MUSCARINIC ACTIVATION OF POTASSIUM AND MIXED CATIONIC CONDUCTANCES

Muscarinic receptor activation can also alter neuronal excitability by activating several types of mixed cationic and K^+ conductances. These changes may include membrane depolarization resulting from the activation of inward cation or Ca^{2+}-activated cation currents or neuronal hyperpolarization due to the activation of a K^+ conductance.

Interneurons in slice preparations of the cerebral cortex and hippocampus, as well as nonpyramidal hippocampal neurons grown in culture, respond to stimulation of muscarinic receptors with a rapid depolarization.[30,61,83] The depolarization was associated with a decrease in membrane resistance and was blocked by atropine, suggesting muscarinic activation of a mixed cation conductance. Activation of this conductance does not appear to be mediated by a G-protein in cultured hippocampal neurons, since activation and deactivation of the inward current was unaffected by intracellular perfusion of GTP-γ-S.[61] Muscarine also activates an inward cation current in rat locus ceruleus neurons.[81] This current was not voltage sensitive, and reversed at approximately +10 mV, indicating that it was carried largely by Na^+. Activation of this current served to potentiate the membrane depolarization produced by the concurrent muscarinic inhibition of an inwardly rectifying K^+ current normally active at rest in these neurons (discussed above). The decreased membrane conductance associated with inhibition of the inward rectifier current was largely offset by the simultaneous increase in conductance due to activation of the mixed cation current, an interaction that altered the I-V relationship and often prevented clear identification of the inward rectifier reversal potential. Similar effects on the I-V relationships of hippocampal and amygdaloid pyramidal neurons have been observed in the presence of muscarinic agonists and have been interpreted as evidence for the muscarinic-induced activation of a nonspecific cation conductance.[51,56]

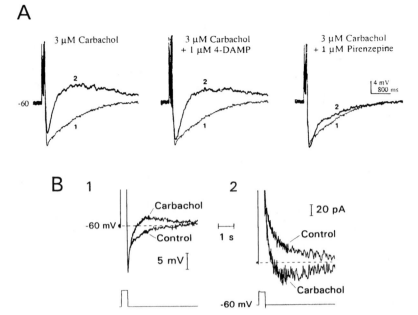

Figure 8-5 Muscarinic receptor activation blocks the slow afterhyperpolarization and induces the appearance of an afterdepolarization in pyramidal neurons from the rat basolateral amygdala. (A) Injection of a 150-ms pulse of depolarizing current elicited a burst of action potentials, which was followed in control medium (1) by a long-lasting AHP. During bath application of carbachol (3 μM), the AHP was replaced by a prolonged ADP (2). The carbachol-induced appearance of the ADP was not prevented by the M_3 antagonist 4-DAMP (1 μM) but was blocked in the presence of pirenzepine (1 μM), suggesting mediation by an M_1 muscarinic receptor subtype. (B) In a different neuron recorded in the presence of TTX (1 μM), bath application of carbachol (5 μM) produced a similar ADP. Hybrid-clamp records revealed that I_{AHP} observed in control medium was replaced in the presence of carbachol by a slowly decaying inward current.

A number of studies have noted that muscarinic blockade of the slow AHP was often accompanied by the appearance of a slowly developing and long-lasting afterdepolarization (ADP) in neurons of the cortex, amygdala, hippocampus, and septal nucleus (Figure 8-5).[30,31,35,40,84-87] The occurrence of ADP-induced repetitive firing contributed to the sustained excitatory response exhibited by these neurons following muscarinic stimulation. More than one mechanism appears to be responsible for generation of the ADP in different central neurons. In neurons of the sensorimotor cortex, septal nucleus, or hippo-campus,[31,84,85,87] the ADP that follows a burst of action potentials in the presence of muscarinic agonists was associated with an increase in membrane conductance, implying that it resulted from the activation of an inward current. Induction of the ADP and its underlying current (I_{ADP}) was prevented by atropine, indicating mediation by muscarinic receptor activation. The estimated reversal potentials for I_{ADP} in these neurons ranged between −30 and −50 mV, and these were altered by changes in extracellular Na^+ or K^+, but not Cl^-, concentrations. The ADP and I_{ADP} were decreased by Ca^{2+} channel blockers or removal of extracellular Ca^{2+}, and I_{ADP} amplitude could be correlated with the magnitude of the depolarization-induced increase in intracellular Ca^{2+} concentration, as revealed by microfluorometric measurements using fura-2.[87] These findings suggest that the ADP arose from the action of a Ca^{2+}-activated nonspecific cation conductance. This current is normally present in septal neurons, where it contributes to the production of spontaneous bursting activity, an action that is augmented by muscarinic receptor activation.[84,85] In sensorimotor cortical and hippocampal neurons, muscarinic stimulation promotes repetitive firing by inhibiting the slow AHP, while at the same time activating the inward cation current responsible for generation of the slow ADP.[31,87]

In contrast, Constanti and Bagetta[86] have reported quite different results in neurons of the guinea pig olfactory cortex. In these cells, muscarinic receptor activation results in generation of an ADP that is associated with a *decrease* in membrane conductance. In addition, this depolarization and its underlying

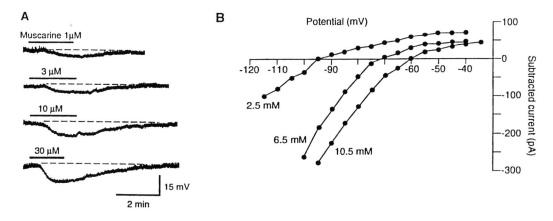

Figure 8-6 Muscarine hyperpolarizes rat nucleus raphé magnus neurons by activating an inwardly rectifying K⁺ current. (A) Superfusion of muscarine during the indicated periods hyperpolarized this neuron in a dose-dependent manner. (B) Current-voltage plots of the steady-state muscarine-induced current in three different external K⁺ concentrations from the same neuron as in (A). The current shows inward rectification, and its reversal potential was shifted positively by increases in extracellular K⁺ concentration. (Modified from Pan, Z. Z. and Williams, J. T., *J. Neurosci.*, 14, 1332, 1994. With permission.)

inward current were dependent on extracellular K⁺ concentration and were blocked by TEA, suggesting that it arose from the blockade of a K⁺ current. However, the depolarization was also sensitive to extracellular Ca²⁺, being blocked by Cd²⁺ or by perfusion of Ca²⁺-free solutions. These authors proposed that calcium entry into the neuron during a burst of action potentials produced a slow Ca²⁺-mediated deactivation of a novel K⁺ current activated by muscarinic receptor stimulation. Inhibition of the outward K⁺ current resulted in production of the ADP. A similar mechanism may underlie production of the muscarinic ADP in pyramidal neurons of the basolateral amygdala. In these neurons, the ADP that is recorded after a burst of action potential in the presence of muscarinic agonists showed a reversal potential near E_K.[35] The ADP of these neurons could be blocked by Cd²⁺ or by 1 μM pirenzepine, but not by 1 μM 4-DAMP, a selective antagonist of M_3 muscarinic receptors, suggesting mediation by the M_1 receptor subtype (Washburn and Moises, unpublished observations) (Figure 8-5A).

The stimulation of muscarinic receptors produces membrane hyperpolarization and neuronal inhibition in neurons from a number of different subcortical and brainstem regions, including neurons of the parabrachial nucleus, thalamic reticular neurons, projection neurons and GABAergic interneurons of the medial and lateral geniculate nuclei, pontine reticular neurons, and neurons of the nucleus raphé magnus.[20,88-92] The inhibition was due to the muscarinic activation of a K⁺ conductance (Figure 8-6), since the response was associated with a marked decrease in input resistance and the presence of an outward current in voltage-clamp, and its reversal potential changed in a Nernstian manner with alterations of extracellular K⁺ concentration.[20,88,91,92] In pontine reticular and raphé nucleus neurons, the conductance was inwardly rectifying, implying that the depressant action of cholinergic stimulation was greater at more negative membrane potentials.[42,91,92] There is a general agreement that the muscarinic receptor subtype that mediates activation of the hyperpolarizing K⁺ conductance is a pirenzepine-insensitive receptor, probably the M_2 receptor.[88,89,91-93] The K⁺ channels responsible for production of the muscarinic hyperpolarization were also opened by activation of receptors for other neurotransmitters, such as mu-opioid and GABA_B receptors in parabrachial nucleus neurons,[93] GABA_B receptors in geniculate neurons,[33] and serotonin receptors in nucleus raphé magnus neurons.[92] These findings indicate that several neurotransmitter systems can have a convergence of action onto the same K⁺ conductance.

C. MUSCARINIC MODULATION OF POTASSIUM CONDUCTANCES

Stimulation of muscarinic receptors can result in changes in central neuron firing patterns via the modulation of several types of voltage-activated K⁺ conductances, including the A-current (I_A) and a delayed rectifier current. Acetylcholine acts on rat neostriatal neurons to decrease their responsiveness to excitatory synaptic inputs, an effect caused in part by the muscarinic modulation of I_A (Figure 8-7).[94,95]

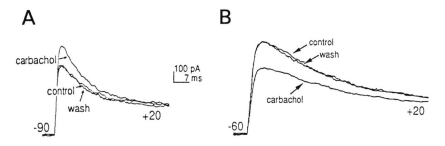

Figure 8-7 Muscarinic modulation of the A-current in cultured embryonic neostriatal neurons. (A) Application of carbachol (100 μM) increased the amplitude of the transient A-current evoked by a voltage step to +20 mV following a 500-ms prepulse to –90 mV. Current amplitude returned to the control level after washout of carbachol. (B) When evoked from a relatively depolarized membrane potential (–60 mV), application of carbachol attenuated I_A amplitude. (From Akins, P. T., Surmeier, D. J., and Kitai, S. T., *Nature (Lond.)*, 344, 240, 1990. Reprinted with permission from *Nature*. Copyright 1990, Macmillan Magazines Limited.)

The A-current is a rapidly inactivating K⁺ current elicited by depolarization from negative membrane potentials that contributes to action potential repolarization and the setting of the firing threshold.[96] In neostriatal neurons, carbachol shifted the voltage dependence of I_A activation and inactivation to more negative membrane potentials, an effect that was blocked by atropine, but not by nicotinic receptor antagonists.[94,95] These neurons tend to maintain two different resting potential levels of –75 or –55 mV, depending on the level of activity of their cortical inputs. Consequently, the 10 to 15 mV negative shift in I_A voltage dependence induced by muscarinic receptor activation had several functional implications. In cells with a hyperpolarized resting potential, the negative shift in I_A activation enabled the A-current to be activated at membrane potentials subthreshold for spike discharge, where it acted to oppose excitatory synaptic inputs, thus maintaining the cell in its quiescent state. In contrast, in cells with a depolarized resting potential, the shift in voltage dependency of I_A inactivation promoted depolarization-induced inactivation and removal of I_A, making these neurons more excitable and allowing them to actively discharge. Thus, the action of ACh on neostriatal neurons was neither excitatory nor inhibitory, but rather served to maintain the present status of the neuron, stabilizing it in either the quiescent or excitable state.

Modulation of I_A has also been observed in cultured rat hippocampal pyramidal neurons.[97] In these cells, ACh or muscarinic agonists produced reductions in I_A amplitude by shifting both the activation and inactivation curves in the depolarizing direction. This change resulted in an increase in the amplitude and duration of individual action potentials. However, similar results have not been observed in recordings from hippocampal neurons in slice preparations taken from adult animals.[5,65]

Muscarinic receptor activation can also potentiate a delayed rectifier-like K⁺ current (I_K) in CA1 hippocampal neurons.[98] This current was normally activated by depolarization, showed slow activation kinetics, did not inactivate, and was sensitive to blockade by TEA. Application of carbachol, in the presence of Na⁺ and Ca²⁺ channel blockers, potentiated I_K amplitude, and atropine blocked this effect. High-frequency stimulation of cholinergic afferents in the hippocampal slice preparation also potentiated I_K, suggesting that the muscarinic modulation of this current may contribute to the functional responses of hippocampal neurons to cholinergic stimulation. The muscarinic potentiation of I_K was mimicked by application of phorbol ester or by intracellular perfusion of GTP-γ-S, and was blocked by inhibitors of protein kinase C. These findings indicate that a G-protein linked to activation of PI hydrolysis may mediate the muscarinic potentiation of I_K. An additional finding was that intracellular perfusion of adenosine nucleotide analogues that do not support kinase-mediated phosphorylation resulted in a time-dependent decrease in I_K amplitude and blockade of carbachol's effect on the current. This raises the possibility that the muscarinic potentiation of I_K required a phosphorylation-dependent process.

D. MUSCARINIC MODULATION OF CALCIUM CURRENTS

An important result of muscarinic receptor activation is the reduction of presynaptic release of neurotransmitter, as measured by a decrease in the amplitudes of the excitatory and inhibitory postsynaptic potentials.[5,6,35,58,99] One mechanism that could contribute to this action is a muscarinic-induced reduction of Ca²⁺ influx through voltage-activated Ca²⁺ channels into the presynaptic nerve terminal.[100] Initial evidence that muscarinic receptor activation can modulate voltage-dependent Ca^{2+} conductances in neurons was provided by studies of Misgeld and co-workers,[101] who showed that ACh or carbachol reduced the Ca^{2+}-dependent plateau potentials recorded in the cell bodies of rat neostriatal neurons. When intracellular recordings were made in the presence of the K^+ channel blocker TEA, depolarization of the neuron produced a sustained action potential with a long-lasting Ca^{2+}-dependent plateau. Cholinergic agonists shortened the duration of these Ca^{2+} action potentials in an atropine sensitive manner. This effect was associated with a decrease in membrane conductance during the plateau phase, suggesting that it resulted from a decrease in Ca^{2+} influx rather than activation of a competing outward current. A similar depressant action of ACh or muscarine on Ca^{2+} action potentials has been reported in CA3 pyramidal cells in hippocampal slice cultures prepared from neonatal rats.[102,103] On the other hand, we were unable to demonstrate any effect of carbachol on Ca^{2+}-action potentials in pyramidal neurons of the adult rat basolateral amygdala.[35]

Voltage-clamp recordings from hippocampal neurons in slice cultures have provided more direct evidence that muscarinic receptor activation inhibits Ca^{2+} currents.[102,103] Muscarinic suppression of Ca^{2+} currents (or channels) also has been described in primary cultures of rat hippocampal neurons[104-106] and in CA1/CA3 neurons of the guinea pig hippocampal slice,[107] as well as in acutely isolated rat magnocellular basal forebrain cholinergic neurons[108] (Figure 8-8). Although not covered in this review, it should be mentioned that inhibition of Ca^{2+} currents via muscarinic receptors has been well documented in rat sympathetic neurons[9,109,110] and in the neuronal-like NG108-15 cell line.[111,112]

Although Gähwiler and co-workers[102,103] demonstrated that muscarinic agonists directly inhibited Ca^{2+} currents in cultured hippocampal neurons, the agonist-sensitive current could not be dissected into components contributed by specific types of Ca^{2+} channels due to the limitations of the single-electrode voltage clamp. Subsequently, a number of groups have used patch-clamp recording techniques to identify several kinetically and pharmacologically distinct components of Ca^{2+} current (and channel types) in hippocampal neurons of different species and at various developmental stages.[113-116] Toselli and his colleagues[104-106,113] reported in cultured hippocampal neurons that activation of muscarinic receptors differentially affected two components of whole-cell Ca^{2+} current, which were distinguished by their voltage dependence of activation, inactivation kinetics, and sensitivities to blockade by Ca^{2+} channel antagonists. Small depolarizing steps (to -50 mV) applied from a relatively negative holding potential (-80 to -100 mV) elicited a rapidly inactivating low voltage-activated (LVA) current that was blocked by nickel, but not by ω-conotoxin GVIA (ω-CgTx), or verapamil. This component corresponds to the low-threshold transient or T-type current described in chick sensory neurons.[117] Larger voltage steps to 0 mV generated an inward current of greater magnitude that could often be resolved kinetically into an inactivating component superimposed on a slowly and incompletely inactivating pedestal current. In addition, two distinct components of the high voltage-activated (HVA) current were resolved pharmacologically, one that was blocked irreversibly by ω-CgTx (thus corresponding to N-type current) and another that was resistant to the toxin but enhanced by the dihydropyridine agonist BAY K8644 (indicative of L-type current).[113] Administration of ACh increased the amplitude of the LVA current in a small faction of cultured rat hippocampal neurons, and this effect was blocked by atropine, but persisted after pretreatment of neurons with PTX. In contrast, in a majority of neurons the HVA current (which was not discriminated into N and L components) was reversibly reduced by ACh, also acting via muscarinic receptors. The muscarinic inhibitory effect was abolished by pretreatment with PTX and mimicked by intracellular application of GTP-γ-S. Moreover, intracellular dialysis of purified G_o- and G_i-type G-proteins or the α subunit of G_o reconstituted the response to ACh in PTX-treated neurons. These data indicate that a PTX-sensitive G-protein, most likely of the G_o type, is involved in the signal transduction mechanism that couples muscarinic receptor activation to inhibition of HVA Ca^{2+} channel activity. However, experiments aimed at identifying second messenger systems responsible for the muscarinic inhibition of HVA current revealed that the effect of ACh was unaffected by intracellular dialysis of neurons either with cyclic-AMP or the protein kinase C inhibitor H7, used at concentrations sufficient to also inhibit cyclic-AMP-dependent and cyclic-GMP protein kinases.[105] In addition, intracellular recordings performed in combination with microfluorometric measurements of cytosolic Ca^{2+} concentration argued against the possibility that the inhibitory effect of ACh resulted from a rise in intracellular free Ca^{2+} and enhanced Ca^{2+}-induced inactivation of the evoked currents.[103] Further studies will be required to determine whether the signaling pathway that couples muscarinic receptors to Ca^{2+} channels in hippocampal neurons involves a novel type of diffusible messenger or might be membrane delimited, possibly mediated via a direct action by G-proteins.

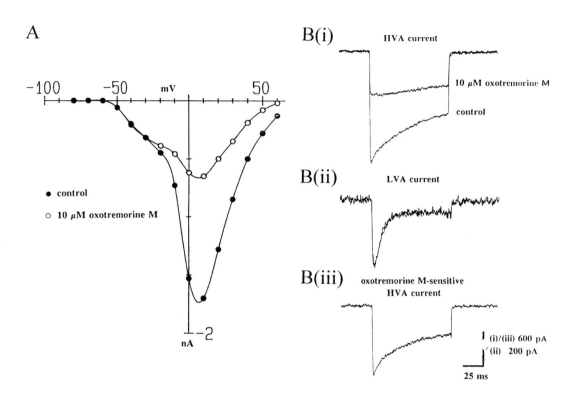

Figure 8-8 The effects of the muscarinic receptor agonist oxotremorine methiodide (oxo-M) on the whole-cell Ca^{2+} current in a magnocellular basal forebrain neuron. (A) Peak current-voltage curves constructed by imposing 100-ms depolarizing voltage steps evoked from a holding potential of −80 mV under control conditions (filled circles) and in the presence of 10 μM oxo-M (open circles). Currents have been corrected for leak by subtracting the current recorded in 200 μM Cd^{2+}. Oxo-M inhibited a component of the high-voltage-activated (HVA) current, but had no effect on the low voltage-activated (LVA) component. (B) Leak subtracted current traces from the neuron shown in (A). B(i) illustrates inhibition of the HVA current by oxo-M at a test potential of +10 mV, while B(ii) shows the lack of effect on the transient LVA current evoked at a test potential of −30 mV and B(iii) the digitally subtracted oxo-M-sensitive component of HVA at a test potential of +10 mV. (From Allen, T. G. J. and Brown, D. A., *J. Physiol. (Lond.)*, 466, 173, 1993. With permission.)

The results obtained from whole-cell recordings in cultured neurons were extended by Fisher and Johnston,[107] who recorded unitary currents through single Ca^{2+} channels in cell-attached patches from CA3 pyramidal neurons in adult guinea pig hippocampal slices. Three types of voltage-gated Ca^{2+} channels were identified in the neurons, based on single-channel conductance, voltage-dependence of activation, and inactivation, and sensitivity to the dihydropyridine agonist BAY K8644.[118] These channel types displayed extensive similarity to those described previously in chick sensory neurons[117] and were therefore given the designation of T-, L-, and N-type. The L- (25 pS, sensitive to BAY K) and N-type (14 pS, BAY K insensitive) channels required large depolarizations to open, and ensemble averages of the single-channel currents yielded sustained and inactivating macroscopic HVA currents, respectively. T-type (8 pS) channels were readily activated by small depolarizations from negative holding potentials and produced a rapidly inactivating LVA whole-cell current. Application of carbachol or muscarine increased the frequency of openings of T-type channels in a patch and reduced the activity of L-type channels, but had no consistent effect on the activity of N-type channels. The effects of carbachol were blocked by low concentrations of atropine, indicating mediation by muscarinic receptors. The fact that unitary current amplitudes were unaffected by the cholinergic agonists suggested that the effect of muscarinic receptor activation was to increase the opening probability of T-type channels and reduce that of L-type channels. However, the possibility could not be ruled out that changes in the number of functional channels within the patch contributed to the observed alterations in channel activity. Nonetheless, these single-channel data are in good agreement with and provide a mechanistic accounting for the muscarinic receptor-induced enhancement of LVA and inhibition of HVA Ca^{2+} currents found in cultured hippocampal neurons.[104-106]

The results of these single-channel experiments are of particular interest in two additional ways. First, because the recordings were from cell-attached patches, drugs applied in the bath did not have direct access to the channels or receptors in the patch. Therefore, the ability of the cholinergic agonists to modify T- and L-type channel activity indicates that muscarinic receptors must be coupled to the Ca^{2+} channels through activation of an intracellular second messenger capable of diffusing into an on-cell patch. Second, the data suggest strongly that the net effect of activation of cholinergic input to CA3 pyramidal neurons is unlikely to translate simply into increases or decreases in Ca^{2+} influx. Rather, it might serve to remodel the profile of Ca^{2+} influx across the neuron so as to favor Ca^{2+} responses to small depolarizing stimuli, while minimizing Ca^{2+} responses to large depolarizations that could lead to excitotoxic damage. This proposed scheme is entirely consistent with the well-documented role that Ca^{2+}-dependent mechanisms and changes in intracellular Ca^{2+} dynamics are thought to play in various forms of synaptic plasticity within the hippocampus, most notably long-term potentiation.

Whole-cell recordings have also been used to examine the modulation of Ca^{2+} currents in acutely dissociated magnocellular cholinergic basal forebrain neurons from neonatal rats.[119] These cells, along with other cholinergic neurons of the basal forebrain complex, provide the major ascending projections of ACh-containing fiber input to large areas of the telencephalon, including the cerebral cortex, hippocampus, and amygdala.[120] The release of ACh into the structures innervated by these neurons is modulated in an inhibitory manner by muscarinic agonists acting on what are thought to be presynaptic autoreceptors.[99] Hence, the characterization of cholinergic action on Ca^{2+} currents of these source neurons, compared to hippocampal target cells, might yield more relevant insights regarding muscarinic mechanisms involved in the modulation of neurotransmitter release. In fact, the effects of cholinergic agonists on Ca^{2+} currents in magnocellular basal forebrain cholinergic neurons (BFCN) are somewhat different from those reported in hippocampal neurons. For example, administration of ACh or oxotremorine methiodide (oxo-M) to BFCN inhibited a transient component of HVA current, but had no effect on LVA T-type currents (Figure 8-8). The agonist-induced inhibition of Ca^{2+} current became irreversible when GTP-γ-S was contained in the patch pipette, and was abolished by dialysis with GDP-β-S or pretreatment of the neurons with PTX. Thus, the muscarinic suppression of HVA current in BFCN was mediated by a PTX-sensitive G-protein, as in hippocampal neurons (see above). However, a large fraction (approximately 55%) of the agonist-sensitive current in BFCN was blocked irreversibly by ω-CgTx, indicating that this component of current was contributed by N-type channels. In addition, the residual ω-CgTx-resistant component of agonist-sensitive current was unlikely to correspond to L-type current, since only a very small proportion of the HVA current in these neurons was inhibited by dihydropyridines.[119] Rather, the ω-CgTx-resistant component of muscarinic-sensitive current exhibited kinetic and pharmacological properties that correspond to those of the P-type current described initially by Llinás et al.[121] It should be noted that the pA_2 values for pirenzepine (6.8) and methoctramine (8.2) obtained by Schild analysis clearly established the involvement of the M_2 subtype of muscarinic receptors in mediating Ca^{2+} current inhibition in these neurons. Recent studies suggest that inhibitory autoreceptors on central cholinergic nerve endings may also be of the M_2 pharmacological subtype.[122] Thus, if the results obtained from these recordings of somal Ca^{2+} currents mirror what occurs at the presynaptic terminal (a currently untestable assumption), then autoreceptor-mediated inhibition of ACh release may occur through a direct suppression of Ca^{2+} influx through N- and possibly P-type Ca^{2+} channels.

IV. CONCLUSION

Acetylcholine elicits a wide variety of responses in neurons of the mammalian central nervous system (CNS). In some areas of the brain, ACh functions as a conventional transmitter in mediating fast, excitatory neurotransmission in a manner analogous to its actions at peripheral ganglionic synapses and the neuromuscular junction. In addition, ACh serves a more widespread role as a neuromodulator in diverse areas of the brain, acting to regulate neuronal excitability by depolarizing or hyperpolarizing neurons, altering neurotransmitter release, or shifting neuronal firing patterns from a bursting behavior to the generation of single spike activity. These responses can result from the activation, inhibition, or modulation of a variety of ionic channels. Understanding the functional consequences of these changes is complicated by the fact that often a single cell type will display more than one cholinergic response. For example, ACh induces a slow depolarization and increases neuronal excitability in hippocampal pyramidal neurons via the muscarinic inhibition of three different K^+ conductances, I_M, I_{Leak}, and I_{AHP}. However, potentiation of a delayed rectifier-like current (I_K), as well as the inhibition and potentiation of different Ca^{2+} currents have also been observed in these neurons. Similarly, electrical stimulation of cholinergic afferents to medial geniculate neurons elicits a fast nicotinic depolarization, followed by a muscarinic hyperpolarization that results from activation of a K^+ conductance, and a subsequent slow muscarinic depolarization generated by the inhibition of a resting K^+ conductance (Figure 8-1).[20,34] This diversity of responses indicates that neurons can have several response patterns, mediated by different receptor subtypes or different intracellular transduction pathways, each of which might be differentially regulated. Studies in which mRNAs for various muscarinic receptor subtypes were transfected into Chinese hamster ovary (CHO) cells demonstrated that different types of G-proteins can couple selectively to particular muscarinic receptors, resulting in significant variations in the activation of PI turnover.[123] Findings such as this indicate that much remains to be learned concerning the role of ACh in the brain and the intracellular mechanisms that underlie cholinergic actions.

REFERENCES

1. Loewi, O., Uber humorale ubertragbarkeit der herznervenwirkung, *Pflugers Arch.*, 189, 239,1921.
2. Dale, H., Chemical transmission of the effects of nerve impulses, *Br. Med. J.*, 12, 835, 1934.
3. Eccles, J. C., Eccles, R. M., and Fatt, P., Pharmacological investigations on a central synapse operated by acetylcholine, *J. Physiol. (Lond.)*, 131, 154, 1956.
4. North, R. A., Muscarinic cholinergic receptor regulation of ion channels, in *The Muscarinic Receptors*, Brown, J. H., Ed., Humana Press, Clifton, NJ, 1989, 341.
5. Halliwell, J. V., Physiological mechanisms of cholinergic action in the hippocampus, *Prog. Brain Res.* 84, 255, 1990.
6. Nicoll, R. A., Malenka, R. C., and Kauer, J. A., Functional comparison of neurotransmitter receptor subtypes in mammalian central nervous system, *Physiol. Rev.*, 70, 513, 1990.
7. Brown, D., A. G-proteins and potassium currents in neurons, *Annu. Rev. Physiol.*, 52, 215, 1990.
8. Brown, A. M. and Birnbaumer, L., Ionic channels and their regulation by G protein subunits, *Annu. Rev. Physiol.*, 52, 197, 1990.
9. Hille, B., G protein-coupled mechanisms and nervous signaling, *Neuron*, 9, 187, 1992.
10. Birnbaumer, L., G proteins in signal transduction, *Annu. Rev. Pharmacol. Toxicol.*, 30, 675, 1990.
11. Simon, M. I., Strathmann, M. P., and Gautam, N., Diversity of G proteins in signal transduction, *Science*, 252, 802, 1991.
12. Bonner, T. I., The molecular basis of muscarinic receptor diversity, *Trends Neurosci.*, 12, 148, 1989.
13. Buckley, N. J., Bonner, T. I., Buckley, C. M., and Brann, M. R., Antagonist binding properties of five cloned muscarinic receptors expressed in CHO-K1 cells, *Mol. Pharmacol.*, 35, 469, 1989.
14. Hulme, E. C., Birdsall, N. J. M., and Buckley, N. J., Muscarinic receptor subtypes, *Annu. Rev. Pharmacol. Toxicol.*, 30, 633, 1990.
15. Agranoff, B. W., Fisher, S. K., Heacock, A. M., and Frey, K. A., The phosphoinositide-linked CNS muscarinic receptor, *Adv. Exp. Med. Biol.*, 236, 195, 1988.
16. El-Fakahany, E. E., Alger, B. E., Lai, W. S., Pitler, T. A., Worley, P. F., and Baraban, J. M., Neuronal muscarinic responses: role of protein kinase C, *FASEB J.*, 2, 2575, 1988.
17. Fisher, S. K., Heacock, A. M., and Agranoff, B. W., Inositol lipids and signal transduction in the nervous system: an update, *J. Neurochem.* 58, 18, 1992.
18. Egan, T. M. and North, R. A., Actions of acetylcholine and nicotine on rat locus ceruleus neurons, *Neuroscience*, 19, 565, 1986.
19. McCormick, D. A. and Prince, D. A., Acetylcholine causes rapid nicotinic excitation in the medial habenular nucleus of guinea pig, *in vitro*, *J. Neurosci.*, 7, 742, 1987.
20. McCormick, D. A. and Prince, D. A., Actions of acetylcholine in the guinea-pig and cat medial and lateral geniculate nuclei, *in vitro*, *J. Physiol. (Lond.)*, 392, 147, 1987.
21. Phelan, K. D. and Gallagher, J. P., Direct muscarinic and nicotinic receptor-mediated excitation of rat medial vestibular nucleus neurons *in vitro*, *Synapse*, 10, 349, 1992.

22. Mulle, C., Choquet, D., Korn, H., and Changeux, J. P., Calcium influx through the nicotinic receptor in rat central neurons: its relevance to cellular regulation, *Neuron,* 8, 135, 1992.

23. Wong, L. A. and Gallagher, J. P., A direct nicotinic receptor-mediated inhibition recorded intracellularly *in vitro, Nature (Lond.),* 341, 439, 1989.

24. Wong, L. A. and Gallagher, J. P., Pharmacology of nicotinic receptor-mediated inhibition in rat dorsolateral septal neurones, *J. Physiol. (Lond.),* 436, 325, 1991.

25. Biscoe, T. J. and Straughn, D. W., Microelectrophoretic studies of neurones in cat hippocampus, *J. Physiol. (Lond.),* 404, 479, 1966.

26. Benardo, L. S. and Prince, D. A., Ionic mechanisms of cholinergic excitation in mammalian hippocampal pyramidal cells, *Brain Res.,* 249, 333, 1982.

27. Cole, A. E. and Nicoll, R. A., Acetylcholine mediates a slow synaptic potential in hippocampal pyramidal cells, *Science,* 221, 1299, 1983.

28. Cole, A. E. and Nicoll, R. A., The pharmacology of cholinergic excitatory responses in hippocampal pyramidal cells, *Brain Res.,* 305, 283, 1984.

29. Cole, A. E. and Nicoll, R. A., Characterization of slow cholinergic post-synaptic potential recorded *in vitro* from rat hippocampal pyramidal cells, *J. Physiol. (Lond.),* 352, 173, 1984.

30. McCormick, D. A. and Prince, D. A., Mechanisms of action of acetylcholine in the guinea-pig cerebral cortex *in vitro, J. Physiol. (Lond.),* 375, 169, 1986.

31. Schwindt, P. C., Spain, W. J., Foehring, R. C., Chubb, M. C., and Crill, W. E. Slow conductances in neurons from cat sensorimotor cortex *in vitro* and their role in slow excitability changes, *J. Neurophysiol.,* 59, 450, 1988.

32. Lacey, M. G., Calabresi, P., and North, R. A., Muscarine depolarizes rat substantia nigra zona compacta and ventral tegmental neurons *in vitro* through M_1-like receptors, *J. Pharmacol. Exp. Ther.* 253, 395, 1990.

33. McCormick, D. A., Cellular mechanisms underlying cholinergic and noradrenergic modulation of neuronal firing mode in the cat and guinea pig dorsal lateral geniculate nucleus, *J. Neurosci.,* 12, 278, 1992.

34. McCormick, D. A., Neurotransmitter actions in the thalamus and cerebral cortex and their role in neuromodulation of thalamocortical activity, *Prog. Neurobiol.,* 39, 337, 1992.

35. Washburn, M. S. and Moises, H. C., Muscarinic responses of rat basolateral amygdaloid neurons recorded *in vitro, J. Physiol. (Lond.),* 449, 121, 1992.

36. Benardo, L. S., Characterization of cholinergic and noradrenergic slow excitatory postsynaptic potentials from rat cerebral cortical neurons, *Neuroscience,* 53, 11, 1993.

37. Wang, Z. and McCormick, D. A., Control of firing mode of corticotectal and corticopontine layer V burst-generating neurons by norepinephrine, acetylcholine, and 1S,3R-ACPD, *J. Neurosci.,* 13, 2199, 1993.

38. Richardson, R. T. and DeLong, M. R., A reappraisal of the functions of the nucleus basalis of Meynert, *Trends Neurosci.,* 11, 264, 1988.

39. Krnjevic, K., Pumain, R. and Renaud, L., The mechanism of excitation by acetylcholine in the cerebral cortex, *J. Physiol. (Lond.),* 215, 247, 1971.

40. Benardo, L. S. and Prince, D. A., Cholinergic excitation of mammalian hippocampal pyramidal cells, *Brain Res.,* 249, 315, 1982.

41. Dodt, H. U. and Misgeld, U., Muscarinic slow excitation and muscarinic inhibition of synaptic transmission in the rat neostriatum, *J. Physiol. (Lond.),* 380, 593, 1986.

42. Greene, R. W., Gerber, U., and McCarley, R. W., Cholinergic activation of medial pontine reticular formation neurons *in vitro. Brain Res.,* 476, 154, 1989.

43. McCormick, D. A. and Williamson, A., Convergence and divergence of neurotransmitter action in human cerebral cortex, *Proc. Natl. Acad. Sci. U.S.A.,* 86, 8098, 1989.

44. Jones, K. A. and Baughman, R. W., Muscarinic M3 receptors inhibit a leak conductance in rat corticocallosal neurons, *NeuroReport,* 3, 889, 1992.

45. Gähwiler, B. H. and Brown, D. A., Functional innervation of cultured hippocampal neurones by cholinergic afferents from co-cultured septal explants, *Nature (Lond.),* 313, 577, 1985.

46. Segal, M., Synaptic activation of a cholinergic receptor in rat hippocampus, *Brain Res.,* 452, 79, 1988.

47. Brown, D. A. and Adams, P. R., Muscarinic suppression of a novel voltage-sensitive K^+-current in a vertebrate neurone, *Nature (Lond.),* 283, 673, 1980.

48. Halliwell, J. V. and Adams, P. R., Voltage-clamp analysis of muscarinic excitation in hippocampal neurons, *Brain Res.,* 250, 71, 1982.

49. Constanti, A. and Galvan, M., M-current in voltage-clamped olfactory cortex neurones, *Neurosci. Lett.,* 39, 65, 1983.

50. Halliwell, J. V., M-current in human neocortical neurones, *Neurosci. Lett.,* 67, 1, 1986.

51. Womble, M. D. and Moises, H. C., Muscarinic inhibition of M current and a potassium leak conductance in neurones of the rat basolateral amygdala, *J. Physiol.,* 457, 93, 1992.

52. Nowak, L. M. and Macdonald, R. L., Muscarinic-sensitive voltage-dependent potassium current in cultured murine spinal cord neurons, *Neurosci. Lett.,* 35, 85, 1983.

53. Constanti, A. and Sim, J. A., Calcium-dependent potassium conductances in guinea-pig olfactory cortex neurones *in vitro, J. Physiol. (Lond.),* 387, 173, 1987.

54. Madison, D. V., Lancaster, B., and Nicoll, R. A., Voltage clamp analysis of cholinergic action in the hippocampus, *J. Neurosci.,* 7, 733, 1987.

55. Madison, D. V. and Nicoll, R. A., Control of the repetitive discharge of rat CA1 pyramidal neurones *in vitro, J. Physiol. (Lond.),* 354, 319, 1984.

56. Benson, D. M., Blitzer, R. D., and Landau, E. M., An analysis of the depolarization produced in guinea-pig hippocampus by cholinergic receptor stimulation., *J. Physiol.,* 404, 479, 1988.

57. Constanti, A. and Sim, J. A., Muscarinic receptors mediating suppression of the M-current in guinea-pig olfactory cortex neurones may be of the M_2-subtype, *Br. J. Pharmacol.,* 90, 3, 1987.

58. Dutar, P. and Nicoll, R. A., Classification of muscarinic responses in hippocampus in terms of receptor subtypes and second-messenger systems, electrophysiological studies *in vitro, J. Neurosci.,* 8, 4214, 1988.

59. Müller, W. and Misgeld, U., Slow cholinergic excitation of guinea pig hippocampal neurons is mediated by two muscarinic receptor subtypes, *Neurosci. Lett.,* 67, 107, 1986.

60. Pitler, T. A. and Alger, B. E., Activation of the pharmacologically defined M_3 muscarinic receptor depolarizes hippocampal pyramidal cells, *Brain Res.,* 534, 257, 1990.

61. Brown, L. D., Kim, K.-M., Nakajima, Y., and Nakajima, S., The role of G protein in muscarinic depolarization near resting potential in cultured hippocampal neurons, *Brain Res.,* 612, 200, 1993.

62. Dutar, P. and Nicoll, R. A., Stimulation of phosphatidylinositol (PI) turnover may mediate the muscarinic suppression of the M-current in hippocampal pyramidal cells, *Neurosci. Lett.,* 85, 89, 1988.

63. Malenka, R. C., Madison, D. V., Andrade, R., and Nicoll, R. A., Phorbol esters mimic some cholinergic actions in hippocampal pyramidal neurons, *J. Neurosci.,* 6, 475, 1986.

64. Lancaster, B. and Adams, P. R., Calcium-dependent current generating the afterhyperpolarization of hippocampal neurons, *J. Neurophysiol.,* 55, 1268, 1986.

65. Lancaster, B. and Nicoll, R. A., Properties of two calcium-activated hyperpolarizations in rat hippocampal neurones, *J. Physiol. (Lond.),* 389, 187, 1987.

66. Schwindt, P. C., Spain, W. J., and Crill, W. E., Calcium-dependent potassium currents in neurons from cat sensorimotor cortex, *J. Neurophysiol.,* 67, 216, 1992.

67. Womble, M. D. and Moises, H. C., Muscarinic modulation of conductances underlying the afterhyperpolarization in neurons of the rat basolateral amygdala, *Brain Res.,* 621, 87, 1993.

68. Baraban, J. M., Snyder, S. H., and Alger, B. E., Protein kinase C regulates ionic conductance in hippocampal pyramidal neurons: electrophysiological effects of phorbol esters, *Proc. Natl. Acad. Sci. U.S.A.,* 82, 2538, 1985.

69. Müller, W., Petrozzino, J. J., Griffith, L. C., Danho, W., and Connor, J. A., Specific involvement of Ca^{2+}-calmodulin kinase II in cholinergic modulation of neuronal responses, *J. Neurophysiol.,* 68, 2264, 1992.

70. Agopyan, N. and Krnjevic, K., Muscarinic actions in hippocampus are probably not mediated by cyclic GMP, *Brain Res.,* 525, 294, 1990.

71. Müller, W. and Connor, J. A., Cholinergic input uncouples Ca^{2+} changes from K^+ conductance activation and amplifies intradendritic Ca^{2+} changes in hippocampal neurons, *Neuron,* 6, 901, 1991.

72. Markram, H. and Segal, M., Long-lasting facilitation of excitatory postsynaptic potentials in the rat hippocampus by acetylcholine, *J. Physiol. (Lond.),* 427, 381, 1990.

73. Malenka, R. C., Kauer, J. A., Zucker, R. S. and Nicoll, R. A., Postsynaptic calcium is sufficient for potentiation of hippocampal synaptic transmission, *Science,* 242, 81, 1988.

74. Malenka, R. C., Kauer, J. A., Perkel, D. J., Mauk, M. D., Kelly, P. T., Nicoll, R. A., and Waxman, M. N., An essential role for postsynaptic calmodulin and protein kinase activity in long-term potentiation, *Nature (Lond.),* 340, 554, 1989.

75. Malinow, R., Schulman, H., and Tsien, R. W., Inhibition of postsynaptic PKC or CaMKII blocks induction but not expression of LTP., *Science,* 245, 862, 1989.

76. Silva, A. J., Stevens, C. F., Tonegawa, S., and Wang, Y., Deficient hippocampal long-term potentiation in α-calcium-calmodulin kinase II mutant mice, *Science,* 257, 201, 1992.

77. Collingridge, G. L., Kehl, S. J., and McLennan, H., Excitatory amino acids in synaptic transmission in the Schaffer collateral-commissural pathway of the rat hippocampus, *J. Physiol. (Lond.),* 334, 33, 1983.

78. Teyler, T. J. and DiScenna, P., Long-term potentiation, *Annu. Rev. Neurosci.,* 10, 131, 1987.

79. Schwindt, P. C., Spain, W. J., and Crill, W. E., Long-lasting reduction of excitability by a sodium-dependent potassium current in cat neocortical neurons., *J. Neurophysiol.,* 61, 233, 1989.

80. Uchimura, N. and North, R. A., Muscarine reduces inwardly rectifying potassium conductance in rat nucleus accumbens neurones., *J. Physiol. (Lond.),* 422, 369, 1990.

81. Shen, K. Z. and North, R. A., Muscarine increases cation conductance and decreases potassium conductance in rat locus ceruleus neurones, *J. Physiol. (Lond.),* 455, 471, 1992.

82. Womble, M. D. and Moises, H. C., Hyperpolarization-activated currents in neurons of the rat basolateral amygdala, *J. Neurophysiol.,* 70, 2056, 1993.

83. Reece, L. J. and Schwartzkroin, P. A., Effects of cholinergic agonists on two non-pyramidal cell types in rat hippocampal slices, *Brain Res.,* 566, 115, 1991.

84. Hasuo, H. and Gallagher, J. P., Facilitatory action of muscarine on slow afterdepolarization of rat dorsolateral septal nucleus neurons *in vitro, Neurosci. Lett.,* 112, 234, 1990.

85. Hasuo, H., Phelan, K. D., and Gallagher, J. P., A calcium-dependent slow afterdepolarization recorded in rat dorsolateral septal nucleus neurons *in vitro, J. Neurophysiol.,* 64, 1838, 1990.

86. Constanti, A. and Bagetta, G., Muscarinic receptor activation induces a prolonged post-stimulus afterdepolarization with a conductance decrease in guinea-pig olfactory cortex neurons *in vitro, Neurosci. Lett.,* 131, 27, 1991.

87. Caeser, M., Brown, D. A., Gähwiler, B. H., and Knöpfel, T. Characterization of a calcium-dependent current generating a slow afterdepolarization of CA3 pyramidal cells in rat hippocampal slice cultures, *Eur. J. Neurosci.,* 5, 560, 1993.

88. Egan, T. M. and North, R. A., Acetylcholine hyperpolarizes central neurones by acting on an M_2 muscarinic receptor, *Nature (Lond.),* 319, 405, 1986.

89. McCormick, D. A. and Prince, D. A., Acetylcholine induces burst firing in thalamic reticular neurones by activating a potassium conductance, *Nature (Lond.),* 319, 402, 1986.

90. McCormick, D. A. and Pape, H. C., Acetylcholine inhibits identified interneurons in the cat geniculate nucleus. *Nature (Lond.),* 334, 246, 1988.

91. Gerber, U., Stevens, D. R., McCarley, R. W., and Greene, R. W., Muscarinic agonists activate an inwardly rectifying potassium conductance in medial pontine reticular formation neurons of the rat *in vitro, J. Neurosci.,* 11, 3861, 1991.

92. Pan, Z. Z. and Williams, J. T., Muscarine hyperpolarizes a subpopulation of neurons by activating an M_2 muscarinic receptor in rat nucleus raphe magnus *in vitro, J. Neurosci.,* 14, 1332, 1994.

93. Christie, M. J. and North, R. A., Agonists at mu-opioid, M_2-muscarinic and $GABA_B$-receptors increase the same potassium conductance in rat lateral parabrachial neurones, *Br. J. Pharmacol.,* 95, 896, 1988.

94. Akins, P. T., Surmeier, D. J., and Kitai, S. T., Muscarinic modulation of a transient K^+ conductance in rat neostriatal neurons, *Nature (Lond.),* 344, 240, 1990.

95. Kitai, S. T. and Surmeier, D. J., Cholinergic and dopaminergic modulation of potassium conductances in neostriatal neurons, *Adv. Neurol.,* 60, 40, 1993.

96. Storm, J. F., Potassium currents in hippocampal pyramidal cells, *Prog. Brain Res.,* 83, 161, 1990.

97. Nakajima, Y., Nakajima, S., Leonard, R. J., and Yamaguchi, K., Acetylcholine raises excitability by inhibiting the fast transient potassium current in cultured hippocampal neurones, *Proc. Natl. Acad. Sci. (U.S.A.,)* 83, 3022, 1986.

98. Zhang, L., Weiner, J. L., and Carlen, P. L., Muscarinic potentiation of I_K in hippocampal neurons: electrophysiological characterization of the signal transduction pathway, *J. Neurosci.,* 12, 4510, 1992.

99. Starke, K., Gothert, M., and Kilbringer, H., Modulation of neurotransmitter release by presynaptic autoreceptors, *Physiol. Rev.,* 69, 864, 1989.

100. Anwyl, R., Modulation of vertebrate neuronal calcium channels by transmitters, *Brain Res. Rev.,* 16, 265, 1991.

101. Misgeld, U., Calabresi, P., and Dodt, H. U., Muscarinic modulation of calcium dependent plateau potentials in rat neostriatal neurons, *Pflügers Arch.,* 407, 482, 1986.

102. Gähwiler, B. H. and Brown, D. A., Muscarine affects calcium-currents in rat hippocampal pyramidal cells *in vitro, Neurosci. Lett.,* 76, 301, 1987.

103. Knöpfel, T., Vranesic, I., Gähwiler, B. H., and Brown, D. A., Muscarinic and β-adrenergic depression of the slow Ca^{2+}-activated potassium conductance in hippocampal CA3 pyramidal cells is not mediated by a reduction of depolarization-induced cytosolic Ca^{2+} transients, *Proc. Natl. Acad. Sci. U.S.A.*, 87, 4083, 1990.

104. Toselli, M. and Lux, H. D., GTP-binding proteins mediate acetylcholine inhibition of voltage dependent calcium channels in hippocampal neurons, *Pflügers Arch.*, 413, 319, 1989.

105. Toselli, M., Lang, J., Costa, T., and Lux, H. D., Direct modulation of voltage-dependent calcium channels by muscarinic activation of a pertussis toxin-sensitive G-protein in hippocampal neurons, *Pflügers Arch.*, 415, 255, 1989.

106. Toselli, M. and Lux, H. D., Opposing effects of acetylcholine on the two classes of voltage-dependent calcium channels in hippocampal neurons, *Experientia Suppl.*, 57, 97, 1989.

107. Fisher, R. and Johnston, D., Differential modulation of single voltage-gated calcium channels by cholinergic and adrenergic agonists in adult hippocampal neurons, *J. Neurophysiol.*, 64, 1291, 1990.

108. Allen, T. G. J. and Brown, D. A., M_2 muscarinic receptor-mediated inhibition of the Ca^{2+} current in rat magnocellular cholinergic basal forebrain neurones, *J. Physiol. (Lond.)*, 466, 173, 1993.

109. Wanke, E., Ferroni, A., Malgaroli, A., Ambrosini, A., Pozzan, T., and Meldolesi, J., Activation of a muscarinic receptor selectively inhibits a rapidly inactivated Ca^{2+} current in rat sympathetic neurons, *Proc. Natl. Acad. Sci. U.S.A.*, 84, 4313, 1987.

110. Mathie, A., Bernheim, L., and Hille, B., Inhibition of N- and L-type calcium channels by muscarinic receptor activation in rat sympathetic neurons, *Neuron*, 8, 907, 1992.

111. Brown, D. A., Docherty, R. J., and McFadzean, I., Calcium channels in vertebrate neurons. Experiments on a neuroblastoma hybrid model, *Ann. N.Y. Acad. Sci.*, 560, 358, 1989.

112. Caulfield, M. P., Robbins, J., and Brown, D. A., Neurotransmitters inhibit the omega-conotoxin-sensitive component of Ca current in neuroblastoma × glioma hybrid (NG 108-15) cells, not the nifedipine-sensitive component, *Pflügers Arch.*, 420, 486, 1992.

113. Toselli, M. and Taglietti, V., Pharmacological characterization of voltage-dependent calcium currents in rat hippocampal neurons, *Neurosci. Lett.*, 112, 70, 1990.

114. Mogul, D. J. and Fox, A. P., Evidence for multiple types of Ca^{2+} channels in acutely isolated hippocampal CA3 neurones of the guinea-pig, *J. Physiol. (Lond.)*, 433, 671, 1991.

115. Thompson, S. M. and Wong, R. K. S., Development of calcium current subtypes in isolated rat hippocampal pyramidal cells, *J. Physiol. (Lond.)*, 439, 671, 1991.

116. O'Dell, T. J. and Alger, B. E., Single calcium channels in rat and guinea-pig hippocampal neurons, *J. Physiol. (Lond.)*, 436, 739, 1991.

117. Fox, A. P., Nowycky, M. C., and Tsien, R. W., Kinetic and pharmacological properties distinguishing three types of calcium currents in chick sensory neurons. *J. Physiol. (Lond.)*, 394, 149, 1987.

118. Fisher, R. E., Gray, R., and Johnston, D., Properties and distribution of single voltage-gated calcium channels in adult hippocampal neurons, *J Neurophysiol.*, 64, 91, 1990.

119. Allen, T. G. J., Sim, J. A., and Brown, D. A., The whole-cell calcium current in acutely dissociated magnocellular cholinergic basal forebrain neurones of the rat, *J. Physiol. (Lond.)*, 460, 91, 1993.

120. Mesulam, M. M., Mufson, E. J., Wainer, B. H., and Levey, A. I., Central cholinergic pathways in the rat, an overview based on an alternative nomenclature (Ch1-Ch6), *Neuroscience*, 10, 1185, 1983.

121. Llinás, R. R., Sugimori, M., Lin, J.-W., and Cherksey, B., Blocking and isolation of a calcium channel from neurons in mammals and cephalopods utilizing a toxin fraction (FTX) from funnel web spider poison, *Proc. Natl. Acad. Sci. U.S.A.*, 86, 1689, 1989.

122. Richards, M. H., Rat hippocampal muscarinic autoreceptors are similar to the M2 (cardiac) subtype, comparison with hippocampal M_1, atrial M_2 and ileal M_3 receptors, *Br. J. Pharmacol.*, 99, 753, 1990.

123. Ashkenazi, A., Peralta, E. G., Winslow, J. W., Ramachandran, J., and Capon, D. J., Functionally distinct G proteins selectively couple different receptors to PI hydrolysis in the same cell, *Cell*, 56, 487, 1989.

Chapter 9

Neurochemical Transduction Processes Associated with Neuronal Muscarinic Receptors

Jesse Baumgold

CONTENTS

I. INTRODUCTION

Muscarinic receptors are cell-surface proteins belonging to the superfamily of proteins having seven transmembrane domains. Five distinct genes for muscarinic receptors have been cloned and sequenced.[1-3] These five genes encode receptors which all use acetylcholine as the endogenous agonist, and are termed m1, m2, m3, m4, and m5. The morphological distribution, biochemical signaling processes, and pharmacological characteristics of these five receptors have been extensively studied and reviewed.[4-6] Mutagenesis studies have identified domains of the muscarinic receptor involved with ligand binding, agonist-induced activation, and G-protein coupling. These studies have recently been reviewed by Wess.[7]

Muscarinic receptors are abundant in the brain, where they mediate a variety of important functions including learning and memory. Loss of brain cholinergic function in Alzheimer's disease (AD) has led to a large effort in developing selective muscarinic agonists as a possible therapeutic treatment for the loss of cognitive function in AD. Muscarinic receptors are also abundant in the peripheral nervous system where they mediate the parasympathetic function of every innervated organ. Their role in the parasympathetic function of the heart, bladder, lungs, and digestive system has been particularly important therapeutically.

Receptor-mediated stimulation of phosphatidylinositol (PI) hydrolysis has been extensively studied and is the subject of several excellent reviews,[5,8] as is muscarinic receptor-mediated inhibition of adenylate cyclase. However, it is now becoming clear that in addition to these well-known signaling pathways, muscarinic receptors also mediate other signaling pathways such as stimulation of phospholipase A2, phospholipase D, and cyclic adenosine 3′,5′-monophosphate (cAMP) levels; regulation of cell growth; and several others. The evidence for muscarinic receptor mediation of each of these signaling pathways, as well as the possible functional relevance of that signaling pathway, is presented in this review.

II. STIMULATION OF PHOSPHOLIPASE C

The availability of cloned receptors has allowed several studies to demonstrate directly that stimulation of odd-numbered muscarinic receptors (m1, m3, and m5) results primarily in increased release of inositol phosphate (InsPs), whereas stimulation of even-numbered receptors results in inhibition of adenylate cyclase.[9] For example, Peralta et al.[9] transfected the m1–m4 muscarinic receptors into cell lines and found that the m1 and m3 receptors preferentially couple to PI hydrolysis, whereas m2 and m4 receptors preferentially couple to the inhibition of adenylate cyclase.[9] The m5 receptor was also shown to couple preferentially to PI hydrolysis.[2] These three odd-numbered muscarinic receptor subtypes, therefore, belong to a large group of G-protein-coupled receptors which elicit this response. This group includes α-1 adrenergic receptors, bradykinin receptors, vasopressin receptors, histamine receptors, thrombin receptors, bombesin receptors, and many others.

Although m2 and m4 receptors couple preferentially to the inhibition of adenylate cyclase, they are also capable of weakly stimulating PI turnover; this was dramatically demonstrated by a study by Ashkenazi et al.[10] who overexpressed recombinant m2 receptors into Chinese hamster ovary (CHO) cells and found a detectable PI hydrolysis response.

Although the availability of cloned receptors allowed for the direct demonstration of the preferential coupling of the various muscarinic receptor subtypes, many elegant studies performed before cloned muscarinic receptors were available demonstrated that muscarinic-mediated PI hydrolysis has a high sensitivity for pirenzepine in some tissues and a lower sensitivity for pirenzepine in others. Gil and Wolfe,[11] for example, found that in the brain, pirenzepine was 15-fold more potent in inhibiting the PI response compared to the cAMP response and concluded that receptors with high affinity for pirenzepine mediate the PI response, whereas receptors with low affinity for pirenzepine mediate the latter.[11] At that time, muscarinic receptors with high affinity were referred to as M1 receptors, whereas those with low affinity for pirenzepine were referred to as M2 receptors.

In some very elegant studies, Ashkenazi et al.[12] demonstrated that activation of m1 receptors transfected into CHO cells elicits PI hydrolysis via two different pathways: a pertussis-sensitive and a pertussis-insensitive pathway. Pertussis toxin selectively adenosine 5'-diphosphate (ADP) ribosylates G_i and G_o and prevents coupling of receptors to these ribosylated G proteins. Two pertussis-toxin insensitive G proteins which stimulate phosphoinositide hydrolysis are known: G_q and G_{11}.[13,14] These G proteins were shown to stimulate phospholipase C (PLC)-β1.[15,16] Consistent with this, was the finding that reconstitution of m1 receptors with either G_q or G_{11}, and phospholipase C-β1 into phospholipid vesicles stimulated PLC activity in the presence of guanosine triphosphate (GTP)τS.[17]

In order to determine which domains of the muscarinic receptor convey selectivity for particular families of G proteins, a variety of chimeric m2/m3 receptors were constructed by Wess et al.[18] Following transfection into mouse A9 L cells, these authors found that substitution of the third cytoplasmic loop from m2 receptors with that from m3 receptors resulted in an m2 chimeric receptor that displayed a functional response similar to an m3 receptor. The converse experiment resulted in an m3 chimeric receptor containing a third cytoplasmic loop from the m2 receptor and displayed a functional response of an m3 receptor. These experiments clearly demonstrate that the third cytoplasmic loop conveys selectivity for G proteins. Other experiments using m2/m3 chimeric receptors showed that only the N-terminal segment (16 or 17 amino acids) of this third cytoplasmic loop conveyed this selectivity,[18,19] although deletion mutants showed that the carboxyl-terminal portion of the third intracellular loop also plays a significant role in G-protein recognition.[20]

Although the third cytoplasmic loop is clearly important in conveying G-protein coupling selectivity, other domains are also important. Mutational analysis of the m1 receptor has shown that a highly conserved Asp residue located at the beginning of the second intracellular loop (Asp 164 in the rat m3 receptor) is similarly important in efficient G-protein coupling.[21]

Receptor-mediated hydrolysis of phosphatidyl inositol-*bis*-phosphate (PIP$_2$) by phospholipase C elicits an extensive cascade of second and third messengers that have been extensively reviewed.[8,22-26] Phospholipase C-catalyzed hydrolysis of PIP$_2$ releases inositoltriphosphate IP$_3$ and 1,2 diacylglycerol (DAG). IP$_3$ binds to its receptor and mobilizes intracellular calcium resulting in activation of many calcium-dependent processes including activation of calcium-dependent potassium channels, activation of receptor-operated calcium channels, and a long list of other responses. Diacylglycerol activates at least seven different protein kinase C isozymes resulting in the phosphorylation of a wide variety of cellular and membranous proteins.[27,28] These protein kinase C isozymes have distinct tissue and cellular distributions,[28] meaning that muscarinic stimulation of this pathway may result in different functional responses in different tissues, depending on which protein kinase C (PKC) isozyme is expressed by that tissue.

III. INHIBITION OF ADENYLATE CYCLASE

In contrast to the odd-numbered receptors, the m2 and m4 muscarinic receptors couple preferentially to the inhibition of adenylate cyclase through pertussis-toxin-sensitive G proteins.[9] Several different palytoxin

(PTX)-sensitive α subunits of G proteins have been identified by molecular cloning techniques, including two α subunits of G_o and three α subunits of G_i.[29] All three of these G_i subunits appear capable of mediating inhibition of adenylyl cyclase.[30] Dell'Acqua et al.[31] recently used agonist-dependent cholera toxin labeling to investigate which of these PTX-sensitive G-protein subunits interact with m2 and m4 muscarinic receptors in transfected CHO cells. These cells were found to express only G_{i2} and G_{i3}, both of which were coupled to m2 receptors upon carbachol activation.

IV. STIMULATION OF ADENYLATE CYCLASE

Stimulation of m1, m3, and m5 receptors expressed in a variety of cells has been shown to elevate intracellular cAMP levels.[9,32,33] Since direct coupling between muscarinic receptors and stimulation of adenylate cyclase has not been observed in membranes from these cells, this response is not mediated by muscarinic receptor coupling to adenylate cyclase through G_s. Several investigators have raised the possibility that the receptor-mediated rise in intracellular calcium can stimulate calmodulin-sensitive adenylate cyclase.[33,34] However, when this hypothesis was tested by pretreating SK-N-SH neuroblastoma cells (which express primarily m3 receptors) with the intracellular calcium chelator BAPTA, muscarinic stimulation of these cells still elicited increased intracellular cAMP levels even though no increase in intracellular calcium was observed.[35] The most likely explanation of this effect comes from recent studies demonstrating that addition of the βτ subunit of G proteins can stimulate type II and type IV adenylate cyclase directly.[30,36] Thus, stimulation of m1, m3 or m5 muscarinic receptors in these cells activates G_q, G_{11}, or G_{14} and liberates a βτ subunit capable of stimulating type II or type IV adenylate cyclase.

Other recent studies on muscarinic receptors in the rat olfactory bulb have found evidence of direct coupling between these receptors and stimulation of adenylate cyclase via G_s.[37] In membranes from this tissue, muscarinic stimulation elicits increased adenylate cyclase activity in the presence of high (10–100 μM) concentrations of GTP and appears to be mediated by m4 receptors.

V. MODULATION OF ION CHANNELS

Stimulation of muscarinic receptors in various tissues has been shown to modulate ion channels both directly through a G protein, and indirectly as a consequence of receptor-mediated calcium mobilization. These receptors have been recently reviewed by Jones.[38] Stimulation of m2 cardiac receptors activates the α subunit of G_i, which then directly activates potassium channels.[39] The βτ subunit has also been shown to modulate potassium channels.[40]

Stimulation of PI-coupled receptors, including m1, m3, and m5 receptors, results in calcium mobilization and a consequent elevation of intracellular calcium levels. These raised intracellular calcium levels have been found to activate several calcium-dependent conductances. For example, in m1-transfected A9 L cells, stimulation of muscarinic receptors activates a calcium-dependent potassium channel resulting in cellular hyperpolarization.[41] However, it is important to note that the electrophysiological consequence of muscarinic receptor activation is not solely dependent on which muscarinic receptor subtype is expressed by particular cells; it is also dependent on which channels are expressed. Thus, although SK-N-SH human neuroblastoma cells express an abundance of PI-coupled m3 receptors which mobilize calcium, only a slight depolarization (probably due to calcium-dependent chloride channels) is observed in response to receptor activation.[42]

Stimulation of m2 and of m4 muscarinic receptors, but not m1 or m3, has been shown to inhibit the M current,[43] a slow voltage-dependent conductance. This M current has been shown to be modulated by a variety of G-protein-coupled receptors.[44]

VI. STIMULATION OF PHOSPHOLIPASE A2

Considerably less is known about muscarinic receptor-mediated stimulation of phospholipase A2 and phospholipase D than the previously mentioned pathways. Stimulation of m1, m3, and m5 receptors has been shown to elicit release of arachidonic acid in a variety of tissues and cells, whereas stimulation of m2 or m4 receptors does not produce this response.[45,46,47] Many of these early reports proposed that this response was secondary to stimulation of phospholipase C.[48,49] However, Conklin et al.,[45] showed that in A9 L cells transfected with m1 or m3 receptors, pretreatment with phenylmercury acetate (PMA) caused a marked enhancement of [³H]arachidonic acid release whereas this treatment inhibited inositol phosphate formation, suggesting that in these cells [³H]arachidonic acid release and inositol phosphate formation are regulated independently. These studies suggest that [³H]arachidonic acid release is not secondary to stimulation of phospholipase C, but instead this response is elicited as a result of receptor coupling, via a G protein, to phospholipase A2.

VII. STIMULATION OF PHOSPHOLIPASE D

In 1985, it was observed that the time course of accumulation of 1,2 diacylglycerol (DAG) in hepatocytes was different from that of IP₃ accumulation, following receptor stimulation.[50,51] It was suggested that the simultaneous hydrolysis of PIP₂ and phospholipase D-mediated breakdown of phosphatidylcholine could account for this apparent discrepancy. This hypothesis was supported by data showing that the fatty acid composition of DAG was different from that of PIP₂, suggesting a different origin.[50] Similarly, muscarinic stimulation of the pancreas had shown that the fatty acid composition of DAG did not resemble that of PIP₂.[52]

Several groups have now demonstrated muscarinic-mediated activation of phospholipase D in neuronal and nonneuronal tissue.[53,54] Phospholipase D catalyzes the hydrolysis of phosphatidylcholine, yielding phosphatidic acid and diacylglycerol. In a more detailed study of muscarinic-mediated activation of phospholipase D, Sandmann et al.[55] used human embryonic kidney cell lines stably transfected with four muscarinic receptors. They found that muscarinic stimulation of cells transfected with m1 and with m3 receptors induced both [³H]phosphatidylethanol and [³H]inositol phosphate formation following preincubation of the cells with [³H]oleic acid and [³H]inositol, suggesting that both phospholipase D (PLD) and PLC were stimulated. In contrast, muscarinic stimulation of cells transfected with m2 or with m4 receptors induced considerably less [³H]phosphatidylethanol and [³H]inositol phosphates, suggesting that these even-numbered receptors were much less effective in eliciting these responses. Furthermore, these authors found that neomycin (1 mM) inhibited m1 and m3 receptor-mediated production of IP by 50% whereas it was considerably less effective in inhibiting phosphatidylethanol formation, suggesting that these two responses may be independently regulated.

VIII. EFFECTS ON CELL GROWTH

In recent years, it has been shown that stimulation of many PI-coupled neurotransmitter receptors can profoundly influence cell growth.[56-58] Stimulation of m1 muscarinic receptors transfected into CHO cells causes increased cell proliferation as measured by [³H]thymidine incorporation.[12] In NIH 3T3 cells transfected with this receptor, muscarinic stimulation even causes foci formation suggesting that this receptor acts like a conditional oncogene in these cells.[59] On the other hand, in other cells such as m1-transfected A9 L cells or m1-transfected PC12 cells, muscarinic receptor stimulation causes the opposite response: inhibition of [³H]thymidine incorporation or induction of neurite outgrowth.[45,60] We recently found that stimulation of m1 muscarinic receptors expressed in either A9 L cells or CHO cells inhibited [³H]thymidine incorporation whether or not serum was present in the assay media.[61] This effect was independent of PKC since PKC inhibitors did not attenuate this response. These diverse effects on cell growth underscore the critical role played by the intracellular machinery on the final signal transduction pathway elicited by receptor stimulation.

Small cell lung carcinoma cells express m3 receptors which, when activated, elicit PI hydrolysis and inhibition of cell proliferation.[62] Upon receptor stimulation, these cells were found to become arrested in the S and G₂/M phases of the cell cycle.

IX. OTHER CELLULAR EFFECTS

Nitsch et al.[63] recently demonstrated that carbachol activation of human embryonic kidney cell lines transfected with the genes for the m1 and the m3 receptor subtypes, increased the basal release of amyloid precursor protein (APP). This increase was blockable by atropine; and it occurred within minutes of treatment, indicating that preexisting APP instead of newly synthesized APP was being released. In m1- and in m3-transfected cells, carbachol (1 mM) stimulated this release by four- to fivefold over control untransfected cells. M2- or m4-transfected cells did not demonstrate this increase, suggesting that the PI turnover pathway may be involved in this effect. Finally, pretreatment of the m1- and m3-transfected cells with staurosporine blocked this effect, suggesting that protein kinases mediate this effect.

In other studies, Haraguchi and Rodbell[64] studied the effect of muscarinic receptor activation on trafficking of endosomes. They found that carbachol stimulation of CHO cells transfected with either the m1 or m3 receptor inhibited the uptake of horseradish peroxidase and Lucifer yellow, both markers of fluid-phase endocytosis. Although the calcium ionophore A23187 mimicked this effect suggesting involvement of intracellular calcium, the mechanism underlying receptor control of endocytic pathways remains to be elucidated.

Along a very different line, Williams et al.[65] recently suggested that muscarinic receptors may regulate processes involving cadherin-mediated adhesion, such as embryonic development, neurogenesis, and cancer metastasis. These authors studied cell-cell adhesion in a small cell carcinoma cell line, and found that stimulation of the m3 receptors endogenously expressed in these cells induces E-cadherin-mediated adhesion. Since phorbol esters also induced this aggregation, they postulate that protein kinase C may mediate this response.

X. CONCLUSION

Muscarinic receptor-mediated transmembrane signaling has indeed become complicated, with five muscarinic receptor subtypes, at least five isozymes of phospholipase C, at least eight isozymes of protein kinase C, an ever-increasing number of phosphoinositides and their kinases, not to mention several inositol lipids, all having extremely diverse cellular and physiological effects. Rather than present a thorough review of each of these areas, clearly a monumental task, this review has attempted to present highlights of direct physiological significance.

In recent years, it has become clear that G-protein-coupled receptors mediate an increasingly diverse series of cellular responses. These responses range from those normally associated with G-protein-coupled receptors, such as mediating changes in intracellular cAMP and calcium levels, to responses with mechanisms yet to be fully elucidated, such as control of cell growth and proliferation and the control of endocytotic and release responses. The varied and cell-specific responses described in this review raise many possibilities of using currently available muscarinic compounds in new and interesting ways.

REFERENCES

1. Bonner, T.I., Buckley, N.J., Young, A.C., and Brann, M.R., Identification of a family of muscarinic acetylcholine receptor genes, *Science*, 237, 527, 1987.
2. Bonner, T.I., Young, A.C., Brann, M.R., and Buckley, N.J., Cloning and expression of the human and rat m5 muscarinic acetylcholine receptor genes, *Neuron*, 1, 403, 1988.
3. Peralta, E.G., Ashkenazi, A., Winslow, J.W., Smith, D.H., Ramachandran, J., and Capon, D.J., Distinct primary structures, ligand-binding properties and tissue-specific expression of four human muscarinic acetylcholine receptors, *EMBO J.*, 6, 3923, 1987.
4. Goyal, R.K., Muscarinic receptor subtypes, *N. Engl. J. Med.*, 321, 1022, 1989.
5. Wolfe, B., Subtypes of muscarinic colinergic receptors: ligand binding, functional studies, and cloning, in *The Muscarinic Receptors,* Brown, J. H., Ed. Humana Press, Clifton, NJ, 1989.
6. Hulme, E.C., Birdsall, N.J.M., and Buckley, N.J., *Annu. Rev. Pharmacol. Toxicol.*, 30, 633, 1990.
7. Wess, J., Molecular basis of muscarinic acetylcholine receptor function, *Trends Pharmacol. Sci.*, 14, 308, 1993.
8. Fisher, S.K., Heacock, A.M., and Agranoff, B.W., Inositol lipids and signal trasduction in the neurvous system: an update, *J. Neurochem.*, 58, 18, 1992.
9. Peralta, E.G., Ashkenazi, A., Winslow, J.W., Ramachandran, J., and Capon, D.J., Differential regulation of PI hydrolysis and adenylyl cyclase by muscarinic receptor subtypes, *Nature (London)*, 334, 434, 1988.
10. Ashkenazi, A., Winslow, J.W., Peralta, E.G., Peterson, G.L., Schimerlik, M.I., Capon, D.J., and Ramachandran, J., An M2 muscarinic receptor subtype coupled to both adenylyl cyclase and phosphoinoistide turnover, *Science,* 238, 672, 1987.
11. Gil, D.W. and Wolfe, B.B., Pirenzepine distinguishes between muscarinic receptor-stimulated phosphoinositide breakdown and inhibition of adenylate cyclase, *J. Pharmacol. Exp. Ther.*, 232, 608, 1985.
12. Ashkenazi, A., Peralta, E.G., Winslow, J.W., Ramachandran, J., and Capon, D.J., Functionally distinct G proteins selectively couple different receptors to PI hydrolysis in the same cells, *Cell*, 56, 487, 1989.

13. Strathman, M. and Simon, M.I., G protein diversity: a distinct class of α subunits is present in vertebrates and invertebrates, *Proc. Natl Acad. Sci. U.S.A.,* 87, 9113, 1990.

14. Strathman, M., Wilkie, T.M., and Simon, M.I., Diversity of the G-protein family: sequences from five additional α subunits in the mouse, *Proc. Natl. Acad. Sci. U.S.A.,* 86, 7407, 1989.

15. Smrcka, A.V., Helpler, J.R., Brown, K.O., and Sternweis, P.C., Regulation of polyphosphoinositide-specific phospholipase C activity by purified G_q, *Science*, 251, 804

16. Taylor, S.J., Chae, H.Z. Rhee, S.G., and Exton J.H., Activation of the β1 isozyme of phospholipase C by α subunits of the Gq class of G proteins, *Nature (London)*, 350, 516.

17. Ross, E.M. and Berstein, G., Regulation of the m_1 muscarinic receptor-G_q-phospholipase C-β pathway by nucleotide exchange and GTP hydrolysis, *Life Sci.*, 52, 413, 1993.

18. Wess, J., Bonner, T.I., Dorje, F., and Brann, M.R., Delineation of muscarinic receptor domains conferring selectivity of coupling to guanine nucleotide-binding proteins and second messengers, *Mol. Pharmacol.*, 38, 517, 1990.

19. Wess, J., Brann, M.R., and Bonner, T.I., Identification of a small intracellular region of the muscairnic m_3 receptor as a determinant of selective coupling to PI turnover. *FEBS Lett.*, 258, 133, 1989.

20. Lameh, J., Philip, M., Sharma, Y.K., Moro, O., Ramachandran, J., and Sadee, W., Hm1 muscarinic cholinergic receptor internalization requires a domain in the third cytoplasmic loop, *J. Biol. Chem.*, 267, 13406, 1992.

21. Fraser, C.M., Wang, C.-D., Robinson, D.A., Gocayne, J.D., and Venter, J.C., Site-directed mutagenesis of m1 muscarinic acetylcholine receptors: conserved asparatic acids play important roles in receptor function, *Mol. Pharmacol.*, 36, 840, 1989.

22. Nishizuka, Y., Turnover of inositol phospholipds and signal transduction, *Science*, 225, 1365, 1984.

23. Berridge, M.J. and Irvine, R.F., Inositol phosphates and cell signalling *Nature (London)*, 341, 197, 1989.

24. Harden, T.K., G prtoein-dependent regulation of phosphlipase C by cell surface receptors, *Am. Rev. Respir. Dis.,* 141, S119, 1990.

25. Sternweiss, P.C., Smrcka, A.V., and Gutowski, S., Hormone signalling via G-protein: regulation of phosphatidylinositol 4,5-bisphosphate hydrolysis by Gq, *Phil. Trans. R. Soc. London B*, 336, 35, 1992.

26. Guillon, G., Mouillac, B., and Savage, A.L., Modulation of hormone-sensitive phospholipase C, *Cell. Signalling*, 4, 11, 1992.

27. Nishizuka, Y., The molecular heterogeneity of protein kinase C and its implications for cellular regulation, *Nature (London)*, 334, 661, 1988.

28. Shearman, M.S., Sekiguchi, K., and Nishizuka, Y., Modulation of ion channel activity: a key function of the protein kinase C enzyme family, *Pharmacol. Rev.*, 41, 211, 1989

29. Simon, M.I., Strathmann, M.P., and Gautam, N., Diversity of G protein in signal transduction, *Science*, 252, 802, 1991.

30. Tang, W.-J. and Gilman, A.G., *Science*, 254, 1500, 1991.

31. Dell'Acqua, M.L., Carrol, R.C., and Peralta, E.G., Transfected m2 muscarinic acetylcholine receptors couple to $G_{\alpha i2}$ and $G_{\alpha i3}$ in Chinese hamster ovary cells, *J. Biol. Chem.*, 268, 5676, 1993.

32. Baumgold, J., Muscarinic receptor-mediated stimulation of adenylyl cyclase. *Trends Pharm. Sci.*, 13, 339, 1992.

33. Felder, C.C., Kanterman, R.Y., Ma, A.L., and Axelrod, J., A transfected m_1 muscarinic acetylcholine receptor stimulates adenylate cyclase via phosphatidylinositol hydrolysis, *J. Biol. Chem.*, 264, 20356, 1989.

34. Jansson, C.C., Kukkonen, J., and Akerman, K.E.O., Muscarinic receptor-linked elevation of cAMP in SH-SY-5Y neuroblastoma cells is mediated by Ca^{2+} and protein kinase C, *Biochim. Biophys. Acta,* 1095, 255, 1991.

35. Baumgold, J., Paek, R., and Fiskum, G., Calcium independence of phosphoinositide hydrolysis-induced increase in cyclic AMP accumulation in SK-N-SH human neuroblastoma cells, *J. Neurochem.*, 58, 1754, 1992.

36. Gao, B. and Gilman, A.G., Cloning and expression of a widely distributed (type IV) adenylyl cyclase, *Proc. Natl. Acad. Sci. U.S.A.,* 88, 10178, 1991.

37. Olianas, M.C. and Onali, P., Properties of muscarinic-stimulated adenylate cyclase activity in rat olfactory bulb, *J. Neurochem.*, 58, 1723, 1992.

38. Jones, S.V.P., Muscarinic receptor subtypes: modulation of ion channels, *Life Sci.*, 52, 457, 1993.

39. Kirsh, G.E., Yatani, A., Codina, J., Burnbaumer, L., and Brown, A.M., α-Subunit of Gk activates atrial K^+ channels of chick, rat and guinea pig., *Am. J. Physiol.*, 254, 1200, 1988.

40. Kim, D., Lewis, D.L., Graziadei, L., Neer, E.J., Bar-Sagi, D. and Clapham, D.E., G-protein βγ-subunit activate the cardia muscarinic K^+-channel via phospholipase A_2. *Nature (London)*, 337, 557, 1989.

41. Jones, S.V.P. , Barker, J.L., Buckley, N.J., Bonner, T.I., and Brann, M.R., Electrophysiological characterization of cloned m_1 muscarinic receptors expressed in A9 L cells, *Proc. Natl. Acad. Sci. U.S.A.*, 85, 4056, 1988.

42. Baumgold, J. and Paek, R., Differences in the functional responses of two cell lines each expressing PI-hydrolysis-coupled muscarinic receptors, *Neurochem. Res.*, 17, 375, 1992.

43. Fukuda, K., Higashida, H., Kubo, T., Maeda, A., Akiba, I., Bujo, H., Mishina, M., and Numa, S., Selective coupling with K^+ currents of muscarinic acetylcholine receptor subtypes in NG108–15 cells, *Nature (London)*, 335, 355, 1988

44. Brown, D.A., G-proteins and potassium currents in neurons, *Annu. Rev. Physiol.*, 52, 215, 1990.

45. Conklin, B.R., Brann, M.R., Buckley, N.J., Ma, A.L., Bonner, T.I., and Axelrod, J., Stimulation of arachidonic acid release and inhibition of mitogenesis by cloned genes for muscarinic receptor subtypes stably expressed in A9 L cells, *Proc. Natl. Acad. Sci. U.S.A.*, 85, 8698, 1988.

46. Abdel-Latif, A.A., Calcium-mobilizing receptors, polyphosphoinostides, and the generation of second messengers, *Pharmacol. Rev.*, 38, 227, 1986.

47. Felder, C.C., Dieter, P., Kinsella, J., Tamura, K., Kanterman, R.Y., and Axelrod, J., A transfected m5 muscarinic acetylcholine receptor stimulates phopholipase A2 by inducing both calcium influx and activation of protein kinase C, *J. Pharmacol. Exp. Ther.* 255, 1140, 1990.

48. Marshall, P.J., Dixon, J.F., and Hokin, L.E., Evidence for a role in stimulus-secretion coupling of prostaglandins derived from release of arachidonoyl residues as a result of phosphotidylinositol breakdown, *Proc. Natl. Acad. Sci. U.S.A.*, 77, 3292, 1980.

49. DeGeorge, J.J., Morell, P., McCarthy, K.D., and Lapetina, E.G., Cholinergic stimulation of arachidonic acid and phosphatidic acid metabolims in C62B glioma cells, *J. Biol. Chem.*, 261, 3428, 1986.

50. Bocckino, S.B., Blackmore, P.F., and Exton, J.H., Stimulation of 1,2-diacylglycerol accumulation in hepatocytes by vasopressin, epinephrine, and angiotensin II, *J. Biol. Chem.*, 260, 14201, 1985.

51. Charest, R., Prpic, V., Exton, J.H., and Blackmore, P.F., Stimulation of inositol trisphosphate formation in hepatocytes by vasopressin, adrenaline and angiotensin II and its relationship to changes in cytosolic free Ca^{2+}, *Biochem. J.*, 227, 79, 1985.

52. Banschbach, M.W., Geison, R.L., and Hokin-Neaverson, M., Effects of cholinergic stimulation on levels and fatty acid composition of diacylglycerols in mouse pancreas, *Biochim. Biophys. Acta*, 663, 34, 1981.

53. Sandmann, J. and Wurtman, R.J., Stimulation of phospholipase D activity in human neuroblastoma (LA-N-2) cells by activation of muscarinic acetylcholine receptors or by phorbol esters: relationship to phosphoinositide turnover, *J. Neurochem.*, 56, 1312, 1991.

54. Qian, Z. and Drewes, L.R., Cross-talk between receptor-regulated phospholipase D and phospholipase C in brain, *FASEB J.*, 5, 315, 1991.

55. Sandmann, J., Peralta, E.G., and Wurtman, R.J., Coupling of transfected muscarinic acetylcholine receptor subtypes to phospholipase D, *J. Biol. Chem.*, 256, 6031, 1991.

56. Meier, E., Herz, L., and Schoushoe, A., *Neurochem. Int.*, 19, 1, 1991.

57. Pouyssegur, J. and Seuwen, K., Transmembrane receptors and intracellular pathways that control cell proliferation, *Annu. Rev. Physiol.*, 54, 195, 1992.

58. Julius, D., Huang, K.N., Livelli, T.J., Axel, R., and Jessell, T.M., The 5HT2 receptor defines a family of structurally distinct but functionally conserved serotonin receptors, *Proc. Natl. Acad. Sci. U.S.A.*, 87, 928, 1990.

59. Gutkind, J.S., Novotny, E.A., Brann, M.R., and Robbins, K.C., Muscarinic acetylcholine receptor subtypes as agonist-dependent oncogenes, *Proc. Natl. Acad. Sci. U.S.A.*, 88, 4703, 1991.

60. Pinkas-Kramarski, R., Stein, R., Lindenboim, L., and Sokolovsky, M., Growth factor-like effects mediated by muscarinic receptors in PC12M1 cells, *J. Neurochem.*, 59, 2158, 1992.

61. Baumgold, J. and Dyer, K., Muscarinic receptor-mediated inhibition of mitogenesis via a protein kinase C-independent mechanism in m1-transfected A9 L cells. *Cell. Signaling*, 6, 103, 1994.

62. Williams, C.L. and Lennon, V.A., Activation of muscarinic acetylcholine receptors inhibits cell cycle progression of small cell lung carcinoma, *Cell Regul.*, 2, 373, 1991.

63. Nitsch, R.M., Slack, B.E., Wurtman, R.J., and Growdon, J.H., Release of Alzheimer amyloid precursor derivatives stimulated by activation of muscarinic acetylcholine receptors, *Science*, 258, 304, 1992.

64. Haraguchi, K. and Rodbell, M., Carbachol-activated muscarinic (M1 and M3) receptors transfected into Chinese hamster ovary cells inhibit trafficking of endosomes, *Proc. Natl. Acad. Sci. U.S.A.*, 88, 5964, 1991.

65. Williams, C.L., Hayes, V.Y., Hummel, A.M., Tarara, J.E., and Halsey, T.J., Regulation of E-cadherin-mediated adhesion by muscarinic acetylcholine receptors in small cell lung carcinoma, *J. Cell Biol.*, 121, 643, 1993.

Chapter 10

Compartmentalization and Release from Neurons in the CNS

Vera Adam-Vizi

CONTENTS

I. INTRODUCTION

One of the greatest challenges for biomedical researchers is to gain greater insight into the function of the brain. To find, however, the best and the most relevant experimental model is sometimes very difficult. This is especially true when the cholinergic system in the mammalian central nervous system (CNS) is concerned. Cholinergic neurons are found dispersed in the brain, although some areas such as the caudate nucleus, hippocampus, and other forebrain regions are relatively rich in cholinergic innervation. Preparations from the brain contain a large number of nerve terminals functioning with other types of transmitters. As revealed by purification of nerve terminals by affinity chromatography[1] or immunoadsorption,[2] cholinergic synaptosomes constitute 5 to 10% of total cerebral cortical synaptosomes.[3] Therefore, preparations from the CNS always suffer from the disadvantage of heterogeneity. With a pure preparation the results are always more unequivocal and can be interpreted more easily. For this reason and because the cholinergic neuron appears to be highly conserved in evolution, studies on nonmammalian but pure preparations have proved to be extremely valuable.

The electric organ of electric fish, having purely cholinergic innervation, is a rich source of cholinergic nerve terminals. The study of this system for the isolation of synaptosomes and vesicles greatly advanced research in this field. In fact, much of our knowledge on the function of the cholinergic nerve terminal was derived from studies on isolated nerve terminals and vesicles from the electric organ of the *Torpedo*. In addition, most of the findings in this system are relevant and hold true for the mammalian central cholinergic neurons. This chapter deals mainly with those findings from different experimental model systems that appear relevant to the problem given in the title, i.e., compartmentalization and release of acetylcholine (ACh) from the cholinergic nerve terminals in the CNS. Various other aspects of the cholinergic nerve teminal have been reviewed elsewhere and can be consulted for detailed morphological, biochemical, and pharmacological information.[4-9]

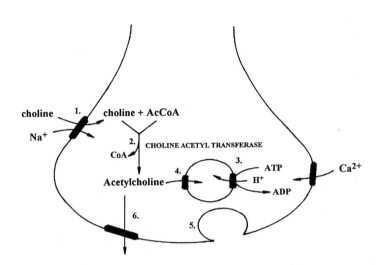

Figure 10-1 The cholinergic nerve terminal. 1. Sodium-dependent high-affinity choline uptake. 2. Choline acetyltransferase. 3. Proton-pumping ATPase of the vesicle. 4. Acetylcholine transporter. 5. Exocytosis of the vesicle. 6. Acetylcholine leakage.

II. OVERVIEW OF ACETYLCHOLINE METABOLISM IN THE PRESYNAPTIC CHOLINERGIC NERVE TERMINAL

Most of the characteristics of the acetylcholine metabolism appear to be similar in the cholinergic nerve terminals from different species (Figure 10-1). Acetylcholine is synthesized in the cytoplasm by choline acetyltransferase, an enzyme specific for cholinergic neurons, from choline and acetyl coenzyme A (CoA).[10] Choline for the synthesis is transported into neurons from the extracellular fluid mainly by the sodium-dependent high-affinity choline uptake system which is located in the presynaptic plasma membrane.[11-13] AcetylCoA is an obligatory intermediary product in the metabolism of carbohydrates, fatty acids, and a number of amino acids; however, for the synthesis of ACh the metabolism of glucose appears to make the largest contribution.[14-16] Acetylcholine from the cytoplasm is transported by an active transport system into the vesicles where it is stored.[9] Upon stimulation, Ca^{2+} enters the nerve terminal and ACh is released from the vesicles by exocytosis. A portion of ACh remains in the cytoplasm from where it leaks out across the presynaptic plasma membrane.

III. CHOLINERGIC VESICLES

A. ACETYLCHOLINE AND ADENOSINE TRIPHOSPHATE (ATP) IN THE VESICLES

Synaptic vesicles located in the cholinergic nerve terminals have a central role in both the compartmentalization and release of ACh. Vesicles were first isolated from brain tissue,[17-19] which was a mixed population containing many different neurotransmitters. Pure cholinergic vesicles were only available when electric organs of electric fish were used as a source.[20,21] Vesicles from the *Torpedo* electric organ, which are much larger than cholinergic vesicles of mammalian origin, contain about 50,000 to 200,000 molecules of ACh[22-24] which corresponds to a concentration of 500 to 880 mM. The ACh content of vesicles from guinea pig cerebral cortex is lower,[25] being in the range of 1600 to 2000 molecules per vesicle.[26] Cholinergic synaptic vesicles contain ATP with a molar ratio to ACh of 1:5,[27] or of 1:10 in another study.[28] In vesicles isolated from affinity-purified cholinergic synaptosomes[1] from the rat brain this ratio is 1:7.[29] The ATP content of vesicles falls during prolonged stimulation and recovers during rest.[30,31] It has been shown that upon activation of nerve terminals, ATP is released along with ACh.[29,32-34] Botulinum and tetanus toxins do not inhibit depolarization-induced release of ATP from electric organ synaptosomes, whereas release of ACh is blocked.[35] This may imply that the storage of ACh and ATP in the vesicles is not closely linked. The functional importance of the costorage and the possible corelease of ACh and ATP from cholinergic vesicles awaits further clarification.

B. pH IN THE CHOLINERGIC VESICLES

From ^{31}P nuclear magnetic resonance (NMR) studies[36] and the self-quenching of the fluorescence of 9-aminoacridine in electric organ vesicles, it has been calculated that the pH in the cholinergic vesicles is

5.2 to 5.5. The proton gradient is essential for the transport of ACh to the vesicles; protonophores or ammonium ions which dissipate the gradient inhibit the storage of ACh.[37-41]

The internally acidic pH is generated by a vesicular ATPase. ATPase activity has been shown to be present in synaptic vesicles,[42,43] and both V-type and P-type ATPases have been demonstrated to be part of the vesicular membrane.[44] The proton-pumping V-type ATPase is activated by Mg^{2+} and Ca^{2+}. The subunit structure of the enzyme is now known.[45-47] It is fully active over a wide pH range, between 5.5 and 9.5. The antibiotic bafilomycin A_1 binds to the ATPase, and in parallel with the inhibition of the enzyme, the ACh active transport of the vesicles is inhibited.[48]

C. TRANSPORT OF ACETYLCHOLINE INTO THE VESICLES

Demonstration of the uptake of ACh into synaptic vesicles was hindered for a long time by experimental drawbacks. Bahr and Parsons[49,50] have described the main difficulties, namely, that vesicular preparations from different sources may vary in the characteristics of the ACh uptake process and that the uptake is dependent on the concentration of vesicles. Initially, only a passive ACh uptake into synaptic vesicles was demonstrated.[51,52] The existence and the characteristics of a specific ACh transport process (AChT) were revealed in the early 1980s mainly by Parsons and co-workers.

The uptake of ACh necessitates that there is a gradient of ACh concentration across the vesicular membrane. However, the extent of this gradient is difficult to determine quantitatively, because it would be necessary to know the exact concentrations of ACh in the cytoplasm and in the vesicles. Because of the heterogeneity of most of the preparations, this is even more difficult for mammalian cholinergic nerve terminals. Nevertheless, both from the attempts to estimate the concentrations and calculations from morphological data, it appears that in most tissues this gradient is about 100-fold.[9]

ACh is taken up by the cholinergic vesicles in the presence of bicarbonate and Mg^{2+}-ATP (or Ca^{2+}-ATP).[53,54] The uptake is driven by proton efflux from the internally acidic vesicles created by the proton-pumping ATPase.[9] The active temperature-dependent and saturable transport of ACh into the cholinergic vesicles is inhibited by the agent *trans*-2(4-phenylpiperidino)cyclohexanol (formerly AH5183), now called vesamicol,[53] without having an effect on the vesicular ATPase activity. The application of vesamicol enabled the direct investigation of the acetylcholine transporter and proved to be very useful in experiments where the subcellular ACh compartments and their contribution to the release were studied. Tetraphenylboron and other hydrophobic anions were also found to be inhibitors of the ACh active transport;[55] however, their binding sites appear to be distinct from that of vesamicol.

Binding studies with vesamicol have shown that its saturable binding site is on the ACh transporter molecule[49] and binding of the drug to the AChT brings about the noncompetitive inhibition of the uptake of radiolabeled ACh to the vesicles.[50,56] A number of ACh analogs were synthesized and also were demonstrated to be transported. Some had higher affinities for the AChT than does ACh;[9] however, choline is not a substrate of the AChT.[57] Cetiedil appears to be a competitive inhibitor of ACh uptake.[56] Vesamicol can permeate the plasma membrane and binds to its receptor, which is on the cytoplasmic surface of the synaptic vesicles.[58] The vesamicol receptor has been purified and solubilized from electric organ vesicles,[59] and it has been demonstrated[60] by photoaffinity labeling with azido-ACh that the AChT is associated with the same structure as the vesamicol receptor. The ACh transporter appears to be associated with proteoglycan[61] in cholinergic vesicles; SV1 and SV2 epitopes both copurify with the vesamicol receptor.[59] Furthermore, the solubilized vesamicol receptor, in addition to the sodium decyl sulfate (SDS)-denatured photoaffinity-labeled AChT, is immunoprecipitated with anti-SV1 monoclonal antibody. Attempts have been made to exploit the unique ability of vesamicol to bind to the cholinergic vesicle in mapping cholinergic neurons in the human brain and assess their function in pathological states such as Alzheimer's disease and Parkinson's disease.[62,63] In agreement with previous animal experiments vesamicol binding in human brain tissues is also correlated with cholinergic nerve terminals and is mainly localized to the synaptic vesicles.

IV. VESICULAR HYPOTHESIS OF THE RELEASE OF ACETYLCHOLINE

A. MORPHOLOGICAL EVIDENCE

A large amount of experimental evidence had been accumulated by the late 1970s which indicated that vesicles were involved in the release of ACh, and it was suggested that fusion of vesicles with the plasma membrane followed by a retrieval is the very essence of the process. The hypothesis that ACh released by nerve stimulation originated from the synaptic vesicles was formulated by Katz.[64] It was assumed that vesicles fused with the synaptic plasma membrane and discharged their content to the synaptic cleft. This apparently simple mechanism could provide a feasible and satisfactory explanation for the occurrence of miniature end plate potentials and the quantal nature of the end plate potentials observed at the neuro-muscular junction. Using this preparation researchers detected a number of morphological alterations and established a correlation of these alterations with the electrophysiological events of the terminal, which appeared to reinforce the hypothesis of vesicular exocytosis. It was predicted that in response to stimulation of the cholinergic nerve, the number of vesicles should be decreased. It was indeed shown that repetitive stimulation and K^+ depolarization depleted synaptic vesicles,[6,65-69] which was usually accompanied by a concomitant increase either in the area of the axolemma or in the number of cisternae in the axoplasm. During release of ACh, plasma membrane infoldings into the presynaptic cytoplasm were seen.[65,67,70] In addition to depletion of vesicles, spectacular exocytotic events could be detected by freeze-fracture studies, such as the occurrence of dimples and protuberances (P-face and E-face) at the active zone of the presynaptic plasma membrane.[6,66,71.] The rapid freezing technique enabled the corre-lation in time of the appearance of dimples and protuberances with the release of ACh evoked by a single-action potential. In the presence of 4-aminopyridine, dimples occurred at the active zone a few millisec-onds after stimulation of the nerve and the increased number of dimples correlated with the increased quantal content of the postsynaptic potential.[71] In quick-frozen samples from neuromuscular junctions, fusion profiles (Ω profile) of vesicles were demonstrated[72] and a good correlation between the rate of occurrence of these figures and the rising phase of the postsynaptic potential was observed. These findings gave strong evidence that fusion of vesicles with the plasma membrane and the release of a quantum of ACh are closely linked.

Recovery of vesicles from the axolemma following electrical stimulation was greatly substantiated by experiments in which large extracellular tracers were found in synaptic vesicles after prolonged stimu-lation.[66,73-76]

B. COMPARTMENTALIZATION OF ACETYLCHOLINE WITHIN THE NERVE TERMINAL

Unequivocal neurochemical evidence for the release of ACh of vesicular origin was more difficult to obtain. Assuming the exocytosis of the vesicles, it could be predicted that ACh in the vesicular pool would be lost upon stimulation. Although this was observed with *Torpedo* electric organ and mammalian brain tissue,[77-79] most of the experimental data permitted alternative interpretations, since ACh within the nerve terminals was found to be present in more than one compartment and alterations in ACh content of these compartments (and not only in that of the vesicles) were reported.

In the early experiments on the cholinergic system, it already appeared that ACh in nerve endings existed in an osmotically resistant (stable-bound) and in an osmotically labile (labile-bound) form. With subcellular fractionation, ACh could be separated into free and vesicle-bound fractions[21,78,80,81,] and the free ACh was thought to be derived mainly from the cytoplasm. The proportion of the cytoplasmic to vesicular ACh proved to be difficult to determine, as experimental conditions greatly influenced the loss of ACh from different compartments during the separation procedure. Estimates of the vesicle-bound ACh range from 10 to 80% of the total tissue ACh.[21,24,82-87] Molenaar et al.[88] and Molenaar and Polak[89] demonstrated that the synaptic activity strongly influenced the amount of ACh in the vesicles. Rat brain slices were incubated with radiolabeled choline, and it was found that at rest the specific radioactivity of ACh in the cytoplasm was higher than in the vesicles; however, in the presence of high K^+ concentration the specific radioactivity of ACh in the two compartments was equal.

Contradictory results emerged from experiments aimed at determining which compartment contrib-uted to the release of ACh. In tissues incubated with radioactive choline the specific radioactivity of the released and the stored ACh was compared. It was shown that released ACh was of a greater specific radioactivity than was ACh isolated from synaptic vesicles.[88,90] ACh appeared to be depleted from the vesicle-bound, but not from the cytoplasmic compartment of mammalian brain tissues,[79,91] indicating the vesicular origin of the released ACh. However, when the slices of cerebral cotex were depolarized by high K^+ concentration, depletion of ACh from both the cytoplasmic and the vesicle-bound pool were observed with no demonstrable difference in the rate of depletion of either of the two fractions.[78] This observation would rather suggest that both compartments could contribute to the release of ACh.

V. CHALLENGES TO THE VESICULAR ORIGIN OF ACETYLCHOLINE RELEASED FROM THE NERVE TERMINAL: ALTERNATIVE SUGGESTIONS

The presence of a large portion of ACh in the cytoplasm is not surprising, but is consistent with the cytoplasmic location of choline acetyltransferase that synthesizes ACh from choline and acetylCoA.

However, it has raised the possibility that ACh quanta are released from the cytoplasm and that vesicles only store ACh within the nerve terminal.

A great challenge to the theory of vesicular release of ACh was the discovery that the most recently synthesized ACh was released in preference to the preformed ACh. This was demonstrated for the superior cervical ganglion,[92] diaphragm,[93] Torpedo electric organ,[94] and cerebral cortex[95] and indicated that the newly synthesized ACh and the preformed ACh were present in different compartments which were not equally available for release of ACh. Mobilization of ACh from these pools responded differently to pharmacological manipulations.[96] Since newly synthesized ACh was found to be accumulated preferentially in the cytoplasm,[97,98] it was suggested that ACh was released mainly from the cytoplasm.[5,7,8,99,100] Several lines of data appeared to support this conclusion. Dunant et al.[101] observed a rapid oscillation of ACh content selective to the cytoplasmic compartment of the Torpedo electric organ during prolonged stimulation. In this organ a decrease of the cytoplasmic ACh content by stimulation was detected, whereas the amount and the specific radioactivity of vesicular ACh were unchanged.[7] The experiment made by Tauc[7] appeared to demonstrate unequivocally a release of ACh of cytoplasmic origin from the presynaptic neuron in Aplysia buccal ganglion. Following the injection of acetylcholinesterase into the neuron, which hydrolyzed the cytoplasmic ACh, the postsynaptic potentials disappeared, whereas the number of the presynaptic vesicles was unchanged and the vesicles were free of acetylcholinesterase.

Israel and co-workers[102-104] made a number of observations which were not suited to the vesicular exocytosis theory. They reported a strictly Ca^{2+}-dependent ACh release from synaptosomal ghosts refilled with ions and soluble ACh.[102] The calcium-dependent ACh release function could be reconstituted in an artificial lipid bilayer that contained presynaptic membrane elements.[103,105] Furthermore, proteoliposomes with the capacity to release ACh in response to Ca^{2+} influx could be made from lyophilized presynaptic membrane.[104] In these experiments stimulation-evoked ACh release was elicited in preparations which did not contain synaptic vesicles. It was suggested that a protein component of the plasma membrane mediated the release of ACh. They solubilized and partly purified this protein from Torpedo electric organ presynaptic membrane and named it mediatophore.[106]

The release of ACh from the cytoplasm of the motor nerve at the neuromuscular junction was hypothesized based on the electrophysiological[84,107,108] and neurochemical[109,110] findings that a fraction of the spontaneous release of ACh was nonquantal. The nonquantal release did not require the presence of Ca^{2+} in the external medium.[108]

These results, especially the preferential release of newly synthesized ACh and the failure to identify the specific radioactivity of the released ACh to that of ACh present in any of the subcellular compartments, did appear inconsistent with the theory of vesicular exocytosis. There could be, however, experimental uncertainties which could not be taken into account in these sudies, e.g., that during the isolation and fractionation procedures ACh might be redistributed between the different subcellular pools altering the original in situ states. It was also a possibility that what was obtained as a vesicular fraction from a density-gradient centrifugation was not homogeneous. The demonstration that synaptic vesicles are, indeed, heterogeneous resolved some of the above contradictions and provided a new interpretation of the results previously not consistent with the vesicular hypothesis of ACh release.

VI. RECENT NEUROCHEMICAL EVIDENCE FOR THE VESICULAR RELEASE OF ACETYLCHOLINE

A. METABOLIC HETEROGENEITY OF SYNAPTIC VESICLES

Metabolic heterogeneity of synaptic vesicles was demonstrated in the mammalian brain as well as in the Torpedo electric organ.[81,111-115] Two populations of vesicles could be separated by density-gradient centrifugation[112] or chromatography;[116] one of them (VP_2) contains smaller and denser vesicles than the other (VP_1). VP_2 vesicles recycle by stimulation and have a high capacity to take up newly synthesized ACh. Upon stimulation the size of these vesicles is decreased and the density is increased, reflecting the loss of ACh and ATP. A fully rested Torpedo electric organ contains a few percent of VP_2 vesicles, whereas after extensive stimulation it can go up to 74%. It has been proposed that VP_2 vesicles near the plasma membrane could undergo repetitive exocytosis and endocytosis. During the rest period, VP_2 vesicles regain the characteristics of the reserve pool; the diameter of the vesicles is increased and the density is decreased. In contrast, VP_1 vesicles form the reserve pool which does not exchange readily with the cytoplasmic ACh and are not involved in exocytosis. By using radiolabeled precursors, rapid incorporation of the newly synthesized ACh into the recycling vesicles has been demonstrated, whereas ACh in the reserve pool is not appreciably labeled. Stimulation of the preparation previously labeled, releases ACh with a specific radioactivity corresponding to that of ACh present in the VP_2 vesicles.[114] This is strong experimental support of the prediction that ACh released from the nerve terminals is derived from the vesicles now identified as being of a VP_2 type. However, it still leaves some room for the assumption that the cytoplasm may be in isotopic equilibrium with the recycling vesicles and that ACh can be (partly or totally) released from there. This uncertainty was resolved by the use of false transmitters and of vesamicol, the inhibitor of the ACh transporter of the vesicles.

Important electrophysiological evidence has been provided by Searl et al.[117] for the presence of a rapidly recycling pool of transmitter at the snake neuromuscular junction. Two classes of miniature end plate currents were produced by prolonged stimulation of the motor nerve. One class having the same mean amplitude as the resting, unstimulated neuromuscular junction was not affected by vesamicol and was thought to reflect the release of ACh from a preexisting vesicular pool. The other class having smaller mean amplitude was only produced by prolonged stimulation and was sensitive to vesamicol. These small-amplitude miniature end plate currents (mepcs) were attributed to the release of recycling vesicles which were created during stimulation, and the effect of vesamicol was to prevent the reloading of ACh into these vesicles.

The discovery of the subclasses of cholinergic vesicles and their distinct role in the storage and release of ACh provided important clues to the understanding of the apparently contradictory reports concerning the preferential release of the most recently synthesized ACh. As a result of this finding and those obtained with false transmitters and the ACh transporter inhibitor vesamicol, the long-lasting debate over the compartmentalization and subcellular origin of ACh released from the cholinergic terminal appears to come to a rest.

B. APPLICATION OF FALSE TRANSMITTERS

False transmitters are analogs of the natural neurotransmitters in that they are taken up by nerve terminals, distributed in the cellular compartments, and can be released by a Ca^{2+}-dependent stimulus. Their pharmacological potencies are usually lower than the endogenous transmitter; however, they were successfully used in experiments aimed at identifying the cellular compartments that are directly involved in the transmitter release.[118]

Noradrenaline analogs greatly contributed to the identification of synaptic vesicles as the source of adrenergic transmitters that are discharged by stimulation of the sympathetic nerve.[119,120] These compounds were taken up by adrenergic nerve terminals, some have the ability to enter the vesicles, and some are distributed only in the cytoplasm. Upon stimulation only those analogs were released which were present in the vesicular compartment.

In the cholinergic nerve terminals the endogenous transmitter acetylcholine, as well as other choline esters, is not taken up readily. Therefore, precursors of false transmitters have been used.[121-124] These can be choline analogs which are taken up by the high-affinity choline uptake of the presynaptic plasma membrane, acetylated in the cytoplasm by the choline acetyl transferase, and to a varying extent can gain entry into the vesicles. False transmitters which have a distribution between the cytoplasm and the vesicles different from that of ACh may help to determine which cellular compartment is discharged upon stimulation.

Triethylcholine and homocholine were successfully used as precursors.[118,125-127] These analogs were taken up by electric organ tissue and were acetylated by ChAT. In a series of experiments acetyltriethylcholine, in addition to the newly synthetized ACh, was labeled by using [³H]acetate. The tissue was stimulated, and the ratio of [³H]ACh to [³H]acetyltriethylcholine released was determined and compared to the ratio found in the total tissue and in the vesicles isolated immediately after the application of the stimulus. Stimulation enhanced the release of both radiolabeled compounds; however, [³H]ACh release was stimulated by a greater extent than the labeled false transmitter. The ratio of released [³H]ACh to [³H]acetyltriethylcholine corresponded to the ratio found in the isolated vesicles. Similar results were obtained when [¹⁴C]choline and [³H]homocholine precursors were used. Again, the ratio of the released ACh and false transmitter represented the ratio in the vesicles. These are clear-cut indicators that stimulation releases acetylcholine from the vesicles. The same conclusion could be drawn from the experiments made by von Schwarzenfeld.[128] She injected [³H]pyrrolidinecholine or [³H]homocholine into a region of the guinea pig cerebral cortex, which was incorporated into the tissue. In two series of experiments, the release of [¹⁴C]ACh and [³H]acetylpyrrolidinecholine or the release of [¹⁴C]ACh and [³H]acetylhomocholine from the surface of the cortex by 30-Hz stimuli was detected. At the end of the stimulations the cortical tissue was subjected to subcellular fractionation. The ratio of the true to the false transmitter in the released samples was similar to that in the membrane-bound vesicular fraction (which may correspond to the fraction of the recycling vesicles), but differed from that of the bulk vesicular fraction and of the supernatant. The vesicular origin of the released ACh is strongly implicated in the experiments by Welner and Collier[129] who used diethylhomocholine on rat brain synaptosomes and cat superior cervical ganglion and found that acetyldiethylhomocholine, which was synthesized in the cytoplasm but did not gain entry to the vesicles, was not released along with ACh by K^+ depolarization. In earlier experiments with false transmitters, Boksa and Collier[130] measured the spontaneous release of transmitters and found that the ratio of true to false transmitter released differed significantly from the ratios measured in the vesicle fraction.

As indicated by the above experiments, cholinergic false transmitters have been successfully used in the subcellular fractionation studies and have provided convincing evidence for a vesicular origin of stimulated ACh release.

C. APPLICATION OF VESAMICOL

The use of vesamicol, a drug that inhibits ACh transport into the cholinergic vesicles, opened a new possibility of studying the subcellular origin of ACh released by stimulation. The drug inhibits reloading of vesicles with ACh; and apart from a slight inhibitory action on the sodium-dependent high-affinity choline uptake system,[131,132] it does not interfere with other presynaptic cholinergic function. It has been used in a variety of tissues[9] including neuromuscular junction, synaptosomes prepared from the *Torpedo* electric organ or mammalian brain tissue, brain slices, superior cervical ganglion, myenteric plexus, and PC12 cells. All these studies provided strong support for the view that synaptic vesicles are reloaded with and release ACh upon nerve stimulation.

At the neuromuscular junction, agents that interfere with the storage of ACh would affect the quantal size, because an individual quanta is thought to reflect the amount of ACh in the vesicle. Indeed it was found that after prolonged stimulation in the presence of vesamicol the quantal size was reduced.[133-135] However, without nerve stimulation vesamicol had no effect on the size of the quanta. This implies that the prepacked ACh available for release is not affected by the inhibitor of the AChT; however, the uptake of ACh into recycling vesicles is inhibited by vesamicol. Vesamicol has no effect either on the frequency of mepcs in unstimulated neuromuscular junction[117,136,137] or at low frequency of nerve stimulation[117,138,139] on the quantal content. The results indicate that the release process, i.e., the vesicular exocytosis is not affected per se by vesamicol. It inhibits the loading of vesicles that are induced to recycle during nerve stimulation, and alteration of the quantal release from neuromuscular junction is the consequence of this effect. Direct measurement of labeled ACh release from the neuromuscular junction[140] confirmed the electrophysiological findings. Vesamicol when applied prior to the loading of the tissue with labeled choline, inhibited the stimulation-evoked release of ACh.

Interestingly, nonquantal, spontaneous release of ACh was also inhibited by vesamicol[136,137,141,142] which was interpreted as indicating the incorporation of the vesicular ACh transporter into the plasma membrane following vesicular exocytosis.[136,143] Consistent with this interpretation was the result that other inhibitors of the ACh transporter such as tetraphenylboron and quinacrine were also effective,[136,143] and increase of the extracellular pH decreased the resting release of ACh.[143] The latter finding implies that the function of the transporter in the plasma membrane, similar to that in the vesicles, would be related to a proton gradient.

Studies with vesamicol on the release of ACh from CNS tissues[144-147] or synaptosomes, both from the mammalian cortex[131,148] and *Torpedo* electric organ, gave essentially the same results. The release of radiolabeled ACh in response to K^+ depolarization was inhibited by vesamicol, only when it was present before and during loading the preparation with the radioactive precursor.[132,148-150] A fraction (15 to 20%) of the release of ACh was not vesamicol sensitive,[149,150] which may represent a nonvesicular, or a vesamicol-insensitive vesicular pool. Roughly the same proportion of total ACh from the perfused superior cervical ganglion could be released in the presence of vesamicol.[151] In rat striatal slices, 20% of the total ACh release could be observed in the presence of vesamicol[152] when slices were challenged with veratridine, an alkaloid which opens potential-dependent Na^+ channels.[153] The vesamicol-insensitive ACh release could be elicited by veratridine in the absence of extracellular calcium.[147,152] The origin of the vesamicol-insensitive ACh release is not quite certain, but it is probably also of a vesicular origin. This is supported by the report that there is a vesamicol-insensitive exchange between the newly synthesized cytoplasmic ACh and vesicular ACh.[154] This remaining ACh pool could be mobilized by α-latrotoxin, the black widow spider venom,[154] which is known to cause vesicle exocytosis from a number of preparations.[155-162] Attempts to determine quantitatively the alterations in the subcellular distribution of ACh in the presence of vesamicol gave somewhat contradictory results. With low concentrations of vesamicol no difference in ACh content of crude subcellular fractions from the mouse brain was reported,[144] whereas in other studies[145,146] the amount of ACh in the cytoplasmic fraction was increased in tissues stimulated in the presence of vesamicol. Ricny and Collier[146] observed a loss of ACh from the subcellular fraction containing synaptic vesicles prepared from rat striatal slices. Gracz et al.[115] found that both VP_1 and VP_2 vesicles for the *Torpedo* electric organ take up radiolabeled ACh and possess binding sites to vesamicol. However, in the VP_2 fraction the ratio of labeled ACh uptake to the concentration of vesamicol receptor was four to seven times higher than in the VP_1 fraction, and it appeared that in the denser fraction (VP_2) the ACh transporter was more active.

The reports on the release of ACh observed in the presence of vesamicol helped to substantiate the opinion that the origin of the ACh released by stimulation is from the cholinergic vesicles. The possibility of the nonvesicular ACh release in response to nerve stimulation appears remote in view of the experiments with vesamicol. Further efforts are now required to elucidate the exact molecular mechanism of exocytosis by which vesicles discharge their content to the extracellular space in response to a Ca^{2+}-dependent stimulus.

REFERENCES

1. Richardson, P. J., Presynaptic distribution of the cholinergic specific antigen Chol-1 and 5′-nucleotidase in rat brain, as determined by complement-mediated release of neurotransmitters, *J. Neurochem.*, 41, 640, 1983.
2. Docherty, M., Bradford, H. F., and Wu, J. Y., The preparation of highly purified GABAergic and cholinergic synaptosomes from mammalian brain, *Neurosci. Lett.*, 81, 232, 1987.
3. Bradford, H. F., Docherty, M., Wu, J. Y., Cash, C. D., Ehret, M., Maitre, M., and Joh, T. H., The immunolysis, isolation, and properties of subpopulations of mammalian brain synaptosomes, *Neurochem. Res.*, 14, 301, 1989.
4. Zimmermann, H., Vesicle recycling and transmitter release, *Neuroscience,* 4, 1773, 1979.
5. Israel, M., Dunant, Y., and Manaranche, R., The present status of the vesicular hypothesis, *Prog. Neurobiol.,* 13, 237, 1979.
6. Ceccarelli, B. and Hurlbut, W. P., Vesicle hypothesis of the release of quanta of acetylcholine, *Physiol., Rev.,* 6, 396, 1980.
7. Tauc, L., Nonvesicular release of neurotransmitter, *Physiol. Rev.*, 62, 857, 1982.
8. Dunant, Y., On the mechanism of acetylcholine release, *Prog. Neurobiol.*, 26, 55, 1986.
9. Parsons, S. M., Prior, C., and Marshall, I. G., Acetylcholine transport, storage, and release, *Int. Rev. Neurobiol.*, 35, 279, 1993.
10. Tucek, S., Choline acetyltransferase and the synthesis of acetylcholine, *Handb. Exp. Pharm.*, 86, 1988, 125.
11. Kuhar, M. J. and Murrin, L. C., Sodium-dependent, high affinity choline uptake, *J. Neurochem.*, 30, 15, 1978.
12. Kuhar, M. J., Sodium-dependent high affinity choline uptake, *Prog. Brain Res.,* 49, 71, 1979.
13. Speth, R. C. and Yamamura, H. I., Sodium dependent high affinity neuronal choline uptake, in *Nutrition and the Brain,* Barbeau, A., Growdon, J. H., and Wurtman, R. J., Eds., Raven, New York, 1979, 129.
14. Quastel, J. H., Source of the acetyl group in acetylcholine, in *Cholinergic Mechanisms and Psychopharmacology*, Jenden, D. J., Ed., Plenum, New York, 1978, 411.
15. Jope, R. S., High affinity choline transport and acetyl-CoA production in brain and their roles in the regulation of acetylcholine synthesis, *Brain Res. Rev.,* 1, 313, 1979.
16. Tucek, S. and Cheng, S. C., Precursors of acetyl groups in acetylcholine in the brain *in vivo*, *Biochim. Biophys. Acta,* 208, 538, 1970.
17. Whittaker, V. P., Michaelson, I. A., and Kirkland, R. J. A., The separation of synaptic vesicles from disrupted nerve ending particles, *Biochem. Pharmacol.*, 12, 300, 1963.
18. Whittaker, V. P., Michaelson, I. A., and Kirkland, R. J. A., The separation of synaptic vesicles from nerve-ending particles ('synaptosomes'), *Biochem. J.*, 90, 293, 1964.

19. De Robertis, E., Rodriguez de Lores Arnaiz, G., Salganicoff, L., Pellegrino de Iraldi, A., and Zieher, L. M., Isolation of synaptic vesicles and structural organization of the acetylcholine system within brain nerve endings, *J. Neurochem.*, 10, 225, 1963.

20. Sheridan, M. N., Whittaker, V. P., and Israel, M., The subcellular fractionation of the electric organ of *Torpedo*, *Z. Zellforsch.*, 74, 291, 1966.

21. Israël, M., Gautron, J., and Lesbats, B., Fractionnement de l'organe électrique de la Torpille: localisation subcellulaire de l'acétylcholine, *J. Neurochem.*, 17, 1441, 1970.

22. Wagner, J. A., Carlson, S. S., and Kelly, R. B., Chemical and physical characterization of cholinergic synaptic vesicles, *Biochemistry*, 17, 1199, 1978.

23. Ohsawa, K., Dowe, G. H. C., Horris, S. J., and Whittaker, V. P., The lipid and protein content of cholinergic synaptic vesicles from the electric organ of *Torpedo marmorata* purified to constant composition: implications for vesicle structure, *Brain Res.*, 161, 447, 1979.

24. Whittaker, V. P., Essman, W. B., and Dowe, G. H. C., The isolation of pure cholinergic synaptic vesicles from the electric organs of elasmobranch fish of the family *Torpedinidae, Biochem. J.*, 128, 833, 1972.

25. Nagy, A., Baker, R. R., Morris, S. J., and Whittaker, V. P., The preparation and characterization of synaptic vesicles of high purity, *Brain Res.*, 109, 285, 1976.

26. Whittaker, V. P. and Sheridan, M. N., The morphology and acetylcholine content of cerebral cortical synaptic vesicles, *J. Neurochem.*, 12, 363, 1965.

27. Dowdall, M. J., Boyne, A. F., and Whittaker, V. P., Adenosine triphosphate, a constituent of cholinergic synaptic vesicles, *Biochem. J.*, 140, 1, 1974.

28. Volknandt, W. and Zimmermann, H., Acetylcholine, ATP, and proteoglycan are common to synaptic vesicles isolated from the electric organs of electric eel and electric catfish as well as from rat diaphragm, *J. Neurochem.*, 47, 1449, 1986.

29. Richardson, P. J. and Brown, S. J., ATP release from affinity-purified rat cholinergic nerve terminals, *J. Neurochem.*, 48, 622, 1987.

30. Zimmermann, H. and Whittaker, V. P., Effect of electrical stimulation on the yield and composition of synaptic vesicles from the cholinergic synapses of the electric organ of *Torpedo*: a combined biochemical, electrophysiological and morphological study, *J. Neurochem.*, 22, 435, 1974.

31. Zimmermann, H. and Whittaker, V. P., Different recovery rates of the electrophysiological, biochemical and morphological parameters in the cholinergic synapses of the *Torpedo* electric organ after stimulation, *J. Neurochem.*, 22, 1109, 1974.

32. Morel, N. and Meunier, F. R., Simultaneous release of acetylcholine and ATP from stimulated cholinergic synaptosomes, *J. Neurochem.*, 36, 1766, 1981.

33. Schweitzer, E., Coordinated release of ATP and ACh from cholinergic synaptosomes and its inhibition by calmodulin antagonists, *J. Neurosci.*, 7, 2948, 1987.

34. Unsworth, C. D. and Johnson, R. G., Acetylcholine and ATP are coreleased from the electromotor nerve terminals of *Narcine brasiliensis* by an exocytotic mechanism. *Proc. Natl. Acad. Sci., U.S.A.*, 87, 553, 1990.

35. Marsal, J., Egea, G., Solsona, C., Rabasseda, X., and Blasi, J., Botulinum toxin type A blocks the morphological changes induced by chemical stimulation on the presynaptic membrane of *Torpedo* synaptosomes, *Proc. Natl. Acad. Sci. U.S.A.*, 86, 372, 1989.

36. Füldner, H. H. and Stadler, H., ^{31}P-NMR analysis of synaptic vesicles. Status of ATP and internal pH, *Eur. J. Biochem.*, 121, 519, 1982.

37. Anderson, D. C., King, S. C., and Parsons, S. M., Uncoupling of acetylcholine uptake from the *Torpedo* cholinergic synaptic vesicle ATP-ase, *Biochem. Biophys. Res. Commun.*, 103, 422, 1981.

38. Anderson, D. C., King S. C., and Parsons, S. M., Proton gradient linkage to active uptake of [^{3}H] acetylcholine by *Torpedo* electric organ synaptic vesicles, *Biochemistry*, 21, 3037, 1982.

39. Michaelson, D. M., Avissar, S., Kloog, Y., and Sokolovski, M., Mechanism of acetylcholine release: possible involvement of presynaptic muscarinic receptors in regulation of acetylcholine release and protein phosphorylation, *Proc. Natl. Acad. Sci., U.S.A.*, 76, 6336, 1979.

40. Toll, L. and Howard, B. D., Evidence that an ATPase and a protonmotive force function in the transport of acetylcholine into storage vesicles, *J. Biol. Chem.*, 255, 1787, 1980.

41. Melega, W. P. and Howard, B. D., Choline and acetylcholine metabolism in PC12 secretory cells, *Biochemistry*, 20, 4477, 1981.

42. Breer, H., Morris, S. J., and Whittaker, V. P., Adenosine triphosphatase activity associated with purified cholinergic synaptic vesicles of *Torpedo marmorata, Eur. J. Biochem.*, 80, 313, 1977.

43. Rothlein, J. E. and Parsons, S. M., Specificity of association of a Ca^{2+}/Mg^{2+} ATPase with cholinergic synaptic vesicles from *Torpedo* electric organ, *Biochem. Biophys. Res. Commun.*, 88, 1069, 1979.

44. Yamagata, S. K. and Parsons, S. M., Cholinergic synaptic vesicles contain a V-type and a P-type ATPase, *J. Neurochem.*, 53, 1354, 1989.

45. Moriyama, Y. and Nelson, N., Lysosomal H^+-translocating ATPase has a similar subunit structure to chromaffin granule H^+-ATPase complex, *Biochim. Biophys. Acta*, 980, 241, 1989.

46. Parry, R. V., Turner, J. C., and Rea, P. A., High purity preparations of higher plant vacuolar H^+-ATPase reveal additional subunits. Revised subunit composition, *J. Biol. Chem.*, 264, 20025, 1989.

47. Perin, M. S., Fried, V. A., Stone, D. K., Xie, X. S., and Sudhof, T. C., Structure of the 116-kDa polypeptide of the clathrin-coated vesicle/synaptic vesicle proton pump, *J. Biol. Chem.*, 266, 3877, 1991.

48. Hicks, B. W. and Parsons, S. M., Characterization of the P-type and V-type ATPases of cholinergic synaptic vesicles and coupling of nucleotide hydrolysis to acetylcholine transport, *J. Neurochem.*, 58, 1211, 1992.

49. Bahr, B. A. and Parsons, S. M., Demonstration of a receptor in *Torpedo* synaptic vesicles for the acetylcholine storage blocker L-*trans*-2-(4-phenyl[3,4-^{3}-H]-piperidino)cyclohexanol), *Proc. Natl. Acad. Sci. U.S.A.*, 83, 2267, 1986.

50. Bahr, B. A. and Parsons, S. M., Acetylcholine transport and drug inhibition in *Torpedo* synaptic vesicles, *J. Neurochem.*, 46, 1214, 1986.

51. Carpenter, R. S., Koenigsberger, R., and Parsons, S. M., Passive uptake of acetylcholine and other organic cations by synaptic vesicles from *Torpedo* electric organ, *Biochemistry*, 19, 4373, 1980.

52. Giompres, P. E. and Luqmani, Y. A., Cholinergic synaptic vesicles isolated from *Torpedo marmorata*: demonstration of acetylcholine and choline uptake in an *in vitro* system, *Neuroscience,* 5, 1041, 1980.

53. Anderson, D. C., King, S. C., and Parsons, S. M., Pharmacological characterization of acetylcholine transport system in purified *Torpedo* electric organ synaptic vesicles, *Mol. Pharmacol.*, 24, 48, 1983.

54. Anderson, D. C., Bahr, B. A., and Parsons, S. M., Stoichiometrics of acetylcholine uptake, release, and drug inhibition in *Torpedo* synaptic vesicles; heterogeneity in acetylcholine transport and storage, *J. Neurochem.*, 46, 1207, 1986.

55. Anderson, D. C., King, S. C., and Parsons, S. M., Inhibiton of [^{3}H] acetylcholine active transport by tetraphenyl-borate and other anions, *Mol. Pharmacol.*, 24, 55, 1983.

56. Diebler, M. F. and Morot Gaudry-Talarmain, Y., AH5183 and cetiedil: two potent inhibitors of acetylcholine uptake into isolated synaptic vesicles from *Torpedo marmorata, J. Neurochem.*, 52, 813, 1989.

57. Parsons, S. M. and Koenigsberger, R., Specific stimulated uptake of acetylcholine by *Torpedo* electric organ synaptic vesicles, *Proc. Natl. Acad. Sci.* U.S.A., 77, 6234, 1980.

58. Kornreich, W. D. and Parsons, S. M., Sidedness and chemical and kinetic properties of the vesamicol receptor of cholinergic synaptic vesicles, *Biochemistry*, 27, 5262, 1988.

59. Bahr, B. A. and Parsons, S. M., Purification of the vesamicol receptor, *Biochemistry*, 31, 5763, 1992.

60. Rogers, G. A. and Parsons, S. M., Photoaffinity labeling of the acetylcholine transporter, *Biochemistry*, 31, 5770, 1992.

61. Carlson, S. S. and Kelly, R. B., A highly antigenic proteoglycan-like component of cholinergic synaptic vesicles, *J. Biol. Chem.*, 258, 11082, 1983.

62. Kish, S. J., Distefano, L. M., Dozic, S., Robitaille, Y., Rajput, A., Deck, J. H., and Hornykiewicz, O., [^{3}H]vesamicol binding in human brain cholinergic deficiency disorders, *Neurosci. Lett.*, 117, 347, 1990.

63. Ruberg, M., Mayo, W., Brice, A., Duyckaerts, C., Hauw, J. J., Simon, H., LeMoal, M., and Agid, Y., Choline acetyltransferase activity and [^{3}H]vesamicol binding in the temporal cortex of patients with Alzheimer's disease, Parkinson's disease, and rats with basal forebrain lesions, *Neuroscience*, 35, 327, 1990.

64. Katz, B., *The Release of Neural Transmitter Substances*, Liverpool University Press, Liverpool, U.K., 1969.

65. Ceccarelli, B., Hurlbut, W. P., and Mauro, A.,Turnover of transmitter and synaptic vesicles at the frog neuromuscular junction, *J. Cell Biol.,* 57, 499, 1973.

66. Ceccarelli, B., Grohovaz, F., Hurlbut, W. P., and Jezzi, N., Freeze-fracture studies of frog neuromuscular junctions during intense release of neurotransmitter. II. Effects of electrical stimulation and high potassium, *J. Cell Biol.*, 81, 178, 1979.

67. Heuser, J. E. and Reese, T. S., Evidence for recycling of synaptic vesicles membranes during transmitter release at the frog neuromuscular junction, *J. Cell. Biol.*, 57, 315, 1973.
68. Perri, V., Sacchi, O., Raviola, E., and Raviola, G., Evaluation of the number and distribution of synaptic vesicles at cholinergic nerve-endings after sustained stimulation, *Brain Res.*, 39, 526, 1972.
69. Reinecke, M. and Walther, C., Aspects of turnover and biogenesis of synaptic vesicles at locust neuromuscular junctions as revealed by zine iodide-osmium tetroxide (ZIO) reacting with intravesicular SH-groups, *J. Cell Biol.*, 78, 839, 1978.
70. Haimann, C., Torri-Tarelli, F., Fesce, R., and Ceccarelli, B., Measurement of quantal secretion induced by ouabain and its correlation with depletion of synaptic vesicles, *J. Cell Biol.*, 101, 1953, 1985.
71. Heuser, J. E., Reese, T. S., Dennis, M. J., Jan, Y., Jan, L., and Evans, L., Synaptic vesicle exocytosis captured by quick freezing and correlated with quantal transmitter release, *J. Cell Biol.*, 81, 275, 1979.
72. Torri-Tarelli, F., Grohovaz, F., Fesce, R., and Ceccarelli, B., Temporal coincidence between synaptic vesicle fusion and quantal secretion of acetylcholine, *J. Cell Biol.*, 101, 1386, 1985.
73. Holtzman, E., Freeman, A. R., and Kashner, A., Stimulation dependent alterations in peroxidase uptake at lobster neuromuscular junctions, *Science*, 173, 733, 1971.
74. Jorgensen, O. S. and Mellerup, E. T., Endocytic formation of rat brain synaptic vesicles, *Nature (London)*, 249, 770, 1974.
75. Ceccarelli, B. and Hurlbut, W. P., The effects of prolonged repetive stimulation in hemicholinium on the frog neuromuscular junction, *J. Physiol. (London)*, 247, 163, 1975.
76. Fried, R. C. and Blaustein, M. P. Synaptic vesicle recycling in synaptosomes *in vitro*, *Nature (London)*, 261, 255, 1976.
77. Zimmermann, H. and Whittaker, V. P., Effect of electrical stimulation on the yield and composition of synaptic vesicles from the cholinergic synapses of the electric organ of *Torpedo*: a combined biochemical, electrophysiological and morphological study, *J. Neurochem.*, 22, 435, 1974.
78. Salehmoghaddam, S. H. and Collier, B., The relationship between acetylcholine release from brain slices and the acetylcholine content of subcellular fractions prepared from brain, *J. Neurochem.*, 27, 71, 1976.
79. Carrol, P. T. and Nelson, S. H., Cholinergic vesicles: ability to empty and refill independently of cytoplasmic acetylcholine, *Science*, 199, 85, 1978.
80. Chakrin, L. W. and Whittaker, V. P., The subcellular distribution of [N-Me-^{3}H] acetylcholine synthetized by brain *in vivo*, *Biochem. J.*, 118, 97, 1969.
81. Barker, L. A., Dowdall, M. J., and Whittaker, V. P., Cholinergic metabolism in the cerebral cortex of giunea pig, *Biochem. J.*, 130, 1063, 1972.
82. Babel-Guérin, E. and Dunant, Y., Entrée de calcium et libération d'acétylcholine dans l'organe électrique de la Torpille, *C. R. Hebd. Séanc. Acad. Sci.*, 275, 2961, 1972.
83. Morel, N., Israel, M., and Manaranche, R., Mastour, P., Isolation of pure cholinergic nerve endings from *Torpedo* electric organ. Evaluation of their metabolic properties, *J. Cell Biol.*, 75, 43, 1977.
84. Katz, B. and Miledi, R., Transmitter leakage from motor nerve endings, *Proc. R. Soc., (London), Ser. B.*, 196, 59, 1977.
85. Suszkiw, J. B., Kinetics of acetylcholine recovery in *Torpedo* electromotor synapses depleted of synaptic vesicles, *Neuroscience*, 5, 1341, 1980.
86. Miledi, R., Molenaar, P. C., and Polak, R. L., An analysis of acetylcholine in frog muscle by mass fragmentography, *Proc. R. Soc. (London) Ser. B*, 197, 285, 1977.
87. Israel, M. and Lesbats, B., Continuous determination of transmitter release and compartmentation in *Torpedo* electric organ synaptosomes, studied with a chemiluminiscent method detecting acetylcholine, *J. Neurochem.*, 37, 1475, 1981.
88. Molenaar, P. C., Nickolson, V. J., and Polak, R. L., Preferential release of newly synthetized [^{3}H] acetylcholine from rat cerebral cortex slices *in vitro*, *Br. J. Pharmacol.*, 47, 97, 1973.
89. Molenaar, P. C. and Polak, R.L., Newly formed acetylcholine in synaptic vesicles in brain tissue, *Brain Res.*, 62, 537, 1973.
90. Richter, J. A. and Marchbanks, R. M., Isolation of [^{3}H] acetylcholine pools by subcellular fractionation of cerebral cortex slices incubated with [^{3}H] choline, *J. Neurochem.*,18, 705, 1971.
91. Rodriguez de Lores Arnaiz, G., Zieher, L. M., and de Robertis, E., Neurochemical and structural studies on the mechanism of action of hemicholinium-3 in central cholinergic synapses, *J. Neurochem.*, 17, 221, 1970.

92. Collier, B., The preferential release of newly synthesized transmitter by a sympathetic ganglion, *J. Physiol. (London)*, 205, 341, 1969.

93. Potter, L.T., Synthesis, storage and release of ^{14}C-acetylcholine in isolated rat diaphragm muscles, *J. Physiol., (London)*, 206, 145, 1970.

94. Dunant, Y., Gautron, J., Israël, M., Lesbats, B., and Manaranche, R., Acetylcholine compartment in stimulated electric organ of *Torpedo marmorata*, *J. Neurochem.*, 19, 1987, 1972.

95. Chakrin, L. W., Marchbanks, R. M., Mitchell, J. F., and Whittaker, V. P., The origin of the acetylcholine released from the surface of the cortex, *J. Neurochem.*, 19, 2727, 1972.

96. Luz, S., Pinchasi, I., and Michaelson, D. M., Newly synthesized and preformed acetylcholine are released from *Torpedo* synaptosomes by different pathways, *J. Neurochem.*, 45, 43, 1985.

97. Marchbanks, R. M. and Israel, M., Aspects of acetylcholine metabolism in the electric organ of *Torpedo marmorata*, *J. Neurochem.*, 18, 439, 1971.

98. Israel, M. and Tucek, S., Utilization of acetate and pyruvate for the synthesis of "total," "bound," and "free" acetylcholine in the electric organ of *Torpedo*, *J. Neurochem.*, 22, 487, 1974.

99. Marchbanks, R. M., The subcellular origin of the acetylcholine released at synapses, *Int. J. Biochem.*, 6, 303, 1975.

100. Israel, M. and Dunant, Y., On the mechanism of acetylcholine release, *Prog. Brain Res.*, 49, 125, 1979.

101. Dunant, Y., Israel, M., Lesbats, B., and Manaranche, R., Oscillation of acetylcholine during nerve activity in the *Torpedo* electric organ, *Brain Res.*, 125, 123, 1977.

102. Israel, M., Lesbats, B., and Manaranche, R., ACh release from osmotically shocked synaptosomes refilled with transmitter, *Nature (London)*, 294, 474, 1981.

103. Israel, M., Lesbats, B., Manaranche, R., and Morel, N., Acetylcholine release from proteoliposomes equipped with synaptosomal membrane constituents, *Biochim. Biophys. Acta*, 728, 438, 1983.

104. Israel, M., Lesbats, B., Morel, M., Manaranche, R., Gulik-Krzywicki, T., and Dedieu, J. C., Reconstitution of a functional synaptosomal membrane possessing the protein constituents involved in acetylcholine translocation, *Proc. Natl. Acad. Sci. U.S.A.*, 81, 277, 1984.

105. Meyer, E. M. and Cooper, J. R., High affinity choline uptake and calcium-dependent acetylcholine release in proteoliposomes derived from rat cortical synaptosomes, *J. Neurosci.*, 3, 987, 1983.

106. Birman, S., Israel, M., Lesbats, B., and Morel, N., Solubilization and partial purification of a presynaptic membrane protein ensuring calcium-dependent acetylcholine release from proteoliposomes, *J. Neurochem.*, 47, 433, 1986.

107. Vyskocil, F. and Illes, P., Non-quantal release of transmitter at mouse neuromuscular junction and its dependence on the activity of Na+-K+ ATPase, *Pfluegers Arch.*, 370, 295, 1977.

108. Vizi, E. S. and Vyskocil, F., Changes in total and quantal release of acetylcholine in the mouse diaphragm during activation and inhibition of membrane ATPase, *J. Physiol., (London)*, 286, 1, 1979.

109. Straugham, D. W., The release of acetylcholine from mammalian motor nerve endings, *Br. J. Pharmacol.*, 15, 417, 1960.

110. Mitchell, J. F. and Silver, A., The spontaneous release of acetylcholine from the denervated hemidiaphragm of the rat, *J. Physiol. (London)*, 165, 117, 1963.

111. Zimmermann, H. and Denston, C. R., Recycling of synaptic vesicles in the cholinergic synapses of the *Torpedo* electric organ during induced transmitter release, *Neuroscience*, 2, 695, 1977.

112. Zimmermann, H. and Denston, C. R., Separation of synaptic vesicles of different functional states from the cholinergic synapses of the *Torpedo* electric organ, *Neuroscience*, 2, 715, 1977.

113. Zimmermann, H., Turnover of adenine nucleotides in cholinergic synaptic vesicles of the *Torpedo* electric organ, *Neuroscience*, 3, 827, 1978.

114. Suszkiw, J. B., Zimmermann, H., and Whittaker, V. P., Vesicular storage and release of acetylcholine in *Torpedo* electroplaque synapses, *J. Neurochem.*, 30, 1269, 1978.

115. Gracz, L. M, Wang, W. C., and Parsons, S. M., Cholinergic synaptic vesicle heterogeneity: evidence for regulation of acetylcholine transport, *Biochemistry*, 27, 5268, 1988.

116. Giompres, P. E., Zimmermann, H., and Whittaker, V. P., Purification of small dense vesicles from stimulated *Torpedo* electric tissue by glass bead chromatography, *Neuroscience*, 6, 775, 1981.

117. Searl, T., Prior, C., and Marshall, I. G., The effects of L-vesamicol, an inhibitor of vesicular acetylcholine uptake, on two populations of miniature endplate currents at the snake neuromuscular junction, *Neuroscience*, 35, 145, 1990.

118. Whittaker, V. P., Cholinergic false transmitters, *Handb. Exp. Pharm.*, 86, 1988, 465.

119. Smith, A. D., Cellular control of the uptake, storage and release of noradrenaline in sympathetic nerves, *Biochem. Soc. Symp.*, 36, 103, 1972.
120. Muscholl, E., Adrenergic false transmitters, *Handb. Exp. Pharm.*, 33, 618, 1972.
121. Barker, L. A., Dowdall, M. J., and Mittag, T. W., Comparative studies on synaptosomes: high affinity uptake and acetylation of N-[Me-³H] *N*-hydroxyethylpyrrolidinum, *Brain. Res.*, 86, 343, 1975.
122. Ilson, D. and Collier, B., Triethylcholine as a precursor to a cholinergic false transmitter, *Nature (London)*, 254, 618, 1975.
123. Collier, B., Barker, L. A., and Mittag, T. W., The release of acetylated choline analogues by a sympathetic ganglion, *Mol. Pharmacol.*, 12, 340, 1976.
124. Zimmermann, H. and Dowdall, M. J., Vesicular storage and release of a cholinergic false transmitter (acetylpyrrolcholine) in the *Torpedo* electric organ, *Neuroscience*, 2, 731, 1977.
125. Whittaker, V. P. and Luqmani, Y. A., False transmitters in the cholinergic system: implications for the vesicle theory of transmitter storage and release, *Gen. Pharmacol.*, 11, 7, 1980.
126. Luqmani, Y. A., Sudlow, G., and Whittaker, V. P., Homocholine and acetylhomocholine: false transmitters in the cholinergic electromotor system of *Torpedo*, *Neuroscience*, 5,153, 1980.
127. Luqmani, Y. A. and Whittaker, V. P., A vesicular site of origin for the release of a false transmitter at the *Torpedo* synapse, in *Cholinergic Mechanisms: Phylogenic Aspects, Central and Peripheral Synapses and Clinical Significance*, Pepeu, G. and Ladinsky, H., Eds., Plenum, New York, 1981, 47.
128. von, Schwarzenfeld, I., Origin of transmitters released by electrical stimulation from a small, metabolically very active vesicular pool of cholinergic synapses in guinea-pig cerebral cortes, *Neuroscience*, 4, 477, 1979.
129. Welner, S. A. and Collier, B., Uptake, metabolism, and releasability of ethyl analogues of homocholine by rat brain, *J. Neurochem*, 43, 1143, 1984.
130. Boksa, P. and Collier, B., Spontaneous and evoked release of acetylcholine and a cholinergic false transmitter from brain slices: comparison to true and false transmitter in subcellular stores, *Neuroscience*, 5, 1517, 1980.
131. Otero, D. H., Wilbekin, F., and Meyer, E. M., Effects of 4-(2-hydroxyethyl)-1-piperazine-ethanesulfonic acid (AH5183) on rat cortical synaptosome choline uptake, acetylcholine storage and release, *Brain Res.*, 359, 208, 1985.
132. Suszkiw, J. B. and Toth, G., Storage and release of acetylcholine in rat cortical synaptosomes: effects of D,L-2-(4-phenylpiperidino)cyclohexanol (AH5183), *Brain Res.*, 386, 371, 1986.
133. Van der Kloot, W., 2-(4-phenylpiperidino) cyclohexanol (AH5183) decreases quantal size at the frog neuromuscular junction, *Pfluegers. Arch.*, 406, 83, 1986.
134. Whitton, P. S., Marshall, I. G., and Parsons, S. M., Reduction of quantal size by vesamicol (AH5183), an inhibitor of vesicular acetylcholine storage, *Brain Res.*, 385, 189, 1986.
135. Lupa, M. T., Effects of an inhibitor of the synaptic vesicle acetylcholine transport system on quantal neurotransmitter release: an electrophysiological study, *Brain Res.*, 461, 118, 1988.
136. Edwards, C., Dolezal, V.,Tucek, S., Zemkova, H., and Vyskocil, F., Is an acetylcholine transport system responsible for nonquantal release of acetylcholine at the rodent myoneural junction?, *Proc. Natl. Acad. Sci. U.S.A.*, 82, 3514, 1985.
137. Smith, D. O., Photoaffinity labeling of the acetylcholine transporter, *Neurosci. Lett.*, 135, 5, 1992.
138. Enomoto, K., Post- and presynaptic effects of vesamicol (AH5183) on the frog neuromuscular junction, *Eur. J. Pharmacol.*, 147, 209, 1988.
139. Searl, T., Prior, C., and Marshall, I. G., Acetylcholine recycling and release at rat motor nerve terminals studied using (-)-vesamicol and troxpyrrolium, *J. Physiol. (London)*, 444, 99, 1991.
140. Vizi, E. S., In favour of the vesicular hypothesis: neurochemical evidence that vesamicol (AH5183) inhibits stimulation-evoked release of acetylcholine from neuromuscular junction, *Br. J. Pharmacol.*, 98, 898, 1989.
141. Vyskocil, F., Inhibition of non-quantal acetylcholine leakage by 2(4-phenylpiperidino)cyclohexanol in the mouse diaphragm, *Neurosci. Lett.*, 59, 277, 1985.
142. Zemkova, H., Vyskocil, F., and Edwards, C., The effects of nerve terminal activity on non-quantal release of acetylcholine at the mouse neuromuscular junction, *J. Physiol. (London)*, 423, 631, 1990.
143. Edwards, C., Dolezal, V., Tucek, S., Zemkova, H., and Vyskocil, F., A possible role for the acetylcholine transport system in non-quantal release of acetylcholine at the rodent myoneural junction, *P.R. Health Sci. J.*, 7, 71, 1988.

144. Carroll, P. T., The effect of the acetylcholine transport blocker 2-(4-phenylpiperidino) cyclohexanol (AH5183) on the subcellular storage and release of acetylcholine in mouse brain, *Brain Res.*, 358, 200, 1985.

145. Jope, R. S. and Johnson, G. V., Quinacrine and 2-(4-phenylpiperidino)cyclohexanol (AH5183) inhibit acetylcholine release and synthesis in rat brain slices, *Mol. Pharmacol.*, 29, 45, 1986.

146. Ricny, J. and Collier, B., Effect of 2-(4-phenylpiperidino)cyclohexanol on acetylcholine release and subcellular distribution in rat striatal slices, *J. Neurochem.*, 47, 1627, 1986.

147. Adam-Vizi, V., External Ca^{2+}-independent release of neurotransmitters, *J. Neurochem.*, 58, 395, 1992.

148. Deri, Z. and Adam-Vizi, V., Parameters not influenced by vesamicol: membrane potential, calcium uptake, and internal calcium concentration of synaptosomes, *Neurochem. Res.*, 17, 539, 1992.

149. Michaelson, D. M. and Burstein, M., Biochemical evidence that acetylcholine release from cholinergic nerve terminals is mostly vesicular, *FEBS Lett.*, 188, 389, 1985.

150. Michaelson, D. M., Burstein, M., and Licht, R., Translocation of cytosolic acetylcholine into synaptic vesicles and demonstration of vesicular release, *J. Biol. Chem.*, 261, 6831, 1986.

151. Collier, B., Welner, S. A., Rícny, J., and Araujo, D. M., Acetylcholine synthesis and release by a sympathetic ganglion in the presence of 2-(4-phenylpiperidino) cyclohexanol (AH5183), *J. Neurochem.*, 46, 822, 1986.

152. Adam-Vizi, V., Deri, Z., Vizi, E. S., Sershen, H., and Lajtha, A., Ca^{2+}-independent veratridine-evoked acetylcholine release from striatal slices is not inhibited by vesamicol (AH5183): mobilization of distinct transmitter pools, *J. Neurochem.*, 56, 52, 1991.

153. Caterall, W. A., Neurotoxins that act on voltage-sinsitive sodium channels in excitable membranes, *Annu. Rev. Pharmacol. Toxicol.*, 20, 15, 1980.

154. Cabeza, R. and Collier, B., Acetylcholine mobilization in a sympathetic ganglion in the presence and absence of 2-(4-phenylpiperidino)cyclohexanol (AH5183), *J. Neurochem.*, 50, 112, 1988.

155. Ceccarelli, B., Grohovaz, F., Hurlbut, W. P., and Jezzi, N., Freeze-fracture studies of frog neuromuscular junctions during intense release of neurotransmitter. I. Effects of black widow spider venom and Ca^{2+} free solutions on the structure of the active zone, *J. Cell Biol.*, 81, 163, 1979.

156. Gorio, A., Hurlbut, W. P., and Ceccarelli, B., Acetylcholine compartments in mouse diaphragm: a comparison of the effects of black widow spider venom, electrical stimulation and high concentrations of potassium, *J. Cell Biol.*, 78, 716, 1978.

157. Tzeng, M. C. and Siekevitz, P., The effect of the purified major protein factor (α-latrotoxin) of black widow spider venom on the release of acetylcholine and norepinephrine from mouse cerebral cortex slices, *Brain Res.*, 139, 190, 1978.

158. Meldolesi, J., Studies on α-latrotoxin receptors in rat brain synaptosomes: correlation between toxin binding and stimulation of transmitter release, *J. Neurochem.*, 38, 1559, 1982.

159. Nicholls, D. G., Rugolo, M., Scott, I. G., and Meldolesi, J., α-Latrotoxin of black widow spider venom depolarizes the plasma membrane, induces massive calcium influx, and stimulates transmitter release in guinea pig brain synaptosomes, *Proc. Natl. Acad. Sci. U.S.A.*, 79, 7924, 1982.

160. Deri, Z., Bors, P., and Adam-Vizi, V., Effect of α-latrotoxin on acetylcholine release and intracellular Ca^{2+} concentration in synaptosomes: Na^+-dependent and Na^+-independent components, *J. Neurochem.*, 60, 1065, 1993.

161. Rosenthal, L. and Meldolesi, J., α-Latrotoxin and related toxins, *Pharmacol. Ther.*, 42, 115, 1989.

Chapter 11

Muscarinic Acetylcholine Receptors and Long-Term Potentiation of Synaptic Transmission

Hendrik W.G.M. Boddeke and Peter H. Boeijinga

CONTENTS

I. INTRODUCTION

Many studies have indicated involvement of the cholinergic system in learning and memory. Muscarinic receptor antagonists disrupt learning tests such as passive and active avoidance and spatial navigation.[1] Also, lesions of the septal nuclei, which constitute the cholinergic innervation of the hippocampus, induce cognitive impairment in rats.[2,3]

Long-term potentiation of synaptic transmission (LTP) is a long-lasting amplification of synaptic strength[4,5] and seems to provide a mechanism for learning which suggests that storage of memory resides in the strengthening of synaptic contacts.

A modulatory role of the cholinergic system in the induction of LTP has been suggested in many *in vitro*[6,7] and *in vivo* studies.[8,9] According to their biochemical and electrophysiological properties, the muscarinic receptors which most likely are involved in induction of LTP are M_1 and M_3 receptors. In this review we discuss the biochemical, electrophysiological, and behavioral effects of muscarinic receptor activation which form the currently accepted mechanistic basis of the enhancement of LTP.

II. DISTRIBUTION OF MUSCARINIC RECEPTOR SUBTYPES IN THE BRAIN

Molecular biological techniques applied to study muscarinic receptors have resulted in the cloning, sequencing, and expression of five different genes which encode proteins with characteristics of muscarinic receptor binding sites.[10] The cloned muscarinic receptors have been named m_1, m_2, m_3, m_4, and m_5, of which the first four correspond to the pharmacologically defined M_1, M_2, M_3, and M_4 muscarinic receptors.[11,12] The muscarinic receptor antagonists pirenzepine and 11[[2-[(diethyl-amino)-methyl]-1-piperidinyl]acetlyl]-5,11-dihydro-6H-pyrido[2,3-6][1,4]-benzodiazepine-6-one (AF-DX 116) were the first M_1 and M_2 receptor subtype selective antagonists, respectively; and provided the basis of the subdivision of muscarinic receptors into M_1 and M_2 receptor subtypes.[13,14] Additionally, an antagonist with a certain selectivity for M_3 receptors, parafluoro-hexahydrosila-difenidol has been described.[15] At present, no selective receptor antagonists for the M_4 and M_5 muscarinic receptor subtypes have been reported. According to *in situ* mRNA hybridization histocytochemistry and radioligand binding studies, all five muscarinic receptor subtypes are found in the brain albeit with specific distribution patterns. In brain areas which have been extensively studied in terms of LTP activity, such as the hippocampus and neocortex, all five muscarinic receptors have been found. In the rat hippocampus, m_1 and m_3 receptor mRNA is found postsynaptically both in the pyramidal cell layer and in the dentate area[16]. Using the radioligands [^{3}H]pirenzepine and 4-[^{3}H]diphenylacetoxy-*N*-methyl-piperidine (DAMP), M_1 and M_3 muscarinic receptor binding sites, respectively, have been found in the hippocampus.[17] m_2 muscarinic receptor mRNA is expressed both pre- and postsynaptically in the pyramidal cell layer but not in dentate granule cells.[18] Also M_2 binding sites have been detected in the hippocampus.[19] m_4 mRNA is expressed as well in the hippocampus; the lack of specific ligands, however, makes it difficult to characterize the regional distribution of this muscarinic receptor subtype in more detail. High levels of m_5 receptor mRNA have been found in the ventral subiculum and the hippocampal pyramidal cell layer of the CA1, whereas lower amounts are found in the CA2 subfield.[20] Very little is currently known, however, on localization of the M_5 receptor protein. In the rat cerebral cortex m_1 receptor mRNA is relatively evenly distributed, whereas [^{3}H]pirenzepine sites are concentrated in the superficial and deep cortical layers.[17] m_3 mRNA is found in superficial and deep layers of the rat cortex and 4-[^{3}H]DAMP sites are found in the same areas.[16,17] m_2 Muscarinic receptor mRNA in the rat neocortex is mainly found in layers IV and VI, whereas binding sites for the selective M_2 muscarinic receptor antagonist [^{3}H](5,11-dihydro-11-{[2-{2-[(dipropyl-amino)methyl]-1-piperidinyl}ethyl)amino]carbonyl}-6H-pyrido(2,3-b)benzodiazepine-6-one (AF-DX 384) are found in the intermediate cortical laminae.[17] A low expression level of m_4 muscarinic receptor mRNA has been observed in the rat cerebral cortex, whereas m_5 receptor mRNA is absent in cortical regions.[20]

At present, most LTP studies concerning muscarinic receptor subtypes consider the M_1 muscarinic receptor most relevant. The expression patterns of both muscarinic receptor mRNAs and protein in the rat hippocampus and cortex, however, show that albeit distributed distinctly all five mRNAs and at least three different muscarinic receptor proteins are present in these brain structures. Accordingly, it is possible that more than one muscarinic receptor subtype is involved in modulation of LTP.

III. ACTIVITY OF MUSCARINIC RECEPTORS IN CNS

Functional expression studies with the five cloned muscarinic receptor subtypes have shown that each muscarinic receptor subtype preferentially couples to one effector system.[21] Thus m_1, m_3, and m_5 muscarinic receptors preferentially stimulate phosphoinositide hydrolysis, thereby generating inositoltriphosphate (IP_3) which mobilizes intracellular calcium and diacylglycerol which activates protein kinase C.[22] Protein kinases and intracellular calcium both are important mediators of LTP,[23] and thus putative effects of M_1, M_3, and M_5 muscarinic receptors on synaptic transmission may, at least partly, be explained by direct activation of these second messengers. m_2 and m_4 muscarinic receptor subtypes primarily couple negatively to adenylate cyclase. In a few cases, however, M_2 and M_4 muscarinic receptors have been reported to activate additional signaling pathways such as phospholipase C and potassium channels.[24-26]

All these muscarinic receptor-mediated second messenger responses have been found in brain tissue suggesting that activation of muscarinic receptors in central nervous system (CNS) neurons modulates a variety of conductance mechanisms, which most likely involve different muscarinic receptor subtypes. Activation of muscarinic receptors results both in excitatory and inhibitory effects, which are summarized in Table 11-1.

A. EXCITATORY EFFECTS
The best known excitatory activity of muscarinic receptors in neurons is the depression of a number of potassium conductances:[27,28]

1. G_K — Carbachol induces decrease in a potassium leak current[29] which leads to depolarization, enhancement of excitatory postsynaptic potentials (EPSPs), and facilitation of cell firing. This effect most likely involves M_1 muscarinic receptors.
2. I_{AHP} — At similar concentrations carbachol mediates blockade of the afterhyperpolarization (AHP). This effect results from a calcium-dependent potassium conductance. Both M_1[30] and M_2[31] receptors have been claimed to be involved in this process. Suppression of AHP causes loss of spike frequency adaptation and thus facilitates afterfiring.

3. I_M — Muscarinic receptor-induced inhibition of the M current, a voltage and time-dependent potassium current, involves activation of protein kinase C and is most likely activated via the M_1 receptor.[32,33] Inhibition of I_M causes a pronounced long-lasting facilitation of excitation.

4. I_A — Inhibition of a fast transient voltage-dependent potassium current by acetylcholine has been observed in cultured hippocampal neurons. This current is responsible for the slow after-hyperpolarization, and inhibition of I_A enhances excitability of cells.[34]

The modulation of the aforementioned potassium currents by muscarinic receptors results in slow synaptic activity. It has been shown that robust stimulation in rat hippocampal slices evokes prolonged slow EPSPs which are enhanced by cholinesterase inhibitors and blocked by muscarinic receptor antagonists.[35] Most likely, blockade of G_K and I_{AHP} are involved in this effect, whereas inhibition of I_M seems to be less prominent.[29] In addition, it has been demonstrated recently that activation of muscarinic receptors reduces a $GABA_B$ receptor-mediated potassium conductance in hippocampal neurons.[36,37] This effect may contribute to the excitatory effects of muscarinic receptor agonists.

Recently, it has been shown that acetylcholine potentiates N-methyl-D-aspartate (NMDA)-receptor mediated currents, presumably via an IP_3-dependent mechanism.[38] Additionally, it has been shown that the NMDA receptor-induced component of the EPSP to θ-burst stimulation is enhanced by activation of M_1 muscarinic receptors.[39,40] Since induction of LTP is critically dependent on activation of NMDA receptors, modulation of these receptors by acetylcholine may be important.

B. INHIBITORY EFFECTS

Depression of calcium currents in CA3 pyramidal cells and in neostriatal neurons has been observed after activation of muscarinic receptors.[41,42] Presently, it is not clear which receptor subtype is involved in the suppression of the calcium conductance. Depression of calcium currents in nerve endings could result in suppression of EPSPs. Indeed, activation of muscarinic receptors suppresses synaptic transmission via a presynaptic mechanism found in CA3 hippocampal neurons[43] and in Schaffer collateral-CA1 synapses.[44] Not only EPSPs but also inhibitory postsynaptic potentials (IPSPs) are suppressed by muscarinic receptor activation.[45,46] Presently, the exact mechanism behind this presynaptic inhibition is unclear. Both M_1 and M_2 receptors have been claimed to be responsible for presynaptic effects in different brain areas.[30,47] Additionally, inhibitory effects resulting from a muscarinic receptor-mediated increase in potassium conductance in thalamic reticular neurons, thalamic interneurons, and the cortex have been reported. This hyperpolarizing effect appears to be mediated by M_2 muscarinic receptors[48,49] and seems, at least in the cortex, to result indirectly from activation of inhibitory neurons.

It is difficult to determine precisely which of the described effects of the cholinergic muscarinic system participate in modulation of LTP. Certainly, excitation mediated through inhibition of G_K, I_{AHP}, I_A and I_M is important. Particularly, suppression of I_M causes a strong depolarization, which has been reported to continue after removal of acetylcholine[50] and accordingly might provide a mechanism for long-term potentiation. Also, as mentioned previously, muscarinic modulation of NMDA currents may be of critical importance for induction of LTP. A fine-tuned muscarinic receptor-mediated modulation of γ-aminobutyric acid (GABA) release but also interneuron synchronization provides the basis of oscillatory activity in hippocampal principal neurons. This oscillatory activity may set the conditions for facilitation by muscarinic receptor activation.

Table 11-1   Effects in CNS mediated by muscarinic receptors

Excitatory

Suppression of potassium conductances
Suppression of GABA-ergic neurotransmission
Reduction of $GABA_B$ receptor-mediated potassium conductance
Enhancement of NMDA responses

Inhibitory

Suppression of calcium current
Supression of glutamatergic EPSPs
Increase in K conductance

IV. MUSCARINIC RECEPTORS AND θ RHYTHM

θ Rhythm or rhythmical slow activity (RSA) has been described in many different species, such as the rat and guinea pig, but also in primates[51] including man.[52] The dominant frequencies of this activity occur in the range of 4 to 7 Hz, and in the rat up to 12 Hz.[53] The origin of oscillatory activity is the result of intrinsic properties of both local networks and individual neurons which are under the control of a "pacemaker"-like input from the medial septum.[54] As we will point out, there is a prominent effect of cholinergic muscarinic drugs on this θ activity. Since in the last few years many reports have appeared on induction of LTP, (notably in the hippocampus) using "patterned" θ-burst stimulation,[55] it is worthwhile to consider the mechanisms underlying the rhythmic behavior of cell populations and the role of muscarinic receptors.

A. θ GENERATORS

Depth profiles of θ activity show a dipole-like generator with a reversal at the proximal parts of the apical dendrites of CA1 pyramidal cells in the hippocampus, as well as in the molecular layer of the fascia dentata.[56-61]

A third generator has been found in the CA3 region using *in vitro* hippocampal slice preparations.[62] Anatomical and biochemical studies have shown that a large number of medial septal cells projecting to the hippocampus are cholinergic.[63,64] More recently, however, evidence was found for GABAergic projections from the septum to the hippocampus as well.[65] Lesions or inactivation of the septal area by procaine abolishes the rhythmic activity in the hippocampal formation.[66-69]

Fox and Ranck[70] have distinguished two types of hippocampal CA1 cell-firing behavior; complex spiking of projection cells and θ-rhythmic firing, representing the majority of interneurons, although principal cells display phasic firing as well.[71,72] Many of these cells respond to stimulation of the septal area with an inhibition for 40 to 140 msec.[69] Moreover, pyramidal cells which are firing more randomly seem to be activated (disinhibited) during or just before a θ epoch.[73] This suggests that inhibitory cells are under control of inputs originating from the septal area, and the net effect of septal activation could be a disinhibition of hippocampal output neurons, resulting in θ synchronization.[74-78] The similarity of θ firing in both septum and hippocampus suggests that, at least in part, the medial septum acts as pacemaker for the hippocampal oscillatory electrical behavior.[54,79-82]

B. MUSCARINIC EFFECTS ON RSA AND HIPPOCAMPAL CIRCUITRY

The best documented pharmacological influences on the hippocampal RSA are the studies with the nonselective muscarinic receptor antagonists atropine and scopolamine. Atropine has been shown to abolish the RSA completely in anaesthetized rats (type 2 θ activity).[83-85] In contrast, type 1 atropine-resistant rhythmical activity has been observed in awake animals.[86-89] These experiments clearly show that activation of muscarinic receptors is important for the induction of RSA. Similar to the *in vivo* situation, carbachol induces rhythmic activity in hippocampal slices, which confirms the crucial role of muscarinic receptors in the generation of θ rhythm.[90,91] Hippocampal θ rhythm or RSA can be abolished not only by anticholinergic compounds, but also by glutamate receptor antagonists. The NMDA receptor antagonists 2-amino-5-phosphonopentanoic acid (APV), 4-(phosphonoprop-2-enyl) piperazine-2-carboxylate (CPP-ene) and (+)-5-methyl-10,11-dihydro-5H-dibenzo (a,d) cyclohepten-5,10-imine (dizocilpine maleate or MK-801) have been shown to abolish RSA.[92,93] It has repeatedly been reported that NMDA receptor antagonists block the induction of LTP. The fact that these compounds abolish θ rhythm in the hippocampus as well strengthens the role of hippocampal circuitry properties for LTP phenomena.

At the cellular level, increased firing of projection cells, associated with θ rhythm was observed after systemic application of muscarinic receptor agonists.[94] The muscarinic receptor antagonist atropine attenuates firing of complex spike cells.[95] These data support the excitatory role of muscarinic tone in the production of θ rhythm. The cholinesterase inhibitor physostigmine shortens the inhibitory responses upon medial septum stimulation.[69] This supports the idea that the cholinergic tone reduces and synchronizes activity of interneurons.[54] It has been shown, however, that interneuron firing is suppressed after iontophoretic application of atropine, suggesting that interneuron activity is supported by a basal cholinergic tone.[96] These data indicate that muscarinic receptors influence the function of interneurons in a complex manner, with simultaneous enhancement of activity on inhibitory interneurons,[97] and a suppression of inhibitory synaptic transmission, possibly due to a reduction in release of GABA (see above).

A descriptive model can be summarized as follows: cells in the medial septum display θ-rhythmic firing; there is a group of atropine-sensitive and a population of atropine-resistant cells,[98] and they can be modulated by ascending pathways from the brain stem (and/or hypothalamic) nuclei.[67,99-102] The septal cells impose their rhythmicity on hippocampal projection cells and interneurons; for the latter a group receiving cholinergic inputs from the septum may depolarize, and another group of interneurons which receive GABAergic inputs respond to septal activity with a periodic inhibition of firing.[54] In this way, the projection cells are phasically activated or inactivated in a feed-forward (or feed-back) mode, resulting in a rhythmic output. Increases in activity of muscarinic receptors not only depolarize pyramidal cells directly, but also induce a net reduction in inhibitory strength of θ cells. These phenomena most likely form the cellular basis of the θ preference of the modulatory enhancement of muscarinic agonists for the induction of LTP.

V. MUSCARINIC RECEPTORS, LEARNING BEHAVIOR, AND LTP

Since LTP has been proposed as a principle for storage of memory, it is of interest to consider the effects of muscarinic agents on learning and memory. It is clear that in most paradigms, the operant tasks require awareness of cues for discrimination, sensitivity for (positive or negative) reinforcement, and locomotion, which are all modulated by muscarinic receptor ligands.

A. LOCOMOTION

It has been reported that systemic injection of the muscarinic receptor antagonists scopolamine and atropine induces a dose-dependent increase in ambulatory behavior.[103-105] Intrahippocampal injection, i.e., in the fascia dentata,[106,107] or in the nucleus accumbens,[108,109] of the muscarinic receptor agonist, however, also induces an increase in locomotion. The different routes of administration and thus distribution in the CNS may provide an explanation for the similar behavior of both muscarinic receptor agonists and antagonists.

Suppression of ambulatory activity (type I behavior) is found after blockade of the septal area with a local anaesthetic, which also abolishes an atropine-sensitive hippocampal rhythmical electroencephalogram (EEG).[68,69,81] These results suggest that the cholinergic system in the hippocampal formation and related structures are involved in spontaneous locomotor activity.

B. PASSIVE AVOIDANCE

In passive avoidance tests, animals learn to avoid a foot shock by suppressing locomotion. Muscarinic receptor antagonists increase the number of punishments, possibly by an impairment of learning.[110,111] Intrastriatal injections of a muscarinic receptor antagonist also impair passive avoidance performance. In these studies, however, motor effects mediated by the muscarinic receptor antagonists cannot be excluded.[112,113] Passive avoidance behavior is also inhibited after lesions of the nucleus basalis of Meynert (nbM) or of the medial septal area.[114] In conclusion, performance of the passive avoidance task seems to be impaired by treatment with muscarinic receptor antagonists. Differences, however, which have been observed in various studies, may be due to motor effects and may complicate the interpretation.[110,111]

C. ACTIVE AVOIDANCE

In active avoidance tests, animals have to perform an instrumental or unlearned response, in order to avoid punishment. In the one-way active avoidance test, according to Anisman[115] neither the muscarinic receptor antagonist scopolamine nor the atropine produced any effect. In contrast, Suits and Isaacson[116] have observed impairment in this test, whereas in the two-way paradigm improvement was observed.[115-117] In mice, it was shown that both subcutaneous and intracerebroventricular injections of muscarinic receptor antagonists impair active avoidance learning. In addition, low doses of a muscarinic receptor agonist led to better scores than with control animals in the same test.[118,119]

Retention tests of initially learned active avoidance are inhibited after treatment with muscarinic receptor antagonists, which is dependent on the time of injection of the antagonist.[119-121] These results indicate that muscarinic receptors may be involved in the consolidation phase for this behavioral task.[105,111]

D. SPATIAL MEMORY

Spatial memory tests, such as the radial and water maze, require recognition of an environment in order to obtain reward.[122-124] Tests with various rodent species (e.g., rats and mice) show that these animals develop a strategy, which is called spatial navigation. Visited locations have to be stored in what is called "working memory", which is reedited upon every new trial.

Impairment of spatial memory has been observed after injection of atropine.[104,125] Muscarinic receptor antagonists applied systemically[126,127] or into well-defined brain structures (amygdala[128]) also increase the number of errors, although cholinergic blockade seems to affect learning more than retention of the task.[104,105]

There is a good deal of evidence that spatial memory requires an intact septo-hippocampal (cholinergic) connection. Lesions of the fornix/fimbria bundle lead to severe impairment in spatial memory tasks.[124,129,130] It is, however, not the only pathway involved. Lesions of the cholinergic cell groups in the basal forebrain, the nbM, also lead to impairment in both types of mazes.[104,130-134] Animals that are impaired in their maze performance due to cholinergic lesions show an improvement after treatment with the cholinesterase inhibitor physostigmine at doses of 0.003 to 0.5 mg/kg.[135,136] Similar doses of physostigmine induce an enhancement of LTP in the fascia dentata *in vivo*.[137,138] Moreover, at these doses, physostigmine induces LTP-like phenomena when low-frequency stimulation is applied *in vivo*.[139] As mentioned previously, muscarinic receptor antagonists impair learning behavior both in animals and in humans.[140] Scopolamine, applied at doses equivalent to those that show effects on learning (1.0 mg/kg), suppresses the induction of LTP *in vivo*.[138]

Thus modulation of both learning behavior and the expression of LTP is observed at similar doses of the same muscarinic agents. This suggests that a change in synaptic strength, which is enhanced by activation of muscarinic receptors, is important in learning behavior. Experimental conditions which cause inhibition of LTP, in most cases, also exert disruptive effects on learning behavior.[141-143] Accordingly, there is evidence for a role of LTP in memory consolidation. Similar to muscarinic receptor antagonists, NMDA receptor antagonists such as APV block the induction of LTP and impair spatial navigation[144] as well as other forms of associative learning.[141]

VI. DIRECT INVOLVEMENT OF MUSCARINIC RECEPTORS IN LTP

Long-term potentiation of synaptic transmission (LTP) is a long-lasting enhancement of synaptic efficacy which is caused by tetanic/high-frequency stimulation of neuronal pathways. Although this phenomenon has mostly been studied in the hippocampus, it has been observed in many other structures of the brain[145,146] and is generally considered as a possible neural mechanism of learning and developmental plasticity.[147,148] In most brain areas, calcium influx via NMDA receptors is imperative for induction of LTP. Since NMDA-mediated currents are strongly voltage dependent, muscarinic receptor-mediated excitation most likely plays an important modulatory role in the induction of LTP. Indeed it has been observed, both in the rat visual cortex[39] and the rat hippocampal slices,[40] that the depolarizing response to the LTP-inducing tetanus stimulation is enhanced upon activation of muscarinic receptors. This enhancement of NMDA current may result from either a muscarinic receptor-induced depolarization or a recently described IP_3-dependent mechanism.[38] In a large number of studies the effect of muscarinic receptor agonists and antagonists on LTP has been investigated.[6-9,39,40,138] Different effects of muscarinic receptor agonists on LTP have been reported. Most studies indicate an enhancement of LTP by muscarinic receptor activation, particularly in the hippocampal CA1 region. In the hippocampal CA3 region, however, activation of muscarinic receptors induces a depression of LTP.[43] In contrast with this finding, however, inhibition of LTP in the CA3 region after treatment with the muscarinic receptor antagonist atropine has been reported as well.[150] In the dentate gyrus, a dual effect of muscarinic agonists has been demonstrated.[6] In this hippocampal region, low concentrations of muscarinic agonist facilitate LTP, whereas high doses induce suppression. This suppression is possibly caused by enhanced M_2 muscarinic receptor-mediated presynaptic inhibition.

As discussed previously, various properties of muscarinic receptors such as modulation of membrane conductances eventually resulting in oscillatory θ-rhythm activity and enhancement of NMDA receptor responses may play a crucial role in enhancement of synaptic plasticity. The aspect of LTP most likely affected by muscarinic receptor activation is its induction.[40] Enhancement of LTP by muscarinic agonists is prevented by pretreatment with muscarinic receptor antagonists. The maintenance phase of LTP, however, is affected neither by muscarinic receptor agonists nor by antagonists.[40]

An important unsolved question concerns the role of various muscarinic receptor subtypes involved in enhancement of LTP. Based on the effects of the M_1 muscarinic receptor antagonist pirenzepine, the M_1 muscarinic receptor is generally assumed to be involved in facilitation of LTP. However, in the hippocampus where M_1 receptor-induced facilitation of LTP has been demonstrated, M_3 and M_5 receptors are present as well. Second, M_1, M_3, and M_5 muscarinic receptors couple to similar G proteins and induce phosphatidylinositol turnover and thus mobilize intracellular calcium and activate protein kinase C, which are both essentially involved in the induction of LTP.[151,152] LTP is a phenomenon that is difficult to analyze in strict pharmacological terms. In addition, the selectivity of most of the muscarinic receptor agonists and antagonists is restricted. Therefore, it is difficult to determine the exact involvement of any of the muscarinic receptor subtypes in the modulation of LTP. Accordingly, it is difficult to discriminate between LTP modulation by M_1 or M_3 receptors, whereas it is at present impossible to recognize putative effects of M_5 receptors. On the other hand, the aforementioned three muscarinic receptor subtypes which have specific distributions in the brain may constitute a fine-tuned system with many possibilities for generating LTP in small specific areas.

As mentioned previously, M_2 muscarinic receptors are located both pre- and postsynaptically.[18] Presynaptically located M_2 muscarinic receptors control acetylcholine release and thereby may exert an effect on LTP.[153] *In vivo* experiments with the lipophilic M_2 muscarinic receptor antagonist BIBN 99 showed improved learning behavior in a Morris maze swim task.[154]

Recently, it has been shown that the presence of rhythmic activity, induced by stimulation of muscarinic receptors, enhances synaptic plasticity.[155] These authors demonstrated that this synaptic enhancement requires presynaptic stimulation, is synapse specific, and is optimal when electrical stimulation is synchronized to the phase of the θ rhythm. Accordingly, under certain circumstances such as cholinergic stimulation, long-term enhancement can be produced without high-frequency stimulation and can be induced by weak stimulation that normally does not induce long-term enhancement of synaptic transmission. Interestingly, the cholinesterase inhibitors such as physostigmine also induce LTP-like phenomena in rats *in vivo*, when low repetition rate stimulation is applied.[139] Taken together, these phenomena may provide an important physiological link between the modulatory role of muscarinic receptor activation on synaptic plasticity and learning, as well as the relevance of θ-burst-patterned induction of LTP.

VII. CONCLUSIONS

As discussed, a clear facilitation of LTP is mediated by the cholinergic muscarinic receptor system. The mechanism of this enhancement of LTP is probably complex and may comprise a number of components. Certainly, excitation through inhibition of a variety of potassium conductances, inhibition of GABA neurotransmission, and positive modulation of NMDA receptor channels play important roles. In addition, a complex mechanism of interneuron firing rate synchronization sets the firing pattern of output neurons during θ activity, a condition which enables incoming inputs to enhance specific synaptic connections. The similarities between the muscarinic activity on both LTP and associative learning suggest that LTP may provide a mechanism for memory processing.

An aspect which is still unclear is the involvement of muscarinic receptor subtypes in facilitation of LTP. Although it is generally assumed that M_1 muscarinic receptors play a predominant role, an effect of M_3, M_5, and eventually presynaptic M_2 muscarinic receptors on LTP, however, cannot be excluded.

REFERENCES

1. Piercey, M. F., Vogelsang, G. D., Franklin, S. R., and Tang, A. H., Reversal of scopolamine-induced amnesia and alterations in energy metabolism by the nootropic piracetam: implications regarding identification of brain structures involved in consolidation of memory traces, *Brain Res.*, 424, 1, 1987.
2. Bartus, R. T., Flicker, C., Dean, R. L., Fisher, S., Pontecorvo, M., and Figuerdo, J., Behavioral and biochemical effects of nucleus basalis lesions: implications and possible relevance to understanding or treating Alzheimer's disease, *Prog. Brain Res.*, 70, 345, 1986.
3. Durkin, T., Central cholinergic pathways and learning and memory processes: presynaptic aspects, *Comp. Biochem. Physiol.*, 93, 273, 1989.
4. Lomo, T., Frequency potentiation of excitatory synaptic activity in the dentate area of the hippocampal formation, *Acta Physiol. Scand.*, 68, 128, 1966.
5. Bliss, T. V. P. and Lomo, T., Long-lasting potentiation of synaptic transmission in the dentate of anaesthetized rabbit following stimulation of the perforant path, *J. Physiol.*, 232, 331, 1973.
6. Burgard, E. C. and Sarvey, J. M., Muscarinic receptor activation facilitates the induction of long-term potentiation (LTP) in the rat dentate gyrus, *Neurosci. Lett.*, 116, 34, 1990.

7. Tanaka, Y., Sakurai, M., and Hayashi, S., Effect of scopolamine and HP 029, a cholinesterase inhibitor, on long-term potentiation in hippocampal slices of the guinea pig, *Neurosci. Lett.*, 98, 179, 1989.

8. Hirotsu, I., Hori, N., Katsuda, and N., Ishihara, T., Effect of anticholinergic drug on long-term potentiation in rat hippocampal slices, *Brain Res.*, 482, 194, 1989.

9. Frank, C., Sagratella, S., Niglio, T., Caporali, M. G., Bronzetti, E., and Scotti De Carolis, A., Hippocampal long-term potentiation in nucleus basalis magnocellularis-lesioned rats, *Brain Res. Bull.*, 29, 847, 1992.

10. Bonner, T. I., The molecular basis of muscarinic receptor diversity, *Trends Neurosci.*, 12, 148, 1989.

11. Doods, H. N., Mathy, M., Davidesko, D., Van Charldorp, K. J., De Jonge, A., and Van Zwieten, P. A., Selectivity of muscarinic antagonists in radioligand and *in vivo* experiments for the putative M_1, M_2 and M_3 receptors, *J. Pharmacol. Exp. Ther.*, 242, 257, 1987.

12. Lazareno, S., Buckley, N. J., and Roberts, F., Characterization of muscarinic M4 binding sites in rabbit lung, chicken heart, and NG 108-15 cells, *Mol. Pharmacol.*, 38, 805, 1990.

13. Hammer, R., Berrie, C. P., Birdsall, N. J. M., Burgen, A. S. V., and Hulme, E.C., Pirenzepine distinguishes between different subclasses of muscarinic receptors, *Nature (London)*, 283, 90, 1980.

14. Hammer, R., Giraldo, E., Schiavi, G.B., Monferini, E., and Ladinsky, H., Binding profile of a novel cardioselective muscarine receptor antagonist, AF-DX 116, to membranes of peripheral tissues and brain in the rat, *Life Sci.*, 31, 2991, 1986.

15. Lambrecht, G., Feifel, R., Forth, B., Strohmann, C., Tacke, R., and Mutschler, E., *p*-Fluoro-hexahydro-sila-difenidol: the first M2β-selective muscarinic antagonist, *Eur. J. Pharmacol.*, 152, 193, 1988.

16. Buckley, N. J., Bonner, T. I., and Brann, M. R., Localization of a family of muscarinic receptor mRNAs in rat brain, *J. Neurosci.*, 8, 4646, 1988.

17. Quirion, R., Aubert, I., Araujo, D. M., Hersi, A., and Gaudreau P., Autoradiographic distribution of putative muscarinic receptor sub-types in mammalian brain, in *Progress in Brain Research*, Cuello, A. C., Ed., 98, 85, 1993.

18. Vilaro, M. T., Wiederholt, K. H., Palacios, J. M., and Mengod, G., Muscarinic M_2 receptor mRNA expression and receptor binding in cholinergic and non-cholinergic cells in the rat brain: a correlative study using *in situ* hybridization histochemistry and receptor autoradiography, *Neuroscience*, 47, 367, 1992.

19. Araujo, D. M., Lapchak, P. A., Robitaille, Y., Gauthier, S., and Quirion, R., Differential alterations of various cholinergic markers in cortical and subcortical regions of the human brain in Alzheimer's disease, *J. Neurochem.*, 50, 1914, 1988.

20. Vilaro, M. T., Palacios, J. M., and Mengod, G., Localization of m_5 muscarinic receptor mRNA in rat brain examined by *in situ* hybridization histochemistry, *Neurosci. Lett.*, 114, 154, 1990.

21. Brann, M. R., Conklin, B., Dean, N. M., Collins, R. M., and Bonner, T. I., Cloned muscarinic receptors couple to different G-proteins and second messengers, *Soc. Neurosci. Abstr.*, 14, 600, 1988.

22. Berridge, M. J. and Irvine, R. F., Inositol phosphates and cell signalling, *Nature (London)*, 341, 197, 1989.

23. Malinow, R., Schulman, H., and Tsien, R. W., Inhibition of postsynaptic PKC or CaMKII blocks induction but not expression of LTP, *Science*, 245, 862, 1989.

24. Ashkenazi, A., Winslow, W. J., Peralta, E. G., Peterson, G. L., Schimerlik, M. I., Capon, D. J., and Ramachandran, J., An M2 muscarinic receptor subtype coupled to both adenylyl cyclase and phosphoinositide turnover, *Science*, 283, 672, 1987.

25. Peralta, F. G., Ashkenazi, A., Winslow, J. W., Ramachandran, J., and Capon, D. J., Differential regulation of PI hydrolysis and adenylyl cyclase by muscarinic receptor subtypes, *Nature (London)*, 344, 434, 1988.

26. Kenakin, T. P. and Boselli, C., Promiscuous or heterogeneous muscarinic receptors in rat atria? I. Schild analysis with competitive antagonists, *Eur. J. Pharmacol.*, 191, 39, 1990.

27. Krnjevic, K., Pumain, R., and Renaud, L., The mechanism of excitation by acetylcholine in the cerebral cortex, *J. Physiol.*, 215, 247 1971.

28. Nicoll, R. A., Malenka, R. C., Kauer, J. A., Functional comparison of neurotransmitter receptor subtypes in mammalian central nervous system, *Phys. Rev.*, 70, 513, 1990.

29. Madison, D. V., Lancaster, B., and Nicoll, R. A., Voltage clamp analysis of cholinergic action in the hippocampus, *J. Neurosci.*, 7, 733, 1987.

30. Dutar, P. and Nicoll, R. A., Classification of muscarinic responses in terms of receptor subtypes and second messenger systems: electrophysiological studies *in vitro*, *J. Neurosci.*, 8, 4214, 1988.

31. Muller, W. and Misgeld, U., Slow cholinergic excitation of guinea pig hippocampal neurons is mediated by two muscarinic receptor subtypes, *Neurosci. Lett.*, 67, 107, 1986.
32. Brown, D. A., M-currents: an update, *Trends Neurosci.*, 11, 294, 1988.
33. Brown, D. A., Marrion, N. V., and Smart, T. G., On the transduction mechanism for muscarine-induced inhibition of M-current in cultured rat sympathetic neurones, *J. Physiol.*, 413, 469, 1989.
34. Nakajima, Y., Nakajima, S., Leonard, R. J., and Yamaguchi, K., Acetylcholine raises excitability by inhibiting the fast transient potassium current in cultured hippocampal neurons, *Proc. Natl. Acad. Sci. U.S.A.*, 83, 3022, 1986.
35. Cole, A. E. and Nicoll, R. A., Characterization of a slow cholinergic postsynaptic potential recorded *in vitro* from rat pyramidal cells, *J. Physiol.*, 325, 173, 1984.
36. Muller, W. and Misgeld, V., Carbachol reduces $I_{K,Baclofen}$ but not $I_{K,GABA}$ in guinea-pig hippocampal slices, *Neurosci. Lett.*, 102, 229, 1989.
37. Worley, P. F., Baraban, J. M., McCareen, M., Snyder, S. H., and Alger, B. E., Cholinergic phosphatidyl inositol modulation of inhibitory, G protein linked, neurotransmitter actions: electrophysiological studies in rat hippocampus, *Proc. Natl. Acad. Sci. U.S.A.*, 84, 3467, 1987.
38. Markram, H. and Segal, M., Long-lasting facilitation of excitatory postsynaptic potentials in the rat hippocampus by acetylcholine, *J. Physiol.*, 427, 381, 1990.
39. Bröcher, S., Artola, A., and Singer, W., Agonists of cholinergic and noradrenergic receptors facilitate synergistically the induction of long-term potentiation in slices of rat visual cortex, *Brain Res.*, 537, 27, 1992.
40. Boddeke, E. W. G. M., Enz, A., and Shapiro, G., SDZ ENS 163, a selective muscarinic M_1 receptor agonist, facilitates the induction of long-term potentiation in rat hippocampal slices, *Eur. J. Pharmacol.*, 222, 21, 1992.
41. Gähwiler, B. H. and Brown, D. A., Muscarine affects calcium currents in rat hippocampal pyramidal cells *in vitro, Neurosci. Lett.*, 76, 301, 1987.
42. Misgeld, U., Calabresi, P., and Dodt, H. U., Muscarinic modulation of calcium dependent plateau potentials in rat neostriatal neurons, *Pfluegers Arch.*, 407, 482, 1986.
43. Williams, S. and Johnston, D., Muscarinic depression of long-term potentiation in CA_3 hippocampal neurons, *Science*, 242, 84, 1988.
44. Valentino, R. J. and Dingledine, R., Presynaptic inhibitory effect of acetylcholine in the hippocampus, *J. Neurosci.*, 1, 784, 1981.
45. Krnjevic, K., Reiffenstein, R.J., and Ropert, N., Disinhibitory action of acetylcholine in the rat's hippocampus: extracellular observations, *Neuroscience*, 12, 2465, 1981.
46. Ben-Ari, Y., Krnjevic, K., Reinhardt, W., and Ropert, N., Intracellular observations on the disinhibitory action of acetylcholine in the hippocampus, *Neuroscience*, 6, 2475, 1981.
47. Hasuo, H., Gallagher, J. P., and Shinnick-Gallagher, P., Disinhibition in the rat septum mediated by M1 muscarinic receptors, *Brain Res.*, 438, 323, 1988.
48. Egan, T. M. and North, R. A., Acetylcholine hyperpolarizes central neurons by acting on an M_2 muscarinic receptor, *Nature (London)*, 319, 405, 1985.
49. McCormick, D. A. and Prince, D. A., Acetylcholine induces burst firing in thalamic reticular neurons by activating a potassium conductance, *Nature (London)*, 319, 402, 1986.
50. Metherate, R., Tremblay, N., and Dykes, R. W., Transient and prolonged effects of acetylcholine on responsiveness of cat somatosensory cortical neurons, *J. Neurophysiol.*, 59, 1253, 1988.
51. Stewart, M. and Fox, S. E., Hippocampal theta activity in monkeys, *Brain Res.*, 538, 59, 1991.
52. Arnolds, D. E. A. T., Lopes da Silva, F. H., Aitink, J. W., Kamp, A., and Boeijinga, P., The spectral properties of hippocampal EEG related to behavior in man, *Electroencephalogr. Clin. Neurophysiol.*, 50, 324, 1980.
53. Lopes da Silva, F. H., Witter, M. P., Boeijinga, P. H., and Lohman, A. H. M., Anatomical organization and physiology of the limbic cortex, *Physiol. Rev.*, 70, 453, 1990.
54. Stewart, M. and Fox, S. E., Do septal neurons pace the hippocampal theta rhythm?, *Trends Neurosci.*, 13, 163, 1990.
55. Rose, G. M. and Dunwiddie, T. V., Induction of hippocampal long-term potentiation using physiologically patterned stimulation, *Neurosci. Lett.*, 69, 244, 1986.
56. Bland, B. H., Andersen, P., Ganes, T., and Sveen, O., Two generators of hippocampal theta activity in rabbits, *Brain Res.*, 94, 199, 1975.
57. Winson, J., Hippocampal theta rhythm. I. Depth profiles in the curarized rat, *Brain Res.*, 103, 57, 1976.

58. Winson, J., Hippocampal theta rhythm. II. Depth profiles in the freely moving rabbit, *Brain Res.*, 103, 71, 1976.

59. Holsheimer, J., Boer, J., Lopes da Silva, F. H., and Van Rotterdam A., The double dipole model of theta rhythm generation: simulation of laminar field potential profiles in dorsal hippocampus of the rat, *Brain Res.*, 235, 31, 1982.

60. Buzsaki, G., Rappelsberger, P., and Kellenyi, L., Depth profiles of hippocampal rhythmic slow activity ('theta rhythm') depend on behavior, *Electroencephalogr. Clin. Neurophysiol.*, 61, 77, 1985.

61. Konopacki, J., Bland, B. H., MacIver, M. B., and Roth, S. H., Cholinergic theta rhythm in transected hippocampal slices: independent CA1 and dentate generators, *Brain Res.*, 436, 217, 1987.

62. Konopacki, J., Bland, B. H., and Roth, S. H., Carbachol-induced EEG 'theta' in hippocampal formation slices: evidence for a third generator in CA3 area, *Brain Res.*, 451, 33, 1988.

63. Lewis, P. R., Shute, C. C. D., and Silver, A., Confirmation from choline acetylase analyses of a massive cholinergic innervation to the rat hippocampus, *J. Physiol.*, 191, 215, 1967.

64. Senut, M. C., Menetrey, D., and Lamour, Y., Cholinergic and peptidergic projections from the medial septum and the nucleus of the diagonal band of Broca to dorsal hippocampus, cingulate cortex and olfactory bulb: a combined wheatgerm agglutinin-apohorseradish peroxidase-gold immunohistochemical study, *Neuroscience*, 30, 385, 1989.

65. Köhler, C., Chen-Palay, V., and Wu, J.Y., Septal neurons containing glutamic acid decarboxylase immunoreactivity project to the hippocampal region in the rat brain, *Anat. Embryol. (Berl.)*, 169, 41, 1984.

66. Sainsbury, R. S. and Bland, B. H., The effects of selective septal lesions on theta production in CA1 and the dentate gyrus of the hippocampus, *Physiol. Behav.*, 26, 1097, 1981.

67. Smythe, J. W., Christie, B. R., Colom, L. V., Lawson, V. H., and Bland, B. H., Hippocampal θ field activity and θ-on/θ-off cell discharges are controlled by an ascending hypothalamo-septal pathway, *J. Neurosci.*, 11, 2241, 1991

68. Lawson, V. H. and Bland, B. H., The role of the septohippocampal pathway in the regulation of hippocampal field activity and behavior: analysis by the intraseptal microinfusion of carbachol, atropine and procaine, *Exp. Neurol.*, 120, 132, 1993.

69. Vinogradova, O. S., Brazhnik, E. S., Stafekhina, V. S., and Kitchigina, V. F., Acetylcholine, theta-rhythm and activity of hippocampal neurons in the rabbit. II. Septal input, *Neuroscience*, 53, 971, 1993.

70. Fox, S. E. and Ranck, J. B., Jr., Electrophysiological characteristics of hippocampal complex-spike and theta cells, *Exp. Brain Res.*, 41, 399, 1981.

71. Colom, L. V. and Bland, B. H., State-dependent spike train dynamics in hippocampal formation neurons: evidence for theta-on and theta-off cells, *Brain Res.*, 422, 277, 1987

72. Bland, B. H. and Colom, L. V., Preliminary observations on the physiology and pharmacology of hippocampal θ-off cells, *Brain Res.*, 505, 333, 1989.

73. Stewart, M., Disinhibition of hippocampal pyramidal cells during the transition into theta rhythm, *Exp. Brain Res.*, 93, 1, 1993.

74. Artemenko, D. P., Role of hippocampal neurons in theta-wave generation, *Neurophysiology*, 4, 531, 1973.

75. Buzsaki, G. and Eidelberg, E., Phase relations of hippocampal projection cells and interneurons with theta activity in the anaesthetized rat, *Brain Res.*, 266, 334, 1983.

76. Leung, L.-W. S. and Yim, C. Y., Intracellular records of theta rhythm in hippocampal CA1 cells of the rat, *Brain Res.*, 367, 323, 1986.

77. Fox, S. E., Membrane potential and impedance changes in hippocampal pyramidal cells during theta rhythm, *Exp. Brain Res.*, 77, 283, 1989.

78. Konopacki, J., Bland, B. H., Colom, L. V., and Oddie, S. D., *In vivo* intracellular correlates of hippocampal formation theta-on and theta-off cells, *Brain Res.*, 586, 247, 1992.

79. Petsche, H., Stumpf, C., and Gogolak, G., The significance of the rabbit's septum as a relay station between the midbrain and the hippocampus. The control of hippocampus arousal activity by septum cells, *Electroencephalogr. Clin. Neurophysiol.*, 14, 202, 1962.

80. Gaztelu, J. M. and, Buño,, W., Septo-hippocampal relationships during EEG theta rhythm, *Electroencephalogr. Clin. Neurophysiol.*, 54, 375, 1982.

81. Stewart, M. and Fox, S. E., Firing relations of medial septal neurons to the hippocampal theta rhythm in urethane anesthetized rats, *Exp. Brain Res.*, 77, 507, 1989.

82. Ford, R. D., Colom, L. V., and Bland, B. H., The classification of medial septum-diagonal band cells as θ-on or θ-off cells in relation to hippocampal EEG states, *Brain Res.*, 493, 269, 1989.

83. Vanderwolf, C. H., Neocortical and hippocampal activation in relation to behavior: effects of atropine, eserine, phenothiazines and amphetamine, *J. Comp. Physiol. Psychol.*, 88, 300, 1975.

84. Vanderwolf, C. H., Kramis, R. C., Gillespie, L. A., and Bland B. H., Hippocampal rhythmic slow activity and neocortical low-voltage fast activity: relations to behaviour, in *The Hippocampus: Neurophysiology and Behaviour*, Vol. 2, Isaacson, R. L. and Pribram, K. H., Eds., Plenum Press, New York, 1975, 101.

85. Kramis, R. C., Vanderwolf, C. H., and Bland, B. H., Two types of hippocampal rhythmical slow activity in both rabbit and the rat: relations of behavior and effects of atropine, diethyl ether, urethane and pentobarbital, *Exp. Neurol.*, 49, 58, 1975.

86. Leung, L.-W. S., Spectral analysis of hippocampal EEG in the freely moving rat: effects of centrally active drugs and relations to evoked potentials, *Electroencephalogr. Clin. Neurophysiol.*, 60, 65, 1985.

87. Buzsaki, G., Czopf, J., Kondakor, I., and Kellenyi, L., Laminar distribution of hippocampal rhythmic slow activity (RSA) in the behaving rat: current-source density analysis, effects of urethane and atropine, *Brain Res.*, 365, 125, 1986.

88. Stewart, M. and Fox, S. E., Detection of an atropine-resistant component of the hippocampal theta rhythm in urethane-anaesthetized rats, *Brain Res.*, 500, 55, 1989.

89. Barnes, J. C. and Roberts, F. F., Central effects of muscarinic agonists and antagonists on hippocampal theta rhythm and blood pressure in the anaesthetized rat, *Eur. J. Pharmacol.*, 195, 233, 1991.

90. Konopacki, J., Bland, B. H., and Roth, S. H., Evidence that activation of *in vitro* hippocampal θ rhythm only involves muscarinic receptors, *Brain Res.*, 455, 110, 1988.

91. Bland, B. H., Colom, L. V., Konopacki, J., and Roth, S. H., Intracellular records of carbachol-induced theta rhythm in hippocampal slices, *Brain Res.*, 447, 364, 1988

92. Leung, L.-W. S. and Desborough, K. A., APV, an *N*-methyl-D-aspartate receptor antagonist, blocks the hippocampal theta rhythm in behaving rats, *Brain Res.*, 463, 148, 1988.

93. Boddeke, H. W. G. M., Wiederhold, K. H., and Palacios, J. M., Intracerebroventricular application of competitive and non-competitive NMDA antagonists induce similar effects upon rat hippocampal electroencephalogram and local cerebral glucose utilization, *Brain Res.*, 585, 177, 1992.

94. Bland, B. H. and Colom, L. V., Responses of phasic and tonic hippocampal theta-on cells to cholinergics: differential effects of muscarinic and nicotinic activation, *Brain Res.*, 440, 167, 1988.

95. Buzsaki, G., Leung, L.-W. S., and Vanderwolf, C. H., Cellular bases of hippocampal EEG in the behaving rat, *Brain Res. Rev.*, 6, 139, 1983.

96. Stewart, M., Luo, Y., and Fox, S. E., Effects of atropine on hippocampal theta cells and complex-spike cells, *Brain Res.*, 591, 122, 1992.

97. Reece, L. J. and Schwartzkroin, P. A., Effects of cholinergic agonists on two non-pyramidal cell types in rat hippocampal slices, *Brain Res.*, 566, 115, 1991.

98. Stewart, M. and Fox, S. E., Two populations of rhythmically bursting neurons in rat medial septum are revealed by atropine, *J. Neurophysiol.*, 61, 982, 1989.

99. Paiva, T., Lopes da Silva, F. H., and Mollevanger, W., Modulating systems of hippocampal EEG, *Electroencephalogr. Clin. Neurophysiol.*, 40, 470, 1976.

100. Vertes, R. P., An analysis of ascending brainstem systems involved in hippocampal synchronization and desynchronization, *J. Neurophysiol.*, 46, 1140, 1981.

101. Vertes, R. P., Brain stem generation of the hippocampal EEG, *Prog. Neurobiol.*, 19, 159, 1982.

102. Brazhnik, E. S. and Vinogradova, O. S., Modulation of the afferent input to the septal neurons by cholinergic drugs, *Brain Res.*, 451, 1, 1988.

103. Hughes, R. N., A review of muscarinic drugs on exploratory choice behavior in laboratory rodents, *Behav. Neural. Biol.*, 34, 5, 1982.

104. Whishaw, I. Q., O'Connor, W. T., and Dunnett, S. B., Disruption of central cholinergic systems in the rat by basal forebrain lesions or atropine: effects on feeding, sensorimotor behavior, locomotor activity and spatial navigation, *Behav. Brain Res.*, 17, 103, 1985.

105. Hagan, J.J. and Morris, R.G.M., The cholinergic hypothesis of memory: a review of animal experiments, in *Handbook of Psychopharmacology*, Iversen, L.L., Iversen, S.D., and Snyder, S.H., Eds., Plenum, New York, 20, 1988, 237.

106. Flicker, C. and Geyer, M., Behaviour during hippocampal microinfusions. II. Muscarinic locomotor activation, *Brain Res. Rev.*, 257, 105, 1982

107. Mogenson G.J. and Nielsen M., A study of the contribution of hippocampal-accumbens-subpallidal projections to locomotor activity, *Behav. Neural. Biol.*, 42, 38, 1984.

108. Austin M.C. and Kalivas, P.W., The effect of cholinergic stimulation in the nucleus accumbens on locomotor behavior, *Brain Res.*, 441, 209, 1988.

109. Jones D.L., Mogenson, G.J., and Wu, M., Injections of dopaminergic, cholinergic, serotoninergic and GABAergic drugs into the nucleus accumbens: effects on locomotor activity in the rat, *Neuropharmacology*, 20, 29, 1981.

110. Bammer, G., Pharmacological investigations of neurotransmitter involvement in passive avoidance responding: a review and some new results, *Neurosci. Biobehav. Rev.*, 6, 247, 1982.

111. Spencer D.G. and Lal, H., Effects of anticholinergic drugs on learning and memory, *Drug Dev. Res.*, 3, 489, 1983.

112. Prado-Alcala, R.A., Fernandez-Samblancat, M., and Solodkin-Herrera, M., Injections of atropine into the caudate nucleus impair the acquisition and the maintenance of passive avoidance, *Pharmacol. Biochem. Behav.*, 22, 243, 1985.

113. Sandberg, K., Sandberg, P. R., and Coyle, J. T., Effects of intrastriatal injections of the cholinergic neurotoxin AF64A on spontaneous nocturnal locomotor behavior in the rat, *Brain Res.*, 299, 339, 1984.

114. Hepler, H. J., Wenk, G. L., Cribbs, B. L., Olton, D. S., and Coyle , J. T., Memory impairments following basal forebrain lesions, *Brain Res.*, 346, 8, 1985.

115. Anisman, H., Cholinergic mechanisms and alterations in behavioral suppression as factors producing time-dependent changes in avoidance performance, *J. Comp. Physiol. Psychol.*, 83, 465, 1973.

116. Suits, E. and Isaacson, R. L., The effects of scopolamine hydrobromide on one-way and two-way avoidance learning in rats, *Int. J. Neuropharmacol.*, 7, 441, 1968.

117. Evangelista, A. M. and Izquierdo, I., Effects of atropine on avoidance condition: interaction with nicotine and comparison with *N*-methyl-atropine, *Psychopharmacology*, 27, 241, 1972.

118. Flood, J. F., Landry, D. W., and Jarvik, M. E., Cholinergic receptor interactions and their effects on long-term memory processing, *Brain Res.*, 215, 177, 1981.

119. Flood, J. F. and Cherkin, A., Scopolamine effects on memory retention in mice: a model for dementia?, *Behav. Neural Biol.*, 45, 169, 1986.

120. Deutsch, J. A., Hamburg, M. D., and Dahl, H., Anti-cholinesterase induced amnesia and its temporal aspects, *Science*, 151, 221, 1966.

121. Deutsch, J. A. and Leibowitz, S. F., Amnesia or reversal of forgetting by anti-cholinesterase, depending simply on time of injection, *Science*, 153, 742, 1966.

122. O'Keefe, J. and Nadel, L., *The Hippocampus as a Cognitive Map*, Clarendon Press, Oxford, 1978.

123. Olton, D. S. and Samuelson, R. J., Remembrance of places passed: spatial memory in rats, *J. Exp. Psychol.: Animal Behav. Process*, 2, 97, 1976.

124. Morris, R. G. M., Garrud, P., Rawlins, J. N. P., and O'Keefe, J. O., Place navigation impaired in rats with hippocampal lesions, *Nature (London)*, 297, 1982.

125. Sutherland, R. J., Whishaw, I. Q., and Regehr, J. C., Cholinergic receptor blockade impairs spatial localization by use of distal cues in the rat, *J. Comp. Physiol. Psychol.*, 96, 563, 1982.

126. Beatty, W. W. and Bierley, R. A., Scopolamine degrades spatial working memory but spares spatial reference memory: dissimilarity of anticholinergic effect and restriction of distal visual cues, *Pharmacol. Biochem. Behav.*, 23, 1, 1985.

127. Riekkinen, Jr., P., Jakala, P., Sirvio, J., and Riekkinen, P., The effects of increased serotoninergic and decreased cholinergic activities on spatial navigation performance in rats, *Pharmacol. Biochem. Behav.*, 39, 25, 1991.

128. Ingles, J. L., Beninger, R. J., Jhamandas, K., and Boegman, R. J., Scopolamine injected into the rat amygdala impairs working memory in the double Y-maze, *Brain Res. Bull.*, 32, 339, 1993.

129. Jarrard, L. E., Selective hippocampal lesions: differential effects on performance by rats of a spatial task with preoperative versus postoperative training, *J. Comp. Physiol. Psychol.*, 92, 1119, 1978.

130. Olton, D. S., Walker, J. A., and Wolf, W. A., A disconnection analysis of hippocampal function, *Brain Res.*, 233, 241, 1982.

131. Bartus, R. T., Flicker, C., Dean, R. L., Pontecorvo, M., Figueiredo, J. C., and Fisher, S. K., Selective memory loss following nucleus basalis lesions: long term behavioral recovery despite persistent cholinergic deficiencies, *Pharmacol. Biochem. Behav.*, 23, 125, 1985.

132. Dunnett, S. B., Whishaw, I. Q., Jones, G. H., and Bunch, S. T., Behavioral, biochemical and histochemical effects of different neurotoxic amino acids injected into the nucleus basalis magnocellularis of rats, *Neuroscience*, 20, 653, 1987.

133. Olton, D. S., Walker, J. A., and Gage, F. H., Hippocampal connections and spatial discrimination, *Brain Res.*, 139, 295, 1978.

134. Hepler, H. J., Olton, D. S., Wenk, B. L., and Coyle , J. T., Lesions in nucleus basalis magnocellularis and medial septal area of rats produce qualitatively similar memory impairments, *J. Neurosci.*, 5, 866, 1985.

135. Murray, C. L. and Fibiger, H. C., Learning and memory deficits after lesions of the nucleus basalis magnocellularis: reversal by physostigmine, *Neuroscience*, 14, 1025, 1985.

136. Matsuoka, N., Maeda, N., Ohkubo, Y., and Yamaguchi, I., Differential effect of physostigmine and pilocarpine on the spatial memory deficits produced by two septo-hippocampal deafferentiations in rats, *Brain Res.*, 559, 233, 1991.

137. Molnar, P. and Gaal, L., Effect of different subtypes of cognition enhancers on long-term potentiation in the rat dentate gyrus *in vivo, Eur. J. Pharmacol.*, 215, 17, 1992.

138. Ito, T., Miura, Y., and Kadokawa, T., Effects of physostigmine and scopolamine on long-term potentiation of hippocampal population spikes in rats, *Can. J. Physiol. Pharmacol.*, 66, 1010, 1988.

139. Ito, T., Miura, Y., and Kadokawa, T., Physostigmine induces in rats a phenomenon resembling long-term potentiation, *Eur. J. Pharmacol.*, 156, 351, 1988.

140. Bartus, R. T., Dean, R. L., Pontecorvo, M., and Flicker, C., The cholinergic hypothesis: a historical overview, current perspective, and future directions, *Ann. N.Y. Acad. Sci.*, 444, 332, 1985.

141. Laroche, S., Doyere, V., and Bloch, V., Linear relation between the magnitude of long-term potentiation in the dentate gyrus and associative learning in the rat. A demonstration using commissural inhibition and local infusion of an *N*-methyl-D-aspartate receptor antagonist, *Neuroscience*, 28, 375, 1989.

142. McNaughton, B. L., Barnes, C. A., Rao, G., Baldwin, R. J., and Rasmussen, M., Long-term enhancement of hippocampal synaptic transmission and the acquisition of spatial information, *J. Neurosci.*, 6, 563, 1986.

143. Cain, D. P., Hargreaves, E. L., Boon, F., and Dennison, Z., An examination of the relations between hippocampal long-term potentiation, kindling, afterdischarge, and place learning in the water maze, *Hippocampus*, 3, 153, 1993.

144. Morris, R. G. M., Anderson, E., Lynch, G. S., and Baudry, M., Selective impairment in learning and blockade of long-term potentiation by an *N*-methyl-D-aspartate receptor antagonist, AP5, *Nature (London),* 319, 774, 1986.

145. Teyler, T. J. and DiScenna, P., Long-term potentiation, *Annu. Rev. Neurosci.*, 10, 131, 1987.

146. Racine, R. J., Milgram, N. W., and Hafner, S., Long-term potentiation phenomena in the rat limbic forebrain, *Brain Res.*, 260, 217, 1983.

147. Frégnac, Y. and Imbert, M., Development of neuronal selection in primary visual cortex of cat, *Physiol. Rev.*, 64, 325, 1984.

148. Singer, W., The role of acetylcholine in use-dependent plasticity of the visual cortex, in *Brain Cholinergic Systems*, Steriade, M. and Biesold, D., Eds., Oxford University Press, Oxford, 1990, 314.

149. Blitzer, R. D., Gil, O., and Landau, M., Cholinergic stimulation enhances long-term potentiation in the CA1 region of rat hippocampus, *Neurosci. Lett.*, 119, 207, 1990.

150. Katsuki, H., Saito, H., and Satoh, M., The involvement of muscarinic, β-adrenergic and metabotropic glutamate receptors in long-term potentiation in the fimbria-CA3 pathway in the hippocampus, *Neurosci. Lett.*, 142, 249, 1992.

151. Hu, G. Y., Hvalby, O., Walaas, I., Albert, K. A., Skjeklo, P., Andersen, P., and Greengard, P., Protein kinase C injection into hippocampal pyramidal cells elicits features of long-term potentiation, *Nature (London)*, 328, 426, 1987.

152. Malenka, R. C., Madison, D. V., and Nicoll, R. A., Potentiation of synaptic transmission in the hippocampus by phorbol esters, *Nature (London)*, 321, 175, 1986.

153. Raiteri, M., Leardi, R., and Marchi, M., Heterogeneity of presynaptic muscarinic receptors regulating neurotransmitter release in the rat brain, *J. Pharmacol. Exp. Ther.*, 228, 209, 1984.

154. Doods, H. N., Quirion, R., Mihm, G., Engel, W., Rudolf, K., Entzeroth, M., Schiavi, G.B., Ladinsky, H., Bechtel, W.D., Ensinger, H.A., Mendla, K.D., and Eberlein, W., Therapeutic potential of CNS-active M_2 antagonists: novel structures and pharmacology, *Life Sci.*, 52, 497, 1993.

155. Huerta, P. T. and Lisman, J. E., Heightened synaptic plasticity of hippocampal CA1 neurons during a cholinergically induced rhythmic state, *Nature (London)*, 364, 723, 1993.

Chapter 12

Nicotinic Receptors and Information Processing

Jennifer M. Rusted and David M. Warburton

CONTENTS

I. INTRODUCTION

Brain nicotine receptors have been mapped by autoradiographic techniques with the most commonly used radioligands being [^{3}H]nicotine and [^{3}H]acetylcholine. Both tritiated nicotine and tritiated acetylcholine bind with high affinity in a saturable and reversible manner to virtually the same sites in the central nervous system when the muscarinic sites are occluded,[1-3] indicating that the nicotinic synapses are predominantly cholinergic.

The autoradiographic maps of the binding sites of [^{3}H]nicotine and [^{3}H]acetylcholine are similar in rats, monkeys, and people.[1-3] Mapping of the regional distribution of high-affinity sites in the human brain with 5 nM of [^{3}H]nicotine shows a large number in the thalamus and parts of the basal ganglia, an intermediate number in areas such as the cerebral cortex (including the sensory areas), and lower numbers in the hippocampus and pons.[4]

Of course, one cannot equate the number of binding sites with the magnitude of psychoactivity. For example, it is clear that although the number of receptors in the hippocampus is relatively small, micromolar concentrations of nicotine will stimulate the release of [^{3}H]acetylcholine from hippocampal synaptosomes which have been preloaded with [^{3}H]choline.[5] Thus, it is the location of the receptors on the neurons that is important.

The site of action for the hippocampal acetylcholine release is thought to be presynaptic, nicotinic autoreceptors on the cholinergic terminals.[5] Similar presynaptic nicotinic autoreceptors are believed to occur on the cortical cholinergic terminals.[6] It is thought that loss of these nicotinic autoreceptors on the cholinergic terminals in the cortical and limbic regions may explain the deficits in nicotinic ligand binding sites in the brains of Alzheimer patients.[4]

Doses of nicotine in the smoking range bind to these high-affinity nicotinic, cholinergic receptor (nAChR) sites and argue for some of the psychopharmacological effects of smoking being due to an interaction with these nAChRs. However, it should be emphasized that nicotine does bind to noncholinergic neurons (e.g., Reference 7) and releases dopamine, norepinephrine, and 5-hydroxytryptamine in the brain.[8] Nevertheless, this review will concentrate on the interactions with the cholinergic system for two reasons. First, there is evidence for the concordance of the binding sites of [^{3}H]nicotine and [^{3}H]acetylcholine. Second, there is independent evidence for a relationship between some cholinergic systems of the brain and information processing (e.g., References 9 to 11).

II. NICOTINE AND INFORMATION PROCESSING

Does nicotine improve cognitive functioning? Despite the wealth of research studies (see References 1 and 2), the extent to which nicotine can influence information processing, and the conditions under which it exerts its effects, continue to be issues about which there are still some doubts.[13]

One of the major problems for researchers is that in trying to reach any conclusion, diverse studies have to be integrated. These studies differ on many critical dimensions, from the route of administration of nicotine to methodological and design factors. With respect to the first issue, the majority of research studies have been conducted with cigarette smoking and the assumption is clearly that the key factor in any observed improvement is the amount of nicotine in the cigarette. There are, however, many other components of the act of smoking and the smoker which could also influence performance: the individual's smoking style and thus the amount of nicotine entering the bloodstream, the preferred smoking context, sensory stimulation from the smoke, smoking history, and personality. All of these amount to an idiosyncratic set of potential modifiers for each smoker.

In addition, there are the more tangible differences across research studies — such as task selection and test sensitivity, choice of procedures, experimental design, and experimental controls — which make the process of extracting conclusions from the literature unusually difficult. This is illustrated by considering a key methodological issue which has received much attention of late — that being against which baseline performance to measure the effect of nicotine.

Many early studies examined the effect of nicotine in overnight-deprived individuals, to ensure minimal residual plasma levels of nicotine prior to the experimental dosing. An obvious problem with this design is that observed performance improvements may reflect relief from the withdrawal effects experienced by regular smokers who have been obliged to abstain, instead of an absolute nicotine-induced improvement in performance. A number of researchers have used this argument to question the validity of earlier claims of direct benefits to cognitive performance of nicotine.[13-17]

These doubts must be seen in the context of the increasing number of experimental studies which demonstrate positive effects of nicotine administration on performance of nonsmokers[18-25] and on the performance of nondeprived or minimally deprived smokers.[22,23,26-29]

Reinstatement may play a role in some of the performance effects observed. It is clear, for example, that on simple performance tasks such as tapping, reaction time, and letter cancellation, nicotine deprivation does impair smokers' performance.[30,31]

In a recent paper, Sherwood et al.[30] provided an elegant demonstration of the combined impact of nicotine reinstatement and direct benefits on measures of critical flicker fusion, tracking, reaction time, and memory scanning time. Using multiple doses of nicotine gum administered over a period of hours to overnight-deprived smokers, they reported significant improvements on all measures in response to the first dose of gum (the reinstatement effect); and continued improvements in performance with subsequent doses on the reaction time, tracking, and memory scanning time measures (primary effects of acute dosing). However, in many of the early studies, we are faced with the additional problem of being unable to differentiate the contribution of each of these elements to the single measure of change in performance reported.

Despite the catalog of difficulties facing those interested in the effects of nicotine, and the multiple contributory influences which ensure that measurable effects on performance are sometimes rather elusive and more often difficult to interpret, there seems to be sufficient depth of evidence for nicotine as an enhancer of cognitive function to stand the scrutiny of even the most hardened of skeptics. Nicotine does not improve cognitive performance in all studies, and thus it may not invariably benefit all aspects of cognition. However, in this review, we will argue that nicotine **is** effective in promoting the individuals capacity for sustained and focused attention and, more critically, that this in turn facilitates the effective encoding and storage of information in memory.

A. NICOTINE AND ATTENTION

The most robust demonstrations of the positive effects of nicotine on human information processing are seen in paradigms designed to measure attention. There is substantial evidence for the beneficial effects of smoking on selective[20] and sustained[19,32,33] attention. Comparable benefits on these tasks can be achieved with nicotine administered as tablets or gum to smokers[18,19,34] and nonsmokers.[25,35] Similarly, positive effects of smoking on smokers[36] and of nicotine gum on both smokers and nonsmokers[27] have been reported for reaction time measures; letter cancellation speed and accuracy is improved following smoking[37] or nicotine gum,[38] as are memory scanning times to correct identification of targets.[23,39] In short, we can conclude that in a wide range of tasks which measure attention, nicotine reliably improves performance.

B. NICOTINE AND MEMORY

In contrast, the evidence for the role of nicotine in promoting memory is rather mixed. Early studies generally reported little or no benefit of nicotine on immediate free recall of lists of items[41-43] while subsequent studies have indicated positive effects under more constrained conditions.

Thus, positive effects of nicotine obtained through smoking have been reported on a paired associate learning task for highly associated word pairs, while no effect[44] or nonsignificant trend for improvement[45] was found when low similarity paired associates were used. According to Craik and Lockhart,[46] associative links between items produce a more distinctive and, hence, more accessible memory trace; and the result would be superior recall performance. Recently, Mangan and Colrain[47] reported that smoking also improved immediate recall of a passage of prose. Since associative processing is necessary to extract meaning from text, these findings are consistent with the notion that nicotine facilitates encoding when the material is intrinsically cohesive, with salient associations between items.

Similarly, if nicotine **is** affecting the nature of the encoded trace, nicotine-induced, state-dependent learning should be observed; the encoding specificity principle[48] argues that if the conditions under which the material is learned affect the nature of the encoded trace, that trace will be better accessed by reinstating those precise conditions at retrieval. Several studies have reported that material learned following a cigarette was recalled better when the smokers were able to smoke at recall as well, that is, when the recall conditions reinstated the learning conditions.[43,49,50] It was also true for these studies that reinstatement of original learning conditions, whether smoking in both sessions or not smoking in either session, resulted in better performance than could be achieved by administration of nicotine at learning without reinstatement at recall.

The fact that performance benefits are not invariably observed in immediate recall measures is obviously a problem for a simple encoding hypothesis of nicotine effects. On the other hand, the fact that studies which fail to report nicotine-induced benefits in immediate recall use conditions which accommodate the state dependency benefits in the not-smoking condition while giving nicotine only at encoding may account in part for the apparent fragility of the nicotine-induced improvements in those studies.

Interestingly, a number of examples can be found in the smoking literature which indicate advantages for long-term recall of material learned under nicotine despite decrements or no effects of nicotine in the immediate recall phase. Andersson and Hockey[41] have suggested that the high level of arousal induced by nicotine administration is not conducive to good performance on immediate recall, since it precludes effective implementation of search strategies for recall. However, this high level of arousal at input does improve attention, and thus delayed recall reflects the benefits of enhanced attention at encoding under optimal retrieval conditions. The fact that cigarettes which deliver high doses of nicotine have been found to impair immediate recall performance[40,51] offers some support for the notion that high levels of arousal may differentially affect the encoding and the retrieval of material. On the other hand, Peeke and Peeke[52] have reported improved free recall following high yield cigarettes in both the immediate and the delayed test phase of their study.

One problem is that some subjects take in too much nicotine with high yield products and feel ill. These early studies do not provide sufficient detail of the smoking characteristics of their subjects to determine the nicotine dose delivered to the smoker from cigarettes of various yields, and the extent to which these doses corresponded to the preferred dose of the person. This is likely to be a significant factor in studies in which smokers had to puff a set number of times, at fixed intervals, and hold the smoke in the lungs for a specified period of time. It is clear from studies of unconstrained smoking that smokers titrate their dose by puffing and inhalation, so that wide variations of plasma nicotine levels are obtained.[53]

A role for electrocortical arousal in the cognitive enhancing effects of nicotine is indicated by electroencephalogram (EEG) studies. Improved cognitive performance following nicotine administration is associated with EEG activity patterns qualitatively different from those producing subjective experience of decreased arousal (relaxation) following nicotine intake.[54] In an earlier study, this change in cortical activity was directly associated with improved performance on a task of sustained attention.[55] More recently, authors have argued for a more complex interaction between EEG activity and the arousing vs. calming effects of smoking.[56-59] Norton et al.[59] reported an interesting result which suggested that the effects of nicotine correlated with the residual nicotine content of the cigarette butt, which is inversely related to the nicotine entering the smoker's mouth. Thus, the arousing effects of nicotine were associated with low doses while increased doses produced an EEG pattern more indicative of a calming action. In addition, Norton and co-workers' data[59] indicate asymmetrical changes in hemispheric EEG measures; increased left hemisphere EEG measures were associated with the stimulant effects and increased right hemisphere EEG with decreased arousal. While the work involved only a small sample of smokers (a total of 17 across two studies), the results are intuitively appealing and provide a parsimonious model for the complex behavioral effects of smoking reported among smokers.

Increased levels of electrocortical arousal produce the subjective experience of increased attention, and it is possible that the attention-enhancing properties of nicotine underlie the observed improvement in memory.[12, 60] Certainly, more robust effects of nicotine on memory have been reported in studies using long word lists which require more sustained attention for effective encoding. Peeke and Peeke[52] reported improved recall on a test of immediate memory for lists of 50 unrelated words; Warburton et al[50] (experiment 2) reported improved recall of 48-item lists following nicotine intake through either smoking or nicotine tablets; Rusted and Eaton-Williams[60] reported that nicotine improves recall of 30-word lists substantially more than it affects 10-word lists. Enhanced attention may also account for the benefits observed in tasks of associative learning, since the formation of links and processing for meaning make heavier demands on the available resources, which would benefit from improved attention.

In contrast to the majority of published studies, Spilich et al.[61] reported that nicotine impaired performance on complex tasks, although it did facilitate "simpler" information processing. In their study, smokers remembered fewer central proposition units from a text passage presented for recall than did nonsmokers, and the authors conclude that nicotine impaired selective attention to main idea units at input. The result is certainly at odds with the substantial body of literature indicating positive effects of nicotine on tasks with high attentional demands; and its relation to previous findings is difficult to assess, since methodological detail was missing from the paper. However, it is significant that subjects were asked "to inhale normally every 25 s and hold that puff for 5 s, for a total of 12 puffs".[61] This would nauseate some subjects with a 1.2-mg cigarette.

In addition, a free recall procedure employed immediately after reading the passage might account for the pattern of results. The acute smoking group was composed of deprived smokers given a cigarette immediately before the experimental session, and the smokers apparent failure to use a systematic recall strategy with primary focus on main idea units may thus be attributed to the high arousal state induced by the reinstatement of nicotine after a period of deprivation. In order to argue the case for impairment by nicotine on this task, Spilich et al.[61] would have had to demonstrate a similar pattern in a delayed recall phase, where the smokers are not disadvantaged by a suboptimal retrieval state.

Human information processing models emphasize the active nature of the memory processes, and thus have tended to focus on the encoding processes and their impact on the way information is stored, with less attention given to the period after the material has been encoded and is being held in storage for subsequent retrieval. The specific locus of nicotine effects in the information processing chain is extremely difficult to determine. Attention-enhancing effects at the encoding phase may induce more distinctive traces through attention to associations, or more established traces through sustained attention to the task of encoding, or both. The outcome of effective encoding is more durable and accessible memory traces, and in this sense the boundary between effects on encoding and effects on storage is blurred. The possibility that nicotine effects are localized at some more discrete, postencoding stage is an interesting alternative which has recently received considerable attention.

In animal studies, a postencoding process which consolidates an unstable memory trace has been indicated by studies demonstrating the disruption[62] or enhancement[63] of memory by chemical agents administered after the learning phase. More importantly, postlearning administration of nicotine to rats appeared to promote consolidation of memory traces (e.g., Reference 64). The process whereby a new and purportedly labile memory trace representing recently stored material is consolidated, or permanently established in memory, has received less attention in the human literature, with the exception of some early clinical studies with head injury patients[65] and ECT patients (e.g., Reference 66). However, some recent studies have examined the potential of nicotine to improve information processing performance when administered after the encoding phase, and have reported significant facilitation by nicotine under these conditions.

Mangan and Golding[45] reported impaired 30-min but improved 1-week and 1-month recall of paired associates when volunteers smoked a cigarette immediately after the paired associates had been learned to criterion. Colrain et al.[67] also report postacquisition enhancement of paired associate list learning. Although Peeke and Peeke[52] failed to obtain effects of posttrial smoking in a free recall study involving lists of unrelated words, we have obtained positive effects of postacquisition nicotine with unrelated word lists both when volunteers were encouraged to form associative links between those items,[68] and when volunteers were undirected as to encoding strategy.[29] These results suggest that an associative learning strategy is not an essential element of the posttrial effects of nicotine. However, in both studies and in the earlier ones cited above, volunteers did have the opportunity for continued rehearsal of the to-be-remembered material during the posttrial smoking phase.

We decided to look at the significance of this point directly, by running a series of studies in which the volunteer was required to complete a secondary distractor task after word list presentation, during the posttrial smoking phase. If posttrial benefits from nicotine are dependent on the opportunity for continued rehearsal, the distractor task, which precludes overt rehearsal, should eliminate the improvement in word list recall previously observed when nicotine is administered in the posttrial phase. In the Rusted and Warburton study,[29] one condition required volunteers to perform a distractor task, in this case, a 10-min rapid visual information processing task (RVIP), after list presentation and during nicotine intake. This task is sufficiently difficult to block any attempt at rehearsal of the word list, and, since it does itself benefit from nicotine intake, it served as a positive control for central nervous system (CNS) effects of nicotine.

The result was clear-cut. Smoking a cigarette after list presentation and concurrently with the RVIP distractor task eliminated the posttrial benefits of nicotine on recall performance, while improving performance on the RVIP task. This suggests that continued processing of words after presentation, and during nicotine administration is the basis of the posttrial effect of nicotine on recall. The fact that the distractor task itself showed a nicotine-induced improvement, however, suggests that some kind of active trade-off or deployment of resources might have occurred, with the focus on the distractor task instead of the word task, in response to concurrent task demands. What happens, then, if the distractor task is not one which normally benefits from concurrent administration of nicotine?

In the next study,[69] volunteer smokers were required to complete a subject-paced, serial sevens task while smoking a cigarette in the postacquisition phase. Confirming previous results with this task in our laboratory, performance on the serial sevens task was not improved by smoking, relative to the sham-smoking control condition. More importantly, despite the fact that this verbal distractor task precluded overt rehearsal of the word list, smoking induced better recall performance than did sham smoking. Thus, posttrial benefits of nicotine on list recall do not appear to be entirely blocked by the requirement to complete a verbal distractor task in the posttrial interval. While implausible, it is possible that the volunteers are somehow managing to covertly rehearse the word list while concurrently completing the distractor task, particularly with the distractor task being self-paced.

In order to address this possibility, we ran a third study in which volunteers completed either a high load or a low load distractor task in the posttrial interval. If they are consciously rehearsing or at least directing some attention to the word list items during the distractor task, then a task with a low processing load should produce a bigger posttrial nicotine effect than a high load task, since less resources need to be committed to the distractor task. We again used serial sevens for the high load task, with the low load task being maintenance repetition of a four-digit number. The results indicated a complex interaction between nicotine and task demands. For both distractor conditions, list recall was considerably reduced relative to the no-distractor baseline measures. In the low load condition, smoking did not improve performance over sham smoking; that is, there was no posttrial facilitation by nicotine. With the high load distractor task, on the other hand, smoking apparently **protects** performance from the detrimental effects of a concurrent distractor task, such that the performance decrement was considerably smaller for the smokers than for the sham smokers. The latter showed a decrement in performance comparable to the low load distractor groups.

Obviously, these results need to be replicated, but the pattern of results is particularly interesting in the context of the literature on nicotine and human information processing for the following reasons. First, these studies provide additional evidence for the positive effects of nicotine on some process acting on the encoded word list during the postacquisition phase, under conditions which eliminate any opportunity for conscious attention to, or rehearsal of, the list items. Second, these effects of nicotine are observed under the more demanding distractor conditions, but not under the low demand condition. This seems at first glance to be counterintuitive. However, suppose the heavy processing demands required by the distractor task are increasing the state of arousal of the system, and this is further boosted by nicotine. Then the end results are direct benefits to performance through some postencoding neuronal activation, which may be equated with the process of consolidation.

A mechanism for the action of nicotine on a consolidation process is proposed in the next section.

III. HYPOTHESIS FOR A CHOLINERGIC ROLE IN THE CONSOLIDATION PROCESS

It has been argued that the central nervous system must have stability and plasticity.[70] Some neural networks must maintain stability, so that there is a relatively unvarying, quantitative relation between the inputs and the outputs. However, there must also be other neurons that can change in response to inputs.

If there were no plastic neurons, then the inputs would merely activate genetically established networks and the outputs would be unvarying. The presence of neurons with plasticity enables outputs to be based on present information *and* information from the more distant past, long-term memory.

In addition, it would seem plausible to have specialized, plasticity-controlling neurons which govern the modification of synaptic connections.[70-73] Specific transmitter systems, such as norepinephrine and acetylcholine, have been suggested as playing a role in plasticity modulation.[70,72-74]

There are many mechanisms of synaptic plasticity, from simple, short-term depletion of transmitter to longer-term structural changes. One of the prolonged changes which has stimulated a great deal of research in recent years has been long-term potentiation, an increase in the strength of synaptic responses which can last anywhere from hours to months.[75]

Long-term potentiation modifies the postsynaptic membrane and occurs as the result of postsynaptic depolarization coinciding with presynaptic activity. The presynaptic activation releases the transmitter glutamate[76] which binds to *N*-methyl-D-aspartate (NMDA) receptors. An essential aspect of the process is an influx of calcium through the NMDA receptor channels.[77] However, the NMDA receptor channels only become activated to allow the influx of calcium when the postsynaptic membrane is adequately depolarized by a strong, cooperative input from other neurons.

The outcome of the calcium influx is an increase in the number of NMDA receptors on the postsynaptic membrane.[75] However, the postsynaptic membrane changes of long-term potentiation are only produced when glutamate binds to the NMDA receptor and when there is depolarization of the membrane by excitation of nonglutamate receptors by other transmitters. In the terminology used above, the nonglutamate neurons act as plasticity controllers for the NMDA receptors.

The major area for the study of long-term potentiation has been the hippocampus,[75] but it has also been observed at the cortex.[78] The hippocampus is of particular interest, because it has long been implicated in mnemonic processes.[79] The hippocampal formation receives converging inputs from high-order areas of the association cortex and has output paths back to the neocortex.

The inputs reach the hippocampal formation through the perforant pathway, ending in a large number of synapses on the CA3 pyramidal cells. The CA3 pyramidal cells project via the Schaffer collaterals to the CA1 pyramidal cells. It is known that many of these synapses are modifiable (long-term potentiation) when there is both activity of the glutamate neurons and strong depolarization of the postsynaptic membrane. Thus, synapses between active input neurons and activated post-synaptic CA3 neurons change so that the neuron responds more the next time it is activated.[80]

The CA3 cells have recurrent collaterals so that the output is fed back and associated with itself, autoassociation. Rolls[81] has hypothesized that the CA3 autoassociation system is able to detect when there is conjoint activation of sets of input fibers from widely separated cortical areas. According to his hypothesis, it allocates output neurons to code economically each complex input event and direct information storage by return pathways to the neocortical areas.

The autoassociation type of memory which is attributed to this hippocampal system is what is required for arbitrary associations of the type which are needed for associative learning. As Rolls[81] has pointed out, the hippocampal formation could provide the substrate for a working memory because the hippocampus sets up a representation which is used to determine how information can best be stored in the neocortex. Impairment of working memory and declarative memory of patients with damage to the hippocampal formation, including Alzheimer patients, may reflect impaired function of the CA3 autoassociation matrix.

Of particular interest is the fact that there are also inputs to the hippocampal pyramidal cells from the cholinergic cells of the medial septal region and the diagonal band of Broca.[82-84] It is known that activation of the septal cholinergic neurons results in a slow depolarization of the hippocampal pyramidal neurons[85] and the same effect can be achieved by the exogenous application of acetylcholine.[86]

The slow time course (a matter of seconds) of exogenously applied and endogenously released acetylcholine in the hippocampus suggests that these cholinergic neurons are involved in modulating the general excitability of the pyramidal neurons, instead of the rapid and precise transfer of high frequency information which would be involved in the CA3 autoassociation processes. This sort of modulatory process would be exactly the sort of specification required for a plasticity controller; and thus the question arises to what extent the membrane actions of cholinergic neurons can function to enhance associative processes.

Supporting evidence comes from the fact that acetylcholine can reduce potassium conductances.[87-89] Singer[90] has pointed out that reduction of potassium conductances will have two consequences. It increases the depolarization in response to excitatory inputs and will facilitate propagation of synaptic potentials. The latter increases temporal and spatial summation of converging inputs, enabling cooperativity among converging excitatory inputs.

In these ways, acetylcholine will increase the likelihood that associated, excitatory inputs produce depolarizations that are of sufficient magnitude to reach the activation threshold of the NMDA receptor to enable calcium influx. Thus, acetylcholine can act as a plasticity controller by influencing the probability that excitatory inputs reach the activation threshold of the NMDA receptor and initiate enhancement of synaptic sensitivity.

The action of acetylcholine on muscarinic, postsynaptic receptors to reduce the potassium conductances[87-89] provides a mechanism whereby muscarinic antagonists impair mnemonic processes[11,90,91] More importantly, as we pointed out in the introductory section, micromolar concentrations of nicotine will stimulate the release of acetylcholine from hippocampal synaptosomes, and if the site of action for the hippocampal acetylcholine release is the presynaptic, nicotinic autoreceptors on the cholinergic terminals,[5] then this would provide a mechanism by which stimulation of nicotinic receptors can modulate the processes of encoding and consolidation.

IV. CONCLUSIONS

First, autoradiographic techniques have demonstrated a concordance of binding sites of [³H]nicotine and [³H]acetylcholine. Pharmacological studies have demonstrated that doses of nicotine in the smoking range bind to high-affinity nicotinic, cholinergic receptor sites and that micromolar concentrations of nicotine will stimulate the release of [³H]acetylcholine from the hippocampal synaptosomes, for example.

Second, psychopharmacological studies with smoking and nicotine have revealed a number of effects of nicotine on human information processing. In this review, we have focused on studies of memory storage. They suggest a positive effect of nicotine on some process acting on encoded material during the postacquisition phase, even when there is no possibility of opportunity for conscious attention to, or rehearsal of, the material. These effects of nicotine are observed under conditions with high processing demands. We suggest that these demands increase the state of arousal of the system; and this state is further boosted by nicotine, with direct benefits to performance through some postencoding neuronal activation. This activation may be equated with the processes of consolidation.

Third, a possible mechanism for the involvement of nicotinic receptors in consolidation processes is that they modulate the membrane excitability of neurons in the hippocampal formation and act as an enabling system which allows temporally associated inputs to produce persisting alterations to the postsynaptic membrane. Consequently, the neuron responds more the next time that it is activated by those inputs. In this system, the nicotinic, cholinergic synapses are not involved in the input processes per se, but have a role in the associative processes.

REFERENCES

1. Clarke, P. B. S., Schwartz, R. D., Paul, S. M., Pert, C. B., and Pert, A. Nicotinic binding in rat brain: autoradiographic comparison of ³H-acetylcholine, ³H-nicotine and ¹²⁵I-alpha-bungarotoxin, *Journal of Neuroscience,* 5, 1307, 1985.
2. Friedman, D. P., Clarke, P. B. S., O'Neill, J.B., and Pert, A., Distributions of nicotinic and muscarinic cholinergic receptors in monkey thalamus, *Society of Neuroscience, Abstracts*, 11, 307, 1985.
3. Adem, A., Jossan, S. S., d'Argy, R., Brandt, R., Winblad, B., and Nordberg, A., Distribution of nicotinic receptors in human thalamus a visualized by ³H-nicotine and ³H-acetylcholine receptor autoradiography, *Journal of Neural Transmission,* 73, 77, 1989.
4. Nordberg, A., Nicotinic receptor loss in brain of Alzheimer patients as revealed by *in vitro* receptor binding and *in vivo* positron emission tomography techniques, in *Effects of Nicotine on Biological Systems*, Adlkofer, F. and Thurau, K., Eds., Birkhäuser Verlag, Basel, 1991, 631.
5. Wonnacott, S., Irons, J., Rapier, C., Thorne, B., and Lunt, G.G., Presynaptic modulation of transmitter release by nicotinic receptors, in *Progress in Brain Research, Vol. 79*, Nordberg, A., Fuxe, K., Holmstedt, B., and Sundwall, A., Eds., Elsevier, Amsterdam, 1989, chap. 15.
6. Araujo, D. M., Lapchak, P. A., Collier, B., and Quirion, R. Characterization of [³H]*N*-methylcarbamylcholine binding sites and effect of *N*-methylcarbamylcholine on acetylcholine release in rat brain, *Journal of Neurochemistry*, 51, 292, 1988.

7. Clarke, P. B. S. and Pert, A., Autoradiographic evidence for nicotine receptors on nigrostriatal and mesolimbic dopaminergic neurons, *Brain Research*, 348, 355, 1985.

8. Chesselet, M. F., Presynaptic regulation of neurotransmitter release in the brain: facts and a hypothesis, *Neuroscience*, 12, 347, 1984.

9. Warburton, D. M., Neurochemistry of behavior, *British Medical Bulletin*, 37, 121, 1981.

10. Rusted, J. M. and Warburton, D. M., Cognitive models and cholinergic drugs, *Neuropsychobiology*, 21, 31, 1989.

11. Warburton, D. M. and Rusted, J. M., Cholinergic control of cognitive resources, *Neuropsychobiology*, 28, 43, 1993.

12. Warburton, D. M., Psychopharmacological aspects of nicotine, in *Nicotine Psychopharmacology*, Wonnacott, S., Russell, M. A. H., and Stolerman, I. P., Eds., Oxford University Press, 1990, chap. 3.

13. West, R., Beneficial effect of nicotine: fact or fiction? (Editorial), *Addiction*, 88, 589, 1993.

14. Ney, T., Gale, A., and Morris, H., A critical evaluation of laboratory studies of the effects of smoking on learning and memory, in *Smoking and Human Behaviour*, Ney, T. and Gale, A., Eds., John Wiley & Sons, Chichester, 1989, chap. 11.

15. Herning, R. I., Brigham, J., Stitzer, M. L., Glover, B. J., Pickworth, W. B., and Henningfield, J. E., The effects of nicotine on information processing: medicating a deficit, *Psychophysiology*, 27, S2, 1990.

16. West, R. J., Nicotine pharmacodynamics: some unresolved issues, in *The Biology of Nicotine Dependence*, Bock, G. and Marsh, J., Eds, John Wiley & Sons, Chichester, 1990, 210.

17. West, R. J., Nicotine addiction: a reanalysis of the arguments, *Psychopharmacology*, 108, 408, 1992.

18. Wesnes, K., Warburton, D. M., and Matz, B., Effects of nicotine on stimulus sensitivity and response bias in a visual vigilance task, *Neuropsychobiology*, 9, 41, 1983.

19. Wesnes, K. and Warburton, D. M., Effects of scopolamine and nicotine on human rapid information processing performance, *Psychopharmacology*, 82, 147, 1984.

20. Wesnes, K. and Revell, A., The separate and combined effects of scopolamine and nicotine on human information processing, *Psychopharmacology*, 84, 51, 1984.

21. West, R. J. and Jarvis, M. J., Effects of nicotine on finger tapping rate in non-smokers, *Pharmacology, Biochemistry and Behaviour*, 25, 727, 1986.

22. Kerr, J. S., Sherwood, N., and Hindmarch, I., Separate and combined effects of the social drugs on psychomotor performance, *Psychopharmacology*, 104, 113, 1991.

23. Sherwood, N., Kerr, J. S., and Hindmarch, I., Effects of nicotine gum on short term memory, in *Effects of Nicotine on Biological Systems: Proceedings of the International Symposium on Nicotine*, Adlkofer, F. and Thurau, K., Eds., Birkhäuser Verlag, Basel, 1990, 531.

24. West, R. and Hack, S., Effects of nicotine cigarettes on memory search rate, in *Effects of Nicotine on Biological Systems: Proceedings of the International Symposium on Nicotine*, Adlkofer, F. and Thurau, K., Eds., Birkhäuser Verlag, Basel, 1990, 547.

25. Provost, S. C. and Woodward, R., Effects of nicotine gum on repeated administration of the Stroop test, *Psychopharmacology*, 104, 536, 1991.

26. Frearson, W., Barrett, P., and Eysenck, H. J., Intelligence, reaction time and the effects of smoking, *Personality and Individual Differences*, 9, 497, 1988.

27. Hindmarch, I., Kerr, J. S., and Sherwood, N., Effects of nicotine gum on psychomotor performance in smokers and non-smokers, *Psychopharmacology*, 100, 535, 1990.

28. Pritchard, W. S., Robinson, J. H., and Guy, T. D., Enhancement of continuous performance task reaction time by smoking in non-deprived smokers, *Psychopharmacology*, 108, 437, 1992.

29. Rusted, J. M. and Warburton, D. M., Facilitation of memory by post-trial administration of nicotine: evidence for an attentional explanation, *Psychopharmacology*, 108, 452, 1992.

30. Sherwood, N., Kerr, J. S., and Hindmarch, I., Psychomotor performance in smokers following single and repeated doses of nicotine gum, *Psychopharmacology*, 108, 432, 1992.

31. Parrott, A. C. and Roberts, G., Smoking deprivation and cigarette reinstatement: effects upon visual attention, *Journal of Psychopharmacology*, 5, 404, 1991.

32. Michel, C., Nil, R., Buzzi, R., Woodson, P.P., and Bättig, K., Rapid information processing and concomitant event-related brain potentials in smokers differing in CO absorption, *Neuropsychobiology*, 17, 161, 1987.

33. Hasenfratz, M., Michel, C., Nil, R., and Battig, K., Can smoking increase attention in rapid information processing during noise? Electrocortical, physiological and behavioral effects, *Psychopharmacology*, 98, 75, 1989.

34. Parrott, A. C. and Winder, G., Nicotine chewing gum (2 mg, 4 mg) and cigarette smoking: comparative effects upon vigilance and heart rate, *Psychopharmacology*, 97, 257, 1989.
35. Wesnes, K. and Warburton, D.M., The effects of cigarette smoking and nicotine tablets upon human attention, in *Smoking Behaviour: Physiological and Psychological Influences,* Thornton, R. E., Ed., Churchill Livingstone, Edinburgh, 1978, 131.
36. Petrie, R. and Deary, I., Smoking and human information processing, *Psychopharmacology*, 99, 393, 1989.
37. Snyder, F. R., Davis, F. C., and Henningfield, J. E., The tobacco withdrawal syndrome: performance deficits assessed on a computerized test battery, *Drug Alcohol Dependence*, 23, 259, 1989.
38. Parrott, A. C. and Craig, D., Cigarette smoking and nicotine gum (0, 2 and 4 mg): effects upon four visual attention tasks, *Neuropsychobiology*, 25, 34, 1992,
39. West, R. J. and Hack, S., Effects of cigaretttes on memory search and subjective ratings, *Pharmacology, Biochemistry and Behaviour*, 38, 281, 1991.
40. Andersson, K. and Post, B., Effects of cigarette smoking on verbal rote learning and physiological arousal, *Scandanavian Journal of Psychology*, 15, 263, 1974.
41. Andersson, K. and Hockey, G. R. J., Effects of cigarette smoking on incidental memory, *Psychopharmacology,* 52, 223, 1977.
42. Williams, D. G., Effects of cigarette smoking on immediate memory and performance in different kinds of smokers, *British Journal of Psychology*, 71, 3, 1980.
43. Peters, R. and McGee, R., Cigarette smoking and state-dependent memory, *Psychopharmacology*, 76, 232, 1982.
44. Mangan, G. L., The effects of cigarette smoking on verbal learning and retention, *Journal of General Psychology*, 108, 203, 1983.
45. Mangan, G. L. and Golding, J. F., The effects of smoking on memory consolidation, *Journal of Psychology*, 115, 65, 1983.
46. Craik, F. I. M. and Lockhart, R. S., Levels of processing: a framework for memory research, *Journal of Verbal Learning and Verbal Behaviour*, 11, 671, 1972.
47. Mangan, G. L. and Colrain, I. M., Relationships between photic driving, nicotine and memory, in *Effects of Nicotine on Biological Systems: Proceedings of the International Symposium on Nicotine*, Adlkofer, F. and Thurau, K., Eds., Birkhäuser Verlag, Berlin, 1990, 537.
48. Tulving, E. and Thomson, D. M., Encoding specificity and retrieval processes in episodic memory, *Psychological Review*, 80, 352, 1973.
49. Kusendorf, R. and Wigner, L., Smoking and memory: state-specific effects, *Perception and Motor Skills*, 61, 158, 1985.
50. Warburton, D. M., Wesnes, K., Shergold, K., and James, M., Facilitation of learning and state dependency with nicotine, *Psychopharmacology,* 89, 55, 1986.
51. Mangan, G. L. and Golding, J. F., The enhancement model of smoking maintenance, in *Smoking Behaviour: Physiological and Psychological Influences*, Thornton, R. E., Ed., Churchill Livingstone, Edinburgh, 1978, 87.
52. Peeke, S. C. and Peeke, V. S., Attention, memory and cigarette smoking, *Psychopharmacology*, 84, 205, 1984.
53. Warburton, D.M., The functional use of nicotine, in *Nicotine, Smoking and the Low Tar Programme*, Wald, N. and Froggatt, P., Eds., Oxford University Press, Oxford, 1989, 182.
54. Pritchard, W., Electroencephalographic effects of cigarette smoking, *Psychopharmacology*, 104, 485, 1991.
55. Edwards, J. A., Wesnes, K., Warburton, D.M., and Gale, A., Evidence of more rapid stimulus evaluation following cigarette smoking, *Addictive Behaviours*, 10, 113, 1985.
56. Norton. R. and Howard, R., Smoking, mood and contingent negative variation in a go-no go avoidance task, *Journal of Psychophysiology*, 2, 109, 1988.
57. Gilbert, D. G., Robinson, J. H., Chamberlin, C. L., and Speilberger, C. D., Effects of smoking/nicotine on anxiety, heart rate, and lateralization of EEG during a stressful movie, *Psychophysiology*, 26, 311, 1989.
58. ONeill, S. T. and Parrott, A. C., Stress and arousal in sedative and stimulant cigarette smokers, *Psychopharmacology*, 107, 442, 1992.
59. Norton, R., Brown, K., and Howard, R., Smoking, nicotine dose and the lateralisation of electrocortical activity, *Psychopharmacology*, 108, 473, 1992.

60. Rusted, J. M. and Eaton-Williams, P., Distinguishing between attentional and amnestic effects in information processing: the separate and combined effects of scopolamine and nicotine on verbal free recall, *Psychopharmacology*, 104, 363, 1991.

61. Spilich, G. J., June, L., and Renner, J., Cigarette smoking and cognitive performance, *British Journal of Addiction*, 87, 1313, 1992.

62. Barondes, S. H., Some critical variables in studies of the effect of inhibitions of protein synthesis on memory, in *Molecular Approaches to Learning and Memory*, Byrne, W. L., Ed., Academic Press, New York, 1970, 27.

63. Greenough, W. T. and McGaugh, J. L., The effect of strychnine sulphate on learning as a function of time of administration, *Psychopharmacologia*, 8, 290, 1965.

64. Garg, M. and Holland, H. C., Consolidation and maze learning. A further study of post-trial injection of a stimulant drug (nicotine), *Neuropharmacology*, 7, 55, 1968.

65. Yarnell, P. R. and Lynch, S., Retrograde amnesia immediately after concussion, *Lancet*, 1, 863, 1970.

66. Cronholm, B. and Lagergren, A., Memory disturbance after electroconvulsive therapy, *Acta Psychiatrica Neurologia Scandinavica*, 34, 283, 1959.

67. Colrain, I. M., Mangan, G. L., Pellett, O. L., and Bates, T. C., The effects of post-learning smoking on memory consolidation, *Psychopharmacology*, 108, 448, 1992.

68. Warburton, D. M., Rusted, J. M., and Fowler, J., A comparison of the attentional and consolidation hypotheses for the facilitation of memory by nicotine, *Psychopharmacology*, 108, 443, 1992.

69. Rusted, J. M., Wang, J., Rodway, P., Warburton, T. S., and Warburton, D. M., unpublished data, 1993.

70. Warburton, D. M., Towards a neurochemical theory of learning and memory, in Gale, R. and Edwards, J., Eds., *Physiological Correlates of Human Behaviour*, Vol. 1, Academic Press, London, 1983, chap. 7

71. Krasne, F. B., Extrinsic control of intrinsic neural plasticity, *Brain Research*, 140, 197, 1978.

72. Kety, S. S., The evolution of concepts of memory, in *The Neural Basis of Behavior*, Beckman, A. L., Ed., SP Medical and Scientific Books, New York, 1982.

73. Kasamatsu, T., Neuronal plasticity maintained by the central norepinephrin system in the cat visual cortex, in *Progress in Psychobiology and Physiological Psychology*, Vol. 10, Sprague, J. M. and Epstein, A. N., Eds., Academic Press, New York, 1983.

74. McGaugh, J. L., Involvement of hormonal and neuromodulatory systems in the regulation of memory storage, *Annual Review of Neuroscience*, 12, 255, 1989.

75. Lynch, G. and Baudry, M., The biochemistry of memory: a new and specific hypothesis, *Science*, 224, 1057, 1984.

76. Dunwiddie, T., Madison, D. V., and Lynch, G., Synaptic transmission is required for the initiation of long-term potentiation, *Brain Research*, 150, 413, 1978.

77. Dunwiddie, T. V. and Lynch, G., The relationship between extracellular calcium concentrations and the induction of hippocampal long-term potentiation, *Brain Research*, 169, 103, 1979.

78. Lee, K. S., Sustained enhancement of evoked potentials following brief, high-frquency stimulation of the cerebral cortex *in vitro*, *Brain Research*, 239, 617, 1982.

79. Scoville, W. B. and Milner, B., Loss of recent memory after bilateral hippocampal lesions, *Journal of Neurology, Neurosurgery and Psychiatry*, 20, 11, 1957.

80. McNaughton, B. L., Activity dependent modulation of hippocampal synaptic efficacy: some implications for memory processes, in *Neurobiology of the Hippocampus*, Seifert, W., Ed., Academic Press, London, 1984, chap. 13.

81. Rolls, E. T., Parallel distributed processing in the brain: implications of the functional architecture of neuronal networks in the hippocampus, in *Parallel Distributed Processing: Implications for Psychology and Neurobiology*, Morris, R.G.M., Ed., Oxford University Press, New York, 1989, 286.

82. Shute, C. C. D. and Lewis, P. R., The use of cholinesterase techniques combined with operative procedures to follow nervous pathways in the brain, in *Bibliotheca Anatomica, Vol. 2, Histochemistry of Cholinesterase*, Schwarzacher, H. G., Ed., Karger, New York, 1961.

83. Shute, C. C. D. and Lewis, P. R., Cholinergic containing systems of the brain of the rat, *Nature (London)*, 199, 1160, 1963.

84. Lewis, P. R., Shute, C. C. D., and Silver, A., Confirmation from choline acetylase analyses of a massive cholinergic innervation to the hippocampus, *Journal of Physiology*, 172, 9, 1964.

85. Gähwiler, B. H. and Brown, D. A., Functional innervation of cultured hippocampal neurones by cholinergic afferents from co-cultured septal explants, *Nature (London)*, 313, 577, 1985.

86. Madison, D. V., Lancaster, B., and Nicoll, R. A. Voltage clamp analysis of cholinergic slow excitation in the hippocampus, *in vitro*, *Journal of Neuroscience*, 7, 733, 1987.
87. Krnjevic, K., Pumain, R., and Renaud, L., The mechanism of excitation by acetylcholine in the cerebral cortex, *Journal of Physiology*, 215, 447, 1971.
88. Halliwell, J. V. and Adams, P. R., Voltage-clamp analysis of muscarinic excitation in hippocampal neurons, *Brain Research*, 250, 71, 1982.
89. McCormick, D. A. and Prince, D. A., Postsynaptic actions of acetylcholine in the mammalian brain *in vitro*, in *Neurotransmitters and Cortical Function*, Avoli, M., Reader, T. A., Dykes, R. W., and Gloor, P., Eds., Plenum Press, New York, 1988, 287.
90. Singer, W., Role of acetylcholine in use-dependent plasticity of the visual cortex, in *Brain Cholinergic Systems,* Steriade, M. and Biesold, D., Eds., Oxford University Press, New York, 1990, 314.
91. Rusted, J. M. and Warburton, D. M., The effects of scopolamine on working memory in healthy young volunteers, *Psychopharmacology*, 105, 145, 1988.
92. Rusted, J. M. and Warburton, D. M., Effects of scopolamine on verbal memory; a retrieval or acquisition deficit?, *Neuropsychobiology*, 21, 76, 1989.

Chapter 13

Cholinergic Neurons and Memory: An Historical Perspective and Overview of Current Research

Rosalee Grette Lydon

CONTENTS

I. INTRODUCTION

The investigation of the relationship between cholinergic neurons of the central nervous system (CNS) and memory has had a rich and diverse history. The understanding of this relationship is accompanied by difficulties in ruling out competing hypotheses, and by disagreement over the types of learning in which the multifaceted neurochemical acetylcholine (ACh) may be involved. Cholinergic neurons are extensively disbursed throughout the central and peripheral nervous system, and encompass numerous and varied behavioral functions.[1] There is little wonder that the quest to determine the role of acetylcholine in memory has presented a challenge.

This chapter is an overview of the literature investigating the role cholinergic neurons play in memory. The first half is a chronological retrospective documenting the history of this investigation, which culminated in the late 1970s with the cholinergic hypothesis of memory dysfunction in senescence and in senile dementia of the Alzheimer type (SDAT). In part, it is meant as a modest tribute to the pioneers and, in part, it is a description of some of the controversy that has spiced up the quest.

The second half is a review of more recent animal and human research. In light of several prior reviews,[1-5] I will focus on a number of specific papers examining the scopolamine model of memory dysfunction in SDAT, and concentrate on studies of maze learning in animals. The animal literature will be examined with respect to the controversy over the claims for a selective involvement of the cholinergic system in working memory and in place learning. Of particular interest is research involving working and reference memory, defined here according to Honig.[6]

II. HISTORICAL PERSPECTIVE

A. ANECDOTAL HISTORY

Agents that block muscarinic receptor sites for ACh have been associated with memory deficit as well as cognitive and other physiological disturbances.[7] The effects are due to tropane alkaloids, including scopolamine or hyoscine, atropine, and hyoscyamine found in various genera of the potato family, *Solanaceae*. Not all effects are due to central psychoactive properties because these alkaloids are also potent blocking agents of the peripheral (autonomic) nervous system and result in various disturbances such as irregular heart rate, dry mouth, temperature increase, and dilation of the pupils leading to blurred vision. A drop of deadly nightshade with its active ingredient, atropine, was used by Roman and Egyptian women to dilate the pupils so as to appear more attractive,[8] thus the species name, *belladonna* or *beautiful woman.*

High doses of anticholinergics result in effects similar to toxic psychosis including confusion, drowsiness, and loss of memory for recent events as well as disturbances of attention, hallucinations, and even death. Poisoners the world over have been aware for centuries of the toxic effects of plants containing anticholinergics such as deadly nightshade; henbane (*Hyosycamus niger*), with active agents scopolamine and hyoscyamine; and mandrake root (*Mandragora officinarum*), which contains all three alkaloids and is found in an area from the Himalayas to the Mediterranean.[7] It was well known in the Middle Ages that the lethal dose for belladonna was 14 berries.[8]

Belladonna and mandrake were important ingredients of the witches' brews concocted in the Dark Ages. *Mandrake, or potent male*, sought after because it was believed to bring fertility as well as sexual stamina, was documented as early as in the book of *Genesis* (30:14–16). Yet even an anecdote is not without controversy. It is thought that the feelings of sexual arousal brought on by belladonna or mandrake were due not to legitimate aphrodesia, but to tachycardia and a placebo effect:[8] the *belief* that the "devil's herb", "apples of Sodom", or "enchanter's nightshade" (Reference 9 in McKim[7]) would ensure that a libidinous time would be had by all around the old coven circle that night. Similarly, the sensation of flying reputedly experienced by witches has been attributed to run-of-the-mill peripheral effects, such as irregular heartbeat, and a central effect of drowsiness[10] in combination with placebo effects.[8]

Psychoactive properties of species of *Datura*, which also contain all three alkaloids, underscore their popularity for medicinal, religious, or magical purposes in both the Old and New Worlds. The priestesses at the Oracle of Delphi were probably under the influence of *Datura* when they uttered prophecies,[11] thought by the ancient Greeks to have been inspired by inhalation of mystic vapors created from burned seeds or ingestion of leaves from the sacred laurel.[8] *Datura metel* was used as a narcotic or sedative in ancient China and Arabia, an aphrodisiac in India, and a poison as well as an intoxicant in Africa. Even in present-day Asia, it is used by thieves to intoxicate victims. *Datura* was widely used by various indigenous people of South America and the U.S. Southwest for spiritual and ceremonial or medicinal purposes.

With respect to properties of memory per se, Ray and Ksir[8] report that Algonquin Indians of North America gave liberal portions of the water of Lethe or wysoccan, made with *Datura stramonium* or jimsonweed, to their youth. This was to help them resolve identity problems of adolescence as they approached manhood, so that they would "lose the remembrance of all former things," and "thus unlive their former lives and commence men by forgetting that they ever had been boys" (Reference 11 in Ray and Ksir.[8]) No doubt, the 18- or 20-day confinement in combination with large quantities of this potent brew as their only sustenance contributed to their amnesia.

This is not to deny the possibility that there was also a genuine effect on memory. It was known in the 1950s and 1960s that scopolamine in combination with a sedative and hypnotic resulted in obstetric twilight sleep, which involved amnesia for events immediately preceding and during delivery.[12] It is interesting, as Bartus et al.[13] note, that this phenomenon was first described in 1906[14] around the same time as Alois Alzheimer[15] published his seminal paper on the "peculiar disease" that was to bear his name.

B. EARLY HISTORY
1. Conceptual and Methodological Issues

The psychopharmacology of memory has been dominated by dual-process theories in which information is thought to enter a short-term store of limited capacity referred to as primary memory, and then is transferred into a permanent long-term store. The latter secondary memory encompasses all information retained beyond the short-term period, and includes recently learned material as well as material from the more remote past. It is common in the literature to speak of memory impairment as resulting from a defect in one of essentially four phases of learning. The first is an initial registration phase of processing during which information is attended to at input. The second phase is acquisition during which associations are formed between stimuli, and information is acquired and encoded into secondary memory. The third phase is storage, and the fourth is retrieval, which involves the recall of information learned previously either from the recent or remote past.

While many of the early researchers found that cholinergic blockade impaired performance of avoidance and discrimination tasks,[16,17] others found no impairment after scopolamine.[18,19] The literature on cholinomimetics (directly acting agonists or acetylcholinesterase inhibitors) is actually more variable, and includes reports of improvement, no improvement, and even impairment in performance. These inconsistencies have continued from the 1960s to the present day. Part of the discrepancy has been attributed to methodological factors such as differences in baseline performance, task complexity, and drug dosages. In general, anticholinergics produce more disruption in animals after partial rather than extended training, in tasks which are more complex, and at higher dosages.[20]

The negative findings on cholinergic drugs do present a problem, but there has also been controversy over whether all of the positive findings reflect disturbances in genuine memory processes. Some findings may reflect nonassociative or performance effects caused by interference with motivation, motor behavior, and the peripheral nervous system. Anticholinergics have been reported to increase the number of incomplete responses[21] as well as run times[22] on a radial maze, and to cause mydriasis and dry mouth effects.[23] Each of these can confound interpretation.[3] Dry mouth effects, for example, can adversely affect procedures that use dry food reward as motivation. (Obviously the latter would present a problem for the interpretation that performance deficits seen in such a procedure were due to forgetting a learned response.) To the extent that performance deficits reflect interference with the peripheral nervous system (and many do), confidence in the results can be increased by appropriate controls to rule out peripheral confounds at least. Researchers have not always used quarternary methylated forms of scopolamine and atropine, which do not readily cross the blood brain barrier, as controls for peripheral effects.

All this notwithstanding, it is that cholinergic agents may exert their effects on attentional mechanisms which has provided the most competition for a memory hypothesis.

2. General Neuroscience and Pharmacology
a. Amnesia, the Hippocampus, and the Source of Cholinergic Innervation

The amnesia caused by neurosurgical lesions stimulated neuropsychological research on the functional roles in learning and memory of particular brain structures, areas, and chemicals. This work has extended into the fields of general neuroscience and neuroanatomy. The first and most influential case of global amnesia was that of H. M., a motor-winder by trade, who suffered from progressive generalized epileptic seizures. In Montreal in 1953, Scoville performed a bilateral resection of the medial temporal lobes in an attempt to control the severe seizure activity. The areas affected by the lesions were the uncus, amygdala, hippocampal gyrus, and much of the anterior hippocampus. The surgery left H. M. with a permanent disability to remember recent events. Through their observations of surgical lesions, Scoville and Milner[24] believed that the critical area for this deficit in recent memory was the hippocampus. This led to the hippocampal or mediotemporal hypothesis of memory.

H. M.'s amnesia has been much studied since its initial description in 1957.[25,26] Two salient features were the profound disturbance in learning new material and a retrograde amnesia for events and features involving a period of 1 to 3 years before surgery. Memory of events during the years prior to this period was normal. The latter findings suggest that the medial temporal lobes and hippocampi are not critical for retrieval processes because an impairment in retrieval should disrupt all past memories including his memory for premorbid events.[27] The former finding may suggest that this region is involved in the formation of new memories; but as Squire[27] points out, there is evidence that the hippocampus is not the actual "site of memory storage."

The hippocampus came to be regarded as critically involved in the formation of new memories, a view not without its own controversy. The controversy began when attempts were made to model human amnesia by lesioning brain areas in animals and particularly in monkeys. Orbach, Milner, and Rasmussen,[28] for example, found no amnesia in monkeys that were given the same lesions to the hippocampus that H. M. had experienced. Horel[29] suggested that the medial temporal lobe with its connecting temporal stem, and not the hippocampus, was the critical area for the amnestic syndrome, while Mishkin[30] argued that it is the combined hippocampal-amygdala lesion that is critical. See References 27 and 29 to 33 for diverging views on memory and hippocampal, amygdaloid, and temporal lobe lesions.)

Let us now turn to the cholinergic system. Some of the early investigators found similarities between the behavioral deficits caused by hippocampal lesions and those resulting from antimuscarinic drugs.[34,35] In 1967, a major cholinergic innervation to the hippocampus from the medial septum and the diagonal band of Broca (dbB) was identified.[36] A relationship was thereby established between the cholinergic system and the hippocampus. Subsequent research focused on the expectation that anticholinergic drugs would produce similar effects to those caused by hippocampal lesions. As Bartus and colleagues[37] report, the relationship between the hippocampus and cholinergic neurons was firmly established 10 years later with the discovery of rich areas of muscarinic receptors in the hippocampus proper.[38,39]

It is now known that the source of cholinergic neurons is located in the gray matter of the basal forebrain (substantia innominata) and specifically within cell bodies in the medial septal nuclei, the dbB, and the nucleus basalis of Meynert (nbM). In addition to the hippocampal innervation identified earlier,[36] a source in the nbM which projects diffusely to the cortex was subsequently identified.[40] (See Chapter 1 in this volume for a description of the distribution of cholinergic perikarya.) While most of the behavioral research on animals has been focused on the projection to the hippocampus, there has recently been substantial interest in the nbM.

Lesion and electrophysiological studies of the hippocampus or associated structures led to the development of two influential theories of hippocampal function. O'Keefe and Nadel[41] maintained that the hippocampus was selectively involved in locale or spatial, but not cue, learning. Olton and Samuelson[42] had demonstrated that rats with fimbria-fornix lesions to the hippocampus were markedly impaired in relearning postoperatively a win-shift strategy on a radial maze. Olton and Papas[43] subsequently reported that the lesioned rats could relearn to avoid the arms of a radial maze that consistently never contained a food reward, a task associated with reference memory. The rats, however, were inaccurate in choosing correctly in a single trial among the remaining arms, each of which was baited with a single food pellet, and Olton and Papas argued that the hippocampus was selectively involved in working memory.

3. Pharmacological Research
a. Rodents

Among the pioneers in the experimental investigation of the cholinergic system and memory were Carlton[44] and Deutsch et al.,[45-47] who studied the effects of injections of antagonists or acetylcholinesterase (AChE) inhibitors on learning tasks in rodents. Carlton proposed that cholinergic neurons were involved in response inhibition or the suppression of behavior that leads to nonreinforcement. Carlton thus thought the function of the cholinergic system was to decrease the likelihood of unrewarded responses (and increase the likelihood of rewarded responses). While Carlton discussed an extinction or habituation process, actually a type of learning, Deutsch claimed a more direct role for the cholinergic system in memory.

Deutsch and his colleagues[46] trained rats to a high criterion of discrimination between the two arms of a Y maze. The rats learned to avoid shock by selecting the arm that was brightly lit. Once the criterion was met (10 out of 10 correct responses), the experimenters then injected the rat with the AChE inhibitors physostigmine or diisopropyl fluorophosphate (DFP) dissolved in peanut oil directly into the hippocampi. (There was a constant 1-day interval between the injection and the retention test.) The rat's retention of the task was examined at varying delays of 30 min to 28 days from original learning, and the rat was retrained to the criterion. Over the course of a series of experiments, it was found that DFP impaired, had no effect, or enhanced retention depending on the time from original learning. (For a review, see Deutsch.[48]) When controls were tested at the different intervals, they easily reacquired the task up to a retention period of 14 days, after which time they took longer and longer to reach criterion. At 28 days posttraining, the control rats appeared to have forgotten the task, while the rats treated with DFP reached criterion in substantially less time.

Deutsch maintained that if a habit was well remembered, enhanced cholinergic neurotransmission would impair recall of the habit, but if the habit was not well remembered, enhanced cholinergic transmission would facilitate recall of the habit. Deutsch[45] proposed that the cholinergic synapse was modified because of learning, and that "the postsynaptic membrane probably becomes increasingly more sensitive to acetylcholine with time after learning, up to a certain point," after which, "sensitivity declines, leading to the phenomenon of forgetting." In other words, there is a definitive level of ACh transmission for the execution of a task learned previously.

The early researchers like Deutsch studied the effects of pharmacological agents on avoidance and discrimination learning based on classical and instrumental conditioning. Because cholinergic antagonists frequently but not consistently improve active avoidance learning, and blockade of peripheral receptors has a pronounced effect on performance, interpretation of active avoidance tasks can be confusing. In appetitive discrimination learning, the animal learns to differentiate between two or more stimuli, one of which is paired with reinforcement. While cholinergic agents were found to affect discrimination learning,[49,50] there are strong indications that the early findings were due to performance effects.[3] Let us therefore examine a sample of the research on passive avoidance in rodents in which the animal is punished for a response and thus learns to inhibit that behavior. Passive avoidance is frequently learned in one trial, and training is easily implemented.

i. Passive Avoidance

There appears to be general agreement in the early and indeed subsequent literature that anticholinergics given before training disrupt performance in retention tests of passive avoidance.[51-54] Traditionally, treatment is thought to affect consolidation or storage of memory if it is found to alter retest performance when given immediately or soon after training. With respect to passive avoidance, Dilts and Berry[53] and Bohdanecký and Jarvik[51] found no effects when scopolamine (0.1 mg/kg and 0.1 to 10.0 mg/kg, respectively) was administered to mice immediately after training (when consolidation would most likely be taking place) and retention was tested 24 hours later. In contrast, Glick and Zimmerberg[54] found that 10 mg/kg scopolamine injected immediately after training impaired retention if a sufficiently strong shock was given during training.

Obviously dose does not account for the discrepancy because the same 10.0-mg/kg dose used by Glick and Zimmerberg[54] was ineffective in the study by Dilts and Berry.[53] There is a compelling reason to suppose that the results by Glick and Zimmerberg[54] were not due to an effect on memory. It has been shown that scopolamine, but not saline, causes an increased reactivity to foot shock in rats, a phenomenon that appears to be particularly pronounced when the shock is in an aversive range.[55] As retention was impaired in Glick and Zimmerberg's mice in the higher but not weaker shock condition, it is possible that this result was due to scopolamine's effects of increasing reactivity to the more aversive stimulus. It is unclear whether this reactivity phenomenon is peripherally or centrally mediated. Feigley et al.[55] did not include a test under methylscopolamine.

With respect to cholinomimetics, physostigmine administered before training appears to cause impairment in passive avoidance tasks similar to that found under scopolamine.[51,53] Dilts and Berry[53] reported no improvement in avoidance learning following posttraining injections of physostigmine or neostigmine over a range of doses, while agonists arecoline and pilocarpine resulted in erratic effects.

On the basis of these results on passive avoidance learning, it is difficult to argue that scopolamine was affecting memory processes, although one could make a case for effects on performance or attentional mechanisms or effects on early phases of information processing. Note that this also presents a difficulty for Deutsch, whose rats were injected long after training, by which time consolidation should have been completed.

ii. Discrimination, Memory and Attention

One method of separating effects of scopolamine on "storage of stimulus information from possible interference with the initial acquisition or discrimination of that information"[56] is to insert a delay after a task is learned and then retest for the memory of the task. Heise and his colleagues[56] trained rats to alternate lever pressing on trials separated by five intertrial intervals, which ranged from durations of 2.5 to 40 sec. Scopolamine (0.13 to 1.0 mg/kg) impaired the alternation behavior after the respective delays, but the degree of impairment did not appear to vary with the length of the delay. Consequently, Heise et al.[56] concluded that the effects of scopolamine were to disrupt attention rather than memory.

An important influence in the development of an attentional hypothesis came from the work of Warburton and his colleagues in the early 1970s.[57,58] In one study, Warburton and Brown[57] trained rats to asymptotic performance on a visual discrimination task. They used the theory of signal detection to investigate whether scopolamine would affect the ability to discriminate, referred to as stimulus sensitivity, or whether it might reflect response bias through interference with response factors such as by increasing perseverative behavior. The results showed no effect on response bias, but scopolamine (0.63 to 0.25 mg/kg) appeared to decrease stimulus sensitivity in a dose-dependent manner. (That scopolamine did not interfere with response behavior is in contradiction to the response disinhibition predictions of Carlton.) Instead of attributing their results to effects on memory (as Deutsch proposed), however, Warburton and Brown suggested that cholinergic neurons were involved in attentional mechanisms during the stimulus input phase of information processing.

With respect to Deutsch, there was some independent support in the 1970s for his hypothesis that ACh is intrinsically involved in the formation of memories.[59-61] Stanes et al.,[61] for example, injected physostigmine (0.05 to 0.20 mg/kg) 30 min before the retention test of a Y-maze discrimination task that was appetitively motivated. They replicated Deutsch's results in showing that an anticholinesterase would impair well-remembered habits, but enhance poorly remembered ones. Deutsch was essentially discussing the retrieval of a memory trace, but he administered the drug 24 hours before the retention test, and so it is difficult to make inferences about retrieval. As Sahakian[5] has pointed out, because Stanes et al. gave the injection only 30 min before the test, one can more comfortably infer that the drug was affecting retrieval processes.

There have been negative findings, however. A troublesome problem has been identified with the method of hippocampal injection used by Deutsch. It has been shown that the amnesia seen in controls, which had previously been attributed to normal forgetting over time, may actually have been due to hippocampal lesions caused by the peanut oil vehicle.[62] This finding has undermined Deutsch's conclusions. In addition, some researchers[63] have not been able to replicate the results of Deutsch and his colleagues.

In an excellent review of the literature on animal learning, Hagan and Morris[3] note that while some of the earliest researchers such as Burešová and her colleagues acknowledged that anticholinergics appeared to disrupt learning, they thought it unlikely that information storage was affected. It is noteworthy that Burešová et al.[52] reported that atropine (6 mg/kg) injected after training did impair performance on a passive avoidance task in rats, but had no effect when the rats were overtrained on the task. Meyers,[19] who found impairment of passive avoidance with posttraining injections of scopolamine (0.1 to 8 mg/kg), attributed his findings to a deficit in response inhibition.

The early research set the stage for what was to follow in terms of a lack of consensus over the involvement of ACh in learning and memory. Themes that continue to appear are impairment of attentional vs. consolidation mechanisms, interference from performance effects, and the importance of baseline performance or level of training in subsequent test results.

b. Nonhuman Primates

As mentioned previously, research with monkeys began with the effort to model human amnesia through lesions of particular brain areas.[28-30,64]

Much of the early, and indeed subsequent, psychopharmacological research in monkeys has used a delayed matching-to-sample (DMS) task in which the subject is trained to match a particular stimulus such as a green or red light with the identical *matching* stimulus presented with one or more distractors.

The presentation of the sample and match stimuli is separated by either a short or longer interval. Impairment after the delay is thought to reflect impairment in memory. While early researchers like Bohdanecký et al.[65] and Glick and Jarvik[66] had found impairment in accuracy after injections of scopolamine, rates of responding and visual discrimination ability were also disrupted, confounding interpretation of the results.[3]

Evans[67] subsequently showed that performance on a form discrimination task in monkeys under scopolamine (7.5 to 60 µg/kg) varied as a function of stimulus luminance. When stimuli with low illumination were used, performance deteriorated to chance levels with 30 µg/kg of scopolamine. When stimuli with high luminance were used, scopolamine had no effects on performance accuracy except at the highest dose (60 µg/kg), where the effect was only minimal. These findings are consistent with the earlier suggestion of Warburton and Brown[57,58] in their studies with rats. It would appear that scopolamine can disrupt performance on discrimination tasks by reducing stimulus sensitivity.

Bartus and Johnson[68] consequently developed a nine-choice delayed response task that appeared less reliant on visual form discrimination. They found an interaction beween delay interval and treatment condition (scopolamine or saline) in young monkeys given intramuscular (i.m.) injections of scopolamine (0.01 to 0.015 or 0.02 to 0.03 mg/kg). Impairment was only slight at short retention intervals, and the magnitude of the deficit increased as the duration of the retention interval increased for the scopolamine-treated animals. Performance was also related to dose, as the highest dose produced greatest impairment. Bartus[69] subsequently reported that physostigmine (0.02 to 0.03 mg/kg i.m.) partially reversed the deficit produced by scopolamine, while methylphenidate (0.013 mg/kg), a catecholaminergic agonist and central nervous system stimulant, potentiated it.

Bartus et al.[70,71] essentially reproduced the same pattern of impairment in aged monkeys compared to young controls, as Bartus and Johnson[68] had found in scopolamine- compared to saline-treated animals. Although unimpaired in performance of the delayed response task at short delay intervals (up to about 5 sec), aged monkeys (18 years and older) showed marked impairment as the retention interval increased (up to 30 sec) in comparison to young controls.[70,71] The deficit became pronounced by the time the delay interval had reached about 15 sec, and appeared unrelated to the duration of stimulus exposure.[70] It may have been related to an increased vulnerability to interference as irrelevant visual distractors presented during the delay intervals seemed to disrupt performance of the aged, but not young, monkeys.[72,73]

The deficit in the aged monkeys did not appear to be due to problems with perceptual sensory processing because visual discrimination learning, or the ability to acquire new associations based on color and pattern discrimination, was unaffected.[72,73] Given this finding and that performance seemed unaffected by stimulus exposure[70] and the task itself is not so dependent on visual discrimination ability, it seems unlikely that Bartus' and Johnson's results were due to interference with sensory processing. The deficits in the scopolamine-treated and aged monkeys might well have been due to an inability to accurately remember the location of the stimulus at the longer retention intervals. The evidenced interaction between delay interval and treatment conditions is certainly suggestive of memory impairment. Further, this important finding has been replicated.

The research by Bartus and colleagues showing similarities in learning and memory impairment in old monkeys and young monkeys treated with scopolamine was prompted by an influential 1974 study on humans.

c. Humans

The 1970s were a particularly fertile period in the fields of experimental psychology, psychopharmacology, and general neuroscience. A major impetus advancing the concept of a key role for the central cholinergic system in memory came from the work of David Drachman and his colleagues. Drachman and Janet Leavitt[74] reported similarities between the memory deficits seen when healthy young humans were administered anticholinergics and the senescent forgetfulness seen in elderly people. In their 1974 study, three groups of young adult students were given one of three drugs subcutaneously (s.c.), 1 mg scopolamine, 1 mg methylscopolamine, or 1 or 2 mg physostigmine. One hour after injection, subjects were given a battery of tests, including standard digit span, thought to measure immediate memory and attention span. Acquisition or storage was measured by supraspan digit span (consisting of three sequences of 15 digits) and by free recall of two lists of 35 words, which, like digit supraspan, was tested in training-test trials. Retrieval was measured by the number of words generated as exemplars (e.g., dog) of familiar categories (animals) in 20 or 60 sec. Subjects were also given the Wecshler Adult Intelligence Scale (WAIS).

Compared to young controls recruited from a previous study, the subjects who received scopolamine showed preserved immediate memory, marked impairment in the supraspan digit and free recall tests, milder but significant impairment in the category naming tests, and impairment of performance, but not verbal, IQ (intelligence quotient). Subjects receiving methylscopolamine or physostigmine did not differ from the controls. These findings supported earlier studies on humans administered scopolamine, which had shown preserved digit span,[75] disruption of the *primacy* (secondary memory), but not *recency* (primary memory), component of a serial position curve, and impaired free recall of a word list and of paired (number-color) associates.[76] (Recall for the terminal items of a word list are recency effects thought to reflect primary memory capacity. Recall for words from the beginning of the list are primacy effects assumed to reflect secondary memory.) Thus, immediate and primary memory were unaffected but secondary memory was disrupted by scopolamine.

Drachman and Leavitt[74] then compared the performance of the scopolamine subjects to that of aged persons from an earlier study,[77] and found that the patterns of performance were similar. The only significant difference between the scores of the scopolamine and aged subjects was for supraspan digits on which the scores of the aged were lower. Drachman and Leavitt concluded that the central cholinergic system was "crucial to the storage of new information," and suggested that impairment of this system might underlie the memory deficits in human aging. Note that their conclusion was based on the impairment found in the supraspan digit and free recall tests. This impairment might actually have reflected effects on earlier acquisitional, not storage, mechanisms, as no attempts were made to separate these hypothetical processes.

Further support for a cholinergic memory hypothesis came from the finding that physostigmine, but not amphetamine (a catecholamine agonist that potentiates the availability of dopamine and norepinephrine and increases the level of arousal), reversed the memory impairment caused by scopolamine in young adults.[78] Drachman[78] believed that the results were due to the effects of scopolamine on memory processes and not on arousal and attention. Electoencephalographic measures in earlier studies[75,79,80] had suggested that scopolamine decreased arousal responses, an effect that appeared independent of cognitive and memory impairment.[79,80] Scopolamine had been shown to produce subjective feelings of drowsiness[75] and a dose-dependent impairment in tests of alertness and selective attention.[81]

Crow and his colleagues[82] in Great Britain provided support for the argument against an arousal hypothesis by showing that scopolamine impaired recall of a word list without affecting performance of a scanning or vigilance test, while a barbiturate, amylobarbitone, and a major tranquilizer, chlorpromazine, impaired scanning, but not memory, of the word list. The effects of scopolamine seemed to be on memory, and those of the barbiturate and tranquilizer, on arousal or attentional mechanisms. The issue is not resolved, however, as diazapam, a benzodiazepine with sedative properties similar to those of barbiturates, has been shown to affect performance on tests of learning and memory in a similar manner to scopolamine.[83] It is noteworthy that one of the first anticholinergic studies on humans[75] reported that subjects had difficulty maintaining attention (during a 1-min task that required pressing a buzzer), as one current hypothesis holds that a major effect of anticholinergics is to impair attention.

4. Neurochemistry of Alzheimer Disease

Senile dementia of the Alzheimer type (SDAT) (DAT in individuals under 65 years of age) is a progressive disease of the brain that results in a profound deterioration of cognitive and memory functions. A deficit in memory for recent events or in new learning ability is one of the most reliable behavioral symptoms and is seen even in early stages of the illness. The neuropathological features of Alzheimer disease (AD), determined only by histological examination at autopsy, are neuritic plaques, neurofibrillary tangles, and granulovacuolar degeneration. In 1968, Blessed et al.,[84] found that the degree of premortem cognitive impairment correlated with the number of senile plaques in persons who had died with AD and, to a lesser extent, in aged persons who died of other causes.

Davies[85] notes that the first study in the field of general neuroscience establishing a link between cholinergic neurons and Alzheimer disease likely appeared in 1965, when Pope and colleagues[249] observed substantial reductions in acetylcholinesterase (AChE) activity in the cerebral cortex of two people who had AD when they died. This finding did not attract much interest probably because AChE is not unique to the cholinergic system. As Davies[85] reports, correlational studies linking AD to choline acetyltransferase (ChAT), the enzyme that synthesizes ACh, did not appear until more than a decade later.

During the mid 1960s to 1970s, progress had been made on the methodology of studying the neurochemistry of human brains removed at autopsy, a development that paved the way for the research that followed. The impetus for the study of the neurochemistry of AD was undoubtedly influenced by the behavioral research on humans and animals that appeared during this period, as reported above. In 1976 and 1977, findings from three independent laboratories[86-88] showed that the activity of ChAT, now known to remain stable in the brain many hours after death, was markedly reduced in the cerebral cortex and hippocampi of persons who had died with AD compared to age-matched controls who died of other causes. A decade after their first correlational study, Blessed and Tomlinson teamed up with the Perrys and others[89] and found correlations between ChAT activity and global ratings of cognitive measures.

The synthetic enzyme ChAT is considered a reliable marker of cholinergic neurons.[90] Initially, findings of reductions in synthesis of ACh among persons with clinical symptoms of dementia appeared to be specific to AD. David Bowen and his colleagues[91] examined ACh synthesis in biopsy samples of brain tissue taken from neurological patients prior to surgery. In patients with clinical symptoms of dementia, ACh synthesis was reduced only in those patients whose samples showed the plaques and tangles characteristic of Alzheimer disease.[91] What this research also suggested, however, was that symptoms of dementia and cognitive impairment could result from disruption in a neurotransmitter other than ACh.

5. Synopsis

The researchers who proposed the cholinergic hypothesis have stimulated volumes of research in the field and interest in the cholinergic system and memory that continues to the present day. The ultimate worth of a hypothesis lies in its heuristic value and not in whether it "is in fact correct."[2] It can thus be said that the cholinergic hypothesis has been immensely successful. We will examine the status of its correctness in our review of current research. The 1970s ended appropriately enough with Drachman and Sahakian[92] suggesting that memory consolidation might involve increased sensitivity in particular synapses and decreased sensitivity in others, rather than "the absolute ACh-sensitivity in a memory synapse" as proposed by Deutsch.*

III. CURRENT RESEARCH

A. ALZHEIMER DISEASE AND DEMENTIA
1. Neurochemistry and Neuropathology

The recent literature on correlations between memory test scores and senile plaques in persons with Alzheimer disease is consonant with the early quantitative study of Blessed et al.,[84] but the findings on choline acetyltransferase (ChAT) activity are contradictory.[93] Neary et al.,[94] for example, reported correlations between plaques as well as tangles and WAIS verbal IQ, between tangles and performance on the token test, and between ACh synthesis and clinical ratings as well as visual reaction time. No significant correlations were found between ACh synthesis and scores on the WAIS or token test, however. Moreover, Neary et al. did not find significant correlations between ChAT activity and any measure of behavioral function. The latter findings are in contrast to Perry et al.[89] who, as reported earlier, found correlations between ChAT activity and global ratings of cognitive measures. Differences in assay procedures, in biopsy of the temporal lobe[94] vs. postmortem assay of the neocortex,[89] or in times of psychological testing relative to cortical assay might account for the discrepancy. See Collerton[93] for a synopsis of correlational findings among various neuropathological indices as well as between these and cognitive measures. He also comments on the difficulties in ascertaining the functional significance of the various correlations reported.

Mann[95] notes that extensive loss of nerve cells from the hippocampus (in persons with AD) had been documented in the 1970s,[96] followed by a report in 1980 of nerve cell loss in the amygdala among other areas. In the early 1980s, postmortem histological examination of brain tissue from a patient with AD showed pronounced degeneration of neurons in the nucleus basalis of Meynert (nbM)[97] and in the septum.[98] It was also noted around this time that Down syndrome, associated with dementia, like AD was also associated with loss of nerve cells within the hippocampus, temporal cortex,[99] and the nbM,[100] as well as with reduction of ChAT activity in the basal forebrain and cortex.[95]

The discovery that the nbM was the source of cholinergic innervation to the cortex prompted many researchers in the early 1980s to hypothesize that cholinergic neurons of the nbM were critically involved

---

* According to Drachman and Sahakian, there is a select level of ACh transmission yielding "optimal signal-trace detectability," and cholinergic drugs are thought to affect this "signal-to-noise ratio."[5]

in cognitive and memory processes. It was shown that degeneration of neurons in the nbM was not specific to AD and occurred in other neurological disorders including Down syndrome, Korsakoff syndrome, Parkinson disease (PD), and Pick's disease. Unlike AD where symptoms of dementia are always seen, however, dementia is not predictable in these other disorders. A natural hypothesis followed that damage to the nbM was critical in those persons with evidence of cognitive impairment, a hypothesis that received subsequent support. For example, loss of neurons in the dbB and nbM had also been implicated in PD (characterized by degeneration of cells in the substantia nigra and loss of dopaminergic neurons passing from the substantia nigra to the striatum). Whitehouse and his colleagues[250] found that what set apart persons with PD who displayed loss of cognitive function from those who did not was damage to the nbM.

It is becoming increasingly apparent that reduced ChAT activity, at least in the neocortex, does not necessarily cause dementia. The former is exemplified in a recent study on people with olivoponto-cerebellar atrophy (OPCA). While these patients showed reductions in ChAT activity in the neocortex comparable to AD rates, they did not present with the dementia syndrome of AD.[101] Bartus and his colleagues[102] similarly document a small number of studies in which healthy elderly subjects with no symptoms of dementia showed reductions in ChAT activity relative to young controls. (The majority of healthy elderly do not show reduced ChAT activity.) The results of animal studies more or less parallel the findings on humans. A few studies have noted small reductions and most report of no reductions in ChAT activity in aged rodents when compared to younger animals.[102]

The literature may be indicating that cholinergic neurons in the nucleus basalis of Meynert are not critical for recent memory. That ChAT activity was not reduced in the hippocampus in the OPCA patients (unlike what is reliably seen in AD) leaves open the possibility that impairment of the cholinergic septo-hippocampal pathway may well underlie some of the memory deficits in Alzheimer disease.[103-105]

Neuronal degeneration is not restricted to cholinergic structures, as a marked reduction of neurons from the locus ceruleus, the source of noradrenergic input to the neocortex, has been noted in AD[106-107] and to a much lesser extent in the normal elderly.[108] Neurons in the ventral tegmentum (a dopaminergic nucleus that projects to the cerebral cortex and amygdala), and to a limited extent in the serotonergic raphé system, have also been found to be reduced in AD.[95] Studies on biochemical markers have shown reduced concentrations of norepinephrine and of serotonin;[109] have suggested the possibility of major deficiencies with putative amino acid neurotransmitters, glutamate, aspartate, and γ–aminobutyric acid (GABA);[95,104] and have shown mixed findings on neuromodulator peptides.[85,110,111] See Mann[95] for a comprehensive discussion of neurochemical and neuropathological aspects of AD.

2. Cholinergic Therapy

The findings on cholinergic receptors have not been consistent, and reductions in density in the cerebral cortex may be a more reliable finding for nicotinic receptors than for muscarinic receptors.[112] That there may be a relative preservation of postsynaptic muscarinic receptors in many patients with AD[95] would seemingly afford the possibility of successful therapeutic intervention with cholinomimetic agents. The optimism has not been borne out in the clinical trials, however, and there are a number of inherent problems with the agents themselves, such as short durations of action and narrow range of effective doses.

The administration of cholinergic precursors such as choline or lecithin has been ineffective.[102] Muscarinic agonists or AChE inhibitors have to date not been very successful in alleviating the memory deficits of SDAT, and any improvement seen has been modest and transitory. This may be in part because increasing muscarinic stimulation appears to result in "downregulation of muscarinic receptors."[113] See References 1, 4, 5, 13, 102, 114, and 115 for a review on cholinergic therapy.

3. Cholinergic Neurons, Memory, and Alzheimer Disease

It is incumbent on proponents of the cholinergic hypothesis of Alzheimer disease to demonstrate that blockade of cholinergic neurons in healthy humans results in deficits similar to those seen in DAT. Such endeavors have yielded only partial success. Recent investigators have maintained that scopolamine administration does not provide an adequate model for either primary memory (PM) or secondary memory deficits in persons with dementia of the Alzheimer type.

a. Primary Memory

For the present discussion, it may be assumed that there are essentially two measures of primary memory. The first is a capacity measure of the immediate span of attention typically assessed by digit and word span tests. This is referred to as immediate memory or what Lishman[116] might call "ultra-short-term memory." The second is a measure of short-term forgetting or the loss of memories from the primary store, which is frequently assessed by the Brown-Peterson short-term memory (STM) test (Kopelman).[117] The Brown-Peterson test measures retention for materials over short periods of up to 30 sec during which a distractor test is performed, and therefore it is a more active measure of PM than span tests.

The studies on humans described earlier reported that initial registration and immediate memory as measured by digit and word span were unaffected by scopolamine.[2,4,74,75,125] While most studies report that both immediate memory and primary memory are impaired in DAT,[4] a small number of investigators have reported digit span[118-120] and word span[121] to be relatively unaffected in earlier stages of the disease. Corkin,[118] for example, found that digit span (an index of both immediate memory and attention) was preserved in patients with mild DAT, but not in those who were moderately or severely impaired. Similarly, with serial position studies, a relatively normal recency (PM) component has been found in patients classified as less severe.[122]

In unpublished observations, the author has found primary memory (as measured by the Tulving and Colotla[123] procedure) to be preserved in subjects with mild SDAT (aged 76 to 94) compared to age and education matched controls. The types of errors made in free recall and cued recall tests suggested that while the SDAT subjects made more errors than controls in free recall, they did not make more of one type of error than another. This was not the case for cued recall, where the SDAT subjects made more extra list (that were not on previous lists), but not prior item (repetitions of words from previous lists) intrusions than the controls. The SDAT group, therefore, showed evidence of proactive interference in the free recall test indicating that previous information was initially processed. These findings are in contrast to those of Wilson et al.,[124] who suggested that persons with DAT might have an "initial processing deficiency." Differences in age as well as in stage of illness may account for the discrepancy. The subjects of Wilson et al.[124] were 15 years younger than our subjects and may have been in a more advanced stage of illness.

It has been reported that performance on the Brown-Peterson STM test is impaired in persons with SDAT, in contrast to what is typically seen in persons under scopolamine.[4] Kopelman and Corn[125] found no impairment on this test in subjects who had received intravenous (i.v.) doses of 0.2 and 0.4 mg of scopolamine. Yet, Caine et al.,[126] Beatty et al.,[127] and Tröster et al.,[128] who used higher doses, did report that scopolamine impaired performance on this test. Thus the literature is inconsistent with respect to scopolamine and the Brown-Peterson test, but at least three studies have reported impairment with higher dosages.

Performance on the Brown-Peterson test in DAT has been found to correlate with the severity of dementia as determined by clinical ratings.[118] In addition, Kopelman[129] found that younger Alzheimer patients performed more poorly on the test than older patients. If it is shown unequivocally that scopolamine disrupts PM, the scopolamine model for the PM deficit in SDAT may have relevance for older patients in the early stages of SDAT. Thus, scopolamine subjects and older patients would both show a mild to moderate deficit in PM (with subjects given smaller doses and patients in the earliest stages showing no impairment). This is in keeping with Kopelman's findings on the higher test performance of the older Alzheimer patients and the relative preservation of PM in our older subjects. As described earlier, most studies suggest that immediate memory is impaired in DAT,[2,4,117,124,129,130] but unaffected after scopolamine. Whether or not immediate memory is unimpaired in SDAT patients at an early stage of illness requires further exploration.

Alternatively, the scopolamine model may have some relevance for SDAT when the processing requirements are more demanding or require more effort to perform.[125] Kopelman and Corn[125] found that scopolamine impaired a visuospatial STM test and a verbal test when a distractor condition made performance of the test more difficult. This is consonant with what is seen in SDAT, and also in Korsakoff syndrome (which is a relatively pure amnestic disorder that develops from Wernicke encephalopathy). Kopelman and Corn's suggestion that the scopolamine model may be an appropriate one for PM impairment in SDAT, when the processing requirements are increased, is compatible with Rusted and Warburton's[131] findings of scopolamine-induced deficits in younger subjects on STM tests in which the processing load was heavier, such as when the subject was required to remember the location of hidden shapes.

b. Secondary Memory

There appears to be general agreement that in measures of recent secondary memory, the scopolamine model does not mirror the memory impairment of SDAT. For example, Collerton[93] notes that when the

WAIS performance of Dat patients was compared to that of scopolamine-treated subjects, performance was similar for only about half of the DAT patients tested. Yet the verbal IQ does appear to be preserved in many persons in the earlier stages of DAT,[132,133] as was found to be the case in both the elderly and the scopolamine-treated young subjects of Drachman and Leavitt.[74] Performance on the WAIS subtests and in particular on the verbal IQ scores may vary as a function of the stage of illness.

Beatty et al.[127] compared the performance of persons with DAT to age and education matched controls treated with scopolamine (0.5 mg i.m.) or the peripherally acting glycopyrrolate (0.1 to 2.0 mg) on a battery of memory tests. Scopolamine, but not glycopyrrolate, impaired the learning and delayed recall of a word list and performance on the Brown-Peterson test (in contrast to the subjects of Kopelman and Corn[125] who had received 0.2 and 0.4 mg of scopolamine as reported above). Unlike the DAT patients, however, who made many prior item or extra list intrusions on these tests, scopolamine did not increase either type of error in the healthy subjects. Surprisingly, scopolamine did not disrupt the learning of a symbol-digit task. This also conflicts with Crow and Grove-White,[76] who found impairment on this paired associate task after a lower dose of scopolamine (0.3 mg).

People with SDAT, even in earlier stages of the disease, have shown pronounced impairment in recalling past public events and faces of famous people,[134] and moderate impairment in locating cities and geographical features on maps.[135] In contrast, scopolamine (0.5 and 0.8 mg) has had no effect on these same measures in healthy middle-aged humans.[128] Tröster et al.[128] found dose-dependent impairment of scopolamine on both immediate and delayed recall of lists of words, on acquisition of new geographical and spatial information, and as above, on the Brown-Peterson test. Consonant with the findings of Beatty et al.,[127] performance on these tests was not accompanied by increases in intrusion errors.

The subjects of Tröster et al.[128] were impaired in a verbal fluency test at the 0.8-mg dose when required to generate exemplars to a broad category (e.g., animals), but not in producing words beginnning with a particular letter. These findings are consonant with those of Caine et al.[126] and also, if you remember, with the results for category fluency reported by Drachman and Leavitt.[74] (Caine et al.[126] found impairment after 0.8 mg scopolamine in a letter fluency test, but only in the second minute of the test, which suggests the effects may have been due to fatigue and not to a problem in memory retrieval.) Recent studies have reported actual improvement in generating words to letter cues after scopolamine, 0.4 mg (s.c.)[136] and 0.6 or 1.2 mg oral dose (o.d.),[137] respectively, relative to saline.

Persons with SDAT, on the other hand, are quite impaired in word fluency tests in which they are required to generate words that are either semantic exemplars or which begin with a particular letter.[120,138] When intrusion errors are examined, the pattern of impairment seems to differ for scopolamine-treated subjects and SDAT patients. The patients make errors of intrusion typically not seen after scopolamine treatment.[128] Neither Drachman and Leavitt[74] nor Caine et al.[126] reported intrusions, but others have not found increases in intrusions under scopolamine.[139]

In summary, there appears to be increasing consensus that the scopolamine model does not adequately approximate the full range of recent or secondary memory deficits seen in dementia of the Alzheimer type.[4,117,125,127,128] As to retrieval from remote verbal memory, Tröster et al.[128] suggested that scopolamine fails "on measures of remote memory that do not emphasize sustained and rapid responding." Studies on diazepam warrant examination as similarities in memory impairment have been reported between elderly people administered diazepam and DAT patients.[140]

B. CHOLINERGIC NEURONS AND ANIMAL RESEARCH
1. Pharmacological Studies

Implicit to the assertion that abnormalities of the cholinergic system account for the memory deficits of AD is the contention that the physiological basis of memory, or at least recent memory, is contained within the cholinergic system. There are obvious restrictions when testing cholinergic hypotheses on human subjects, although researchers have blocked muscarinic receptors as reviewed previously. Blocking cholinergic receptors is not always innocuous. Tröster et al.[128] reported that one of their subjects experienced toxic delirium under 0.5 mg scopolamine. While animal models do not provide appropriate analogs for human memory, they do permit manipulations not possible in human subjects, and the results are suggestive.

In the literature, working memory (WM) is often used to describe all recent memory or new learning, although theoretically WM does not include learning that subsequently becomes a stable base of knowledge. It is common in the animal research to refer to this relatively stable knowledge base acquired through learning discriminations as reference memory (RM). According to Honig,[6] WM is a type of recent memory for events that change from trial to trial, while RM is a type of memory that remains relatively stable from trial to trial. Because WM involves choice behaviors dependent on choices made in previous trials (within a test session) and because RM involves choice behaviors independent of previous choices, the terms *trial-dependent* and *trial-independent* are often used to describe tasks based on WM and RM, respectively.

a. Nonhuman Primates

Bartus and Johnson's 1976 finding of an interaction between (scopolamine vs. saline) treatment and retention intervals was subsequently replicated in DMS[141] as well as in delayed nonmatching-to-sample[142] procedures. In the delayed nonmatching-to-sample task, objects unique to each trial are used, and the subject is rewarded for choosing the novel object. These interactions do suggest impairment of memory processes.

Ridley et al.[143] were able to determine which phase of the learning of an object discrimination task was affected by scopolamine. Marmosets were treated immediately before and after training, and immediately before retention testing 24 hours later. Scopolamine (0.06 mg/kg i.m.) impaired performance when given immediately before training and before the retention test, but not when given immediately after learning. The effects did not appear due to state dependency. Ridley et al. also found some evidence of retrieval impairment, as scopolamine administered before retention testing was disruptive whether or not the marmosets were under scopolamine when they acquired the discrimination. This effect, however, did not hold for well-trained animals, which maintained their training (90% correct) performance level. That the overtrained animals performed so well under scopolamine would seem to rule out interference from perceptual and motor mechansims. Ridley et al.[143] concluded that scopolamine disrupts acquisition and may have a mild effect on retrieval, but does not affect storage. This conclusion contradicts that of Drachman and Leavitt,[74] who had postulated effects on storage, but unlike Ridley and colleagues, did not actually vary the times of scopolamine administration.

Dean and Bartus[144] noted that impairment of spatial delayed response tasks was not found in adult monkeys after drug treatment that affected either dopaminergic or serotonergic neurotransmitters.[145] Furthermore, consonant with Drachman's[78] finding on young humans, physostigmine[69] and the nootropic drug aniracetam[146] attenuated some of the impairment caused by scopolamine on both delayed response[69] and DMS tasks.[146] Also in keeping with Drachman's[78] finding for amphetamine, Dean and Bartus[144] reported that methylphenidate, a CNS stimulant of catecholaminergic neurons, was not found to improve performance.

Two additional findings on delayed responding reported by Dean and Bartus[144] are noteworthy. First, amitriptyline, a tricyclic antidepressant with strong anticholinergic properties, but not desipramine with only weak anticholinergic properties, impaired performance in young monkeys. That much higher doses of desipramine (16 times that of amitriptyline) were used indirectly supports the possibility that performance impairment was due to the actions of amitriptyline on the cholinergic system. The tricyclics potentiate the availability of adrenergic and serotonergic neurotransmitters, however, which precludes ruling out involvement of these other systems. The same (delayed response) task was also markedly impaired after tetrahydrocannabinol (THC),[145] the psychoactive agent in marijuana. Whether the effects of THC were due to cholinergic mechanisms such as reduced ACh synthesis[147] or to other neurotransmitters[148] awaits future research.[144]

Not so easy to attribute to the cholinergic system are the findings that diazepam (a benzodiazepine receptor agonist) impaired performance on delayed response tasks in aged and young monkeys,[145] and in a manner reportedly similar to what is seen in older humans given diazepam.[140,144] Yet some of the effects of benzodiazepines may be due to indirect action on cholinergic neurons through disinhibition of GABA and benzodiazepine receptor complexes. See Sarter et al.[149] for discussion of GABA and benzodiazepine-mediated inhibition of cholinergic neurons.

The tasks used in the above monkey research involve working memory. Most of the lesion research has focused on lesions to the amygdala and/or hippocampus, as mentioned previously, with mixed results. (See References 144 and 150 for reviews of the lesion research in monkeys.) One lesion study relevant herein relates to the finding of Olton and Papas[43] that fimbria-fornix lesions in rats appeared to selectively disrupt the WM component of a radial maze task. Specifically, Zola-Morgan et al.[33] found that combined amygdala-hippocampal lesions in monkeys profoundly disrupted a delayed nonmatching-to-sample task when the WM requirements of the task were increased, but had no effect on a visual discrimination task that required only RM. Santi et al.[151] were unable to replicate this finding in a study of cholinergic blockade on a DMS task in pigeons.

b. Pigeons
Santi et al.[151] adapted the DMS task to examine the effects of scopolamine (0.01, 0.03, 0.05, 0.1 mg/kg) on WM and RM simultaneously. The WM task consisted of a delayed visual discrimination in which the correct response was to select a comparison key that matched the sample presented previously. The RM task consisted of a simultaneous visual discrimination in which the correct response was to choose a horizontal or vertical line irrespective of the previous sample. Even though there was a ceiling effect in the RM component, both RM and WM were disrupted under scopolamine relative to saline conditions, in contrast to the lesion studies of Zola-Morgan et al.[33] and Olton and Papas.[43] Santi et al.[151] also manipulated the delay intervals (and used delays of 1, 2, 4, or 8 sec) between presentation of the sample and match stimuli. Although WM, but not RM, accuracy decreased as the delay interval increased, there was nonetheless a significant effect of drug treatment on RM. As there was no interaction between drug dose and delay interval, the authors concluded that the decreased accuracy on WM trials was not due to interference with the active maintenance of information during the delay (memory).

c. Rodents
There has been substantial interest in maze learning since the early 1980s. Researchers have used radial and water maze as well as T-maze and Y-maze procedures. Individual tasks on these mazes typically involve either WM or RM. Some procedures have been adapted to include both types of memory in one experiment. Radial mazes have been particularly popular in rodent studies.

i. Radial Maze Studies
Radial maze procedures are thought to have high ecological validity for rats because they are similar to the natural food-seeking behavior in this species.[42] A radial maze consists of a central platform with arms fanning outward from the center. Typically, a cup containing a food reward is placed at the end of an arm. The food is not visible when the animal is placed in the central platform. Behavior is thought to be under the control of distal extramaze cues, and the positions of the arms relative to the salient cues are kept constant. Controls are implemented so that rats do not use local intramaze or olfactory cues to find the reward. When each arm of the maze contains a food reward, a win-shift task is required in which the rat uses primarily working memory because the correct response to each arm (enter once and retrieve the reward) changes within a trial. This task is not a pure measure of WM because it does contain components of RM,[152] but it likely reflects a much stronger requirement on working memory.

The literature has been consistent in reporting reductions in choice accuracy following the intraperitoneal injection of scopolamine prior to training.[153-155] Two studies in the early 1980s reported impairment in choice accuracy when scopolamine was administered after training and prior to testing.[21,156] Burešová and Bureš,[157] however, found that the performance of well-trained rats was not disrupted by 0.1 mg/kg scopolamine unless a 5-min delay was inserted between choices, midway through the procedure. The rats used by Eckerman et al.[21] had only 5 to 7 days of training, which points to the level of training as a cause of the inconsistency. The low dose alone used by Burešová and Bureš does not appear responsible for the discrepancy, because Eckerman et al.[21] found the same dose to be effective at impairing performance. A later study of a water maze task by Burešová and her colleagues[158] suggested that level of training and dose may interact. Burešová et al.[158] reported that while acquisition was disrupted by pretrial administration of 0.1 and 0.2 mg/kg scopolamine, a higher dose (1.0 mg/kg) was needed to impair performance in rats that had been previously well trained in the task.

Olton and Papas[43] adapted the radial maze procedure to examine both types of memory in one session. Particular arms of the radial maze are either always or never baited. The correct response to an unbaited arm, to avoid entering it, remains constant and thus reflects reference memory. The dependent measures are types of errors, with first entries into unbaited arms scored as errors in RM, and reentries into the baited arms recorded as errors in WM. While rather unwieldy probability calculations were used earlier to define the respective measures, it has become popular to simply count the errors.

Researchers in the 1980s examined the effects of anticholinergics on radial maze performance in light of Olton and Papas'[43] findings on fimbria-fornix lesions. Rats were trained prior to testing to retrieve food from a baited subset of arms of a multiarm radial maze. The results were contradictory. Rats subsequently tested under scopolamine (i.p. injections) were reported to display selective impairment of WM[159,160] on the one hand, and impairment of both WM and RM[161] on the other hand.

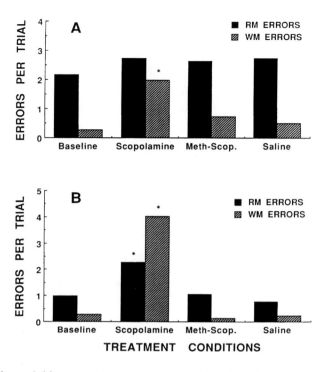

Figure 13-1 (A) Means of two types of errors per trial under each condition in Experiment 1. (B) Means of two types of errors per trial under each condition in Experiment 2. *Significant difference from control conditions. (Reprinted from *Pharmacol. Biochem. Behav.*, Vol. 43, R. Grette Lydon and S. Nakajima, Differential effects of scopolamine on working and reference memory depend upon level of training, p. 647, Copyright 1992, with kind permission from Elsevier Science Ltd, The Boulevard, Langford Lane, Kidlington OX5 1GB, UK.)

Grette Lydon and Nakajima[20] recently addressed this controversy. In Experiment 1, rats trained to a criterion of at least 75% correct responding made significantly more reentries into baited arms, but not first entries into unbaited, arms under scopolamine (0.5 mg/kg i.p.) relative to control conditions. In other words, scopolamine appeared to increase WM but not RM errors (Figure 13-1A.) The results did not appear due to peripheral effects as there were no differences among control conditions that included methylscopolamine. Our findings were congruent with those of Beatty and Bierley[159] and Wirsching et al.,[160] but not with Okaichi et al.[161]

We noted that even in the last days of training, our animals made more RM errors (1.67) per trial than WM errors (0.43). The means of the baseline or control scores for the other three studies showed the same pattern, lower WM and substantially higher RM errors for all but the Okaichi et al.[161] study where the RM means were also low. We then did a second experiment with the animals trained to a higher criterion of learning (92% correct performance), and like Okaichi et al.,[161] found that both WM and RM were impaired under scopolamine relative to control conditions (Figure 13-1B).

A graphic depiction of the means of RM errors under control conditions for the five studies is instructive. Figure 13-2 shows the respective means of RM baseline or control errors plotted from highest to lowest. Note that the learning criteria for the five studies, which ranged from a low of 75% for our Experiment 1 to the 95% correct responding used by Okaichi et al.,[161] fall in inverse order to the error means of the respective studies. It would appear that Okaichi et al.[161] and Grette Lydon and Nakajima (Experiment 2),[20] did not find differential effects of scopolamine because they used higher criteria of learning. Differences in dosages did not account for the discrepancies, as contradictory outcomes were reported for the same dosages. We proposed that the differences in baseline errors for WM and RM made it "easier to detect an impairment in WM" when animals are trained to a lower criterion of learning. A higher level of training lowers baseline rates sufficiently to also reveal a deficit in RM. This problem with baseline errors might account for the selective impairment in WM reported in a radial maze study of a different species: mice.[162]

ii. T-Maze and Y-Maze Studies

It is possible that differences in baseline errors or in task difficulty underlie reports of selective WM impairment for other maze tasks including T-maze procedures after intrahippocampal[163] or systemic[164,165] injections of scopolamine.

A T-maze usually contains a food cup at the distal end of each arm, and the starting box is separated from the stem by a guillotine door. The standard T-maze alternation procedure requires WM because the rat must remember the choice made on the previous trial to choose the correct arm on the next trial. One

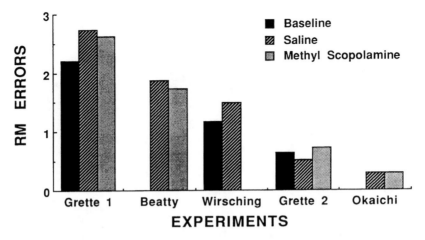

Figure 13-2 Means of RM errors from highest to lowest under control conditions for five studies. (Reprinted from *Pharmacol. Biochem. Behav.*, Vol. 43, R. Grette Lydon and S. Nakajima, Differential effects of scopolamine on working and reference memory depend upon level of training, p. 648, Copyright 1992, with kind permission from Elsevier Science Ltd, The Boulevard, Langford Lane, Kidlington OX5 1GB, UK.)

version of the task consists of two runs, a forced run and a free choice. In the forced run, only one arm is open, and the rat retrieves the reward from the food cup. The rat is then removed from the goal arm and placed in the start box for the second run. In the second run both arms are open, and the correct choice is to enter the arm not visited on the forced run.

T-maze tasks have been modified for examining both WM and RM. A trial-independent RM procedure can be used on a second T-maze in which the rat learns a left/right discrimination useful for all trials. The stem is divided in half, with one side open to the goal arms and the other side blocked. The correct choice (the open side of the stem) always remains the same. The WM alternation and RM discrimination procedures just described were used by Bartolini et al.[164] In this study and others that reported that scopolamine disrupted the WM task selectively, the impaired WM task was also the more difficult task[164,165] and/or the one for which baseline performance was lower.[163-165]

Bartolini et al.[164] noted that their rats reached criterion (of 90% correct) in only six sessions for the stem discrimination task, but required 18 sessions for the WM alternation task. Four months after training, while stem discrimination was maintained (at about 98% correct as rats benefitted from the additional training needed to reach criterion on alternation), alternation arm responses fell to chance levels. Scopolamine (given about a week after criterion was reached) obviously impaired the more difficult task. Equating the baseline measures for WM and RM by further training would enhance the argument for a selective effect on WM, were a dissociation of the WM and RM tasks subsequently found.

A similar problem exists for a relatively new procedure, the double Y-maze.[166] The maze consists of two Y-shaped arms connected by a single stem. RM is assessed in the first half of the maze and involves a task, which remains constant across trials. The rat is randomly placed in one arm and must traverse the stem for a food reward. WM is assessed in the second half of the maze. To choose the correct arm in the WM task, the rat must remember in which arm it was placed during the RM task. (See Mallet and Beninger[166] for a description.) Mallet and Beninger found, however, that the acquisition rates for WM and RM differ in this procedure, and that substantially more sessions of training are required to reach criterion on the WM component.

This presents a difficulty for the interpretation that injections of scopolamine into the amygdala[167] or of muscimol, a GABAergic agonist, into the nbM,[168] selectively impaired the WM tasks. In the scopolamine study,[167] accuracy in baseline performance was about 97% for the RM task, but only about 88% for the alternation task. In the muscimol experiment,[168] baseline performance was 100% for the RM task, but only about 92% for the WM task. Further, while the tasks were differentially affected at a lower dose of scopolamine (24 μg) or muscimol (0.1 μg), a higher dose (72 μg scopolamine, 1.0 μg muscimol) affected both WM and RM tasks relative to saline.

The problem is compounded by the fact that WM and RM may be differentially sensitive to interference. As scopolamine causes the animal to move slowly in the mazes, it takes longer to complete a trial. It is possible that the prolonged run time (latency) increases WM errors more than RM errors. This problem could be addressed by calculating a correlation between WM errors and response latency. Were these measures not found to be correlated, it would increase confidence that increased WM errors were not due to interference from the slowed movement.

iii. Water Maze Studies

Performance deficits do appear in studies of the effects of anticholinergics on tasks that require only spatial RM. Most of these studies used a swimming pool task[169] in which the rat learns to locate the position of a platform just below the surface, which permits escape from the water in the pool. The water in the tank is usually made opaque by adding powdered milk, and the task requires RM because the location of the platform relative to the extramaze cues remains constant. (Technically, this apparatus is not a maze. Because it has become known as the Morris water maze, its popular usage is retained here.) A number of studies have shown that 50 mg/kg atropine was effective at disrupting the acquisition and performance of this water escape task.[170-174] While 50 mg/kg is high relative to the doses of scopolamine typically used on the radial maze or T-maze, the escape task in the Morris water maze does not require eating behavior, and thus has an advantage over appetitive maze tasks that can be affected by interference from dry mouth effects.[3]

Three points are notable with respect to the water maze studies. The first point is that if this procedure does in fact depend on a stable reference memory, then RM is affected by cholinergic blockade. The second relates to O'Keefe and Nadel's[41] theory of hippocampal function that the hippocampus is selectively involved in spatial, but not cue, learning. It was reported in the above studies[170-174] that the rats were more impaired when acquiring a spatial strategy reliant on the location of the platform relative to distal cues than on a cue strategy more dependent on local cues.

Hagan et al.,[175] however, found that atropine (10 or 50 mg/kg) disrupted acquisition of both a spatial version of the task (in which the escape platform was distinguishable from a float only on the basis of location) and a nonspatial (cued) version where the platform and float were discriminable on the basis of appearance. In the cued version, the platform was striped and the float was a solid color; the use of a spatial strategy was discouraged by the concealment of room cues with a curtain. Both the rigid platform, which permitted escape, and the float, which sank if the rat climbed up on it, were raised 1.5 cm above the water surface.

Whishaw and Petrie[176] also found that 50 mg/kg of atropine impaired the acquisition of a visual discrimination (black/white- or horizontal/vertical-painted platform) version of the cued task. A subsequent study showed that atropinized rats would acquire a pattern discrimination as well as control rats if they were given pretraining experience with pattern discriminations different to those required by the test trials. (If they pretrained on a horizontal/vertical discrimination, they subsequently would not be impaired relative to controls on a black/white discrimination.) The main point here, however, is that cholinergic blockade does not selectively impair place learning. This has also been found to be the case on radial maze tasks where both place and cue versions were equally disrupted.[177]

Finally, Whishaw and Petrie[176] questioned whether it was even memory per se that was impaired by cholinergic blockade. They argued that what was disrupted was not discrimination learning, but the "ability to call up and experiment with various movements or strategies" needed for the successful learning of discriminations. Whishaw and Petrie[176] noted that their atropinized rats were impaired in the early stages of acquisition as shown on the plotted acquisition curves. When the investigators moved the acquisition curves of the drugged rats to the left, the curves resembled those of the controls. The results did not appear due to a tendency to inhibit responding or to peripheral effects, as methylatropine was not disruptive.

d. Effects of Delay

As noted earlier, Burešová and Bureš[157] found that scopolamine impaired radial maze performance only after a delay interval in their highly trained rats. To counter problems of anticholinergic-induced dry mouth, Burešová et al.[178] developed a radial maze task based on aversive motivation and adapted it for use in a water tank. In this procedure, the rat is first placed on the central platform. The platform is lowered after 15 sec, and the rat must swim to a bench at the end of each of the eight channels. The bench is collapsed after 20 sec, and the rat must return to the raised platform. This is repeated until all eight arms are visited. Visits to previously entered arms are counted as errors.

Burešová et al.[158] examined the effects of anticholinergics on uninterrupted vs. delayed performance of this radial water maze. Rats received 60 training trials on the eight-arm water maze and were injected

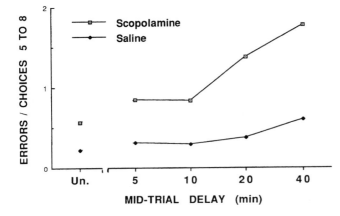

Figure 13-3 Effects of scopolamine and saline on uninterrupted or delayed performance of a radial water maze task. (From O. Burešová, J. J. Bolhuis, and J. Bures, Differential effects of cholinergic blockade on performance of rats in the water tank navigation task and in a radial water maze, *Behav. Neurosci.*, 100, 478, 1986. Copyright 1986 by the American Psychological Association. Reprinted by permission.)

with scopolamine 20 min before the test trial. Scopolamine had no effect on a continuous test, but increased errors in choices 5 to 8 when delays of 5, 10, 20, and 40 min were introduced between choices 4 and 5. Further, there was a significant interaction between treatment (scopolamine and saline) conditions and delay intervals, with errors under scopolamine increasing as the duration of the delay increased. (See Figure 13-3.) This finding does suggest a problem with working memory. Yet these results conflicted with those of Gooding et al.[179] that scopolamine (1.0 to 5.0 mg/kg) did not substantially increase errors in the second half of an eight-arm radial maze task when a 5-hour delay was inserted between choices 4 and 5.

Burešová et al.[158] next included an additional trial in which they injected scopolamine immediately after choice 4 during a midsession 40-min delay. In contrast to the results when scopolamine was injected 20 min before the trial began (at choice 1), errors did not increase in the trial following the delay. Thus, scopolamine did not affect retention of memory for the arms visited before the delay (and hypothetical storage mechanisms). The authors concluded that scopolamine affected or interfered with the acquisition of the delayed task. This study by Burešová et al.[158] is important because it showed that retention and retrieval were unaffected, and that the effects of scopolamine appeared to be limited to disruption in the acquisitional stages of information processing.

Beatty and Bierley[180] attempted to examine the effects of scopolamine on radial maze performance by timing injections (i.p.) within a trial so that inferences could be made about encoding, storage, or retrieval mechanisms purported to underlie WM. A 4-hour delay was inserted between choices 4 and 5 on an eight-arm maze, and the rats were trained to a baseline level of at least 85% correct responding on choices after the delay interval. Scopolamine injections (0.25 mg/kg) were given at specific times relative to the to-be-remembered event (TBRE, the first 4 choices made). The injection times were 15 min before the TBRE as an indicator of encoding, immediately after choice 4 as an indicator of storage, and 15 min before choice 5 as an indicator of retrieval.

The effect of scopolamine was to increase errors when administered before the TBRE and 15 min before choice 5, but not when administered immediately after the TBRE. Beatty and Bierley[180] concluded "that scopolamine impairs encoding and retrieval, but does not affect storage." These findings are consonant with those of Burešová et al.[158] and Gooding et al.[179] on rats in radial mazes, and of Ridley et al.[143] on delayed performance in marmosets. Taken together they suggest that scopolamine disrupts acquisition or encoding, but does not affect storage.

e. Aging

Nilsson and Gage[181] found that atropine (50 mg/kg i.p.) disrupted the search behavior of aged rats compared to young controls on both cued (visible platform) and spatial (nonvisible platform) versions of the Morris water maze task. In Experiment 1, the aged animals were divided into two groups, impaired or unimpaired, based on their performance on the pretest. (This is an important control procedure because Stanes et al.[61] showed that fast and slow learners were differentially affected by an anticholinesterase.) For half the trials, a cover was placed over the platform to make it visible to the rat. After drug testing, a spatial probe trial was given with the platform removed.

The effect of atropine was to increase response latencies in the aged rats relative to young controls on both visible and nonvisible platform trials. It should be noted that the young rats were also impaired in

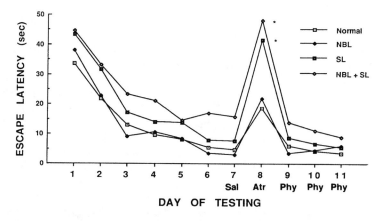

Figure 13-4 Escape latencies for the normal, septal-lesioned (SL), nbM-lesioned (NBL), and combined-lesioned (NBL + SL) rats. Days 1 to 6 = acquisition without drugs. The rats were given saline on Day 7, atropine on Day 8, and physostigmine on Days 9 to 11. *Significant difference from control conditions. (Reprinted from *Neurobiol. Aging*, Vol. 14, O. G. Nilsson and F. H. Gage, Anticholinergic sensitivity in the aging rat septohippocampal system as assessed in a spatial memory task, p. 489, Copyright 1993, with kind permission from Elsevier Science Ltd, The Boulevard, Langford Lane, Kidlington 0X5 1GB, UK.)

response latency and in their spatial navigation under atropine compared to control conditions. The effects of atropine appeared to be centrally controlled, as methylatropine was not disruptive. Examination of swim speeds suggested that atropine's effects on the aged rats were not secondary to swimming ability or motor performance. The spatial probe trial showed that relative to young controls the aged rats made fewer crossings over the site of the previous platform location, suggesting perhaps that both their spatial memory and their overall search behavior were impaired.

Nilsson and Gage's[181] study showed then that the aged rats were particularly sensitive to the blockade of central cholinergic receptors, an effect that led to pronounced impairment on both cued and spatial trials of a task thought to reflect reference memory. Deficits in RM and indeed in working memory have also been noted for other tasks or in aged animals of other species. Nondrugged, aged rats have been reported to be impaired on a more complex RM task, the 14-choice-point Stone maze.[182] As described earlier, both aged[70,71] and scopolamine-treated[68,69] monkeys were found to be impaired in a WM task (delayed responding).

2. Lesion Studies
a. Water Maze

Nilsson and Gage[181] (Experiment 2) compared the effects of atropine (50 mg/kg i.p.) on rats given septal lesions, nbM (nuceus basalis magnocellularis, analagous to the nbM in humans) lesions, and combined septal and nbM lesions. Four to six months after surgery the rats were tested for acquisition of the water maze task over days 1 to 7 and were given saline on day 7, atropine on day 8, and physostigmine on days 9 to 12. The combined-lesioned group were a little slower in acquiring the task than those in the other groups, which included an unoperated control group. All groups under atropine showed increased latencies at finding the hidden platform in comparison to day 7. The escape latencies of the septal- and combined-lesioned groups were significantly increased relative to day 7, and both groups of septal-lesioned rats were more impaired than the nbM and control groups. Physostigmine did not affect the latencies of any group (see Figure 13-4). In a spatial probe trial with the platform removed after day 7, all groups were impaired under atropine, but the septal-lesioned groups seemed to be more random in their search strategies than the normal group. The nbM (NBL) group did not differ from the controls.

The pattern of severe impairment in the young rats with septal lesions (relative to controls and nbM-lesioned animals) under atropine was similar to that for the aged rats (relative to young controls) when challenged by atropine in Experiment 1 of Nilsson and Gage.[181] In Experiment 2, the (radiofrequency) lesions of the septal area resulted in reduced ChAT activity of 67 to 73% in the hippocampus; and the (quisqualic acid) lesion to the nbM and to the combined group reduced ChAT in the neocortex by 29 to

51%. These two experiments by Nilsson and Gage suggest that the septo-hippocampal system may play a more important role in the navigation task of the water maze than the basalocortical system. That atropine resulted in similar patterns of impairment in aged rats and in young rats given septal lesions may indicate that the performance deficit in the aged rats was due to disruption of the septo-hippocampal system.

Similar results had earlier been reported by Hagan et al.,[183] who found that acquisition of the water maze task was disrupted by bilateral ibotenic lesions of the medial septum (MS) and diagonal band of Broca (dbB), but not of the nbM. The impairment was evident across the 24 training or acquisition trials, but there were no retention deficits in a subsequent spatial probe trial. ChAT was reduced in the hippocampus and posterior, but not anterior, cortex by the MS/dbB lesions and in the cortex, but not in the hippocampus, by the nbM lesions. The results of Nilsson and Gage[181] and Hagan et al.[183] contrast with those of other investigators who reported impairment in this task following ibotenic lesions of the nbM.[173,184-186]

Page et al.[187] recently suggested that impairment in acquisition of the water maze task due to ibotenate-induced nbM lesions may be unrelated to (magnocellular) cholinergic neuronal damage in the nbM and cholinergic deafferentation of the neocortex. Page et al. found that the excitotoxin AMPA (α-amino-3-hydroxy-5-methyl-4-isoxazole proprionic acid), which decreased cortical ChAT by 70%, had no effect on water maze acquisition and performance, while ibotenic acid, which produced marked impairment, reduced cortical ChAT by only 50%. The authors stated that while retention deficits of passive avoidance behavior (also tested in the study) appeared due to destruction of cholinergic neurons in the nbM, deficits in acquisition of the water maze may have been due to injury to pallidal or other neurons in the basal forebrain.

There have been similar reports for different excitotoxins and behaviors. Lesions from quisqualic acid, which caused greater reductions in neocortical ChAT than ibotenic acid, were found to produce less impairment in both acquisition of the water maze and passive avoidance,[185,188] as well as in delayed matching and delayed alternation.[189] These observations are troublesome for the hypothesis that cholinergic neurons in the basalocortical projection are involved in memory.

b. T-Maze and Y-Maze

It was outlined earlier how differences in baseline responding might underlie the findings of selective WM effects in pharmacological (WM/RM) dissociation studies. Also discussed was how differential effects of scopolamine on WM and RM might be confounded with task difficulty, or with differential sensitivity to interference, in T-maze and Y-maze studies. These factors are also relevant to the lesion literature. As was noted, the acquisition rates for WM and RM differ in the double Y-maze procedure.[166] The selective impairment in the WM component of the double Y-maze task seen after quisqualate lesions to the nbM[190] may, therefore, reflect differences in task difficulty of the two memory components. Similarly, differences in task difficulty or baseline performance might have contributed to findings of impairment on WM alternation responding, but not on RM stem or spatial discrimination, in a T-maze procedure after nbM lesions.[165,191-193]

Hepler et al.[194] did find selective impairment of WM performance when preoperative baseline responding was equated, however. Two T-mazes were used in which the stem was divided in half with one side open and leading to the goal arms, and the other blocked. Each maze was placed in a different location relative to distal extramaze cues. The rat had to determine the location in which it was being tested in order to choose the correct side (left on one maze, right on the other), an RM discrimination that remained constant across trials. The WM arm-alternation task was the same forced run and free choice procedure used by Bartolini et al.[164] described earlier. Preoperative baseline performance at the end of ten sessions of training was similar for the two memory components (about 95 and 97% correct for the respective arm and stem discriminations). This is in marked contrast to the baselines of Bartolini et al.[164]

Hepler et al.[194] gave the rats bilateral ibotenic acid or radiofrequency lesions to the MS area, the nbM, and combined MS/nbM areas. All lesioned groups were impaired in arm discrimination relative to controls for 14 postoperative sessions, after which choice accuracy reached that of preoperative levels. The groups with medial septal lesions were more impaired at arm discrimination than the nbM-lesioned groups on sessions 6 through 14. In contrast, choice accuracy of the stem task was unimpaired in all groups. The combined MS/nbM-lesioned group was impaired at stem discrimination only when a 10-min delay was inserted, and when a more difficult RM task, stem reversal, was used. As baseline performance of the respective tasks was equated, the selective impairment of the WM task did not seem to be due to task difficulty.

c. Radial Maze

Knowlton et al.[195] have argued that task difficulty does not account for the reported dissociations between WM and RM tasks. Because T-maze procedures involve relatively simple discriminations, Knowlton and her colleagues used more complex tasks, a standard radial maze procedure and a Stone maze task, as WM and RM measures, respectively. Combined ibotenic lesions of the nbM and medial septal areas resulted in marked impairment of acquisition of the radial maze task, but lesioned rats were not impaired relative to sham-operated controls in acquiring the RM task on the Stone maze. Because the Stone maze involves a relatively complex task of 14 choice-point discriminations, Knowlton et al. do not think task difficulty accounts for the selective impairment of WM tasks in radial maze and other dissociation studies. (In this study by Knowlton et al.,[195] ChAT activity in the hippocampus and cortex were reduced by 21 and 33%, respectively.)

A number of radial maze studies have found increases in both WM and RM errors after unilateral quinolinic acid[196,197] and bilateral ibotenic acid[198] injections into the nbM, and bilateral ibotenic acid lesions to the nbM and medial septal areas.[199,200] In fact, at least three studies suggested predominant or selective deficits in reference memory after bilateral ibotenate nbM lesions.[198,201,202]

In contrast to Knowlton et al.,[195] Bartus and colleagues[203] did not find a postoperative effect on uninterrupted performance of a similar radial maze procedure. They did, however, report an interesting finding. Rats were trained preoperatively and given bilateral ibotenic lesions of the nbM, or sham operations. When tested 3 to 7 weeks later, the ibotenic-lesioned rats were impaired only after a delay interval compared to the sham controls. In this regard, the finding is similar to the radial maze studies of Burešová et al.[158] and Burešová and Bureš[157] in their well-trained animals where scopolamine exerted an effect only after a delay, and to the T-maze study of Hepler et al.[194] where a combined lesion affected RM only after a delay. When tested 17 to 20 weeks after surgery, however, Bartus et al.'s rats showed a gradual and complete recovery of performance. This occurred despite the fact that postmortem examinations showed neuronal degeneration in the nbM, and reductions in both cortical ChAT (44%) and high-afinity choline uptake (33%) in the frontal cortex. As there appeared to be no neurochemical changes in cholinergic markers in the terminal field or histological differences around the site of the lesion, compensatory changes in the brain did not account for the results.

Bartus et al.'s[203] results question the functional role that cholinergic neurons in the nbM play in memory for the radial maze task. They do not preclude a role for the septo-hippocampal neurons (in studies where memory effects are found) because there appeared to be no effects of the lesion on the hippocampus. This also leaves open the possibility that the impairment evidenced by Knowlton et al.[195] was due to septo-hippocampal disruption. In addition to a reduction in hippocampal ChAT, Knowlton et al.[195] had found extensive cell loss in the medial septum and dbB.

Alternatively, months of practice following the lesions attenuated the memory deficit in the Bartus et al.[203] rats. Studies in the field of information processing have shown that tasks that initially require effort to perform can, with extensive practice, become more automatic.[204] (The more an event is practiced the less attention it requires.) It may be that the effect of practice in this experiment was to enable performance of the task to become more automatic and thus require less attention or mental effort to perform. This complete recovery of function and attenuation of memory impairment is not seen in persons with AD.

3. Synopsis

The lesion studies question the role that cholinergic neurons in the basalocortical projection play in acquisition or retention of the radial maze, water maze, and other maze procedures, as well as of delayed matching, delayed responding, and passive avoidance. Kesner et al.[205] used a modified radial maze procedure to examine item and order recognition memory for a sequence of baited arms visited in a predetermined order. The authors concluded that the nbM may be involved with order, but not item, memory for spatial locations, perhaps through cholinergic afferents to limbic (amygdaloid) but not neocortical projection areas. Boegman et al.[206] recently found that lesions from excitotoxins such as ibotenic acid, which tend to produce greatest impairment in memory tasks, caused the greatest reduction of ChAT in the amygdala. The basaloamygdaloid projection may thus have some involvement in memory or cognition. See Reference 205 for discussion of functional dissociations between the medial septum and nbM, and between the hippocampus and amygdala.

The literature provides no strong justification for ruling out the involvement of the cholinergic septo-hippocampal pathway in memory and attentional processes. Consistent with the pharmacological and lesion studies reviewed above, however, the lesion research on the hippocampus is not unequivocal. For example, lesions of cholinergic neurons in the MS/dbB areas, which resulted in pronounced decreases in hippocampal and frontal ChAT, were found to produce greater deficits in acquisition of the water maze than hippocampal lesions themselves.[187]

Moreover, subsequent investigators have been unable to replicate Olton and Papas'[43] findings of selective effects on working memory, and reported that hippocampectomy impaired trial-independent RM as well as trial-dependent WM on the partially baited radial maze.[207,208] There is no doubt that reducing cholinergic transmission impairs performance on a radial maze at least in naive rats. The evidence is not convincing, however, that the deficit is selective for working memory. Olton's paradigm has had heuristic value nonetheless, and the foraging hypothesis does offer an improved face validity over more traditional learning paradigms.

4. Methodological Issues Revisited

It was noted earlier that anticholinergics increase run times and abortive responses, and result in dry mouth effects and pupillary dilation, any of which can cause performance deficits. Yet as Hagan and Morris[3] point out, with respect to peripheral effects at least, the authors who have used methylated antagonists as controls in the radial maze studies invariably report that these do not affect choice accuracy. Peripheral effects do not appear to account for the results of many of the pharmacological studies reviewed here on the radial maze,[20-22,156,159,161,162,200] T-maze,[49,50,164] and water maze,[158,171,174-176,181] as well as in some of the studies on avoidance, discrimination learning, and delayed matching.[19,51,53,65,67,143,151]

Grette Lydon and Nakajima[20] showed that scopolamine appeared to affect WM selectively when a lenient criterion of learning was used, but that it impaired both WM and RM in well-trained animals. Using this study as a basis to discuss methodological and conceptual issues, impaired choice accuracy in the task did not result from an increase in response rates or an impairment of peripheral motor mechanisms. Response latency and WM errors were not correlated, for example, and therefore the increased WM errors under scopolamine were not due to increased running speed. The same argument could be made for RM errors: that the impairment in RM in the well-trained animals was not due to similar motoric interference. Had we not paid attention to the level of training, we would have concluded erroneously that scopolamine impaired WM selectively, and further, we would have been confident that the effects were on memory and not on performance factors. And our confidence would not have been warranted.

It is generally assumed that when a differential effect is found within experimental conditions, the results likely reflect the effects of the independent variable (drug treatment conditions here) on the dependent measures (one type of memory and not another). All conditions being equal, interference with performance arising from confounding variables should affect each drug condition *equally*. Movement problems due to the blockade of peripheral motor activity, for example, should affect the number of errors on a task thought to measure RM to the same extent as on a task thought to measure WM. A clear effect on only one type of error could well be due to the difference in the type of memory required in that task. If there is no differential effect, if both RM errors and WM errors increase under scopolamine, for example, one is then unable to rule out possible interference from perceptual or attentional mechanisms.

It would appear that interpreting the results of drug or lesion studies is precarious even when differential effects within experimental conditions are found.

C. ATTENTION

Memory storage is assumed necessarily to involve a temporal component. Results of memory experiments can, in fact, be due to effects on discriminative processes such as confusion over the locations of the arms in the case of a radial maze, for example.[21] As Eckerman et al.[21] note, "Scopolamine may reduce discriminative control without changing the loss of this stimulus control over time (i.e., memory)." As discussed earlier, investigators have attempted to differentiate scopolamine's effects on storage from effects on discrimination processes by the insertion of delays. Such attempts often indicate that the degree of impairment seen does not vary with the length of the delay, suggesting that scopolamine may be affecting attention rather than memory processes.[56] Warburton and Brown[57,58] also found an effect of scopolamine on stimulus sensitivity.

Cheal[209] and Spencer et al.[210] maintain that stimulus sensitivity per se does not fully account for the effects of anticholinergics. Cheal observed investigative behavior in gerbils. Under 0.01 to 10 mg/kg scopolamine, the gerbils showed appropriate investigative behavior to novel stimuli (a pen or a paper cup holder), and appeared to have no problems in selectively attending to odor stimuli. When tested either 60 sec or 24 hours after a prior 60-sec exposure, the animals spent less time investigating the stimulus. This indicated normal habituation. Yet the frequency of approaching the object did not decrease. The shorter duration of each investigation, but not frequency of approach, suggested a problem maintaining attention. Cheal[209] proposed that animals under scopolamine can discriminate appropriate stimuli, but have difficulty maintaining attention when the task requires "more than a minimal amount of time and effort." The results did not indicate a clear role for ACh in either behavioral inhibition as suggested earlier by Carlton,[44] or in stimulus selectivity as proposed more recently by Warburton and Brown.[57,58]

It is noteworthy that in the experiments by Evans,[67] Cheal,[209] and Spencer et al.,[210] methylscopolamine exerted little effect, suggesting that the attentional effects of scopolamine were centrally mediated.

While most of the cholinergic research has focused on muscarinic receptors, there has been a recent interest in nicotinic receptors. It has been reported that mecamylamine, which blocks nicotinic receptors, and scopolamine given in combination produce greater deficits in the radial maze performance of rats than either drug alone.[113] Nicotine has also been found to reverse WM impairment on a radial maze caused by septo-hippocampal and basalocortical knife-cut lesions,[113] and by colchicine lesions, of the nbM.[211] Both arecoline and nicotine, but not amphetamine, have been reported to attenuate WM and RM deficits in a radial and in a water maze produced by quisqualic lesions to the hippocampus and the nbM.[212]

Evidence is growing that nicotinic receptors are important in cognitive processes. Yet it has also become apparent that the amelioration of lesion-induced deficits, attributed in some earlier reports to effects of agonists on memory, may have been due to effects on performance.[149,213] Some of the human research indicates that the facilitatory effects of nicotine in patients with SDAT is restricted to improvement of attentional mechanisms,[214-216] although there have been reports of effects on memory, as well as on attention, in healthy humans.[217-219] The finding of facilitation of attentional and not memory mechanisms that predominates in the nicotine research may also apply to muscarinic enhancing agents in SDAT patients.[220]

Nicotine has not been found to affect retrieval, and there is a state-dependent effect.[219] Whether cholinergic agents exert independent effects on attention and memory[221] encoding, or the effects on memory are secondary to effects on attentional mechanisms, cannot be ascertained at present.[5,222] While there are clear difficulties with testing these hypotheses in human subjects,[5] Warburton and Rusted have made some progress in teasing out the two variables (see Chapter 12 in this volume).

Warburton[223] has proffered a functional state model in which performance on a cognitive task is viewed as a product of the energy requirements of the task and the state of arousal in the brain. Some tasks are performed best under high states of arousal, and others under low states of arousal. Dunne et al.[136] recently provided support for this hypothesis by showing that scopolamine impaired performance on tasks requiring selective attention and psychomotor speed such as symbol-digit substitution (in contrast to Beatty et al.[127]). As reported earlier, scopolamine actually improved the generation of words to letter cues, perhaps because verbal fluency tasks benefit from a low state of arousal, "where the mind is free to wander."[136]

The suggestion that scopolamine might impair effortful but not automatic processing[204] is appealing.[224,225] Dunne[226] tested the effects of scopolamine (0.9 mg o.d.) on the ability of adult subjects to generate members of natural categories. While scopolamine increased repetitions, it did not disrupt item retrieval, even when the task was made more demanding by restricting subjects to one retrieval strategy or by extending the duration of testing. Similarly, the recall of abstract words, which because of their inherent low imagery are thought to require more effortful processing requirements,[204] vs. concrete high-imagery words, was not differentially affected by scopolamine.[227,228] The hypothesis on effortful processing failed then as far as retrieval for items on word lists is concerned. Dunne[226] concluded that scopolamine's effects were restricted to information acquisition.

IV. GENERAL DISCUSSION

Cholinergic hypotheses of memory, first introduced in the 1960s, have stimulated research ever since. To say that many of the findings support these hypotheses may be to overinterpret the results.[2] The evidence is not convincing that the cholinergic system is *selectively* involved in working memory or in spatial learning, or that the critical structure in the brain for recent or working memory is the nucleus basalis of Meynert. Many of the early and recent investigators have failed to find a role for cholinergic neurons in memory storage.[143,157,158,179,180] By the same token, these and other studies have suggested that anticholinergic drugs, or lesions to structures rich in cholinergic neurons, disrupt acquisition or attention.

The pharmacological and lesion studies on maze learning in animals do not support the argument for a selective involvement of the cholinergic system in working memory or in spatial learning. Differences in baseline performance, task difficulty, or sensitivity to interference confound some of the findings of selective WM impairment. Moreover, anticholinergics or lesions do have a deleterious effect on choice accuracy in various maze tasks that rely primarily on RM. Similarly, rats with fimbria-fornix lesions to the hippocampus also show a deficit, albeit a smaller one, on radial maze tasks that require trial-independent RM.

It may well be that tasks involving working memory are more sensitive to cholinergic blockade or lesions than are tasks involving reference memory. In the T-maze study by Brito et al.,[163] for example, rats under scopolamine were impaired at both alternation, a WM task, and visual cue discrimination relative to saline conditions, but they still performed the cue discrimination task significantly better than the alternation task. (In this discrimination task,[163] the presence or absence of a light placed at the distal end of the arm signaled the correct choice. The visual cue remained the same across trials, and so reflected RM.)

Similarly, while cue learning in radial and water maze tasks is also disrupted, cholinergic blockade does impair place learning more than tasks based on nonspatial discriminations or relations among local cues. Working memory on mazes is dependent on spatial relations among distal cues, whereas the cues in taxon learning are nearby, or underfoot, or unique. Perhaps learning discriminations dependent on spatial relations place greater demands on attentional or processing capacity than learning discriminations to proximal cues. It may be noteworthy that each working memory trial as defined here is a trial of acquisition.

Although cholinergic neurons in the basaloamygdaloid projection may have some involvement in memory, many of the animal studies question the role that cholinergic neurons in the basalocortical projection play in the learning and memory of diverse tasks. Similarly, research on human subjects has shown that recent memory can be relatively unaffected when cholinergic activity is reduced in the neocortex. As noted earlier, patients with olivopontocerebellar atrophy[101] and a small number of healthy elderly[102] having no symptoms of dementia have shown significant reductions in neocortical ChAT activity.

The animal literature does not offer sufficient evidence, however, to preclude a possible role for the cholinergic septo-hippocampal pathway in memory and attentional processes. In like manner, the human findings leave open the possibility that impairment of cholinergic neurons in the septo-hippocampal pathway may well underlie some of the memory deficits in Alzheimer disease. It may be noteworthy that the hippocampal/amygdaloid complex has been proposed as the loci where Alzheimer disease may first strike.[95]

It is noteworthy that some researchers found a scopolamine-induced or postoperative effect on performance of a radial maze procedure[157,158,203] and of an RM task on a T-maze[194] only after delay intervals in well-trained animals. Moreover, in the radial maze study by Bartus and his colleagues,[203] preoperatively trained rats gradually recovered full performance (even when interrupted by delays) when tested a few months after surgery. In a similar vein, Whishaw and Petrie[176] found that atropinized rats could learn to discriminate patterns on a water maze if they had prior experience with pattern discriminations different to those of the test trials. It is intriguing that practice or overtraining offers protection from disruption by cholinergic blockade, as naive animals are quite sensitive to a scopolamine challenge when performing these various maze tasks.

One effect of practice might be to enable performance of the task to become more automatic, and so require less mental effort to perform. One wonders what effect prior experience or practice has on the variablility of memory impairment seen among persons in the early stages of Alzheimer disease. There has been the suggestion that automatic processing is relatively unaffected in the earlier stages of SDAT, while effortful processing is quite impaired.[229] Along a similar vein, it has frequently been shown that anticholinergics or fimbria-fornix lesions have more pronounced effects on the more difficult tasks.[230] Could the scopolamine model have some relevance for the primary memory deficits in the early stages of SDAT, when the processing requirements are made more effortful? Given that the involvement of cholinergic neurons appears limited to acquisitional phases of processing and to retrieval processes only in restricted circumstances, there is little wonder that the scopolamine model of secondary memory deficits in SDAT has been found inadequate.

Some prudence is in order with respect to cholinomimetic agents or neural grafts. The changes evoked by cholinergic enhancing agents, such as those for scopolamine, are frequently related to baseline performance.[5] Sitaram et al.,[231] for example, found that good performers (on baseline measures) showed less improvement after arecoline and choline, and less impairment under scopolamine, than poor performers. As Sahakian[5] notes, it is important to be mindful of this baseline dependency effect when administering cholinergic agonists or anticholinesterases in clinical settings. It would seem that at best facilitation would occur in very restricted circumstances. It is notable that the findings for the cholinergic enhancing agents are consonant with Deutsch's prediction, respective facilitation and impairment of well-remembered vs. poorly remembered information.

While most of the deficit reversal research has focused on muscarinic agonists, it is becoming evident that nicotinic receptors may also be important in cognition. We can thus expect that research on nicotine will intensify. It would be wise to examine the nicotinic research for performance effects,[149,213] however, before nicotine gum is liberally dispensed in dementia clinics. This is especially true in light of nicotine's adverse effects,[232] and of the actual impairment in memory trials that has been noted after even relatively low doses of nicotine.[153,200]

Taken along with the well-documented problems of cholinomimetics including side effects, the known involvement of acetylcholine in diverse functions and in complex interactions with other transmitters, and the number of unknowns, these considerations underscore the need for careful individual evaluation before recommending a course of cholinergic therapy. One should also keep in mind that to date the facilitatory effects of nicotinic as well as of muscarinic cholinomimetics seem limited to the improvement of attentional processes, at least in patients with symptoms of Alzheimer disease.[215,216,220]

The roles of other neurotransmitter systems have not been extensively researched, but a glutamatergic hypothesis of AD appears to be emerging, with the dementing symptoms of AD attributed to loss of glutamatergic receptors in the hippocampus and neocortex.[233-235] There has been some recent interest in GABA and benzodiazepine receptors. Intraseptal muscimol, which appears to decrease ACh turnover in the hippocampus in rats,[236] has recently been found to disrupt retention of a radial maze task[237] and acquisition of a water maze task,[238] for example, while injections into the nbM impaired choice accuracy on a double Y-maze.[168]

As was found to be the case in the 1970s and 1980s, in the 1990s we are witnessing occasional suggestions of similarities between scopolamine and benzodiazepines in terms of their amnestic as well as sedative effects.[239,240] (See Curran[241] for a review of benzodiazepines and memory.) It has been suggested that these effects are due in part to GABA/benzodiazepine-mediated inhibition of cholinergic neurons.[149] As Sarter et al.[149] also note in a cautionary vein, the enhancement of compromised corticocortical pyramidal neurons may accelerate cell death, one of several important considerations that require further study before GABAergic disinhibitory treatment is advocated for SDAT.

It has become increasingly clear that the cholinergic system is not the only neurotransmitter system involved in memory. Given the complex circuitry in the brain and the fact that ACh is so widely distributed there,[36] this is not surprising. It follows that acetylcholine is not the only transmitter involved in the cognitive impairment of dementia of the Alzheimer type. Given the breadth of impairment seen in Alzheimer disease and the widespread neuronal damage, this also is not surprising. According to David Mann,[95] "There is in fact no transmitter-based selectivity of neuronal loss in Alzheimer's disease. What has presumed to be transmitter selective is in actuality an apparent rather than a real selectivity based on anatomical rather than biochemical grounds." That notwithstanding, authors have also noted that while there exists a suggestive relationship between cholinergic markers and cognitive impairment, no such relationship has yet to be established for any other transmitter.[149]

The present review underscores the need for prudence when interpreting the results of studies attributing either a selective role in a particular type of learning and memory, or a role for memory storage, to the cholinergic neuron or structure. As a research community we foster a tendency to overinterpret our results. Nonetheless, researchers have been able to rule out interference from a number of performance and nonassociative variables.

Where we have not been particularly successful is in identifying the contribution of attention to performance on tests of learning and memory. Research is clearly needed in somehow teasing out attentional and memory mechanisms. Otherwise we must restrict ourselves to definitions of learning that are sufficiently broad so as to include attentional, and, if Whishaw[172,243] would have his way, motor processes as well. The field of human information processing (with attentional components built into the models) may offer parsimonious alternatives for looking at cognition and memory. The model of working memory developed by Baddeley and Hitch[244,245] does include an attentional-controlling system. Building on Baddeley and Hitch's model, Warburton and Rusted[222] have proposed that memory and attentional processes "operate through a common, limited capacity" mechanism. In fact, Honig's[6] model of working memory necessarily involves an attentional component because the animal must pay attention to the various cues or information needed for the successful performance of foraging and avoidance behaviors.

To their credit, Deutsch and subsequent proponents of his hypothesis have sparked an interest in the cholinergic system and memory, which continues to the present day. Deutsch's claim that the engram is found at the site of the cholinergic synapse may no longer be a tenable concept, as stored memories are no longer viewed as "static entities".[2] Yet evocative findings that continue to appear, such as the reported alteration in the functioning of muscarinic synapses during associative learning in mollusks,[242] suggest that it would be premature to dismiss a possible role for the cholinergic neuron in learning. It will be interesting to see what implications these findings on invertebrates have for vertebrates and higher species.

The neural substrate for the locus of the engram is likely not the cholinergic neuron. Cholinergic neurons are active during many states that do not include obvious learning behaviors, such as in simple motor tasks or grooming,[246,247] as well as in species-specific and more complex emotional behaviors.[1] The cholinergic system does not seem to have a role in immediate memory or in the earliest registration phase of information processing, and may be involved in retrieval mechanisms in some circumstances. Whishaw[243] notes that because ACh is not fast acting and appears to play a more modulatory role at most synapses,[248] it is unlikely that it is "primarily involved in encoding new memories." That being said, there is substantial evidence that scopolamine appears to disrupt acquisitional as well as attentional processes. The challenge awaiting future researchers is to determine the relative involvement of cholinergic neurons in either process.

ACKNOWLEDGMENTS

I thank Shinshu Nakajima of Dalhousie University for his assistance, the Psychology Department of Saint Mary's University for photocopying privileges, and the Alzheimer Society of Nova Scotia for its support.

REFERENCES

1. Bartus, R. T., Dean, R., L., and Flicker, C., Cholinergic psychopharmacology: an integration of human and animal research on memory, in *Psychopharmacology: The Third Generation of Progress*, Meltzer, H. Y., Ed., Raven Press, New York, 1987, 219.
2. Collerton, D., Cholinergic function and intellectual decline in Alzheimer's disease, *Neuroscience*, 19, 1, 1986.
3. Hagan, J. J. and Morris, R. G. M., The cholinergic hypothesis of memory: a review of animal experiments, in *Handbook of Psychopharmacology*, Vol. 20, *Psychopharmacology of the Aging Nervous System,* Iversen, L. L., Iversen, S. D., and Snyder, S. H., Eds., Plenum Press, New York, 1988, 237.
4. Kopelman, M. D., The cholinergic neurotransmitter system in human memory and dementia: a review, *Q. J. Exp. Psychol.*, 38A, 535, 1986.
5. Sahakian, B. J., Cholinergic drugs and human cognitive performance, in *Handbook of Psychopharmacology*, Vol. 20, *Psychopharmacology of the Aging Nervous System,* Iversen, L. L., Iversen, S. D., and Snyder, S. H., Eds., Plenum Press, New York, 1987, 393.
6. Honig, W. K., Studies on working memory in the pigeon, in *Cognitive Processes in Animal Behavior*, Hulse, S. H., Fowler, H., and Honig, W. K., Eds., Lawrence Erlbaum Press, Hillsdale, NJ, 1978, 211.
7. McKim, W. A., *Drugs and Behavior*, Prentice-Hall, Englewood Cliffs, NJ, 1986, chap. 14.
8. Ray, O. and Ksir, C., *Drugs, Society, and Human Behavior*, 4th ed., Times Mirror/Mosby, St. Louis, 1987, chap. 15 and 295.
9. Le Strange, R., *A History of Herbal Plants*, Angus and Robertson, London, 1977.
10. Langdon-Brown, W., *From Witchcraft to Chemotherapy*, Cambridge University Press, Cambridge, 1941.
11. Schultes, R. E., The plant kingdom and hallucinogens (Part lll), *Bull. Narcotics*, 22 (1), 25, 1970.
12. Lambrechts, W. and Barkhouse, J., Postoperative amnesia, *Br. J. Anesthesia*, 33, 397, 1961.
13. Bartus, R. T., Dean, R., L., and Fisher, S. K., Cholinergic treatment for age-related memory disturbances, in *Treatment Development Strategies for Alzheimer's Disease*, Crook, T., Bartus, R. T., Ferris, S., and Gershon, S., Eds., Mark Powley, Madison, CT, 1986, chap. 17.

14. Gauss, C. J., Geburten im kunstlichem dammerschlaf, *Arch. Gynekol.*, 78, 579, 1906.

15. Alzheimer, A., Uber eine eigenartige erkrankung der hirnrinde, *Allg. Z. Psychiatrie*, 64, 146, 1907.

16. Hertz, A., Die gedentung der gahung fur die wirkung von scopolamin und ahnlichen substanzen auf gedengte reaktionen, *Z. Biol.*, 112, 104, 1960.

17. Meyers, B. and Domino, E. F., The effect of cholinergic blocking drugs on spontaneous alternation in rats, *Arch. Int. Pharmacodyn.*, 150, 3, 1964.

18. Leaf, R. C. and Muller, S. A. Effects of scopolamine on operant avoidance acquisition and retention, *Psychopharmacologia*, 9, 101, 1966.

19. Meyers, B., Some effects of scopolamine in a passive avoidance response in rats, *Psychopharmacologia*, 8, 111, 1965.

20. Grette Lydon, R. and Nakajima, S., Differential effects of scopolamine on working and reference memory depend upon level of training, *Pharmacol. Biochem. Behav.*, 43, 645, 1992.

21. Eckerman, D. A., Gordon, W. A., Edwards. J. D., MacPhail, R. C., and Gage, M. I., Effects of scopolamine, pentobarbitol, and amphetamine on radial arm maze performance in the rat, *Pharmacol. Biochem. Behav.*, 12, 595, 1980.

22. Kasckow, J. W., Thomas, G. J., and Herndon, R. M., Performance factors in regard to impaired memory and tolerance induced by atropine sulphate, *Physiol. Psychol.*, 12, 111, 1984.

23. Gilman, A. G., Goodman, L. S., and Gilman, A., Eds., *Goodman and Gilman's The Pharmacological Basis of Therapeutics*, 6th ed., Macmillan, New York, 1980.

24. Scoville, W. B. and Milner, B., Loss of recent memory after bilateral hippocampal lesions, *J. Neurol. Neurosurg. Psychiatry*, 20, 11, 1957.

25. Milner, B., Amnesia following operation on the temporal lobes, in *Amnesia*, Whitty, C. W. M. and Zangwill, O. L., Eds., Buttersworth, London, 1966, 109.

26. Cohen, N. J. and Corkin, S., The amnesic patient H. M.: learning and retention of a cognitive skill, *Soc. Neurosci. Abstr.*, 7, 235, 1981.

27. Squire, L. R., The hippocampus and the neuropsychology of memory, in *Neurobiology of the Hippocampus*, Seifert, W., Ed., Academic Press, London, 1983, 491 and 497.

28. Orbach, J., Milner, B., and Rasmussen, T., Learning and retention in monkeys after amygdala-hippocampus resection, *Arch. Neurol.*, 3, 230, 1960.

29. Horel, J. A., The neuroanatomy of amnesia: a critique of the hippocampal memory hypothesis, *Brain*, 101, 403, 1978.

30. Mishkin, M., Memory in monkeys severly impaired by combined but not by separate removal of amygdala and hippocampus, *Nature (London)*, 273, 297, 1978.

31. Seifert, W., Ed., *Neurobiology of the Hippocampus*, Academic Press, London, 1983.

32. Squire, L. R., *Memory and Brain*, Oxford University Press, Oxford, 1987, 193.

33. Zola-Morgan, S., Squire, L. R., and Mishkin, M., The neuroanatomy of amnesia: amygdala-hippocampus vs. temporal stem, *Science*, 218, 1337, 1982.

34. Douglas, R. J., The hippocampus and behavior, *Psychol. Bull.*, 67, 416, 1966.

35. Douglas, R. J. and Truncer, P. C., Parallel but independent effects of pentobarbital and scopolamine on hippocampus-related behavior, *Behav. Biol.*, 18, 359, 1976.

36. Lewis, P. R. and Shute, C. C. D., The cholinergic limbic system: projections to hippocampal formation, medial cortex, nuclei of the ascending reticular system, and the subfornical organ in supra-optical crest, *Brain*, 90, 521, 1967.

37. Bartus, R. T., Dean, R. L., Pontecorvo, M. J., and Flicker, C., The cholinergic hypothesis: a historical overview, current perspective, and future directions, in *Annals of the New York Academy of Sciences*, Vol. 444: *Memory Dysfunctions: An Integration of Animal and Human Research from Preclinical and Clinical Perspectives*, Olton, D. S., Gamzu, E., and Corkin, S., Eds., The New York Academy of Sciences, NY, 1985a, 332.

38. Yamamura, H. I., Kuhar, M. J., Greenberg, D., and Snyder, S. H., Muscarinic cholinergic receptor binding: regional distribution in monkey brain, *Brain Res.*, 66, 541, 1974.

39. Yamamura, H. I., Kuhar, M. J., and Snyder, S. H., *In vivo* identification of muscarinic cholinergic receptor binding in rat brain, *Brain Res.*, 80, 170, 1974.

40. Johnston, M. V., McKinney, M., and Coyle, J. F., Evidence for a cholinergic projection to neocortex from neurons in basal forebrain, *Proc. Nat'l. Acad. Sci. U.S.A.*, 76, 5392, 1979.

41. O'Keefe, J. and Nadel., L., The Hippocampus as a Cognitive Map, Oxford University Press, Oxford, 1978.

42. Olton, D. S. and Samuelson, R. J., Remembrance of places past: spatial memory in rats, *J. Exp. Psychol.: Anim. Behav. Processes*, 2, 97, 1976.

43. Olton, D. S. and Papas, B. C., Spatial memory and hippocampal function, *Neuropsychologia*, 17, 669, 1979.

44. Carlton, P. L., Cholinergic mechanisms in the control of behavior by the brain, *Psychol. Rev.*, 70, 19, 1963.

45. Deutsch, J. A., The cholinergic synapse and the site of memory, *Science*, 174, 788, 1971.

46. Deutsch, J. A., Hamburg, M. D., and Dahl, H., Anticholinesterase-induced amnesia and its temporal aspects, *Science*, 151, 221, 1966.

47. Deutsch, J. A. and Rogers, J. B., Cholinergic excitability and memory: animal studies and their clinical implications, in *Brain Acetylcholine and Neuropsychiatric Disease*, Davis, K. L. and Berger, P. A., Eds., Plenum Press, New York, 1979, 175.

48. Deutsch, J. A., The cholinergic synapse and the site of memory, in *The Physiological Basis of Memory*, 2nd ed., Deutsch, J. A., Ed., Academic Press, New York, 1983, 367.

49. Whitehouse, J. M., Effects of atropine on discrimination learning in the rat, *J. Comp. Physiol. Psychol.*, 57, 13, 1964.

50. Whitehouse, J. M., The effects of physostigmine on discrimination learning, *Psychopharmacologia*, 9, 183, 1966.

51. Bohdanecký, Z. and Jarvik, M. E., Impairment of one trial passive avoidance learning in mice by scopolamine, scopolamine methyl bromide and physostigmine, *Int. J. Neuropharmacol.*, 6, 217, 1967.

52. Burešová, O., Bureš, J., Bohdanecký, Z., and Weiss, T., Effect of atropine on learning, extinction, retention and retrieval in rats, *Psychopharmacologia*, 5, 255, 1964.

53. Dilts, S. L. and Berry, C. A., Effect of cholinergic drugs on passive avoidance in the mouse, *J. Pharmacol. Exp. Ther.*, 158, 279, 1967.

54. Glick, S. D. and Zimmerberg, B., Amnesic effects of scopolamine, *Behav. Biol.*, 7, 245, 1972.

55. Feigley, D. A., Beakey, W., and Saynisch, M. J., Effect of scopolamine on the reactivity of the albino rat to footshock, *Pharmacol. Biochem. Behav.*, 4, 255, 1976.

56. Heise, G. A., Conner, R., and Martin, R. A., Effects of scopolamine on variable intertrial interval spatial alternation and memory in the rat, *Psychopharmacology*, 49, 131, 1976.

57. Warburton, D. M. and Brown, K., Attenuation of stimulus sensitivity induced by scopolamine, *Nature (London)*, 230, 126, 1971.

58. Warburton, D. M. and Brown, K., The facilitation of discrimination performance by physostigmine sulfate, *Psychopharmacologia*, 27, 275, 1972.

59. Biederman, G. B., Forgetting of an operant response: physostigmine produced increases in escape latency in rats as a function of time of injection, *Q. J. Exp. Psychol.*, 22, 384, 1970.

60. Squire, L. R., Glick, S. D., and Goldfarb, J., Relearning at different times after training as affected by centrally and peripherally acting cholinergic drugs in the mouse, *J. Comp. Physiol. Psychol.*, 74, 41, 1971.

61. Stanes, M. D., Brown, C. P., and Singer, G., Effect of physostigmine on Y maze discrimination retention in the rat, *Psychopharmacologia*, 46, 269, 1976.

62. George, G. and Mellanby, J., A further study on the effect of physostigmine on memory in rats, *Brain Res.*, 81, 133, 1974.

63. George, G., Mellanby, H., and Mellanby, J., When does inhibition of brain acetylcholinesterase cause amnesia in rats?, *Brain Res.*, 122, 568, 1977.

64. Gaffan, D., Recognition impaired and association intact in the memory of monkeys after transection of the fornix, *J. Comp. Physiol. Psychol.*, 86, 1100, 1974.

65. Bohdanecký, Z., Jarvik, M. E., and Carley, J. C., Differential impairment of delayed matching in monkeys by scopolamine and scopolamine methylbromide, *Psychopharmacologia*, 11, 293, 1967.

66. Glick, S. D. and Jarvik, M. E., Differential effects of amphetamine and scopolamine on matching performance of monkeys with lateral frontal lesions, *J. Comp. Physiol. Psychol.*, 73, 307, 1970.

67. Evans, H. L., Scopolamine effects on visual discrimination: modifications related to stimulus control, *J. Pharmacol. Exp. Ther.*, 195, 105, 1975.

68. Bartus, R. T. and Johnson, H. R., Short term memory in the rhesus monkey: disruption from the anticholinergic scopolamine, *Pharmacol. Biochem. Behav.*, 5, 39, 1976.

69. Bartus, R. T., Evidence for a direct cholinergic involvement in the scopolamine-induced amnesia in monkeys: effects of concurrent administration of physostigmine and methylphenidate with scopolamine, *Pharmacol. Biochem. Behav.*, 9, 833, 1978.

70. Bartus, R. T., Fleming, D., and Johnson, H. R., Aging in the rhesus monkey: debilitating effects on short term memory, *J. Gerontol.*, 33, 858, 1978.

71. Bartus, R. T., Dean, R. L., and Fleming, D., Aging in the rhesus monkey: effects on visual discrimination learning and reversal learning, *J. Gerontol.*, 34, 209, 1979.

72. Bartus, R. T. and Dean, R. L., Recent memory in aged non-human primates: hypersensitivity to visual interference during retention, *Exp. Aging Res.*, 5, 385, 1979.

73. Davis, R. T., Old monkey behavior, *Exp. Gerontol.*, 13, 237, 1978.

74. Drachman, D. A. and Leavitt, J., Human memory and the cholinergic system: a relationship to aging?, *Arch. Neurol.*, 30, 113, 1974.

75. Ostfeld, A. M. and Aruguete, A., Central nervous system effects of hyoscine in man, *J. Pharmacol. Exp. Ther.*, 137, 133, 1962.

76. Crow, T. J. and Grove-White, I. G., An analysis of the learning deficit following hyoscine administration to man, *Br. J. Pharmacol.*, 49, 322, 1973.

77. Drachman, D. A. and Leavitt, J., Memory impairment in the aged: storage versus retrieval deficit, *J. Exp. Psychol.*, 93, 302, 1972.

78. Drachman, D. A., Memory and cognitive function in man: does the cholinergic syystem have a specific role?, *Neurology,* 27, 783, 1977.

79. Dundee, J. W. and Pandit, S. K., Anterograde amnesic effects of pethidine, hyoscine, and diazepam in adults, *Br. J. Pharmacol.,* 44, 140, 1972.

80. Crow, T. J. and Grove-White, I. G., A differential effect of atropine and hyoscine on human learning capacity, *Br. J. Pharmacol.,* 43, 464, 1971.

81. Safer, D. J. and Allen, R. P., The central effects of scopolamine in man, *Biol. Psychiat*ry, 3, 347, 1971.

82. Crow, T. J., Grove-White, I. G., and Ross, D. G., The specificity of the action of hyoscine on human learning, *Br. J. Clin. Pharmacol.*, 2, 367, 1975.

83. Ghoneim, M. M. and Mewaldt, S. P., Studies on human memory: the interaction of diazepam, scopolamine and physostigmine, *Psychopharmacology*, 52, 1, 1977.

84. Blessed, G., Tomlinson, B. E., and Roth, M., The association between quantitative measures of dementia and of senile change in the cerebral grey matter of elderly subjects, *Br. J. Psychiatry*, 114, 797, 1968.

85. Davies, P., An update on the neurochemistry of Alzheimer's disease, in *The Dementias*, Mayeux, R. and Rosen, W. G., Raven Press, New York, 1983, 75.

86. Bowen, D. M., Smith, C. B., White, P., and Davison, A. N., Neurotransmitter-related enzymes and indices of hypoxia in senile dementa and other abiotrophies, *Brain* , 99, 459, 1976.

87. Davies, P. and Maloney, A. J. R., Selective loss of central cholinergic neurons in Alzheimer's disease (letter to the editor), *Lancet*, 2, 1403, 1976.

88. Perry, E. K., Perry, R. H., Blessed, G., and Tomlinson, B. E., Necropsy evidence of central cholinergic deficits in senile dementia, *Lancet*, 1, 189, 1977.

89. Perry, E. K., Tomlinson, B. E., Blessed, G., Bergman, K., Gibson, P. H., and Perry, R. H., Correlation of cholinergic abnormalities with senile plaques and mental test scores in senile dementia, *Br. Med. J.*, 2, 1457, 1978.

90. Coyle, J. T., Price, D. L., and DeLong, M. D., Alzheimer's disease: a disorder of cortical cholinergic innervation, *Science*, 219, 1184, 1983.

91. Bowen, D. M., Smith, C. B., White, P., Flack, R. H., Carrassco, L. H., Gedye, J. L., and Davison, A. N., Chemical pathology of the organic dementias. II. Quantitative estimation of cellular changes in post-mortem brains, *Brain*, 100, 427, 1977.

92. Drachman, D. A. and Sahakian, B. J., Effects of cholinergic agents on human learning and memory, in *Nutrition and the Brain*, Vol. 5, Barbeau, A., Growdon, J. H., and Wurtman, R. J., Eds., Raven Press, New York, 1979, 351 and 362.

93. Collerton, D., Problems in the cognitive neurochemistry of Alzheimer's disease, in *Cognitive Neurochemistry*, Stahl, S. M., Iversen, S. D., and Goodman, E. C., Eds., Oxford University Press, Oxford, 1987, 272.

94. Neary, D., Snowdon, J. S., Mann, D. M. A., Bowen, D. M., Sims, N. R., Northen, B., Yates, P. O., and Davison, A. N., Alzheimer's disease: a correlative study, *J. Neurol., Neurosurg. Psychiatry*, 49, 163, 1986.

95. Mann, D. M. A., Neuropathological and neurochemical aspects of Alzheimer's disease, in *Handbook of Psychopharmacology*, Vol. 20, *Psychopharmacology of the Aging Nervous System,* Iversen, L. L., Iversen, S. D., and Snyder, S. H., Eds., Plenum Press, New York, 1987, 1 and 48.

96. Ball, M. J., Neurofibrillary tangles and the pathogenesis of dementia. A quantitative study, *Neuropathol. Appl. Neurobiol.*, 2, 395, 1976.

97. Whitehouse, P. J., Price, D. L., Clark, A. W., Coyle, J. T., and DeLong, M. R., Alzheimer's disease: evidence for selective loss of cholinergic neurons in the nucleus basalis, *Ann. Neurol.*, 10, 122, 1981.

98. Nakano, I. and Hirano, A., Loss of large neurons of the medial septal nucleus in an autopsy case of Alzheimer's disease, *J. Neuropath. Exp. Neurol.*, 41, 341, 1982.

99. Ball, M. J. and Nuttall, K., Neurofibrillary tangles, granulovacuolar degeneration and neuron loss in Down's syndrome. Quantitative comparison with Alzheimer's dementia, *Ann. Neurol.*, 7, 462, 1980.

100. Price, D. L., Whitehouse, P. J., Strubble, R. G., Coyle, J. T., Clark, A. W., DeLong, M. R., Cork, L. C., and Hedreen, J. C., Alzheimer's disease and Down's syndrome, *Ann. N.Y. Acad. Sci.*, 396, 145, 1982.

101. Kish, S. J., Munir, E.-A., Schut, T., Leach, L., Oscar-Berman, M., and Freedman, M., Cognitive deficits in olivopontocerebellar atrophy; implications for the cholinergic hypothesis of Alzheimer's dementia, *Ann. Neurol.*, 24, 200, 1988.

102. Bartus, R. T., Dean, R. L., Beer, B., and Lippa, A. S., The cholinergic hypothesis of geriatric memory dysfunction, *Science*, 217, 408, 1982.

103. Fibiger, H. C., Cholinergic mechanisms in learning, memory and dementia: a review of recent evidence, *Trends Neurosci.*, 14, 220, 1991.

104. Francis, P. T., Pangalos, M. N., and Bowen, D. M., Animal and drug modelling for Alzheimer synaptic pathology, *Prog. Neurobiol.*, 39, 517, 1992.

105. Perry, E. K., Irving, D., and Perry, R. H., Letters to the editor, cholinergic controversies, *Trends Neurosci.*, 14, 483, 1991.

106. Mann, D. M. A., Lincoln, J., Yates, P. O., Stamp, J. E., and Toper, S., Changes in monoamine containing neurons of the human CNS in senile dementia, *Br. J. Psychiatry*, 136, 533, 1980.

107. Perry, E. K., Tomlinson, B. E., Blessed, G., Perry, R. H., Cross, A. J., and Crow, T. J., Neuropathological and biochemical observations on the noradrenergic system in Alzheimer's disease, *J. Neurol. Sci.*, 51, 279, 1981a.

108. Vijayashankar, N. and Brody, H., A quantitative study of the pigmented neurons in the nuclei locus coeruleus and subcoeruleus in man as related to aging, *J. Neuropathol. Exp. Neurol.*, 38, 490, 1979.

109. Adolfsson, R., Gottfries, C. G., Roos, B. E., and Winblad, B., Changes in brain catecholamines in patients with dementia of Alzheimer type, *Br. J. Psychiatry*, 135, 216, 1979.

110. Francis, P. T., Bowen, D. M., Lowe, S. L., Neary, D., Mann, D. M. A., and Snowdon, J. S., Somatostatin content and release measured in cerebral biopsies from demented patients, *J. Neurol. Sci.*, 78, 1, 1987.

111. Perry, R. H., Dockray, G. J., Dimaline, R., Perry, E. K., Blessed, G., and Tomlinson, B. E., Neuropeptides in Alzheimer's disease, depression and schizophrenia. A post mortem analysis of vasointestinal polypeptide and cholecystokinin in cerebral cortex, *J. Neurol. Sci.*, 51, 465, 1981b.

112. Perry, E. K., Cortical neurotransmitter chemistry in Alzheimer's disease, in *Psychopharmacology: The Third Generation of Progress*, Meltzer, H. Y., Ed., Raven Press, New York, 1987, 887.

113. Levin, E. D., Nicotinic systems and cognitive function, *Psychopharmacology*, 108, 417, 1992.

114. Bagne, C. A., Pomara, N., Crook, T., and Gershon, S., Alzheimer's disease: strategies for treatment and research, in *Treatment Development Strategies for Alzheimer's Disease*, Crook, T., Bartus, R. T., Ferris, S., and Gershon, S., Eds., Mark Powley, Madison, CT, 1986, chap. 26.

115. Whitehouse, P. J., Development of neurotransmitter-specific therapeutic approaches in Alzheimer's disease, in *Treatment Development Strategies for Alzheimer's Disease*, Crook, T., Bartus, R. T., Ferris, S., and Gershon, S., Eds., Mark Powley, Madison, CT, 1986, chap. 20.

116. Lishman, W. A., *Organic Psychiatry, the Psychological Consequences of Cerebral Disorder*, The Alden Press, Oxford, 1978.

117. Kopelman, M. D., How far could cholinergic depletion account for the memory deficits of Alzheimer-type dementia or the alcoholic Korsakoff syndrome?, in *Cognitive Neurochemistry*, Stahl, S. M., Iversen, S. D., and Goodman, E. C., Eds., Oxford University Press, Oxford, 1987, 303.

118. Corkin, S., Some relationships between global amnesia and the memory impairments in Alzheimer's disease, in *Alzheimer's Disease: A Report of Progress in Research. Aging,* 19, Corkin, S., Davis, K. L., Growdon, J. H., Usdin, E., and Wurtman, R. J., Eds., Raven Press, New York, 1982, 149.

119. Fuld, P. A., Psychometric differentiation of the dementias: an overview, in *Alzheimer's Disease: The Standard Reference*, Reisberg, B., Ed., Free Press, New York, 1983, 201.

120. Weingartner, H., Kaye, W., Smallberg, S. A., Ebert, H., Gillin, J. C., and Sitaram, N., Memory failures in progressive idiopathic dementia, *J. Abnorm. Psychol.*, 90, 187, 1981.

121. Warrington, E. K., The selective impairment of semantic memory, *Q. J. Exp. Psychol.*, 27, 635, 1975.

122. Harris, S. J. and Dowson, J. H., Recall of a 10-word list in the assessment of dementia in the elderly, *Br. J. Psychiatry*, 141, 524, 1982.

123. Tulving, E. and Colotla, V. A., Free recall of trilingual lists, *Cognit. Psychol.*, 1, 86, 1970.

124. Wilson, R. S., Bacon, L. D., Fox, J. H., and Kaszniak, A. W., Primary memory and seconday memory in dementia of the Alzheimer type, *J. Clin. Neuropsychol.*, 5, 337, 1983.

125. Kopelman, M. D. and Corn, T. H., Cholinergic 'blockade' as a model for cholinergic depletion, *Brain*, 111, 1079, 1988.

126. Caine, E. D., Weingartner, H., Ludlow, C. L., Cudahy, E. A., and Wehry, S., Qualitative analysis of scopolamine-induced amnesia, *Psychopharmacology*, 74, 74, 1981.

127. Beatty, W. W., Butters, N., and Janowsky, D., Patterns of memory failure after scopolamine treatment: implications for cholinergic hypotheses of dementia, *Behav. Neural Biol.*, 45, 196, 1986.

128. Tröster, A. I., Beatty, W. W., Staton, R. D., and Rorabaugh, A. G., Effects of scopolamine on anterograde and remote memory in humans, *Psychobiology*, 17, 12, 1989.

129. Kopelman, M. D., Rates of forgetting in Alzheimer-type dementia and Korsakoff's syndrome, *Neuropsychologia*, 23, 623, 1985.

130. Morris, R. G. M., Short-term forgetting in senile dementia of the Alzheimer's type, *Cognit. Neuropsychol.*, 3, 77, 1986.

131. Rusted, J. M. and Warburton, D. M., The effects of scopolamine on working memory in healthy young volunteers, *Psychopharmacology*, 96, 145, 1988.

132. Ferris, S. H., Crook, T., Clark, E., McCarthy, M., and Rae, D., Facial recognition memory deficits in normal aging and senile dementia, *J. Gerontol.*, 35, 707, 1980.

133. Gutzmann, H., Klimitz, H., and Avdaloff, W., Correlations between psychopathology, psychological test results and computerized tomography changes in senile dementia, *Arch. Gerontol. Geriatr.*, 1, 241, 1982.

134. Flicker, C., Ferris, S. H., Crook, T., and Bartus, R. T., Implications of memory and language dysfunction in the naming deficit of senile dementia, *Brain Language*, 31, 187, 1987.

135. Beatty, W. W., Salmon, D. P., Butters, N., Heindel, W. C., and Granholm, E. L., Retrograde amnesia in patients with Alzheimer's disease or Huntington's disease, *Neurobiol. Aging*, 9, 181, 1988.

136. Dunne, M. P., Statham, D., Raphael, B., Kemp, R., and Kelly, B., Further evidence that scopolamine can improve verbal fluency, *J. Psychopharmacol.*, 7, 159, 1993.

137. Lines, C. R., Preston, G. C., Broks, P., and Dawson, C., The effects of scopolamine on retrieval from semantic memory, *J. Psychopharmacol.*, 5(3), 234, 1991.

138. Martin, A., Brouwers, P., Cox, C., and Fedio, P., On the nature of the verbal memory deficit in Alzheimer's disease, *Brain Language*, 25, 323, 1985.

139. Mewaldt, S. P. and Ghoneim, M. M., The effects and interactions of scopolamine, physostigmine and methamphetamine on human memory, *Pharmacol. Biochem. Behav.*, 10, 205, 1979.

140. Block, R. I., DeVoe, M., Stanley, B., Stanley, M., and Pomara, M., Memory performance in individuals with primary degenerative dementia: its similarity to diazepam-induced impairments, *Exp. Aging Res.*, 11, 151, 1985.

141. Penetar, D. M. and McDonough, J. H., Jr., Effects of cholinergic drugs on delayed match-to-sample performance of rhesus monkeys, *Pharmacol. Biochem. Behav.*, 19, 963, 1983.

142. Aigner, T. G. and Mishkin, M., The effects of physostigmine and scopolamine on recognition memory in monkeys, *Behav. Neural Biol.*, 45, 81, 1986.

143. Ridley, R. M., Bowes, P. M., Baker, H. F., and Crow, T. J., An involvement of acetylcholine in object discrimination learning and memory in the marmoset, *Neuropsychologia*, 22, 253, 1984.

144. Dean, R. L. and Bartus, R. T., Behavioral models of aging in nonhuman primates, in *Handbook of Psychopharmacology*, Vol. 20, *Psychopharmacology of the Aging Nervous System,* Iversen, L. L., Iversen, S. D., and Snyder, S. H., Eds., Plenum Press, New York, 1987, chap. 8.

145. Dean, R. L., Beer, B., and Bartus, R. T., Drug induced memory impairments in nonhuman primates, *Soc. Neurosci. Abstr.*, 8, 322, 1982.

146. Pontecorvo, M. J. and Evans, H. C., Effects of aniracetam on delayed matching-to-sample performance of monkeys and pigeons, *Pharmacol. Biochem. Behav.*, 22, 745, 1985.

147. Miller, L. L. and Branconnier, R. J., Cannabis: effects on memory and the cholinergic limbic system, *Psychol. Bull.*, 93, 441, 1983.

148. Martin, B. R., Cellular effects of cannabinoids, *Pharmacol. Rev.*, 38, 45, 1986.

149. Sarter, M., Bruno, J. P., and Dudchenko, P., Activating the damaged basal forebrain cholinergic system: tonic stimulation vs. signal amplification, *Psychopharmacology*, 101, 1, 1990.

150. Squire, L. R. and Zola-Morgan, S., The neurology of memory: the case for correspondence between the findings for human and nonhuman primate, in *The Physiological Basis of Memory,* Deutsch, J. A., Ed., Academic Press, New York, 1983, 199.

151. Santi, A., Hanemaayer, C., and Reason, W., The effect of scopolamine on reference and working memory in pigeons, *Anim. Learn. Behav.*, 15, 395, 1987.

152. Morris, R. G. M., An attempt to dissociate "spatial-mapping" and "working memory" theories of hippocampal function, in *Neurobiology of the Hippocampus*, Seifert, F., Ed., Academic Press, New York, 1983, 405.

153. Mundy, W. R. and Iwamoto, E. T., Nicotine impairs acquisition of radial maze performance in rats, *Pharmacol. Biochem. Behav.*, 30, 119, 1988.

154. Stevens, R., Scopolamine impairs spatial maze performance in rats, *Physiol. Behav.*, 27, 385, 1981.

155. Watts, J., Stevens, R., and Robinson, C., Effects of scopolamine on radial maze performance in rats, *Physiol. Behav.*, 26, 845, 1981.

156. Hiraga, Y. and Iwasaki, T., Effects of cholinergic and monoaminergic antagonists and tranquilizers upon spatial memory in rats, *Pharmacol. Biochem. Behav.*, 20, 205, 1984.

157. Burešová, O. and Bureš, J., Radial maze as a tool for assessing the effect of drugs on the working memory of rats, *Psychopharmacology*, 77, 268, 1982.

158. Burešová, O., Bolhuis, J. J., and Bureš, J., Differential effects of cholinergic blockade on performance of rats in the water tank navigation task and in a radial water maze, *Behav. Neurosci.*, 100, 476, 1986.

159. Beatty, W. W. and Bierley, R. A., Scopolamine degrades spatial working memory but spares spatial reference memory: dissimilarity of anticholinergic effect and restriction of distal visual cues, *Pharmacol. Biochem. Behav.*, 23, 1, 1985.

160. Wirsching, B. A., Beninger, R. J., Jhamandas, K., Boegman, R. J., and El-Defrawy, S. R., Differential effects of scopolamine on working and reference memory of rats in the radial maze, *Pharmacol. Biochem. Behav.*, 20, 659, 1984.

161. Okaichi, H., Oshima, Y., and Jarrard, L. E., Scopolamine impairs both working and reference memory in rats: a replication and extension, *Pharmacol. Biochem. Behav.*, 34, 599, 1989.

162. Levy, A., Kluge, P. B., and Elsmore, T. F., Radial arm maze performance of mice: acquisition and atropine effects, *Behav. Neural Biol.*, 39, 229, 1983.

163. Brito, G. N. O., Davis, B. J., Stopp, L. C., and Stanton, M. E., Memory and the septo-hippocampal cholinergic system in the rat, *Psychopharmacology*, 81, 315, 1983.

164. Bartolini, L., Risaliti, R., and Pepeu, G., Effect of scopolamine and nootropic drugs on rewarded alternation in a T-maze, *Pharmacol. Biochem. Behav.*, 43, 1161, 1992.

165. Beninger, R. J., Jhamandas, K., Boegman, R. J., and El-Defrawy, S. R., Effects of scopolamine and unilateral lesions of the basal forebrain on T-maze spatial discrimination and alternation in rats, *Pharmacol. Biochem. Behav.*, 24, 1353, 1986a.

166. Mallet, P. E. and Beninger, R. J., The double Y-maze as a tool for assessing memory in rats, *Neurosci. Prot.*, 93-010-02, 1, 1993.

167. Ingles, J. L., Beninger, R. J., Jhamandas, K., and Boegman, R. J., Scopolamine injected into the rat amygdala impairs working memory in the double Y-maze, *Brain Res. Bull.*, 32, 339, 1993.

168. Beninger, R. J., Ingles, J. L., Mackenzie, P. J., Jhamandas, K., and Boegman, R. J., Muscimol injections into the nucleus basalis magnocellularis of rats: selective impairment of working memory in the double Y-maze, *Brain Res.*, 597, 66, 1992.

169. Morris, R. G. M., Spatial localization does not require the presence of local cues, *Learn. Motiv.*, 12, 239, 1981.

170. Ellen, P., Taylor, H. S., and Wages, C., Cholinergic blockade effects on spatial integration vs. cue discrimination performance, *Behav. Neurosci.*, 100, 611, 1986.

171. Sutherland R. J., Whishaw, I. Q., and Regehr, J. C., Cholinergic receptor blockade impairs spatial localization by use of distal cues in the rat, *J. Comp. Physiol. Psychol.*, 96, 563, 1982.

172. Whishaw, I. Q., Cholinergic receptor blockade in the rat impairs locale but not taxon strategies for place navigation in a swimming pool, *Behav. Neurosci.*, 99, 979, 1985.

173. Whishaw, I. Q., O'Connor, W. T., and Dunnett, S. B., Disruption of central cholinergic systems in the rat by basal forebrain lesions or atropine: effects on feeding, sensorimotor behavior, locomotor activity and spatial navigation, *Behav. Brain Res.*, 17, 103, 1985.

174. Whishaw, I. Q. and Tomie, J.-A., Cholinergic receptor blockade produces impairments in a sensorimotor subsystem for place navigation in the rat: evidence from sensory, motor, and acquisition tests in a swimming pool, *Behav. Neurosci.*, 101, 603, 1987.

175. Hagan, J. J., Tweedie, F., and Morris, R. G. M., Lack of task specificity and absence of posttraining effects of atropine on learning, *Behav. Neurosci.*, 100, 483, 1986.

176. Whishaw, I. Q. and Petrie, B. F., Cholinergic blockade in the rat impairs strategy selection but not learning and retention of nonspatial visual discrimination problems in a swimming pool, *Behav. Neurosci.*, 102, 662, 1988.

177. Okaichi, H. and Jarrard, L. E., Scopolamine impairs performance of a place and cue task in rats, *Behav. Neural Biol.*, 35, 319, 1982.

178. Burešová, O., Bureš, J., Oitzl, M. S., and Zahálka, A., Radial maze in the water tank: an aversely motivated spatial working memory task, *Physiol. Behav.*, 34, 1003, 1985.

179. Gooding, P. R., Rush, J. R., and Beatty, W. W., Scopolamine does not disrupt spatial working memory in rats, *Pharmacol. Biochem. Behav.*, 16, 919, 1982.

180. Beatty, W. W. and Bierley, R. A., Scopolamine impairs encoding and retrieval of spatial working memory in rats, *Physiol. Psychol.*, 14, 82, 1986.

181. Nilsson, O. G. and Gage, F. H., Anticholinergic sensitivity in the aging rat septo-hippocampal system as assessed in a spatial memory task, *Neurobiol. Aging*, 14, 487, 1993.

182. Goodrick, C., Learning by mature-young and aged Wistar rats as a function of test complexity, *J. Gerontol.*, 27, 353, 1972.

183. Hagan, J. J., Salamone, J. D., Simpson, J., Iversen, S. D., and Morris, R. G. M., Place navigation in rats is impaired by lesions of medial septum and diagonal band but not nucleus basalis magnocellularis, *Behav. Brain Res.*, 27, 9, 1988.

184. Dunnett, S. B., Toniolo, G., Fine, A., Ryan, C. N., Bjorklund, A., and Iversen, S. D., Transplantation of embryonic ventral forebrain neurons to the neocortex of rats with lesions of nucleus basalis magnocellularis. II. Sensorimotor and learning impairments, *Neuroscience,* 16, 787, 1985.

185. Dunnett, S. B., Whishaw, I. Q., Jones, G. H., and Bunch, S. T., Behavioural, biochemical and histochemical effects of different neurotoxic amino acids injected into nucleus basalis magnocellularis of rats, *Neuroscience,* 20, 653, 1987.

186. Mandel, R. J. and Thal, L. J., Physostigmine improves water maze performance following nucleus basalis magnocellularis lesions in the rat, *Psychopharmacology*, 96, 421, 1988.

187. Page, K. J., Everitt, B. J., Robbins, T. W., Marston, H. M., and Wilkinson L. S., Dissociable effects on spatial maze and passive avoidance acquisition and retention following AMPA- and ibotenic acid-induced excitotoxic lesions of the basal forebrain in rats: differential dependence on cholinergic neuronal loss, *Neuroscience*, 43, 457, 1991.

188. Riekkinen, M., Riekkinen, P., and Riekkinen, P., Jr., Comparative effects of quisqualic and ibotenic acid nucleus basalis magnocellularis lesions on water maze and passive avoidance performance, *Brain Res. Bull.*, 27, 119, 1991.

189. Etherington, R., Mittleman, G., and Robbins, T. W., Comparative effects of nucleus basalis and fimbria-fornix lesions on delayed matching and alternation tests of monkeys, *Neurosci. Res. Commun.*, 1, 135, 1987.

190. Biggan, S. L., Beninger, R. J., Cockhill, J., Jhamandas, K., and Boegman, R. J., Quisqualate lesions of rat NBM: selective effects on working memory in a double Y-maze, *Brain Res. Bull.*, 26, 613, 1991.

191. Markowska, A. L., Wenk, G. L., and Olton, D. S., Nucleus basalis magnocellularis and memory: differential effects of two neurotoxins, *Behav. Neural Biol.*, 54, 13, 1990.

192. Wenk, G. L., Markowska, A. L., and Olton, D. S., Basal forebrain lesions and memory: alterations in neurotensin, not acetylcholine, may cause amnesia, *Behav. Neurosci.*, 103, 765, 1989.

193. Beninger, R. J., Wirsching, B. A., Jhamandas, K., Boegman, R. J., and El-Defrawy, S. R., Effects of altered cholinergic function on working and reference memory in the rat, *Can. J. Physiol. Pharmacol.*, 64, 376, 1986b.

194. Hepler, D. J., Olton, D. S., Wenk, G. L., and Coyle, J. T., Lesions in nucleus basalis magnocellularis and medial septal area of rats produce qualitatively similar memory impairments, *J. Neurosci.*, 5, 866, 1985.

195. Knowlton, B. J., Wenk, G. L., Olton, D. S., and Coyle, J. T., Basal forebrain lesions produce a dissociation of trial-dependent and trial-independent memory performance, *Brain Res.*, 345, 315, 1985.

196. Beninger, R. J., Jhamandas, K., Boegman, R. J., and El-Defrawy, S. R., Kynurenic acid-induced protection of neurochemical and behavioral deficits produced by quinolinic acid injections into the nucleus basalis of rats, *Neurosci. Lett.*, 68, 317, 1986c.

197. Wirsching, B. A., Beninger, R. J., Jhamandas, K., Boegman, R. J., and Bialik, M., Kynurenic acid protects against the neurochemical and behavioral effects of unilateral quinolinic acid injections into the nucleus basalis of rats, *Behav. Neurosci.*, 103, 90, 1989.

198. Murray, C. L. and Fibiger, H. C. Learning and memory deficits after lesions of the nucleus basalis magnocellularis: reversal by physostigmine, *Neuroscience*, 14, 1025, 1985.

199. Hodges, H., Allen, Y., Kershaw, T., Lantos, P. L., Gray, J. A., and Sinden, J., Effects of cholinergic-rich neural grafts on radial maze performance of rats after excitotoxic lesions of the forebrain cholinergic projection system. I. Amelioration of cognitive deficits by transplants into cortex and hippocampus but not into basal forebrain, *Neuroscience*, 45, 587, 1991a.

200. Hodges, H., Allen, Y., Sinden, J., Lantos, P. L., and Gray, J. A., Effects of cholinergic-rich neural grafts on radial maze performance of rats after excitotoxic lesions of the forebrain cholinergic projection system. II. Cholinergic drugs as probes to investigate lesion-induced deficits and transplant-induced functional recovery, *Neuroscience*, 45, 609, 1991b.

201. Murray, C. L. and Fibiger, H. C., Pilocarpine and physostigmine attenuate spatial memory impairments produced by lesions of the nucleus basalis magnocellularis, *Behav. Neurosci.*, 100, 1986.

202. Kesner, R. P., DiMattia, B. V., and Crutcher, K. A., Evidence for neocortical involvement in reference memory, *Behav. Neural Biol.*, 47, 40, 1987.

203. Bartus, R. T., Flicker, C., Dean, R. L., Pontecorvo, M., Figueiredo, J. C., and Fisher, S. K., Selective memory loss following nucleus basalis lesions: long term behavioral recovery despite persistent cholinergic deficiencies, *Pharmacol. Biochem. Behav.*, 23, 125, 1985b.

204. Hasher, L. and Zacks, R. T., Automatic and effortful processes in memory, *J. Exp. Psychol. Gen.*, 108, 356, 1979.

205. Kesner, R. P., Crutcher, K. A., and Omana, H., Memory deficits following nucleus basalis magnocellularis lesions may be mediated through limbic, but not neocortical, targets, *Neuroscience*, 38, 93, 1990.

206. Boegman, R. J., Cockhill, J., Jhamandas, K., and Beninger, R. J., Excitotoxic lesions of rat basal forebrain: differential effects of choline acetyltransferase in the cortex and amygdala, *Neuroscience*, 51, 129, 1992.

207. Jarrard, L. E., Selective hippocampal lesions and behavior: effect of kainic acid lesions on performance of place and cue tasks, *Behav. Neurosci.*, 97, 873, 1983.

208. Oades, R. D., Impairments of search behavior in rats after haloperidol treatment, hippocampal or neocortical damage suggest a mesocorticolimbic role in cognition, *Biol. Psychol.*, 12, 77, 1981.

209. Cheal, M. L., Scopolamine disrupts maintenance of attention rather than memory processes, *Behav. Neural Biol.*, 33, 163, 1981.

210. Spencer, D. G., Pontecorvo, M. J., and Heise, G. A., Central cholinergic involvement in working memory: effects of scopolamine on continuous nonmatching and discrimination performance in the rat, *Behav. Neurosci.*, 99, 1049, 1985.

211. Tilson, H. A., McLamb, R. L., Shaw, S., Rogers, B. C., Pediaditakis, P., and Cook, L., Radial-arm maze deficits produced by colchicine administered into the area of the nucleus basalis are ameliorated by cholinergic agents, *Brain Res.*, 438, 83, 1988.

212. Hodges, H., Sinden, J., Turner, J. J., Netto, C. A., Sowinski, P., and Gray, J. A., Nicotine as a tool to characterise the role of the forebrain cholinergic projection system in cognition, in *The Biology of Nicotine: Current Research Issues*, Lippiello, P. M., Collins, A. C., Gray, J. A., and Robinson, J. H., Eds., Raven Press, New York, 1992, 157.

213. Robbins, T. W., Everitt, B. J., Ryan, C. N., Marston, H. M., Jones, G. H., and Page, K. J., Comparative effects of quisqualic and ibotenic acid-induced lesions of the substantia innominata and globus pallidus on the acquisition of a conditional discrimination: differential effects on cholinergic mechanisms, *Neuroscience*, 28, 337, 1989.

214. Jones, G. M. M., Sahakian, B. J., Levy, R., Warburton, D. M., and Gray, J. A., Effects of acute sub-cutaneous nicotine on attention, information processing, and short-term memory in Alzheimer's disease, *Psychopharmacology*, 108, 485, 1992.

215. Sahakian, B. J. and Jones, G. M. M., The effects of nicotine on attention, information processing, and working memory in patients with dementia of the Alzheimer type, in *Proc. of the Int'l. Symp. on Nicotine: Effects of Nicotine on Biological Systems*, Birkhäuser, Berlin, 1990.

216. Sahakian, B., Jones, G., Levy, R., Gray, J., and Warburton, D., The effects of nicotine on attention, information processing, and short-term memory in patients with dementia of the Alzheimer type, *Br. J. Psychiatry*, 154, 797, 1989.

217. Rusted, J. M. and Eaton-Williams, P., Distinguishing between attentional and amnestic effects in information processing: the separate and combined effects of scopolamine and nicotine on verbal free recall, *Psychopharmacology*, 104, 363, 1992.

218. Rusted, J. M. and Warburton, D. M., Facilitation of memory by post-trial administration of nicotine: evidence for an attentional explanation, *Psychopharmacology*, 108, 452, 1992.

219. Warburton, D. M., Rusted, J. M., and Fowler, J., A comparison of the attentional and consolidation hypotheses for the facilitation of memory by nicotine, *Psychopharmacology*, 108, 443, 1992.

220. Sahakian, B. J., Owen, A. M., Morant, N. J., Eagger, S. A., Boddington, S., Crayton, L., Crockford, H. A., Crooks, M., Hill, H., and Levy, R., Further analysis of the cognitive effects of tetrahydroaminoacridine (THA) in Alzheimer's disease: assessment of attentional and mnemonic function using CANTAB, *Psychopharmacology*, 110, 395, 1993.

221. Petersen, R. C., Scopolamine-induced learning failures in man, *Psychopharmacologia*, 52, 283, 1977.

222. Warburton, D. M. and Rusted, J. M., Cholinergic control of cognitive resources, *Neuropsychobiology*, 28, 43, 1993.

223. Warburton, D. M., A state model for mental effort, in *Energetics and Human Information Processing*, Hockey, G. R. J., Gaillard, A. W. K., and Coles, M. G. H., Eds., Martinus Nijhoff Publishers, Dordrecht, 1986, 217.

224. Dunne, M. P. and Hartley, L. R., The effects of scopolamine upon verbal memory: evidence for an attentional hypothesis, *Acta Psychol.*, 58, 205, 1985.

225. Dunne, M. P. and Hartley, L. R., Scopolamine and the control of attention in humans, *Psychopharmacology*, 89(1), 94, 1986.

226. Dunne, M. P., Scopolamine and sustained retrieval from semantic memory, *J. Psychopharmacol.*, 4(1), 13, 1990.

227. Frith, C. D., Richardson, J. T. E., Samuel, M., Crow, T. J., and McKenna, P. J., The effects of intravenous diazepam and hyoscine upon human memory, *Q. J. Exp. Psychol.*, 36A, 133, 1984.

228. Rusted, J. M. and Warburton, D. M., Effects of scopolamine on verbal memory; a retrieval or acquisition deficit?, *Neuropsychobiology*, 21, 76, 1989.

229. Jorm, A. F., Controlled and automatic information processing in senile dementia: a review, *Psychol. Med.*, 15, 1985.

230. Bresnahan, E. L., Wiser, P. R., Muth, N. J., and Ingram, D. K., Delayed matching-to-sample performance by rats in a new avoidance-motivated maze: response to scopolamine and fimbria-fornix lesions, *Physiol. Behav.*, 51, 735, 1992.

231. Sitaram, N., Weingartner, H., and Gillin, J. C., Human serial learning: enhancement with arecholine and choline and impairment with scopolamine, *Science*, 201, 274, 1978.

232. Newhouse, P. A., Sunderland, T., Tariot, P. N., Blumhardt, C. L., Weingartner, H., Mellow, A., and Murphy, D. L., Intravenous nicotine in Alzheimer's disease: a pilot study, *Psychopharmacology*, 95, 171, 1988.

233. Greenamyre, J. T., The role of glutamate in neurotransmission and neurologic disease, *Arch. Neurol.*, 43, 1058, 1986.

234. Maragos, W. F., Greenamyre, J. T., Penny, J. B., and Young, A. B., Glutamate dysfunction in Alzheimer's disease: an hypothesis, *Trends Neurosci.*, 10, 65, 1987.

235. Myhrer, T., Animal models of Alzheimer's disease: glutamatergic denervation as an alternative approach to cholinergic denervation, *Neurosci. Biobehav. Rev.*, 17, 195, 1993.

236. Blaker, W. D., Peruzzi, G., and Costa, E., Behavioral and neurochemical differentiation of specific projections in the septal-hippocampal cholinergic pathway of the rat, *Proc. Natl. Acad. Sci. U.S.A.*, 81, 1880, 1984.

237. Chrobak, J. J., Stackman, R. W., and Walsh, T. J., Intraseptal administration of muscimol produces dose-dependent memory impairments in the rat, *Behav. Neural Biol.*, 52, 357, 1989.

238. Brioni, J. D., Decker, M. W., Gamboa, L. P., Izquierdo, and McGaugh, J. L., Muscimol injections in the medial septum impair spatial learning, *Brain Res.*, 522, 227, 1990.

239. Curran, H. V., Schifano, F., and Lader, M., Models of memory dysfunction? A comparison of the effects of scopolamine and lorazepam on memory, psychomotor performance and mood, *Psychopharmacology*, 103, 83, 1991.

240. Rusted, J. M., Eaton-Williams, P., and Warburton, D. M., A comparison of the effects of scopolamine and diazepam on working memory, *Psychopharmacology*, 105, 442, 1991.

241. Curran, H. V., Benzodiazepines, memory, and mood: a review, *Psychopharmacology*, 105, 1, 1991.

242. Morielli, A. D., Matera, E. M., Kovac, M. P., Schrum, R. G., McCormack, K. J., and Davis, W. J., Cholinergic suppression: a postsynaptic mechanism of long-term associative learning, *Proc. Natl. Acad. Sci. U.S.A.*, 83, 4556, 1986.

243. Whishaw, I. Q., Dissociating performance and learning deficits on saptial navigation tasks in rats subjected to cholinergic muscarinic blockade, *Brain Res. Bull.*, 23, 347, 1989.

244. Baddeley, A. D., *Working Memory*, Clarendon Press, Oxford, 1986.

245. Baddeley, A. D. and Hitch, G., Working memory, in *Recent Advances in Learning and Motivation*, Vol. 8, Bower, G., Ed., Academic Press, New York, 1974.

246. Leung, L.-W. S., Pharmacology of theta phase shift in the hippocampal CA1 region of freely moving rats, *Electroencephalogr. Clin. Neurophysiol.*, 58, 457, 1984.

247. Vanderwolf, C. H., Near-total loss of "learning" and "memory" as a result of combined cholinergic and serotinergic blockade in the rat, *Behav. Brain Res.*, 23, 43, 1987.

248. Krnjevic, K., Acetycholine as transmitter in the cerebral cortex, in *Proc. Symp. Neurotransmitters and Cortical Function: From Molecules to Mind* (Montreal, 1986), Avoli, M., Reader, T. A., Dykes, R., and Gloor, P., Eds., Plenum Press, New York, 1988, 227.

249. Pope, A., Hess, H. H., and Lewin, E., Microchemical pathology of the cerebral cortex in pre-senile dementias, *Trans. Am. Neurol. Assoc.*, 89, 15, 1965.

250. Whitehouse, P. J., Hedreen, J. C., White, C. L., III, and Price, D. L., Basal forebrain neurons in the dementia of Parkinson disease, *Ann. Neurol.*, 13, 243, 1983.

Chapter 14

Nerve Growth Factor: Influence on Cholinergic Neurons in the CNS

Carola Eva

CONTENTS

I. INTRODUCTION

Target-derived growth factors that regulate neuronal survival and differentiation, and maintain neuronal phenotype may be of critical importance for normal development and mature function of the central nervous system (CNS). The classical studies of Hamburger and Levi-Montalcini on naturally occurring neuronal cell death, followed by the discovery of nerve growth factor (NGF) provided the first experimental support for the neurotrophic factor hypothesis.[1,2] In the past few years there has been tremendous progress in the study of neurotrophic factors and their receptors. The NGF gene family, collectively named neurotrophins, has been expanded to include at least three additional members which are structurally and functionally related to NGF: brain-derived neurotrophic factor (BDNF), neurotrophin 3 (NT-3), and neurotrophin 4 (NT-4).[3]

The physiological role of NGF on the sympathetic and neuronal crest-derived sensory neurons has been well characterized.[4,5] It is well established that NGF supports the development and survival of mammalian peripheral sympathetic and neural crestal-derived sensory neurons from the embryonic and postnatal periods throughout adulthood.[6-9] NGF is synthesized and released in limited amounts from target tissues, which in the periphery are nonneuronal cells. During development, a retrograde flow of NGF from the target into the nerve terminal, and up to the axon and the cell body, is established. The retrograde flow is initiated by the binding of NGF to its receptors located on the nerve terminal and, following internalization the NGF-NGF receptor complex, is retrogradely transported to the cell soma. The interaction of NGF with its receptors also initiates signal transduction events; whether signal transduction events occur only at the level of the nerve terminals or also during transport of the NGF-NGF receptor complex, still remains to be determined.

On the basis of findings obtained on sympathetic neurons, it was first anticipated that NGF may also affect catecholaminergic neurons in the CNS. However, findings reported by several laboratories clearly demonstrated that central dopaminergic and noradrenergic neurons do not respond to NGF.[10-12] More recently, evidence has been provided that NGF also regulates survival and development of certain populations of cholinergic neurons in the CNS. Central cholinergic neurons include the ascending projection neurons of the basal forebrain, striatal interneurons, neurons of the pontomesencephalic reticular formation located in the pedunculopontine and dorsolateral tegmental neurons, and motoneurons of the spinal cord.[13-18] The most extensive cholinergic pathways in the rat brain arise from cholinergic cell bodies in the basal forebrain. The rostromedial cells (medial septum and vertical limb of the diagonal band) project primarily to both sides of the hippocampus, while more caudal cells (horizontal limb of the diagonal band and nucleus basalis of Meynert [nbM]) project to the ipsilateral amygdala and to the ipsilateral regions of the neocortex. A variety of data obtained during the last 10 years indicate that, in the CNS, NGF acts as a neurotrophic factor for cholinergic neurons located in basal forebrain and striatum.

This review attempts to describe the main physiological features of NGF in the CNS. Since the recent cloning and characterization of the NGF receptors have expanded our knowledge of the molecular action of this neurotrophic factor, the largest part of the first section is the description of the molecular structure of the NGF receptors and their functional properties and interaction. The second part of the review summarizes current knowledge on the effect of NGF on basal forebrain and striatal cholinergic neurons. Emphasis is given to functional aspects of NGF in developing and adult brain, and to its pharmacological potential in the treatment of neurodegenerative diseases.

II. STRUCTURE OF NGF

NGF was originally purified from mouse submandibular glands.[19] This extremely rich source allowed the purification of NGF, the generation of antibodies, and the determination of the primary structure of the NGF precursor. NGF occurs as a complex of three subunits (alpha, beta and gamma; stoichiometry $\alpha_2\beta\gamma_2$) with a molecular weight of approximately 140 kDa and a sedimentation coefficient of 7 S (for a review see Greene and Shooter[6] and Thoenen and Barde[7]). The biological activity of NGF is carried out entirely by the β subunit, which exists as a 26-KDa dimer of two identical 118 amino acid chains held together by noncovalent forces. Each chain has three disulfide bonds that are important for biological activity. β-NGF is synthesized as a prepromolecule that undergoes subsequent proteolytic cleavage at both the amino and carboxyl termini to liberate the mature peptide. The nucleotide sequences of β-NGF have been determined in several species, including man, and appear to be highly conserved during evolution.[20-23] More recently, the crystal structure of NGF has been solved.[24] The protomer structure consists of three antiparallel pairs of strands that together form a flat surface. In the dimer, two subunits associate through this surface. Four loop regions, which contain many of the variable residues observed between different NGF-related molecules, may confer binding specificities to the receptors.

III. THE RECEPTORS FOR NGF

A. THE LOW-AFFINITY NGF RECEPTOR

Early binding studies with ^{125}I-NGF revealed two distinct kinetic classes of NGF receptors in chick embryonic sensory neurons.[25] The major population of receptors, the low-affinity NGF receptors (LNGFR), have a K_d of $10^{-9}M$, while the high-affinity NGF receptors (HNGFR) have a K_d of $10^{-11}M$. The two receptor types can also be distinguished by their dissociation rate constants (fast vs. slow), trypsin sensitivity, and detergent solubility.[26-28] The availability of monoclonal antibodies against the human and rat LNGFR permitted cloning of these receptors.[29,30] The isolated cDNA encodes for a 75-kDa transmembrane glycoprotein containing a single membrane spanning domain separating a slightly longer extracellular domain from a shorter cytoplasmic region. The LNGFR is expressed in all NGF-responsive cells, including sympathetic, sensory, and basal forebrain cholinergic neurons.[31 32] The extracellular domain of the LNGFR contains four cysteine-rich regions (loops) which were shown to determine the NGF binding domain. The C-terminal region contains a single mastoparan-like domain, a consensus sequence for the binding of G proteins.[33] The cloned LNGFR receptor displays all the properties of the low-affinity receptor, as defined by binding and kinetic studies. It binds to NGF with a Kd of $10^{-9}M$, NGF dissociates from it rapidly, and the binding of NGF to it is rapidly released by mild trypsin treatment.[30] The LNGFR was renamed p75[NGFR], based on the approximate molecular weight of the human receptor of 75 kDa. The p75[NGFR] is a member of a family of related peptides which includes the two receptors for tumor necrosis factor, TNFRI and TNFRII, and the Fas antigen.[34-38] This family of molecules, which exhibits a striking homology in the pattern of cysteine-rich repeats in the extracellular domain, can mediate apoptotic cell death, suggesting that p75[NGFR] may play a role in determining neuronal cell survival.

B. THE HIGH-AFFINITY NGF RECEPTOR

Studies on the cell biology of NGF action demonstrated that NGF stimulates cellular tyrosine and serine/threonine phosphorylation, revealing similarities to growth factors acting on tyrosine kinase receptors.[39-41] These signaling events were not reproduced by transfection of p75[NGFR] into nonneuronal cells, suggesting that a second neuron-specific protein, possibly a tyrosine kinase, was required for NGF action (for review see Chao[42]). The identity of the HNGFR was first suggested by chemical cross-linking studies that revealed two NGF-receptor complexes of 100 and 160 kDa, representing NGF bound to separate proteins with a molecular weight of approximately 80 kDa (p75[NGFR]) and 140 kDa, respectively. Concomitant studies revealed that the proto-oncogene *trk* encoded for a 140-kDa phosphoprotein tyrosine kinase (p140[trkA]) that might be a target of NGF activation. The *trk* gene was originally detected as an oncogene in a tumorigenicity assay of NIH 3T3 cells.[43] The identification of p140[trkA] as an NGF receptor was made after it was noted that the distribution of p140[trkA] was restricted to embryonic dorsal root ganglia and to other sensory ganglia such as the jugular and trigeminal ganglia.[44] Findings of *trk* mRNA in sympathetic ganglia and basal forebrain cholinergic neurons have confirmed that p140[trkA] expression is directly associated with classical targets of NGF.[45 46] The observation that p140[trkA] could be specifically cross-linked to NGF and immunoprecipitated with antisera specific for the *trk* product clearly identified the 160-kDa HNGFR as p140[trkA].[47,48] Moreover, the demonstration that NGF directly binds to p140[trkA] and stimulates its intrinsic tyrosine kinase activity and autophosphorylation in PC12 cells suggested that p140[trkA] may play an important role in the signal transduction events induced by NGF.[49]

The p140[trkA] displays some of the properties of the HNGFR: the NGF binding to p140[trkA] is stable to trypsin treatment and NGF dissociates slowly from the receptor expressed in COS cells.[28] The p140[trkA] receptor, like p75[NGFR], is a heavily glycosilated protein with a single membrane spanning domain. Its cytoplasmatic domain contains the tyrosine kinase activity that puts it in the large family of tyrosine kinase receptors.[50]

The identification of p140[trkA] as the receptor for NGF reveals that receptor tyrosine kinase mediates the initial signal transduction of NGF, and suggests that many short-term actions of NGF exist in addition to its better characterized long-term actions. The p140[trkA] receptor is rapidly activated, and autophosphorylation of p140[trkA] induced by NGF occurs within a minute.[49] Studies using inhibitors of phosphorylation have demonstrated that p140[trkA] autophosphorylation is a specific requirement for NGF function. The alkaloid K-252a, which is a potent inhibitor of NGF-induced biological responses, is also a potent and highly selective inhibitor of the protein kinase activity of p140[trkA].[51,52] The activation of p140[trkA] leads to the phosphorylation of other proteins including a number of immediate early gene products, and several cellular signaling proteins, such as phospholipase C-τ_1; and a number of serine/threonine protein kinases, including mitogen activated protein (MAP) kinases, mitogen-sensitive MAP kinase kinase (MEK), and the *raf* protooncogene product.[53-55] The current dogma for NGF action indicates a possible linear pathway from p140[trkA] − → p21[ras] − → raf-1 − → MEK − → MAP kinase − → neurite extension. Paradoxically, the steps in signaling for NGF appear to involve the same molecules used for transduction of mitogenic signals (for a review see Chao[56]). For example, stimulation of NGF and EGF receptors in PC12 cells activates the same initial signal events but leads to opposite biological effects, that is, cell differentiation and proliferation, respectively. Thus none of the metabolic activities identified for signal transduction of NGF appears to be characteristic and distinctive. Later steps must be more obligatory, and further studies are necessary to resolve the enigma of growth factors signaling specificity.

C. POSSIBLE ROLES OF p75[NGFR] AND p140[trkA] IN NGF ACTION

Findings reported above strongly indicate that p140[trk] and p75[NGFR] are separate receptors for NGF. The relative role of each of these receptors in NGF action, however, is still controversial. Critical questions are why two separate receptors exist for NGF and what the relevance of p75[NGFR] to NGF function is.

Several lines of evidence suggest that the expression of p140[trkA] alone is sufficient for signal generation. The use of antisera against the extracellular domain of p75[NGFR] effectively blocked low-affinity NGF binding, but had no effect on NGF responsiveness in neuronal cells.[57] Moreover, mutant recombinant

NGFs that display negligible binding affinity for p75[NGFR] still retain specific biological activity, as measured by neurite outgrowth in PC12 cells.[58] *In vivo* studies also suggested that p140[trkA] alone is sufficient for NGF biological activity. MAH sympathoadrenal progenitor cells differentiated in response to NGF without requiring the expression of p75[NGFR].[59] In addition, dorsal root ganglia expressing p140[trkA] were selectively sensitive to NGF deprivation, whereas p75[NGFR] expression was found in both NGF sensitive and insensitive neurons.[60]

What is, therefore, the role of p75[NGFR]? This protein, which belongs to a functionally diverse family of cell surface receptors, is expressed in most NGF-responsive neurons. A number of studies have shown that NGF binding and responsiveness are influenced by the expression, activity, and stoichiometry of both NGF receptors. Gene knockout in transgenic animals demonstrated that p75[NGFR] plays an important role in the development of sensory neurons. Homozygous animals lacking p75[NGFR] are viable, but show a pronounced deficit in thermal sensitivity due to a loss of sensory neurons and smaller dorsal root ganglia than those in wild-type mice.[61] Moreover, antisense oligonucleotides against p75[NGFR] were shown to inhibit maturation of sensory neurons.[62] Hempstead and co-workers[63] proposed that p75[NGFR] and p140[trkA] form a stable complex that is required for high-affinity NGF-binding and subsequent biological activity, citing as evidence the difference in affinity measurements of cells expressing either the proteins alone or in combination. Reconstitution studies using gene transfer and membrane fusion techniques in COS, in NIH 3T3, in melanoma cells, and in a mutant derivative of PC12 cells (NR18) have indicated that high-affinity binding requires coexpression of both receptors.[63,64] However, experiments performed in fibroblasts expressing only the p140[trkA] receptor suggest that high-affinity binding can be formed in the absence of p75[NGFR].[48] Antibodies against p75[NGFR] do not coprecipitate 140[trkA], and antibodies against 140[trkA] do not coprecipitate p75[NGFR] from PC12 cells.[27,47] In addition, under conditions in which NGF binding is restricted to the HNGFR, only an NGF-140[trkA] complex can be observed following cross-linking with various agents.[26,65] Thus, although we cannot currently exclude that p75[NGFR] may modulate p140[trkA] by inducing higher affinity binding, the definition of the precise role played by p75[NGFR] in NGF action still requires further studies. Several other possible functions for p75[NGFR] have been suggested. First, p75[NGFR] might play a role in controlling the local concentration of NGF.[66] Second, p75[NGFR] might participate in a signal transduction mechanism, through activation of a G protein.[33] Third, p75[NGFR] might play a role in retrograde transport of NGF, although p75[NGFR] is internalized relatively slowly in tissue culture cells.[67] Finally, since binding studies have revealed that p75[NGFR] binds with low-affinity NGF, BDNF, and NT-3, this receptor could be considered more appropriately as a neurotrophin receptor (and not merely an NGF receptor) and might play a role in providing binding selectivity among neurotrophins.[68,69]

IV. INFLUENCES OF NGF ON BASAL FOREBRAIN CHOLINERGIC NEURONS

A. EFFECTS OF NGF ON DEVELOPING BASAL FOREBRAIN CHOLINERGIC NEURONS

In vitro and *in vivo* studies suggest that NGF stimulates the survival of basal forebrain cholinergic neurons during development. NGF affects neurite extension and/or survival of cholinergic neurons isolated from the medial septum, diagonal band, and nbM, depending on the age of animals and the conditions used for culturing the cells.[70-73] NGF was found to stimulate the growth of cholinergic neurons from cultured septal slices to cocultured slices of hippocampal tissue.[74] Exogenous administration of NGF stimulates high K+-evoked acetylcholine (ACh) release from basal forebrain cholinergic neurons in culture[75] and selectively increases presynaptic cholinergic markers, including choline acetyltransferase (ChAT) activity, acetylcholine esterase (AChE) activity, and ACh endogenous levels in primary cultures containing cholinergic neurons from septum or nbM.[70,76-85] NGF increases the expression of the ChAT mRNA and stimulates the transcriptional activity of the ChAT promoter in these neuronal cultures, suggesting that the NGF-mediated increase in ChAT activity in these cultures is regulated at the transcriptional level.[86,87]

The notion that NGF affects development of basal forebrain cholinergic neurons gets further support from experiments performed *in vivo*. Measurements of levels of NGF and NGF mRNA in the hippocampus and neocortex demonstrated that there is a temporal association of increases in ChAT activity and NGF expression during early postnatal development.[88-91] Since low levels of NGF are found in the basal forebrain of neonatal rats, NGF that is found in this region is likely to be transported retrogradely from the hippocampus and the cortex.[88-90] Moreover, expression of NGF mRNA and protein in the hippocampus and the cortex, and expression of p140[trkA] mRNA, p75[NGFR] mRNA, and protein in basal forebrain cholinergic neurons appear to be developmentally regulated in a similar fashion.[31,92-95] Intraventricular administration of NGF increases ChAT activity and ChAT mRNA in the basal forebrain of neonatal rats.[77,80,96-98] Moreover, the neutralization of endogenous NGF was shown to cause alterations of NGF-responsive cholinergic neurons. The intracerebroventricular injection of immunoaffinity-purified rabbit anti-NGF antibody and their Fab fragments decreases the ChAT activity in the hippocampus, septal area, and cortex of newborn rats suggesting that endogenous NGF regulates the ChAT activity in forebrain cholinergic neurons, thereby performing a physiological role in the developing CNS.[99] The treatment of telencephalic neuronal cultures from newborn rat brain induces an early transient increase in the number of muscarinic cholinergic receptors that is followed by a downregulation of the receptor number and of mRNA encoding m1 and m3 muscarinic receptor subtypes.[85] This was confirmed by *in vivo* experiments that showed the intracerebroventricular administration of NGF to neonatal rats altering in a biphasic and reversible fashion the muscarinic binding sites in the cerebral cortex during postnatal development.[100] Thus, the NGF-induced stimulation of central cholinergic neurons during development also results in changes at the synaptic site and may thus modulate muscarinic cholinergic receptor function at the target neurons.

In neonatal rats an earlier forebrain cholinergic system maturation induced by NGF might have some behavioral implications. Alleva et al.[101] reported that in young mice a single intracerebroventricular injection of NGF potentiates the effects of scopolamine on locomotor activity, a behavioral response which is under cholinergic muscarinic control.[102] Furthermore, Hess and Blozovski[103] showed that in newborn rats intrahippocampal NGF accelerates the ontogenesis of spontaneous alternation, an effect which may relate to earlier maturation of the septo-hippocampal cholinergic system.

B. EFFECTS OF NGF ON ADULT BASAL FOREBRAIN CHOLINERGIC NEURONS

Several studies support the role of NGF as a neurotrophic molecule for adult basal forebrain cholinergic neurons.

The levels of both NGF mRNA and protein correlate with the density of cholinergic innervation in the adult rat brain.[104-109] NGF levels are highest in the hippocampus, cortex, and basal forebrain, and NGF mRNA is localized primarily in the cerebral cortex, dentate gyrus, and pyramidal neurons of the hippocampal formation.[110-114] The distribution of NGF and NGF mRNA indicates that NGF is synthesized by the synaptic targets of cholinergic neurons and retrogradely transported to the cell bodies of responding neurons in the basal forebrain. However, neurochemical studies have detected low levels of NGF mRNA within basal forebrain areas of normal and experimentally lesioned animals, thus suggesting that some NGF synthesis may actually occur within the cell body of the responsive cholinergic cells.[115]

Adult cholinergic neurons contain receptors for NGF and retrogradely transport NGF.[116-118] Binding studies with iodinated murine and recombinant human NGF have revealed that there is an excellent correlation between the distribution of high-affinity NGF binding sites and that of cholinergic cell bodies and terminals.[119-121] The low-affinity receptor for NGF (p75[NGFR]) is synthesized both in the rat and the primate basal forebrain cholinergic cell bodies and transported to the hippocampal formation through the fimbria fornix.[31,122-128] Septal cholinergic neurons possess p75[NGFR]-immunoreactive somata, dendrites, axons, and terminals projecting to the hippocampus.[129] In both the human and nonhuman primate basal forebrain, virtually all NGF receptor-containing neurons are also immunoreactive for ChAT.[126,130,131] A similar percentage was found in the basal forebrain of the adult rat.[127,132] In addition, the mRNA for the tyrosine receptor kinase high-affinity NGF receptor (p140[trk]) is highly restricted in its distribution in the adult rat forebrain, it is present in cholinergic neurons, and most of cholinergic neurons contain p140[trk] mRNA.[46,133] Thus the binding site and the functional component of NGF are associated with adult basal forebrain cholinergic cells.

A variety of data indicate that the exogenous administration of nerve growth factor affects uninjured adult rat septo-hippocampal cholinergic neurons. NGF produces a dose-dependent and long-lasting increase of ChAT activity and ChAT mRNA in septo-hippocampal neurons of adult rats.[98,134,135] Moreover, NGF was found to have excitatory effects on the spontaneous firing of neurons in septal grafts, which is blocked by a low-affinity receptor antibody.[136] A recent study by Lindefors et al.[137] demonstrates that the activation of septo-hippocampal afferents by an injection of quisqualate is followed by a rapid increase in the level of NGF mRNA in the hippocampus that is prevented by scopolamine pretreatment. Thus, an increased release of ACh from cholinergic neurons of the medial septal nucleus can regulate the expression of NGF in the target area. NGF treatment in normal adult animals increases the expression of p75[NGFR] and p140[trk] mRNA and results in cellular hypertrophy of NGF-responsive cholinergic neurons.[46,138-140] Via induction of p75[NGFR] and p140[trk] NGF may provide a means to enhance or maintain receptor density on neurons enlarging under influence of NGF. Upregulation of NGF receptors may play a role in mediating NGF effects on mature cholinergic neurons.

Evidence summarized in this section suggests that in the adult brain septo-hippocampal and basalocortical cholinergic neurons respond to NGF and that NGF may be required for the functional integrity of these cholinergic pathways. However, results from Sofroniew and co-workers[141] have questioned this hypothesis. These authors demonstrated that 6 months following the excitotoxin lesions of the hippocampus, which destroy most endogenous NGF-producing hippocampal neurons, the hippocampal projecting cholinergic septal/diagonal band neurons had still not degenerated. These findings were also confirmed by Kordower et al.[142] who showed that removal of hippocampal target neurons by ibotenic acid injections does not alter the number, morphology, or projections of cholinergic septo-hippocampal neurons. These neurons still project ipsilaterally, and sprouting to the NGF-rich controlateral side does not occur. These data strongly suggest that although NGF is essential for the viability of cholinergic neurons during development,[143,144] it may not be necessary for the maintenance of cholinergic neurons in the intact adult organism. An interesting possibility is that only small levels of NGF are required for the maintenance of basal forebrain cholinergic neurons under normal conditions and that levels of NGF within the intact hippocampus might serve to regulate other cellular functions such as promotion of the expression of cholinergic phenotypes within the septo-hippocampal neurons, neuronal differentiation, or plasticity. Alternatively, low levels of NGF following lesion of the hippocampus might be provided by astrocytes, or NGF synthesized within the septal region may become available for local trophic support after removal of the target neurons.[115]

C. EFFECTS OF NGF IN ADULT ANIMALS WITH EXPERIMENTAL LESIONS
1. Experimental Models

Indirect evidence for a role of NGF in the maintenance and function of cholinergic neurons of septo-hippocampal and basalocortical pathways is further provided by studies on rats and primates with experimental lesions. Animal models have assumed critical importance in the effort to elucidate the mechanisms involved in neurodegenerative diseases and in the assessment of new therapeutic interventions aimed at preventing or alleviating neurological impairments. The findings that basal forebrain cholinergic neurons express NGF receptors and respond to exogenous NGF and that a dramatic loss of these neurons occurs in neurodegenerative diseases such as Alzheimer's disease, prompted several groups to investigate the effects of NGF in animals with experimentally induced lesions of forebrain cholinergic neurons.[145,146]

Both full and partial transections of the fimbria, which sever the cholinergic septo-hippocampal pathway, have been used by many investigators during the past decade. This pathway was chosen because it represents the best characterized cholinergic projection from the basal forebrain and because the axons are easily accessible for lesioning. Partial fimbria transections offer the advantage of leaving a natural bridge between septum and hippocampus through which severed axons may regenerate. This system, given the appropriate conditions, allows the investigator to study the effect of NGF on these surviving axons; however, axonal regeneration cannot be easily distinguished from sprouting in this experimental model. Full transections combined with tissue bridge have been used extensively as a model to investigate the pattern of regeneration of severed axons and reinnervation of the target hippocampus.

Animal models that induce cellular atrophy and degeneration of nbM cholinergic neurons have also attracted considerable interest because the dramatic loss of cholinergic neurons is detected in the nbM in pathologies like Alzheimer's disease. Lesions of the nmB have been carried out by various methods, including excitotoxins, i.e., ibotenic acid, kainic acid, and quisqualic acid. Although these lesions destroy the majority of cholinergic neurons, a small population of cells usually survives.[147] The biochemical and morphological damage which occurs in the nbM after a cortical devascularizing lesion, which causes an infarct of the fronto-parietal cortex, represents a well-established experimental tool to study the processes of transsynaptic retrograde degeneration. In this lesion model, axons are not transected; instead cholinergic fibers ascending from the nbM degenerate gradually due to the infarct and cause significant decrease in ChAT activity and cholinergic cell shrinkage in the nbM.[148]

2. Methods of NGF Administration

Single intraventricular injections of NGF were found to be ineffective in enhancing cholinergic function following experimental lesions of basal forebrain cholinergic neurons, and chronic treatment with NGF seems to be required to prevent neurodegenerative changes of these severed neurons.[149] Chronic NGF treatments have been performed by means of repeated intraventricular injections (daily or biweekly), or by chronic delivery systems including microspheres,[150,151] osmotic minipumps,[152-154] or genetically modified cells.[155-158] Large doses of NGF (microgram amounts) were used for intraventricular injections, because distribution characteristics of injected NGF and its final concentration in the septal, hippocampal, and cortical area cannot be easily determined. However, dose-response relationships for biochemical and behavioral action of intraventricularly administered NGF have been recently reported.[159-161] Studies performed in cell culture systems suggest that there is a dose-dependent effect of NGF and that NGF affects cholinergic neurons *in vivo* at concentrations similar to those mediating the well-known actions on sympathetic or sensory neurons.[70,76-79,81] Exogenous NGF has also been delivered by means of miniosmotic pumps releasing small amounts of NGF into the ventricular cavity or directly into the parenchyma. There are, however, several drawbacks to the long-term use of these miniosmotic pumps. First, the biological activity of NGF stored within the reservoir may diminish over extended periods. Further, only a limited amount of NGF can be stored within the reservoir, and thus new NGF must be added periodically. Genetically modified fibroblasts that produce NGF have offered an alternative approach to achieve a local delivery of this trophic factor to rescue lesioned neurons. Grafts of immortalized cell lines, however, often give rise to tumors following implantation within the brain, exhibiting robust proliferative activity. Thus it is likely that higher rate of cell growth may directly influence levels of NGF production by these intracerebral grafts. That is, higher rate of cell growth may result in higher levels of NGF that, in turn, may induce an overcompensation of lesion-induced degenerative changes of basal forebrain cholinergic neurons.[152,158,162] Moreover, histocompatibility of donor cells with the recipient nervous system may be an important factor determining the long-term survival of grafted cells not derived from the host. Primary cells have been successfully used for gene transfer and intracerebral grafting to obviate concerns of tumor formation and immune rejections often associated with immortalized cells grafted within the CNS.[158]

3. Effects of NGF in Adult Animals with Fimbria/Fornix Transections

In several studies, administration of NGF to rats with fimbria transections was found to prevent the lesion-induced loss of ChAT and AChE activity in the medial septal nucleus/diagonal band of Broca and at the level of cholinergic projections fibers in the hippocampal formation.[149,159,163-165] Chronic NGF also attenuates lesion-induced deficit of hippocampal cholinergic function measured in hippocampal slices *in vitro*, where it increases [³H]ACh synthesis and release.[149] Moreover, chronic NGF treatment enhances *in vivo* synthesis, storage, and release of endogenous ACh in adult rats with unilateral partial fimbria transections.[166] The NGF-mediated increase of *in vivo* cholinergic function becomes manifest only following treatment with CNS stimulants which induce seizures and decrease endogenous hippocampal ACh content. This suggests that surviving cholinergic axons may be able to functionally compensate for the partial degeneration of the septo-hippocampal pathway.

Recent data also suggest that trophic effects of NGF treatment on presynaptic cholinergic function correlate with functional changes at the level of postsynaptic cholinergic receptors.[167] In animals receiving full fimbria transections, which by themselves did not alter muscarinic receptor density, chronic intraventricular NGF treatment increases the density of [³H]quinuclidinyl benzilate binding sites by approximately 40% in the CA1 region. Moreover, in partially fimbriectomized rats, chronic NGF treatment prevents the lesion-induced supersensitivity of the oxotremorine-induced inositol triphosphate production by hippocampal slices.

Histochemical analysis revealed that intracerebroventricular administration of NGF prevents the disappearance of cholinergic septal/diagonal band neurons induced by fimbria-fornix transection in rats and nonhuman primates.[153,154,159,162-165,168-170] Whether the apparent loss of ChAT and NGF receptor is the result of the actual death of these neurons, or instead it is due to shrinkage associated with a loss of expression of the cholinergic marker enzyme still remains an unresolved issue. Although several authors have demonstrated that most of the axotomized cholinergic neurons die after these lesions,[163,171,172] others have shown that downregulation of marker proteins and cellular atrophy precede cell death and that exogenous NGF prevents both the atrophic and degenerative processes in rats and nonhuman primates.[152,173]

It has been shown that a significant retrieval of ChAT-immunoreactive neurons occurs in the medial septum after delayed NGF treatment of rats with fimbria-fornix transections, therefore suggesting that fimbrial transection results in shrinkage of cholinergic neurons.[169] However, other investigators reported that delayed NGF administration is ineffective in preventing the loss of cholinergic neurons.[174]

A neurite-promoting action of NGF also has been evidenced in the adult septo-hippocampal model. The intraventricular NGF administrations result in a dramatic increase in ChAT immunoreactive fibers in the ipsilateral septum that is likely to represent sprouting of the axotomized cholinergic neurons.[160,162,164,165,175] Moreover, Junard et al.[176] reported that the long-term NGF administration also stimulates neurite growth that is restricted to limited growth close to the lesion site. There is now much evidence indicating that true axonal regeneration can occur, given an appropriate substratum for growth. A number of different types of tissues have been successfully used to bridge new axon growth from septal neurons to the deafferentated hippocampus, including segments of autologous peripheral nerves, embryonic rat hippocampus, and genetically engineered immortalized fibroblasts (rat 208F and mouse NIH-3T3) or primary fibroblasts expressing the transgene for NGF.[155-158,177,178] Results from these studies indicate that NGF-sensitive axons arising from perturbed septal neurons require NGF and a permissive graft environment for new growth. The availability of NGF appears to be a necessary requirement to sustain axotomized cholinergic septal neurons and to promote axonal regeneration and cholinergic reinnervation of hippocampal neurons. Kawaja and co-workers[158] recently reported that septal axons that regenerate across NGF-rich grafts of collagen and fibroblasts reestablish at least some aspects of the normal synaptic organization within the dentate gyrus.

4. Effect of NGF in Adult Animals with Lesions of the Basalocortical Cholinergic Neurons

Experimental evidence has shown that both biochemical and morphological degenerative changes observed in the nbM and the remaining cortex following devascularizing lesions could be prevented and even overcompensated with immediate intracerebroventricular application of NGF. The early administration of NGF prevents the decrease of ChAT activity, high-affinity choline uptake and ACh, and release in rats and nonhuman primates with cortical devascularizing lesions.[150,151,179-181] In addition, intraventricular administration of NGF to decorticated rats and nonhuman primates increases the number and size of cortical cholinergic synapses and prevents cell body shrinkage and neurite loss of cholinergic neurons of the nbB.[151,179,182,183] Recovery of nbM cholinergic neurons was also induced by grafting NGF secretor fibroblasts in rats bearing unilateral cortical devascularizing lesions.[184]

In rats with lesions of the nbM, the intraventricular administration of NGF counteracts reductions of ChAT and AChE activities and ACh release in the neocortex.[161,179,185-189] Moreover, NGF was found to increase the size of nbM neurons and the area of ChAT positive fiber staining in the neocortex of the same animals.[190,191] The increased area of fiber staining may be due to sprouting and/or increased ChAT expression in terminals from nbM neurons, since NGF did not affect the ChAT staining or the size of cortical cholinergic interneurons.[191]

V. EFFECTS OF NGF DURING AGING

There is biochemical, electrophysiological, pharmacological, and behavioral evidence that cholinergic functions in the brain decline with age and that these changes are associated with cognitive disturbance.[192,193] It has been suggested that deficient synthesis or metabolism of NGF during aging could underlie or contribute to the cholinergic deficits in old animals, although discrepancies exist in the literature. Levels of NGF and its mRNA,[194] as well as receptors for NGF,[195-197] were found to be reduced in the basal forebrain of old rats and in the human brain. Moreover, the decrease of NGF levels and of the number of NGF receptor-positive basal forebrain cholinergic neurons correlates with the decline in spatial learning ability during aging.[198-200] Studies showing improvement of neuronal deficits after administering exogenous NGF to senescent rats support the hypothesis that the marked atrophy and cell loss in the forebrain cholinergic system of behaviorally impaired aged rats may be caused by a reduced availability of NGF in the cholinergic target areas. The chronic administration of NGF partially restores cholinergic functions in aged rates. NGF increases the activity of high-affinity choline uptake and ChAT in brain of senescent rats.[201-203] Intraventricular administration of NGF also prevents the loss of NGF receptor-positive neurons and stimulates collateral sprouting of intact cholinergic axons in the septo-hippocampal system in aged rats with fimbria transections.[204,205] In addition, NGF ameliorates cholinergic neuron atrophy and spatial memory impairment in aged rats.[192,206,207]

VI. BEHAVIORAL EFFECTS OF NGF

Several lines of evidence indicate that the NGF-induced increase of presynaptic cholinergic functions in animals with experimental lesions of basal forebrain cholinergic neurons is associated with the attenuation and normalization of lesion-induced deficits in behavioral function.[208] Administration of NGF to animals with lesions of septal cell bodies or fimbria transections was found to normalize responses in multiple tasks associated with memory formation and consolidation, such as the T maze and the eight-arm radial maze.[209-212] Single intraventricular injections of NGF at a concentration approximately equal to the total amount of NGF used in the repeated or continuous infusion schedule, ameliorates performances of rats tested in a maze task after a delay of several months.[213,214] After mild septal damage NGF induces important and long-lasting behavioral improvement; however, NGF was ineffective after large lesions, irrespective of the dosage used.

In rats with lesions of nbM, early or delayed treatment with NGF was also found to improve acquisition of spatial tasks.[186,190,215,216] These behavioral effects are thought to be mediated by basal forebrain cholinergic neurons that have been shown to be involved in memory processes.[217,218] However, no correlation between the behavioral findings and cholinergic markers was found in most of the studies described above.[190-219] Although the explanation for this lack of correlation is not simple, these authors have suggested that NGF might ameliorate functional performance either in a way not manifest in changes of cholinergic parameters, or instead by affecting systems other than the cholinergic one.[190,219]

VII. NGF AND ALZHEIMER'S DISEASE

Degenerative changes in the basal forebrain cholinergic system have been suggested to play a critical role in the development of dementia in Alzheimer's disease. The finding that NGF plays a role in trophic support of basal forebrain cholinergic neurons has provoked speculation that a decline in NGF-related mechanisms may be involved in the pathogenesis of Alzheimer's disease and that agents mimicking its function may be beneficial for the clinical treatment of this disease.[208,220,221] However, evidence indicating that Alzheimer's disease is associated with a significant deficiency in NGF function so far is lacking. NGF content and its mRNA were found unchanged in the CNS of patients affected by Alzheimer's disease.[105,222] In addition, Lorigados and co-workers[223], by using a two-site immunoassay for the determination of serum levels of β-NGF, recently demonstrated that sera from patients with Alzheimer's disease show only slight reductions of NGF immunoreactivity. Reports concerning NGF receptor expression in Alzheimer's disease have been somewhat conflicting. Several studies demonstrated a decrease of the density of NGF receptors in the nbM of patients with Alzheimer's disease that occurs in conjunction with the loss of cholinergic neurons and their processes.[196,224-228] Thus, the decrease of NGF receptor immunoreactive neurons in Alzheimer's disease seems to reflect neuron loss and not the failure of viable neurons to synthesize NGF receptors. Interestingly, the decrease of the NGF receptor is not supported by *in situ* hybridization experiments which show normal or increased expression of NGF receptor mRNA in the nbM of Alzheimer's disease patients.[229,230]

Alzheimer's disease is known to be a multisystem disorder with the primary insult probably originating in the cerebral cortex. The recent discovery of a pathogenic mutation in the β-amyloid precursor protein (APP) gene on chromosome 21 suggested that altered metabolism of APP and β-amyloid deposition are the primary events in the disease process (for review, see Hardy and Allsop[231]). On the other hand, the association between the loss of basal forebrain cholinergic neurons and the loss of memory in Alzheimer's disease has been taken to suggest that agents which maintain the function of these cells might be used to slow the deterioration of cognitive function that occurred in this disease. The fact that NGF receptors and ChAT are colocalized in the nbM of normal human brain and in the remaining cholinergic neurons in Alzheimer's disease, suggests that in Alzheimer's disease these neurons may still be responsive to the beneficial effect of NGF and supports a rationale for the use of NGF in the treatment of Alzheimer's disease. Preliminary results recently obtained from a clinical trial of NGF infusion into the brain of Alzheimer's disease patients indicate that NGF may counteract the cholinergic deficit in Alzheimer's disease.[232] These authors report that the intraventricular infusion of 6.6 mg of NGF for 3 months into the brain of one 68-year-old woman with Alzheimer's disease resulted in a marked increase in uptake and binding of [^{11}C]nicotine in the frontal and temporal cortex, a persistent increase in cortical blood flow, and an improvement in tests of verbal episodic memory. Although in this study no adverse effects of NGF were observed, the potential toxicity of this neurotrophic factor should be considered (for review see Butcher and Woolf[233]). Yankner and co-workers[234] reported that NGF markedly increases the toxicity of β-amyloid toward differentiated hippocampal neurons in culture. This effect may be physiologically relevant, because reactive astrocytes that proliferate in response to a lesion begin to produce NGF, which could thus exacerbate the loss of neuronal perikarya surrounding plaques containing β-amyloid. Thus, the suggested use of NGF replacement therapy for Alzheimer's disease patients could be counterproductive.

Recent results reported by Hagg and Varon[235] have also raised some concerns about the potential use of NGF in pathological situations. These authors have demonstrated that the NGF-induced sprouting of axotomized adult rat cholinergic neurons is oriented toward the site of administration of NGF, resulting in an aberrant location of the cholinergic axons. This neurotropic action of NGF occurs at doses needed to prevent or reverse degenerative changes of lesioned axons, and was demonstrated both in rats that received only fimbria transections but no nerve grafts and in rats that received both unilateral or bilateral fimbria transections and sciatic nerve grafts. These results will have to be taken into account for planning treatments with NGF of patients with Alzheimer's disease and suggest the development of accurately planned treatment protocols.

The findings reported above strongly suggest that before NGF treatment can be considered beneficial in the treatment of Alzheimer's disease further studies are needed to elucidate the exact physiological role of NGF in the CNS and to better understand its pharmacological actions on lesioned cholinergic neurons.

VIII. INFLUENCES OF NGF ON CHOLINERGIC NEURONS IN THE CORPUS STRIATUM

Although considerable interest has been directed at the actions of NGF on basal forebrain cholinergic neurons, several lines of evidence demonstrate that NGF acts as a neurotrophic factor also for caudate-putamen cholinergic neurons. The neurotrophic dependence of these neurons is of particular interest in that, in contrast to other NGF-responsive populations, these cells are interneurons that form anatomically and functionally distinct populations of neurons confined to this area.[15,236] Thus, the caudate putamen contains these cells and serves as their target.

Data consistent with the view that NGF is a trophic factor during development for striatal cholinergic neurons include robust, dose-dependent responses to NGF administration. Exogenously administered NGF increases ChAT activity both in primary fetal striatal cultures and in neonatal rats *in vivo*.[81,96,97,237] The treatment with NGF of fetal striatal neurons increases ChAT mRNA levels, suggesting that the observed increase of ChAT activity in NGF-treated neurons probably reflects an increase in gene transcription.[238] Moreover, intracerebroventricular administration of the anti-NGF antibody decreases ChAT activity in the striatum of rat pups, indicating that endogenous NGF regulates striatal ChAT activity during postnatal development.[99]

NGF mRNA and protein are present in the developing striatum, and during development increases in NGF high-affinity binding in this tissue parallel increasing ChAT activity.[239] Moreover, p75[NGFR], p75[NGFR] mRNA, and p140[trkA] mRNA are expressed in the developing caudate-putamen cholinergic neurons.[91,94,95,239,240] The time course for the expression of NGF, NGF binding, and ChAT activity suggests that NGF influences cholinergic differentiation in the caudate putamen.

The following evidence demonstrates that NGF may continue to act on caudate-putamen neurons also in the adult. Exogenous NGF, chronically infused intraventricularly or intrastriatally, increases ChAT immunoreactivity, size of cholinergic cell bodies, and ChAT activity in intact rat neostriatum.[159,160,241-243] Moreover, NGF selectively prevents excitotoxin-induced degeneration of striatal cholinergic neurons.[243,244] The induction of ChAT gene expression in striatum demonstrated by these studies indicates that NGF acts vigorously on these cells. However, a discrepancy between NGF binding and p75[NGFR] expression in the mature striatum has been reported. While it is detected in the damaged adult caudate putamen,[241] little if any p75[NGFR] expression has been found by *in situ* hybridization or by immunohistochemistry in the intact adult caudate putamen.[92,94,108,123,241,245-248] In contrast, tissue autoradiographic studies indicate that high-affinity NGF binding is present in the striatum of both developing and mature rats, and the levels in adults are actually greater than in rats in the early postnatal period.[119-121,239] The amount of high-affinity binding in the adult striatum is similar to or even greater than that found in the basal forebrain. More recently, high expression of p140[trk] mRNA has been found in the striatum of adult rats.[46,113] An interesting possibility is that in these neurons p140[trk] may be responsible for creating both high-affinity binding and signal transduction by itself. However, other interpretations should also be considered. p75[NGFR] mRNA is present in the intact adult caudate putamen, though its level in this tissue is much lower than that found in the adult basal forebrain.[239] It is therefore possible that in the striatum, p75[NGFR] participates in mediating NGF high-affinity binding and signal transduction as in the basal forebrain; if this is so, much more p75[NGFR] would be expressed in basal forebrain cholinergic neurons than is required for producing NGF high-affinity binding and signaling. This latter possibility is supported by the observation that in chick sensory neurons occupancy of only a few hundred of HNGFRs is required for a half maximal biological response.[249] NGF effects on striatal neurons differ in other respects from those in the basal forebrain, possibly reflecting specific features related to the "interneuronal" character of these cells. The temporal patterns for NGF mRNA and protein expression during development also are different from those in other brain regions. Moreover, quite low levels for NGF and its mRNA were found in the caudate putamen; indeed each was only about 10% of that detected in the hippocampus. Given the substantial level of cholinergic innervation on these two tissues and a comparable number of innervating cholinergic neurons, it is possible that relatively less NGF is required to support striatal interneurons. This possibility is in line with the observation that the hypertrophic response of adult striatal neurons requires less NGF than does that of medial septum neurons.[160] The treatment of neonatal rats during postnatal development does not affect the muscarinic cholinergic receptor density in the striatum.[100] Thus, the NGF-induced changes in muscarinic cholinergic receptor density appear to be restricted to, or more selective for, the basal forebrain cholinergic system. The fact that developing cholinergic neurons of basal forebrain and striatum differentially respond to the NGF treatment suggests that differences in morphology and physiology may account for a different response at the target brain areas. Moreover, distribution of muscarinic cholinergic receptors within brain regions, as detected by *in situ* hybridization, reveals a regional selectivity and may also account for the different responses to NGF treatment.[250,251] Neostriatal cholinergic neurons are lost in aged rats but also in Alzheimer's disease, Parkinson's disease, and Huntington's chorea (for review see Altar[252]). Future studies need to address whether these losses result from a decline in NGF-related mechanisms and whether NGF administration may retard or reverse these cholinergic losses.

IX. CONCLUSIONS

In this review we have attempted to highlight some of the current advances in the study of the mechanism of action of NGF and its physiological role on forebrain cholinergic neurons. For several decades NGF has been the prototypic neurotrophic factor; and its roles in promoting neuronal survival and differentiation, and in maintaining neuronal phenotype are the best characterized. However, the functional role of NGF in development and maintenance of forebrain cholinergic neurons needs to be reevaluated in light of the recent discovery of the family of NGF-related neurotrophic factors. BDNF, NT-3, and NT-4 have been recently identified as members of the NGF-related products, now termed neurotrophins.[253-259] These neurotrophins all bind to p75[NGFR] with similar affinity; molecules related to *trk*A, such as *trk*B and *trk*C, bind and respond preferentially to BDNF and NT-3, respectively, but not to NGF.[68,69] All members of the neurotrophin family differ functionally in promoting survival of different subsets of neurons. For example, BDNF and NT-3, but not NGF, support survival of sensory neurons of the nodose ganglion;[253,260] cultured retinal ganglion cells and cultured dopaminergic neurons of the embryonic ventral mesencephalon appear to respond to BDNF, but not to NGF.[261,262] Recent observations suggest the possibility that BDNF may also serve as a trophic factor for basal forebrain cholinergic neurons. In cholinergic cultures, BDNF, similarly to NGF, increases ChAT activity and intensity of ChAT immunoreactivity staining.[263,264] BDNF was found to reduce axotomy-induced degenerative changes, although the protective effect of BDNF treatment on adult septal hippocampal neurons was only partial and less significant than that produced by NGF.[265] Moreover, BDNF mRNA levels were found to be reduced in the brain of patients with Alzheimer's disease.[266] Thus, it is possible that BDNF is involved in the maintenance of a subpopulation of cholinergic neurons in the basal forebrain, which may express the *trkB* receptor, and that a deficiency of BDNF expression might contribute to the alteration of cholinergic neurons in the basal forebrain in Alzheimer's disease.

These and other findings indicate wide-ranging roles for the neurotrophins and a system of neurotrophic regulation of development and maintenance of the nervous system that is much more complex than previously postulated. Understanding the neuronal specificities of each neurotrophin will bring insights to the specific role of each member of the family in the process of development, regeneration, and degeneration of the nervous system; and may provide clues for the therapeutic application of neurotrophic molecules.

ACKNOWLEDGMENTS

I would like to thank Dr. Guido Vantini, Istituto di Ricerca sulla Senescenza, Sigma-Tau, Roma, Italy, for helpful discussions throughout the writing of this review and his careful revision of the manuscript. I also express my appreciation to Prof. Francesco De Matteis, Istituto di Farmacologia e Terapia Sperimentale, Facoltà di Medicina e Chirurgia, Università di Torino, Italy, and Dr. Renzo Levi, Istituto di Fisiologia Generale, Dipartimento di Biologia Animale, Università di Torino, Italy, for their critical comments of the manuscript.

REFERENCES

1. Levi-Montalcini, R., The nerve growth factor 35 years later, *Science*, 237, 1154, 1987.
2. Levi-Montalcini, R. and Calissano P., Nerve growth factor as a paradigm for other polypeptide growth factors, *Trends Neurosci.*, 9, 437, 1986.
3. Thoenen, H., The changing scene of neurotrophic factors, *Trends Neurosci.*, 14(5), 165, 1991.
4. Levi, A., Biocca, S., Cattaneo, A., and Calissano, P., The mode of action of nerve growth factor in PC12 cells, *Mol. Neurobiol.*, 2, 201, 1988.
5. Snider, W. D. and Johnson, E. M., Jr. Neurotrophic molecules, *Ann. Neurol.*, 26, 489, 1989.
6. Greene, L. A. and Shooter, E. M., The nerve growth factor: biochemistry, synthesis and mechanism of action, *Annu. Rev. Neurosci.*, 3, 353, 1980.
7. Thoenen, H. and Barde, Y.-A., Physiology of nerve growth factor, *Physiol. Rev.*, 60, 1285, 1980.
8. Levi Montalcini R., and Angeletti, P. U., Nerve growth factor, *Physiol. Rev.*, 48, 534, 1968.
9. Davies, A. M., The emerging generality of the neurotrophic hypothesis, *Trend Neurosci.*, 11, 243, 1988.
10. Konkol, R. J., Mailman, R., Bendeich, E. G., Garrison, A. M., Mueller, R. A., and Breese, G. R., Evaluation of the effects of nerve growth factor anti-nerve growth factor on the development of central catecholamine-containing neurons, *Brain Res.*, 144, 277, 1978.
11. Schwab, M. E., Otten, U., Agid, Y., and Thoenen, H., Nerve growth factor (NGF) in the rat CNS: absence of specific axonal transport and tyrosine hydrolase induction in locus ceruleus and substantia nigra, *Brain Res.*, 168, 473, 1979.
12. Dreyfus, C. F., Peterson, E. R., and Crain, S. M., Failure of nerve growth factor to affect fetal mouse brainstem cathecolaminergic neurons in culture, *Brain Res.*, 194, 540, 1980.
13. Mesulam, M. M., Mufson, B. H., Wainer, B. H., and Levey, A. I., Central cholinergic pathways in the rat: an overview based on an alternative nomenclature (Ch1-Ch6), *Neuroscience*, 10, 1185, 1983.
14. Mesulam, M. M., Mufson, E. J., Levey, A. I., and Weiner, B. H., Cholinergic innervation of cortex by the basal forebrain: cytochemistry and cortical connections of the septal area, diagonal band nuclei, nucleus basalis (substantia innominata), and hypothalamus in the rhesus monkey, *J. Comp. Neurol.*, 214, 170, 1983.
15. Schwaber, J. S., Rogers, W. T., Satoh, K., and Fibiger, H. C., Distribution and organization of cholinergic neurons in the rat forebrain demonstrated by computer-aided data acquisition and three-dimensional reconstruction, *J. Comp. Neurol.*, 263, 309, 1987.
16. McGeer, P.L., McGeer, E. G, and Peng Peng, J. H., Choline acetyltransferase purification and immunohistochemical localization, *Life Sci.*, 34, 2319, 1984.
17. Wainer, B. H., Levey, A. I., Mufson, E. F., and Mesulam, M. M., Cholinergic systems in mammalian brain identified with antibodies against choline acetyltransferase, *Neurosci. Int.*, 6, 163, 1984.
18. Woolf, N. J., Eckenstein, F., and Butcher, L. L., Cholinergic system in the rat brain. I. Projections to the limbic telencephalon, *Brain Res. Bull.*, 13, 751, 1984.
19. Bocchini, V. and Angeletti, P. U., The nerve growth factor purification as a 30,000-molecular weight protein, *Proc. Natl. Acad. Sci. U.S.A.*, 64, 787, 1969.
20. Scott, J., Selby, M., Urdea, M., Quiroga, M., Bell, G. I., and Rutter, W. J., Isolation and nucleotide sequence of a cDNA encoding the precursor of mouse nerve growth factor, *Nature (London)*, 302, 538, 1983.
21. Ullrich, A., Gray, A., Berman, C., and Dull, T. J., Human beta-nerve growth factor gene sequence highly homologous to that of mouse, *Nature (London)*, 303, 821, 1983.
22. Meier, R., Becker-Andre, M., Gotz, R., Heumann, R., Shaw, A., and Thoenen, H., Molecular cloning of bovine and chick nerve growth factor (NGF): delineation of conserved and unconserved domains and their relationship to the biological activity and antigenicity of NGF, *EMBO J.*, 5, 1489, 1986.

23. Inoue, S., Oda, T., Koyama, J., Ikeda, R., and Hayashi, K., Amino acid sequences of nerve growth factors derived from cobra venoms, *FEBS Lett.*, 279, 38, 1991.

24. McDonald, N. Q., Lapatto, R., Murray-Rust, J., Gunning, J., Wlodawer, A., and Blundell, T. L., A new protein fold revealed by a 2.3 Å resolution crystal structure of nerve growth factor, *Nature (London)*, 354, 411, 1991.

25. Sutter, A., Riopelle, R. J., Harris-Warrich, R. M., and Shooter, E. M., NGF receptors: characterization of two distinct classes of binding sites on chick embryo sensory ganglia cells, *J. Biol. Chem.*, 254, 5972, 1979.

26. Hosang, M. and Shooter, E. M., Molecular characteristics of nerve growth factor receptors on PC12 cells, *J. Biol. Chem.*, 260, 655, 1985.

27. Meakin, S. O. and Shooter, E. M., Molecular investigations on the high affinity nerve growth factor receptor, *Neuron*, 6, 153, 1991.

28. Meakin, S. O., Suter, U., Drinkwater, C. C., Welcher, A. A., and Shooter, E. M., The rat *trk* protooncogene product exhibits properties characteristic of the slow NGF receptor, *Proc. Natl. Acad. Sci. U.S.A.*, 89, 2374, 1992.

29. Johnson, D., Lanahan, A., Buck, C. R., Sehgal, A., Morgan, C., Mercer, E., Bothwell, M., and Chao, M., Expression and structure of the human NGF receptor, *Cell*, 47, 545, 1986.

30. Radeke, M. J., Misko, T. P., Hsu, C., Herzenberg, L. A., and Shooter, E. M., Gene transfer and molecular cloning of the rat nerve growth factor receptor, *Nature (London)*, 325, 593, 1987.

31. Buck, C. R., Martinez, H., Black, I. B., and Chao, M. V., Developmentally regulated expression of the nerve growth factor receptor gene in the periphery and brain, *Proc. Natl. Acad. Sci. U.S.A.*, 84, 3060, 1987.

32. Patil, N., Lacy, E., and Chao, M. V., Specific neuronal expression of human NGF receptors in the basal forebrain and cerebellum of transgenic mice, *Neuron*, 4, 437, 1990.

33. Feinstein, D. and Larhammer, D., Identification of a conserved protein motif in a group of growth factor receptors, *FEBS Lett.*, 272, 7, 1990.

34. Loetscher, H., Pan, Y.-C. E., Lahm, H.-W., Gentz, R., Brockhaus, M., Tabuchi, H., and Lesslauer, W., Molecular cloning and expression of the human 55kd tumor necrosis factor receptor, *Cell*, 61, 351, 1990.

35. Mallet, S., Fossum, S., and Barclay, A. N., Characterization of the MRC OX40 antigen of activated CD4 positive T lymphocytes — a molecule related to nerve growth factor receptor, *EMBO J.*, 9, 1063, 1990.

36. Schall, T. J., Lewis, M., Koller, K. J., Lee, A., Rice, G. C., Wong, G. H. W., Gatanaga, T., Granger, G. A., Lentz, R., Raab, H., Kohr, W. J., and Goeddel, D. V., Molecular cloning and expression of a receptor for human tumor necrosis factor, *Cell*, 61, 361, 1990.

37. Itoh, N., Yonehara, S., Ishii, A., Yonehara, M., Mizushima, S.-I., Sameshima, M., Hase, A., Seto, Y., and Nagata, S., The polypeptide encoded by the cDNA for human cell surface antigen Fas can mediate apoptosis, *Cell*, 66, 233, 1991.

38. Dürkop, H., Latza, U., Hummel, M., Eitelbach, F., Seed, B., and Stein, H., Molecular cloning and expression of a new member of the nerve growth factor receptor family that is characteristic for Hodgkin's disease, *Cell*, 68, 421, 1992.

39. Maher, P. A., Nerve growth factor induces protein-tyrosine phosphorylation, *Proc. Natl. Acad. Sci. U.S.A.*, 85, 6788, 1988.

40. Miyasaka, T., Chao, M. V., Sherline, P., and Saltier, A. R., Nerve growth factor stimulates a protein kinase in PC12 cells that phosphorylates microtubule-associated protein-2, *J. Biol. Chem.*, 265, 4730, 1990.

41. Miyasaka, T., Sternberg, I., Miyasaka, J., Sherline, P., and Saltier, A. R., Nerve growth factor stimulates protein tyrosine phosphorylation in PC12 pheochromocytoma cells, *Proc. Natl. Acad. Sci. U.S.A.*, 88, 2653, 1991.

42. Chao, M. V., The membrane receptor for nerve growth factor, *Curr. Top. Microbiol. Immunol.*, 165, 39, 1991.

43. Martin-Zanca, D., Hughes, S. H., and Barbacid, M., A human oncogene formed by the fusion of truncated tropomyosin and protein tyrosine kinase sequences, *Nature (London)*, 319, 743, 1986.

44. Martin-Zanca, D., Barbacid, M., and Parada, L. F., Expression of the *trk* proto-oncogene is restricted to the sensory cranial and spinal ganglia of neural crest origin in mouse development, *Genes Dev.*, 4, 683, 1990.

45. Schecterson, L. C. and Bothwell, M., Novel roles for neurotrophins are suggested by BDNF and NT-3 mRNA expression in developing neurons, *Neuron*, 9, 449, 1992.

46. Holtzman, D. M., Li, Y., Parada, L. F., Kinsman, S., Chen, C.-K., Valletta, J. S., Zhou, J., Long, B. J., and Mobley, W. C., p140^{trk} mRNA marks NGF-responsive forebrain neurons: evidence that *trk* gene expression is induced by NGF, *Neuron*, 9, 465, 1992.

47. Kaplan, D. R., Hemstead, B. L., Martin-Zanca, D., Chao, M. V., and Parada, L. F., The *trk* protooncogene product: a signal transducting receptor for nerve growth factor, *Science*, 252, 552, 1991.

48. Klein, R., Jing, S., Nanduri, V., O'Rourke, E., and Barbacid, M., The *trk* protooncogene encodes a receptor for nerve growth factor, *Cell*, 65, 189, 1991.

49. Kaplan, D. R., Martin-Zanca, D., and Parada, L. F., Tyrosine-phosphorylation and tyrosine kinase activity of the *trk* protooncogene product induced by NGF, *Nature (London)*, 350, 158, 1991.

50. Martin-Zanca, D., Oskam, R., Mitra, G., Copeland, T., and Barbacid, M., Molecular and biochemical characterization of the human *trk* proto-oncogene, *Mol. Cell Biol.*, 9, 24, 1989.

51. Koizumi, S., Contreras, M., Matsuda, Y., Hama, T., Lazarovici, P., and Guroff, G., K-252a: a specific inhibitor of the action of nerve growth factor on PC12 cells, *J. Neurosci.*, 8, 751, 1988.

52. Berg, M. M., Sternberg, D. W., Parada, L. F., and Chao M. V., K-252a inhibits nerve growth factor-induced trk proto-oncogene tyrosine phosphorylation and kinase activity, *J. Biol. Chem.*, 267(1), 13, 1992.

53. Ohimichi, M., Decker, S. J., Pang, L., and Saltiel, A. R., Nerve growth factor binds to the 140-kDa *trk* proto-oncogene product and stimulates its association with src homology domain of phospholipase C-τ1, *Biochem. Biophys. Res. Commun.*, 179, 217, 1991.

54. Ohmichi, M., Pang, I., Decker, S. J., and Saltier, A. R., Nerve growth factor stimulates the activities of the raf-1 and the mitogen activated protein kinases via the *trk* protooncogene, *J. Biol. Chem.*, 267, 14604, 1992.

55. Saltier, A. R. and Ohimichi, M., Pleiotropic signaling from receptor tyrosine kinases, *Curr. Opinion Neurobiol.*, 3, 352, 1993.

56. Chao, M. V., Growth factor signaling: where is the specificity?, *Cell*, 68, 995, 1992.

57. Weskamp, G. and Reichardt, L. F., Evidence that biological activity of NGF is mediated through a novel subclass of high affinity receptors, *Neuron*, 6, 649, 1991.

58. Ibàñez, C. F., Ebendal, T., Barbany, G., Murray-Rust, J., Blundell, T. L., and Persson, H., Disruption of the low affinity receptor binding site in NGF allows neuronal survival and differentiation by binding to the *trk* gene product, *Cell*, 69, 329, 1992.

59. Birren, S. J., Verdi, J. M., and Anderson, D. J., Membrane depolarization induces p140^{trk} and and NGF responsiveness but not p75 (LNGFR), in MAH cells, *Science*, 257, 395, 1992.

60. Carrol, S. L., Silos-Santiago, I., Frese, S. E., Ruit, K. G., Milbrandt, J., and Snider, W. D., Dorsal root ganglion neurons expressing *trk* are selectively sensitive to NGF deprivation *in utero*, *Neuron*, 9, 779, 1992.

61. Lee, K.-F., Li, E., Huber, L. J., Landis, S. C., Sharpe, A. H., Chao, M. V., and Jaenisch, R., Targeted mutation of the gene encoding the low affinity NGF receptor p75 leads to deficits in the peripheral sensory nervous system, *Cell*, 69, 737, 1992.

62. Wright, E. M., Vogel, K. S., and Davies, A. M., Neurotrophic factor promote the maturation of developing sensory neurons before they become dependent of these factors for survival, *Neuron*, 9, 139, 1992.

63. Hempstead, B. L., Martin-Zanca, D., Kaplan, D. R., Parada, L. F., and Chao, M. V., High affinity NGF binding requires co-expression of the *trk* proto-oncogene product and the low affinity NGF receptor, *Nature (London)*, 350, 678, 1991.

64. Hempstead, B. L., Patil, N., Theil, B., and Chao, M. V., Deletion of cytoplasmatic sequences of the NGF receptor leads to loss of high affinity ligand binding, *J. Biol. Chem.*, 265, 9595, 1990.

65. Radeke, M. J. and Feinstein, S., Analytical purification of the slow, high affinity NGF receptor: identification of a novel 135 kd polypeptide, *Neuron*, 7, 141, 1991.

66. Taniuchi, M., Clark, H. B., Schwertzer, J. B., and Johnson, E. M., Expression of nerve growth factor receptors by Schwann cells of axotomized peripheral nerves: ultrastructural location, suppression by axonal contact and binding properties, *J. Neurosci.*, 8, 664, 1988.

67. Kahle, P. and Hertel, C., NGF receptor in rat glial cell lines, *J. Biol. Chem.*, 267, 13917, 1992.

68. Rodriguez-Tébar, A., Dechant, G., and Barde, Y.-A., Binding of brain-derived neurotrophic factor to the nerve growth factor receptor, *Neuron*, 4, 487, 1990.

69. Rodriguez-Tébar, A., Dechant, G., Götz, R., and Barde, Y.-A., Binding to neurotrophin-3 to its neuronal receptors and interactions with nerve growth factor and brain-derived neurotrophic factor, *EMBO J.*, 11, 917, 1992.

70. Hartikka, J. and Hefti, F., Development of septal cholinergic neurons in culture: plating density and glial cells modulate effects of NGF on surviving fiber growth, and expression of transmitter-specific enzymes, *J. Neurosci.*, 8, 2967, 1988.

71. Hatanaka, H., Tsukui, H., and Nihonmatsu, I., Developmental change in the nerve growth factor action from induction of choline acetyltransferase to promotion of cell survival in cultured basal forebrain cholinergic neurons from postnatal rats, *Dev. Brain Res.*, 39, 88, 1988.

72. Alderson, R.F., Hua, Z. W., and Hersh, L. B., Nerve growth factor and phorbol esters increase the number of choline acetyltransferase-positive cells in two morphologically distinct classes of basal forebrain neurons in primary cultures, *Brain Res. Dev. Brain Res.*, 48(2), 229, 1989.

73. Arimatsu, Y. and Miyamoto, M., Survival-promoting effect of NGF on *in vitro* septohippocampal neurons with cholinergic and GABAergic phenotypes, *Brain Res. Dev. Brain Res.*, 58(2), 189, 1991.

74. Gähwiler, B.H., Enz, A., and Hefti, F., Nerve growth factor promotes development of the rat septohippocampal cholinergic projection *in vitro*, *Neurosci. Lett.*, 75, 6, 1987.

75. Takei, N., Tsukui, H., and Hatanaka, H., Intracellular storage and evoked release of acetylcholine from postnatal rat basal forebrain cholinergic neurons in culture with nerve growth factor, *J. Neurochem.*, 53(5), 1405, 1989.

76. Honneger, P. and Leinor, D., Nerve growth factor (NGF) stimulation of cholinergic telencephalic neurons in aggregating cell cultures, *Dev. Brain Res.*, 3, 229, 1982.

77. Gnahn, H., Hefti, F., Heumann, R., Schwab, M. E., and Thoenen, H., NGF mediated increase of choline acetyltransferase (ChAT) in the neonatal rat forebrain: evidence for a physiological role of NGF in the brain, *Dev. Brain Res.*, 9, 45, 1983.

78. Hefti, F., Hartikka, J., Eckenstein, F., Gnahn, H., Heumann, R., and Schawb, M., Nerve growth factor (NGF) increases choline acetyltransferase but not survival or fiber outgrowth of cultured fetal septal cholinergic neurons, *Neuroscience*, 14, 55, 1985.

79. Hatanaka, H. and Tsukui, H., Differential effects of nerve growth factor and glioma-conditioned medium on neurons cultured from various regions of fetal rat central nervous system, *Dev. Brain Res.*, 30, 47, 1986.

80. Mobley, W.C., Rutkowski, J. L., Tennekoon, G., Gemski, J., Buchanan, K., and Johnston, H.V., Nerve growth factor increases choline acetyltransferase activity in developing basal forebrain neurons, *Mol. Brain Res.*, 1, 53, 1986.

81. Martinez, H. J., Dreyfus, C. F., Jonakait, G. M., and Black, I. B., Nerve growth factor selectively increases cholinergic markers but not neuropeptides in rat basal forebrain in culture, *Brain Res.*, 412, 295, 1987.

82. Takei, N., Tsukui, H., and Hatanaka, H., Nerve growth factor increases the intracellular content of acetylcholine in cultured septal neurons from developing rats, *J. Neurochem.*, 51, 1118, 1988.

83. Hatanaka, H., Nishio, C., Kushima, Y., and Tsukui, H., Nerve growth factor dependent and cell density independent survival of septal cholinergic neurons in culture from postnatal rats, *Neurosci. Res.*, 8(2), 69, 1990.

84. Knusel, B., Burton, L. E., Longo, F. M., Mobley, W. C., Koliatsos, V. E., Price, D. L., and Hefti, F., Trophic actions of recombinant human nerve growth factor on cultured rat embryonic CNS cells, *Exp. Neurol.*, 110(3), 274, 1990.

85. Eva, C., Fusco, M., Bono, C., Tria, M. A., Ricci Gamalero, S., Leon, A., and Genazzani, E., Nerve growth factor modulates the expression of muscarinic cholinergic receptor messenger RNA in telencephalic neuronal cultures from newborn rat brain, *Mol. Brain Res.*, 14, 344, 1992.

86. Bejanin, S., Habert, E., Berrard, S., Edwards, J. B., Loeffler, J. P., and Mallet, J., Promoter elements of the rat choline acetyltransferase gene allowing nerve growth factor inducibility in transfected primary cultured cells, *J. Neurochem.*, 58(4), 1580, 1992.

87. Lorenzi, M. V., Knusel, B., Hefti, F., and Strauss, W. L., Nerve growth factor regulation of choline acetyltransferase gene expression in rat embryo basal forebrain cultures, *Neurosci. Lett.*, 140(2), 185, 1992.

88. Large, T. H., Bodary, S. C., Clegg, D. O., Weskamp, G., Otten, U., and Reichardt, L. F., Nerve growth factor gene expression in the developing rat brain, *Science*, 234, 352, 1986.

89. Whittemore, S. R., Ebendal, T., Lärkfors, L., Olson, L., Seiger, A., Strömberg, I., and Persson, H., Developmental and regional expression of β nerve growth factor mRNA and protein in rat central nervous system, *Proc. Acad. Sci. U.S.A.*, 83, 817, 1986.

90. Auburger, G. R., Heumann, R., Hellweg, S., Korsching, S., and Thoenen, H., Developmental changes of nerve growth factor and its mRNA in the rat hippocampus: comparison with choline acetyltransferase, *Dev. Biol.*, 120, 322, 1987.

91. Lu, B., Buck, C. R., Dreyfus, C. F., and Black, I. B., Expression of NGF and NGF receptor mRNA in the developing rat brain: evidence for local delivery and action of NGF, *Exp. Neurol.*, 104, 191, 1989.

92. Eckenstein, F., Transient expression of NGF-receptor-like immunoreactivity in postnatal brain and spinal cord, *Brain Res.*, 446, 149, 1988.

93. Yan, Q. and Johnson, E. M., An immunohistochemical study of the nerve growth factor receptor in developing rats, *J. Neurosci.*, 8, 3481, 1988.

94. Koh, S. and Higgins, G. A., Differential regulation of the low-affinity nerve growth factor receptor during postnatal development of the rat brain, *J. Comp. Neurol.*, 313(3), 494, 1991.

95. Ringstedt, T., Lagercrantz, H., and Persson, H., Expression of members of the trk family in the developing postnatal rat brain, *Brain Res. Dev. Brain Res.*, 72(1), 119, 1993.

96. Johnston, M. V., Rutkowski, J. L., Wainer, B. H., Long, J. B., and Mobley, W. C., NGF effects on developing forebrain cholinergic neurons are regionally specific, *Neurochem. Res.*, 12, 985, 1987.

97. Mobley, W. C., Rutkowski, J. L., Tennekoon, G. I., Buchanan, K., and Johnston, M. V., Choline acetyltransferase activity in striatum of neonatal rats increased by nerve growth factor, *Science*, 229, 284, 1985.

98. Cavicchioli, L., Flanigan, T. P., Dickson, J. G., Vantini, G., Dal Toso, R., Fusco, M., Walsh, F. S., and Leon, A., Choline acetyltransferase messenger RNA expression in developing and adult rat brain: regulation by nerve growth factor, *Brain Res.*, 9, 319, 1991.

99. Vantini, G., Schiavo, A., Di Martino, A., Polato, P., Triban, C., Callegaro, L., Toffano, G., and Leon, A., Evidence for a physiological role of nerve growth factor in the central nervous system of neonatal rats, *Neuron*, 3, 267, 1989.

100. Eva, C., Fusco, M., Brusa, R., Schiavo, N., Ricci Gamalero, S., Vantini, G., and Genazzani, E., Intracerebroventricular administration of nerve growth factor affects muscarinic cholinergic receptors in the cerebral cortex of neonatal rats, *Neurochem. Int.*, 24(1), 57, 1994.

101. Alleva, E., Aloe, L., and LaViola, G., Pretreatment of young mice with nerve growth factor enhances scopolamine-induced hyperactivity, *Dev. Brain Res.*, 28, 1278, 1986.

102. Alleva, E. and Bignami, G., Development of mouse activity, stimulus reactivity, habituation, and response to amphetamine and scopolamine, *Physiol. Behav.*, 34, 519, 1985.

103. Hess, C. and Blozovski, D., Le NGF en injection intrahippocampique accélère l'ontogenèse de l'alternance spontanée et la maturation de l'innervation cholinergique septo-hippocampique chez le rat, *Neurobiologie*, 310, 533, 1990.

104. Korsching, S., Auburger, G., Heumann, R., Scott, J., and Thoenen, H., Levels of nerve growth factor and its mRNA in the central nervous system of the rat correlate with cholinergic innervation, *EMBO J.*, 4, 1389, 1985.

105. Goedert, M., Fine, A., Hunt, S. P., and Ullrich, A., Nerve growth factor mRNA in peripheral and central rat tissue and in the human central nervous system: lesion effects in the brain and level in Alzheimer's disease, *Mol. Brain Res.*, 1, 85, 1986.

106. Shelton, D. L. and Reichardt, L. F., Studies on the expression of the β-nerve growth factor (NGF) mRNA suggest that NGF functions as a trophic factor for several distinct populations of neurons, *Proc. Natl. Acad. Sci. U.S.A.*, 83, 2714, 1986.

107. Whittemore, S. R., Friedman, P. L., Larhammar, D., Persson, H., Gonzalez-Carvajarl, M., and Holets, V. R., Rat beta-nerve growth factor sequence and site of synthesis in the adult hippocampus, *J. Neurosci. Res.*, 20, 403, 1988.

108. Pioro, E. P. and Cuello, A. C., Distribution of nerve growth factor receptor-like immunoreactivity in the adult rat central nervous system. Effect of colchicine and correlation with the cholinergic system. I. Forebrain, *Neuroscience*, 34(1), 57, 1990.

109. Conner, J. M., Muir, D., Varon, S., Hagg, T., and Manthorpe, M., The localization of nerve growth factor-like immunoreactivity in the adult rat basal forebrain and hippocampal formation, *J. Comp. Neurol.*, 319(3), 454, 1992.

110. Rennert, P. D. and Heinrich, G., Nerve growth factor mRNA in brain: localization by *in situ*-hybridization, *Biochem. Biophys. Res. Comm.*, 138, 813, 1986.
111. Ayer-LeLievre, C., Olson, L., Ebendahl, T., Seiger, A., and Persson, H., Expression of the beta-nerve growth factor gene in hippocampal neurons, *Science*, 240, 1339, 1988.
112. Gall, C. M. and Isackson, P. J., Limbic seizures increase neuronal production of messenger RNA for nerve growth factor, *Science,* 245, 758, 1989.
113. Ernfors, P., Wetmore, C., Olson, L., and Persson, H., Identification of cells in rat brain and peripheral tissues expressing mRNA for members of the nerve growth factor family, *Neuron,* 5, 511, 1990.
114. Conner, J. M. and Varon, S., Distribution of nerve growth factor-like immunoreactive neurons in the adult rat brain following colchicine treatment, *J. Comp. Neurol.*, 326(3), 347, 1992.
115. Lauterborn, J. C., Isackson, P. J., and Gall, C. M., Nerve growth factor mRNA-containing cells are distributed within regions of cholinergic neurons in the rat basal forebrain, *J. Comp. Neurol.*, 306(3), 439, 1991.
116. Schwab, M. N., Otten, U., Agid, Y., and Thoenen, H., Nerve growth factor (NGF) in the rat CNS: absence of specific retrograde axonal transport and tyrosine hydroxylase induction in locus ceruleus and substantia nigra, *Brain Res.*, 168, 473, 1979.
117. Seiler, M. and Schwab, M. E., Specific retrograde transport of nerve growth factor (NGF) from neocortex to nucleus basalis in the rat, *Brain Res.*, 300, 34, 1984.
118. Ferguson, I. A., Schweitzer, J. B., Bartlett, P. F., and Johnson, E. M., Jr., Receptor-mediated retrograde transport in CNS neurons after intraventricular administration of NGF and growth factors, *J. Comp. Neurol.*, 313(4), 680, 1991.
119. Richardson, P. M., Verge Issa, V. M. K., and Riopelle, R. J., Distribution of neuronal receptors for nerve growth factor in the rat, *J. Neurosci.*, 6, 2312, 1986.
120. Riopelle, R. J., Verge, V. M., and Richardson, P. M., Properties of receptors for nerve growth factor in the mature rat nervous system, *Mol. Brain Res.*, 3, 45, 1987.
121. Altar, C. A., Burton, L. E., Bennet, G. L., and Dugich-Djordjevic, M., Recombinant human nerve growth factor is biologically active and labels novel high affinity binding sites in rat brain, *Proc. Natl Acad. Sci. U.S.A.*, 88, 281, 1991.
122. Taniuchi, M. and Johnson, E. M., Characterization of the binding properties and retrograde transport of a monoclonal antibody directed against the rat nerve growth factor receptor, *J. Cell Biol.*, 101, 1100, 1985.
123. Hefti, F., Hartikka, J., Salvatierra, A., Weiner, W. J., and Mash, D. C., Localization of nerve growth factor receptors in cholinergic neurons of the human basal forebrain, *Neurosci. Lett.*, 20, 275, 1986.
124. Johnson, E. M., Jr., Taniuchi, M., Clark, B. H., Springer, J. E., Koh, S., Tayrien, M., and Loy, R., Demonstration of the retrograde transport of nerve growth factor receptor in the peripheral and central nervous system, *J. Neurosci.*, 7, 923, 1987.
125. Springer, J. E., Koh, B., Tavrien, M. W., and Loy, R., Basal forebrain magnocellular neurons stain for nerve growth factor receptor: correlation with cholinergic cell bodies and effects of axotomy, *J. Neurosci. Res.*, 17, 111, 1987.
126. Kordower, J. H., Bartus, R. T., Bothwell, M., Schatteman, G., and Gash, D. M., Nerve growth factor receptor immunoreactivity in the non-human primate (*Cebus apella*): distribution, morphology and colocalization with cholinergic enzymes, *J. Comp. Neurol.*, 277, 465, 1988.
127. Batchelor, P. E., Amstrong, D. M., Blaker, S. N., and Gage, F., H., Nerve growth factor receptor and choline acetyltransferase colocalization in neurons within the rat forebrain: response to fimbria-fornix transection, *J. Comp. Neurol.*, 72, 248, 1989.
128. Kerwin, J., Morris, C., Oakley, A., Perry, R., and Perry, E., Distribution of nerve growth factor receptor immunoreactivity in the human hippocampus, *Neurosci. Lett.*, 121(1–2), 178, 1991.
129. Kawaja, M. D. and Gage, F. H., Nerve growth factor receptor immunoreactivity in the rat septohippocampal pathway: a light and electron microscope investigation, *J. Comp. Neurol.*, 307(3), 517, 1991.
130. Allen, S. J., Dawbarn, D., Spillantini, M. G., Goedert, M., Wilcock, G. K., Moss, T. H., and Semenenko, F. M., Distribution of beta-nerve growth factor receptors in the human basal forebrain, *J. Comp. Neurol.*, 289(4), 626, 1989.
131. Mufson, E. J., Bothwell, M., Hersh, L. B., and Kordower, J. H., Nerve growth factor receptor immunoreactive profiles in the normal, aged human basal forebrain: colocalization with cholinergic neurons, *J. Comp. Neurol.*, 285(2), 196, 1989.

132. Woolf, N. J., Gould, E., and Butcher, L. L., Nerve growth factor receptor is associated with cholinergic neurons of the basal forebrain but not the pontomesencephalon, *Neuroscience*, 30(1), 143, 1989.

133. Vazquez, M. E. and Ebendal, T., Messenger RNAs for trk and the low affinity NGF receptor in the rat basal forebrain, *Neuroreport*, 2, 593, 1991.

134. Fusco, M., Oderfeld-Nowak, B., Vantini, G., Schiavo, N., Gradkowska, M., Zaremba, M., and Leon, A., Nerve growth factor affects uninjured, adult rat septohippocampal cholinergic neurons, *Neuroscience*, 33, 47, 1989.

135. Fusco, M., Tria, M. A., Schiavo, N., Leon, A., and Vantini, G., Nerve growth factor induces a marked and long-lasting enhancement of choline acetyltransferase activity in septohippocampal neurons of uninjured adult rats, *Neurosci. Res. Commun.*, 7, 97, 1990.

136. Palmer, M. R., Eriksdotter-Nilsson, M., Henschen, A., Ebendal, T., and Olson, L., Nerve growth factor-induced excitation of selected neurons in the brain which is blocked by a low-affinity receptor antibody, *Exp. Brain. Res.*, 93(2), 226, 1993.

137. Lindefors, N., Ernfors, P., Falkenberg, T., and Persson, H., Septal cholinergic afferents regulate expression of brain-derived neurotrophic factor and beta-nerve growth factor mRNA in rat hippocampus, *Exp. Brain Res.*, 88(1), 78, 1992.

138. Cavicchioli, L., Flanigan, T. P., Vantini, G., Fusco, M., Polato, P., Toffano, G., Walsh, F. S., and Leon, A., NGF amplifies expression of NGF receptor messenger RNA in forebrain cholinergic neurons of rats, *Eur. J. Neurosci.*, 1, 258, 1989.

139. Higgins, G. A., Koh, S. K., Chen, K. S., and Gage, F. H., NGF induction of NGF receptor gene expression and cholinergic neuronal hypertrophy within the basal forebrain of the adult rat, *Neuron*, 3, 247, 1989.

140. Fusco, M., Polato, P., Vantini, G., Cavicchioli, L., Bentivoglio, M., and Leon, A., Nerve growth factor differentially modulates the expression of its receptor within the CNS, *J. Comp. Neurol.*, 312(3), 477, 1991.

141. Sofroniew, M. V., Galletly, N. P., Isacson, O., and Svendsen, C. N., Survival of adult basal forebrain cholinergic neurons after loss of target neurons, *Science*, 247, 338, 1990.

142. Kordower, J. H., Burke-Watson, M., Roback, J. D., and Wainer, B. H., Stability of septohippocampal neurons following excitotoxic lesions of the rat hippocampus, *Exp. Neurol.*, 117(1), 1, 1992.

143. Hsiang, J., Wainer, B. H., Shalaby, I. A., Hoffmann, P. C., Heller, A., and Heller, B. R., Neurotrophic effects of hippocampal target cells on developing septal cholinergic neurons in culture, *Neuroscience*, 21, 333, 1987.

144. Hsiang, J., Price, S. D., Heller, A., Hoffmann, P. C., and Wainer, B. H., Ultrastructural evidence for hippocampal target cell-mediated effect on developing septal cholinergic neurons in reaggregating cell cultures, *Neuroscience*, 26, 417, 1988.

145. Whitehouse, P. J., Price, D. L., Struble, R. G., Clark, A. W., Coyle, J. T., and DeLong, M. R., Alzheimer's disease and senile dementia: loss of neurons in the basal forebrain, *Science,* 215, 1237, 1982.

146. Arendt, T., Bigl, V., Arendt, A., and Tennstedt, A., Loss of neurons in the nucleus basalis of Meynert in Alzheimer's disease, paralysis agitants and Korsakoff's disease, *Acta Neuropathol.*, 61, 101, 1983.

147. Dunnet, S. B., Whishaw, I. Q., Jones, G. H., and Bunch, S. T., Behavioral, biochemical and histochemical effects of different neurotoxic aminoacids injected into the nucleus basalis magnocellularis of rats, *Neuroscience*, 20, 653, 1987.

148. Sofroniew, M. W., Pearson, R. C. A., Eckenstein, F., Cuello, A. C., and Powell, T. P. S., Retrograde changes in cholinergic neurons in the basal forebrain of the rat following cortical damage, *Brain Res.*, 289, 370, 1983.

149. Lapchak, P. A. and Hefti, F., Effect of recombinant human nerve growth factor on presynaptic cholinergic function in rat hippocampal slices following partial septohippocampal lesions: measures of [3H]acetylcholine synthesis, [3H]acetylcholine release and choline acetyltransferase activity, *Neuroscience*, 42(3), 639, 1991.

150. Maysinger, D., Herrera-Marschitz, M., Ungerstedt, U., and Cuello, A.C., Effects of treatment with microencapsulated monosialoganglioside GM1 on cortical and striatal acetylcholine release in rats with cortical devascularizing lesions, *Neurosci. Lett.*, 118, 252, 1990.

151. Maysinger, D., Jalsenjak, I., and Cuello, A. C., Microencapsulated nerve growth factor: effects on the forebrain neurons following devascularizing cortical lesions, *Neurosci Lett.*, 140(1), 71, 1992.

152. Koliatsos, V. E., Nauta, H. J., Clatterbuck, R. E., Holtzman, D. M., Mobley, W. C., and Price, D. L., Mouse nerve growth factor prevents degeneration of axotomized basal forebrain cholinergic neurons in the monkey, *J. Neurosci.*, 10(12), 3801, 1990.

153. Tuszynski, M. H., U, H. S., Amaral, D. G., and Gage, F. H., Nerve growth factor infusion in the primate brain reduces lesion-induced cholinergic neuronal degeneration, *J. Neurosci.*, 10(11), 3604, 1990.

154. Koliatsos, V. E., Clatterbuck, R. E., Nauta, H. J., Knusel, B., Burton, L. E., Hefti, F. F., Mobley, W. C., and Price, D. L., Human nerve growth factor prevents degeneration of basal forebrain cholinergic neurons in primates, *Ann. Neurol.*, 30(6), 831, 1991.

155. Rosenberg, M. B., Friedmann, T., Robertson, R. C., Tuszynski, M., Wolff, J. A., Breakefield, X. O., and Gage, F. H., Grafting genetically modified cells to damaged brain: restorative effects of NGF expression, *Science*, 242, 1575, 1988.

156. Strömberg, I., Wetmore, C. J., Ebendal, T., Ernfors, P., Persson, H., and Olson, L., Rescue of basal forebrain cholinergic neurons after implantation of genetically modified cells producing recombinant NGF, *J. Neurosci. Res.*, 25(3), 405, 1990.

157. Tuszynski, M. H., Buzsaki, G., and Gage, F. H., Nerve growth factor infusions combined with fetal hippocampal grafts enhance reconstruction of the lesioned septohippocampal projection, *Neuroscience*, 36(1), 33, 1990.

158. Kawaja, M. D., Rosenberg, M. B., Yoshida, K., and Gage, F. H., Somatic gene transfer of nerve growth factor promotes the survival of axotomized septal neurons and the regeneration of their axons in adult rats, *J. Neurosci.*, 12(7), 2849, 1992.

159. Williams, L. R., Jodelis, K. S., and Donald, M. R., Axotomy-dependent stimulation of choline acetyltransferase activity by exogenous mouse nerve growth factor in adult rat basal forebrain, *Brain Res.*, 498(2), 243, 1989.

160. Vahlsing, H. L., Hagg, T., Spencer, M., Conner, J. M., Manthorpe, M., and Varon, S., Dose-dependent responses to nerve growth factor by adult rat cholinergic medial septum and neostriatum neurons, *Brain Res.*, 552(2), 320, 1991.

161. Dekker, A. J. and Thal, L. J., Effect of delayed treatment with nerve growth factor on choline acetyltransferase activity in the cortex of rats with lesions of the nucleus basalis magnocellularis: dose requirements, *Brain. Res.*, 584(1–2), 55, 1992.

162. Gage, F. H., Armstrong, D. M., Williams, L. R., and Varon, S., Morphological response of axotomized septal neurons to nerve growth factor, *J. Comp. Neurol.*, 269, 147, 1988.

163. Hefti, F., Nerve growth factor (NGF) promotes survival of septal cholinergic neurons after fimbrial transections, *J. Neurosci.*, 6, 2155, 1986.

164. Williams, L. R., Varon, S., Peterson, G. M., Wictorin, K., Fischer, W., Bjorklund, A., and Gage, F. H., Continuous infusion of nerve growth factor prevents basal forebrain neurons death after fimbria fornix transection, *Proc. Natl Acad. Sci. U.S.A.*, 83, 9231, 1986.

165. Hagg, T., Fass Holmes, B., Vahlsing, H. L., Manthorpe, M., Conner, J. M., and Varon, S., Nerve growth factor (NGF) reverses axotomy-induced decreases in choline acetyltransferase, NGF receptor and size of medial septum cholinergic neurons, *Brain Res.*, 505(1), 29, 1989.

166. Lapchak, P. A., Jenden, D. J., and Hefti, F., Pharmacological stimulation reveals recombinant human nerve growth factor-induced increases of *in vivo* hippocampal cholinergic function measured in rats with partial fimbrial transections, *Neuroscience*, 50(4), 847, 1992.

167. Lapchak, P. A., Araujo, D. M., and Hefti, F., Regulation of hippocampal muscarinic receptor function by chronic nerve growth factor treatment in adult rats with fimbrial transections, *Neuroscience*, 53(2), 379, 1993.

168. Kromer, L. F., Nerve growth factor treatment after brain injury prevents neuronal death, *Science*, 235, 214, 1987.

169. Hagg, T., Mantorphe, M., Vahlsing, H. L., and Varon, S., Delayed treatment with nerve growth factor reverses the apparent loss of cholinergic neurons after acute brain damage, *Exp. Neurol.*, 101, 303, 1988.

170. Koliatsos, V. E., Applegate, M. D., Knusel, B., Junard, E. O., Burton, L. E., Mobley, W. C., Hefti, F. F., and Price, D. L., Recombinant human nerve growth factor prevents retrograde degeneration of axotomized basal forebrain cholinergic neurons in the rat, *Exp. Neurol.*, 112(2), 161, 1991.

171. Gage, F. H., Wictorin, K., Fischer, W., Williams, L. R., Varon, S., and Bjorklund, A., Life and death of cholinergic neurons: in the septal and diagonal band region following complete fimbria fornix transections, *Neuroscience*, 19, 2431, 1986.

172. Tuszynski, M. H., Armstrong, D. M., and Gage, F. H., Basal forebrain cell loss following fimbria/fornix transection, *Brain Res.*, 508(2), 241, 1990.

173. Fischer, W. and Bjorklund, A., Loss of AChE- and NGFr-labeling precedes neuronal death of axotomized septal-diagonal band neurons: reversal by intraventricular NGF infusion, *Exp. Neurol.*, 113(2), 93, 1991.

174. Montero, C. N. and Hefti, F., Rescue of lesioned septal cholinergic neurons by nerve growth factor: specificity and requirement for chronic treatment, *J. Neurosci.*, 8, 2986, 1988.

175. Hagg, T., Quon, D., Higaki, J., and Varon, S., Ciliary neurotrophic factor prevents neuronal degeneration and promotes low affinity NGF receptor expression in the adult rat CNS, *Neuron*, 8(1), 145, 1992.

176. Junard, E. O., Montero, C. N., and Hefti, F., Long-term administration of mouse nerve growth factor to adult rats with partial lesions of the cholinergic septohippocampal pathway, *Exp. Neurol.*, 110(1), 25, 1990.

177. Hagg, T., Vahlsing, H. L., Manthorpe, M., and Varon, S., Nerve growth factor infusion into the denervated adult rat hippocampal formation promotes its cholinergic reinnervation, *J. Neurosci.*, 10(9), 3087, 1990.

178. Hagg, T., Vahlsing, H. L., Manthorpe, M., and Varon, S., Septohippocampal cholinergic axonal regeneration through peripheral nerve bridges: quantification and temporal development, *Exp. Neurol.*, 109, 153, 1990.

179. Cuello, A. C., Garofalo, L., Kenigsberg, A. L., and Maysinger, D., Gangliosides potentiate in vivo and in vitro effects of nerve growth factor in central cholinergic neurons, *Proc. Natl. Acad. Sci. U.S.A.*, 86, 2056, 1989.

180. Maysinger, D., Herrera-Marschitz, M., Goiny, M., Ungerstedt, U., and Cuello, A. C., Effects of nerve growth factor on cortical and striatal acetylcholine and dopamine release in rats with cortical devascularizing lesions, *Brain Res.*, 577(2), 300, 1992.

181. Cuello, A. C., Maysinger, D., and Garofalo, L., Trophic factor effects on cholinergic innervation in the cerebral cortex of the adult rat brain, *Mol. Neurobiol.*, 6(4), 451, 1992.

182. Garofalo, L., Ribeiro da Silva, A., and Cuello, A. C., Nerve growth factor-induced synaptogenesis and hypertrophy of cortical cholinergic terminals, *Proc. Natl. Acad. Sci. U.S.A.*, 89(7), 2639, 1992.

183. Liberini, P., Pioro, E. P., Maysinger, D., Ervin, F. R., and Cuello, A. C., Long-term protective effects of human recombinant nerve growth factor and monosialoganglioside GM1 treatment on primate nucleus basalis cholinergic neurons after neocortical infarction, *Neuroscience*, 53(3), 625, 1993.

184. Piccardo, P., Maysinger, D., and Cuello, A. C., Recovery of nucleus basalis cholinergic neurons by grafting NGF secretor fibroblasts, *Neuroreport*, 3(4), 353, 1992.

185. Haroutunian, V., Kanof, P. D., and Davies, K. L., Partial reversal of lesion-induced deficits in cortical cholinergic markers by nerve growth factor, *Brain Res.*, 386, 397, 1986.

186. Casamenti, F., Di Patre, P. L., Milan, F., Petrelli, L., and Pepeu, G., Effects of nerve growth factor and GM1 ganglioside on the number and size of cholinergic neurons in rats with unilateral lesion of the nucleus basalis, *Neurosci. Lett.*, 103, 87, 1989.

187. Di Patre, P. L., Casamenti, F., Cenni, A., and Pepeu, G., Interaction between nerve growth factor and GM1 monosialoganglioside in preventing cortical choline acetyltransferase and high affinity choline uptake decrease after lesion of the nucleus basalis, *Brain Res.*, 480, 219, 1989.

188. Haroutunian, V., Kanof, P. D., and Davies, K. L., Attenuation of nucleus basalis of Meynert lesion-induced cholinergic deficits by nerve growth factor, *Brain Res.*, 487(1), 200, 1989.

189. Dekker, A. J., Langdon, D. J., Gage, F. H., and Thal, L. J., NGF increases cortical acetylcholine release in rats with lesions of the nucleus basalis, *Neuroreport*, 2(10), 577, 1991.

190. Dekker, A. J., Gage, F. H., and Thal, L. J., Delayed treatment with nerve growth factor improves acquisition of a spatial task in rats with lesions of the nucleus basalis magnocellularis: evaluation of the involvement of different neurotransmitter systems, *Neuroscience*, 48(1), 111, 1992.

191. Dekker, A. J. and Thal, L. J., Nerve growth factor increases cortical choline acetyltransferase-positive fiber staining without affecting cortical cholinergic neurons, *Brain Res.*, 601(1–2), 329, 1993.

192. Gage, F. H., Chen, K. S., Buzsachi, G., and Amstrong, D., Experimental approaches to age-related cognitive impairments, *Neurobiol. Aging*, 9, 645, 1988.

193. Hefti, F., Hartikka, J., and Knusel, B., Function of neurotrophic factors in the adult and aging brain and their possible use in the treatment of neurodegenerative diseases, *Neurobiol. Aging*, 10, 515, 1989.

194. Lärkfors, L., Ebendal, T., Whittemore, S. R., Persson, H., Hoffer, B., and Olson, L., Decreased level of nerve growth factor (NGF) and its messenger RNA in the aged rat brain, *Mol. Brain Res.*, 3, 55, 1987.

195. Koh, S. and Loy, R., Age-related loss of nerve growth factor sensitivity in rat basal forebrain cholinergic neurons, *Brain Res.*, 440, 396, 1988.

196. Hefti, F. and Mash, D. C., Localization of nerve growth factor receptors in the normal human brain and in Alzheimer's disease, *Neurobiol. Aging*, 10(1), 75, 1989.

197. Alberch, J., Perez-Navarro, E., Arenas, E., and Marsal, J., Involvement of nerve growth factor and its receptor in the regulation of the cholinergic function in aged rats, *J. Neurochem.*, 57(5), 1483, 1991.

198. Koh, S., Chang, P., Collier, T. J., and Loy, R., Loss of NGF receptor immunoreactivity in basal forebrain neurons of aged rats: correlation with spatial memory impairment, *Brain Res.*, 498(2), 397, 1989.

199. Fischer, W., Chen, K. S., Gage, F. H., and Bjorklund, A., Progressive decline in spatial learning and integrity of forebrain cholinergic neurons in rats during aging, *Neurobiol. Aging*, 13(1), 9, 1992.

200. Henriksson, B. G., Soderstrom, S., Gower, A. J., Ebendal, T., Winblad, B., and Mohammed, A. H., Hippocampal nerve growth factor levels are related to spatial learning ability in aged rats, *Behav. Brain Res.*, 48(1), 15, 1992.

201. Williams, L. R. and Rylett, R. J., Exogenous nerve growth factor increases the activity of high affinity choline uptake and choline acetyltransferase in brain of Fisher 344 male rats, *J. Neurochem.*, 55, 1042, 1990.

202. Williams, L. R., Exogenous nerve growth factor stimulates choline acetyltransferase activity in aging Fischer 344 male rats, *Neurobiol. Aging*, 12(1), 39, 1991.

203. Hadjiconstantinou, M., Karadsheh, N. S., Rattan, A. K., Tejwani, G. A., Fitkin, J. G., and Neff, N. H., GM1 ganglioside enhances cholinergic parameters in the brain of senescent rats, *Neuroscience*, 46(3), 681, 1992.

204. Montero, C. N. and Hefti, F., Intraventricular nerve growth factor administration prevents lesion-induced loss of septal cholinergic neurons in aging rats, *Neurobiol. Aging*, 10(6), 739, 1989.

205. Yunshao, H., Zhibin, Y., and Yici, C., Effect of nerve growth factor on the lesioned septohippocampal cholinergic system of aged rats, *Brain Res.*, 552(1), 159, 1991.

206. Fischer, W, Wictorin, K., Bjorklund, A., Williams, L. R., Varon, S., and Gage, F. H., Amelioration of cholinergic neurons atrophy and spatial memory impairment in aged rats by nerve growth factor, *Nature (London)*, 329, 65, 1987.

207. Gage, F. H. and Bjorklund, A., Cholinergic septal grafts into the hippocampal formation improve spatial learning and memory in aged rats by an atropine-sensitive mechanism, *J. Neurosci.*, 6, 2837, 1986.

208. Lapchak, P. A., Therapeutic potential for nerve growth factor in Alzheimer's disease: insights for pharmacological studies using lesioned central cholinergic neurons, *Rev. Neurosci.*, 3, 109, 1992.

209. Will, B. and Hefti, F., Behavioral and neurochemical effects of chronic intraventricular injections of nerve growth factor in adult rats with fimbria lesions, *Behav. Brain Res.*, 17, 17, 1985.

210. Pallage, V., Toniolo, G., Will, B., and Hefti, F., Long-term effects of nerve growth factor and neural transplants on behavior of rats with medial septal lesions, *Brain Res.*, 386, 197, 1986.

211. Will, B., Hefti, F., Pallage, V., and Toniolo, G., Nerve growth factor. Effects on CNS neurons and on behavioral recovery from brain damage, in *Pharmacological Approaches to the Treatment of Brain and Spinal Cord Injury*, Stein D. and Sabel B., Eds., Plenum Press, New York, 1988, 339.

212. Will, B., Pallage, B., and Eclancher, F., Nerve growth factor and behavioral recovery after brain damage in rats, in *Neurotrophic Factors and Alzheimer's Disease*, Springer, New York, 1990, 117.

213. Pallage, V., Effects of nerve growth factor on functional recovery in rats receiving grafts after septal lesions: a preliminary report, *Med. Sci. Res.*, 18, 511, 1990.

214. Pallage, V., Orenstein, D., and Will, B., Nerve growth factor and septal grafts: a study of behavioral recovery following damage to the septum in rats, *Behav. Brain Res.*, 47, 1, 1992.

215. Mandel, R. J., Gage, F. H., and Thal, L. J., Spatial learning in rats: correlation with cortical choline acetyltransferase and improvement with NGF following NBM damage, *Exp. Neurol.*, 104(3), 208, 1989.

216. Casamenti, F., Milan, F., and Pepeu, G., Lesions of the nucleus basalis magnocellularis in the rat: morphological, biochemical and behavioral reparative effect of nerve growth factor and ganglioside GM1, *Acta Neurobiol. Exp. (Warsz.)*, 50(4–5), 461, 1990.

217. Bartus, R. T., Dean, R. L., Pontecorvo, M. J., and Lippa, A. S., The cholinergic hypothesis: an historical overview, current perspective and future directions, in *Memory Dysfunctions; An Integration of Animal and Human Research from Preclinical and Clinical Perspectives*, Vol. 444, Olton D. S., Gamzu E., and Corkin S., Eds., The New York Academy of Sciences Press, New York, 1985, 332.

218. Olton, D. S. and Wenk, G. L., Dementia: animal models of the cognitive impairments produced by degeneration of the basal forebrain cholinergic system, in *Psychopharmacology: The Third Generation of Progress*, Meltzer, H. Y., Eds., Raven Press, New York, 1987, 941.

219. Pallage, V., Knusel, B., Hefti, F., and Will, B., Functional consequences of a single nerve growth factor administration following septal damage in rats, *Eur. J. Neurosci.*, 5, 669, 1993.

220. Hefti, F. and Schneider, L. S., Rationale for the planned clinical trials with nerve growth factor in Alzheimer's disease, *Psychiatry Dev.*, 7(4), 297, 1989.

221. Phelphs, C. H., Gage, F. H., Growdown, J. H., Hefti, F., Harbaugh, R., Johnston, M. V., Khachaturian, Z. S., Mobley, W. C., Price, D. L., Raskind, M., Simpkins, J., Thal, L. J., and Woodcock, J., Potential use of nerve growth factor to treat Alzheimer's disease, *Neurobiol. Aging*, 190, 205, 1989.

222. Allen, S. J., MacGowan, S. H., Treanor, J. J., Feeney, R., Wilcock, G. K., and Dawbarn, D., Normal beta-NGF content in Alzheimer's disease cerebral cortex and hippocampus, *Neurosci. Lett.*, 131(1), 135, 1991.

223. Lorigados, L., Söderström, S., and Ebendal, T., Two-site enzyme immunoassay for βNGF applied to human patient sera, *J. Neurosci. Res.*, 32, 329, 1992.

224. Kordower, J. H., Gash, D. M., Bothwell, M., Hersh, L., and Mufson, E. J., Nerve growth factor receptor and choline acetyltransferase remain colocalized in the nucleus basalis (Ch4) of Alzheimer's patients, *Neurobiol Aging*, 10(1), 67, 1989.

225. Mufson, E. J., Bothwell, M., and Kordower, J. H., Loss of nerve growth factor receptor-containing neurons in Alzheimer's disease: a quantitative analysis across subregions of the basal forebrain, *Exp. Neurol.*, 105(3), 221, 1989.

226. Allen, S. J., Dawbarn, D., MacGowan, S. H., Wilcock, G. K., Treanor, J. J. S., and Moss, T. H., Quantitative morphometric analysis of basal forebrain neurons expressing β-nerve growthfactor receptors in normal and Alzheimer's disease brains, *Dementia*, 1, 125, 1990.

227. Kerwin, J. M., Morris, C. M., Perry, R. H., and Perry, E. K., Hippocampal nerve growth factor receptor immunoreactivity in patients with Alzheimer's and Parkinson's disease, *Neurosci Lett.*, 143(1–2), 101, 1992.

228. Strada, O., Hirsch, E. C., Javoy-Agid, F., Lehericy, S., Ruberg, M., Hauw, J.-J., and Agid, Y., Does loss of nerve growth factor receptors precede loss of cholinergic neurons in Alzheimer's disease? An autoradiographic study in the human striatum and basal forebrain, *J. Neurosci.*, 12(12), 4766, 1992.

229. Goedert, M., Fine, A., Dawbarn, D., Wilcock, G. K., and Chao, M. V., Nerve growth factor receptor mRNA distribution in human brain: normal levels in basal forebrain in Alzheimer's disease, *Mol. Brain Res.*, 5(1), 1, 1989.

230. Ernfors, P., Linderfors, N., Chan-Palay, V., and Persson, H., Cholinergic neurons in the nucleus basalis express elevated levels of nerve growth factor receptor mRNA in senile dementia of the Alzheimer's type, *Dementia*, 1, 138, 1990b.

231. Hardy, J. and Allsop, D., Amyloid deposition as the central event in the aetiology of Alzheimer's disease, *Trends Pharmacol. Sci.*, 12, 383 1991.

232. Olson, L., Nordberg, A., von-Holst, H., Backman, L., Ebendal, T., Alafuzoff, I., Amberla, K., Hartvig, P., Herlitz, A., Lilja, A., Lundqvist, H., Langström, B., Meyerson, B., Persson, A., Viitanen, M., Winblad, B., and Seiger, A., Nerve growth factor affects ^{11}C-nicotine binding, blood flow, EEG, and verbal episodic memory in an Alzheimer patient (case report), *J. Neural Transm.*, 4(1), 79, 1992.

233. Butcher, L. L. and Woolf, N. J., Neurotrophic agents may exacerbate the pathologic cascade of Alzheimer's disease, *Neurobiol. Aging*, 10, 557, 1989.

234. Yankner, B. A., Caceres, A., and Duffy, L. K., Nerve growth factor potentiates the neurotoxicity of beta amyloid, *Proc. Natl. Acad. Sci. U.S.A.*, 87(22), 9020, 1990.

235. Hagg, T. and Varon, S., Neurotropism of nerve growth factor for adult rat septal cholinergic axons in vivo, *Exp. Neurol.*, 119(1), 37, 1993.

236. Bolam, J. P., Wainer, B. H., and Smith, A. D., Characterization of cholinergic neurons in the rat neostriatum. A combination of choline acetyltransferase immunocytochemistry, Golgi-impregnation and electron microscopy, *Neuroscience*, 12 (3), 711, 1984.

237. Hartikka, J. and Hefti, F., Comparison of nerve growth factor's effects on development of septum, striatum, and nucleus basalis cholinergic neurons *in vitro, J. Neurosci. Res.*, 21, 352, 1988.

238. Ebstein, R. P., Bennet, E. R., Sokoloff, and Shoham, S., The effect of nerve growth factor on cholinergic cells in primary fetal striatal cultures: characterization by *in situ* hybridization, *Dev. Brain Res.*, 73, 165, 1993.

239. Mobley, W. C., Woo, J. E., Edwards, R. H., Riopelle, R. J., Longo, F. M., Weskamp, G., Otten, U., Valletta, J. S., Johnston, M. V., Developmental regulation of nerve growth factor and its receptor in the rat caudate-putamen, *Neuron*, 3(5), 655, 1989.

240. Ernfors, P., Hallböök, F., Ebendal, T., Shooter, E. M., Radeke, M. J., Misko, T. P., and Persson, H., Developmental and regional expression of β-nerve growth factor receptor mRNA in the chick and rat neuron, *Neuron*, 1, 983, 1988.

241. Gage, F. H., Batchelor, P., Chen, K. S., Chin, D., Higgins, G. A., Koh, S., Deputy, S., Rosenberg, M. B., Fischer, W., and Björklund, A., NGF receptor reexpression and NGF-mediated cholinergic neuronal hypertrophy in the damaged adult neostriatum, *Neuron*, 2, 1177, 1989.

242. Hagg, T., Hagg, F., Vahlsing, H. L., Manthorpe, M., and Varon, S., Nerve growth factor effects on cholinergic neurons of neostriatum and nucleus accumbens in the adult rat, *Neuroscience*, 30(1), 95, 1989.

243. Altar, C. A., Armanini, M., Dugich-Djordjevic, M., Bennet, G. L., Williams, R., Feinglass, S., Anicetti, V., Sinicropi, D., and Bakhit, C., Recovery of cholinergic phenotype in the injured rat neostriatum: roles for endogenous and exogenous nerve growth factor, *J. Neurochem.*, 59(6), 2167, 1992.

244. Davies, S. W. and Beardsall, K., Nerve growth factor selectively prevents excitotoxin induced degeneration of striatal cholinergic neurones, *Neurosci. Lett.*, 140(2), 161, 1992.

245. Gomez-Pinilla, F., Cotman, C. W., and Nieto-Sampedro, M., NGF receptor immunoreactivity in rat brain: topographical distribution and response to etorhinal ablation, *Neurosci. Lett.*, 82, 260, 1987.

246. Schatteman, G. C., Gibbs, L., Lanahan, A. A., Claude, P., and Bothwell, M., Expression of NGF receptor in the developing and adult primate central nervous system, *J. Neurosci.*, 8, 860, 1988.

247. Kiss, J., McGovern, J., and Patel, A. J., Immunohistochemical localization of cells containing nerve growth factor receptor in the different regions of the adult rat forebrain, *Neuroscience*, 27, 731, 1988.

248. Springer, J. E., Robbins, E., Meyer, S., Baldino, F.J., and Lewis, M.E., Localization of nerve growth factor receptor mRNA in the rat basal forebrain with *in situ* hybridization histochemistry, *Cell. Mol. Neurobiol.*, 10, 33, 1990.

249. Meakin, S. O. and Shooter, E. M., The nerve growth factor family of receptors, *Trends Neurosci.*, 15(9), 323, 1992.

250. Buckley, N. J., Bonner, T. I., and Brann, M. R., Localization of a family of muscarinic receptor mRNAs in rat brain, *J. Neurosci.*, 8, 4646, 1988.

251. Weiner, D. M. and Brann, M.R., Distribution of m_1-m_5 muscarinic receptor mRNAs in rat brain, *FEBS Lett.*, 253, 207, 1989.

252. Altar, C. A., Nerve growth factor and the neostriatum, *Prog. Neuropsychopharmacol. Biol. Psychiatry*, 15(2), 157, 1991.

253. Leibrock, J., Lottspeich, F., Hohn, A., Hofer, M., Hengerer, B., Masiakowski, P., Thoenen, H., and Barde, Y.-A., Molecular cloning and expression of brain-derived neurotrophic factor, *Nature (London)*, 341, 149, 1989.

254. Ernfors, P., Ibàñez, C. F., Ebendal, T., Olson, L., and Persson, H., Molecular cloning and neurotrophic activities of a protein with structural similarities to nerve growth factor: developmental and topographical expression in the brain, *Proc. Natl. Acad. Sci. U.S.A.*, 87, 5454, 1990.

255. Hohn, A., Leibrock, J., Bailey, K., and Barde, Y.-A., Identification and characterization of a novel member of the nerve growth factor/brain-derived neurotrophic factor family, *Nature (London)*, 344, 339, 1990.

256. Jones, K. R. and Reichardt, L. F., Molecular cloning of a human gene that is a member of the nerve growth factor family, *Proc. Natl. Acad. Sci. U.S.A.*, 87, 8060, 1990.

257. Maisonpierre, P. C., Belluscio, L., Squinto, S., Ip, N. Y., Furth, M. E., Lindsay, R. M., and Yancopuloulos, G. D., Neurotrophin-3: a neurotrophic factor related to NGF and BDNF, *Science*, 247, 1466, 1990.

258. Rosenthal, A., Goeddel, D. V., Nguyen, T., Lewis, M., Shih, A., Laramee, G. R., Nikolics, K., and Winslow, J. W., Primary structure and biological activity of a novel human neurotrophic factor, *Neuron*, 4, 767, 1990.

259. Hallböök, F., Ibàñez, C. F., and Persson, H., Evolutionary studies of the nerve growth factor family reveal a novel member abundantly expressed in *Xenopus* ovary, *Neuron*, 6, 845, 1991.

260. Lindsay, R. M., Thoenen, H., and Barde Y.-A., Placode and neural crest-derived sensory neurons are responsive at early developmental stages to brain-derived neurotrophic factor, *Dev. Biol.*, 112, 319, 1985.

261. Johnson, J. E., Barde Y.-A., Schwab, M., and Thoenen, H., Brain-derived neurotrophic factor supports the survival of cultured rat retinal ganglion cells, *J. Neurosci.*, 6, 3031, 1986.

262. Hyman, C., Hofer, M., Barde, Y.-A., Juhasz, M., Yancopoulos, G. D., Squinto, S. P., and Lindsay, R. M., BDNF is a neurotrophic factor for dopaminergic neurones of the substantia nigra, *Nature (London)*, 350, 230, 1991.

263. Alderson, R. F., Alterman, A. L., Barde, Y.-A., and Lindsay, R. M., Brain-derived neurotrophic factor increases survival and differentiated functions of rat septal cholinergic neurons in culture, *Neuron*, 5(3), 297, 1990.

264. Knusel, B., Winslow, J. W., Rosenthal, A., Burton, L. E., Seid, D. P., Nikolics, K., and Hefti, F., Promotion of central cholinergic and dopaminergic neuron differentiation by brain-derived neurotrophic factor but not neurotrophin 3, *Proc. Natl. Acad. Sci. U.S.A.*, 88(3), 961, 1991.

265. Knusel, B., Beck, K. D., Winslow, J. W., Rosenthal, A., Burton, L. E., Widmer, H. R., Nikolics, K., and Hefti, F., Brain-derived neurotrophic factor administration protects basal forebrain cholinergic but not nigral dopaminergic neurons from degenerative changes after axotomy in the adult rat brain, *J. Neurosci.*, 12(11), 4391, 1992.

266. Phillips, H. S., Hains, J. M., Armanini, M., Laramee, G. R., Johnson, S. A., and Winslow, J. W., BDNF mRNA is decreased in the hippocampus of individuals with Alzheimer's disease, *Neuron*, 7, 695, 1991.

INDEX